W0245771

Ecological responses to environmental stresses

Tasks for vegetation science 22

Series Editors

HELMUT LIETH

University of Osnabrück, Germany

HAROLD A. MOONEY

Stanford University, Stanford, Calif., U.S.A.

Ecological responses to environmental stresses

edited by

J. ROZEMA & J.A.C. VERKLEIJ

Kluwer Academic Publishers

Dordrecht / Boston / London

Library of Congress Cataloging-in-Publication Data

Ecological responses to environmental stresses / edited by J. Rozema
   and J.A.C. Verkleij.
         p.   cm. -- (Tasks for vegetation science ; 22)
      Includes index.

      1. Plants, Effect of stress on.   2. Botany--Ecology.   I. Rozema,
   J. II. Verkleij, J. A. C. (Jos A. C.)   III. Series.
   QK754.E36   1990
   581.5'22--dc20   90-4762

ISBN-13: 978-94-010-6757-7   e-ISBN-13: 978-94-009-0599-3
DOI: 10.1007/978-94-009-0599-3

Published by Kluwer Academic Publishers,
P.O. Box 17, 3300 AA Dordrecht, The Netherlands.

Kluwer Academic Publishers incorporates
the publishing programmes of
D. Reidel, Martinus Nijhoff, Dr W. Junk and MTP Press.

Sold and distributed in the U.S.A. and Canada
by Kluwer Academic Publishers,
101 Philip Drive, Norwell, MA 02061, U.S.A.

In all other countries, sold and distributed
by Kluwer Academic Publishers Group,
P.O. Box 322, 3300 AH Dordrecht, The Netherlands.

Printed on acid-free paper

Contents

List of contributors

Andel, J. van, Department of Plant Ecology, University of Groningen, Biological Centre, P.O. Box 14, 9750 AA Haren (Gn), The Netherlands.

Arp, W.J., Smithsonian Environmental Research Center, Box 28, Edgewater MD 21037, Maryland, USA.

Assche, F. van, Limburgs Universitair Centrum, Universitaire Campus, B-3610 Diepenbeek, Belgium.

Baalen, J. van, Ministerie van Landbouw en Visserij, Directie NMF, Postbus 20401, 2500 EK 's-Gravenhage, The Netherlands.

Beckhoven, K. van, Department of Ecology and Ecotoxicology, Vrije Universiteit, De Boelelaan 1087, 1081 HV Amsterdam, The Netherlands.

Beeftink, W.G., Delta Institute for Hydrobiological Research, Vierstraat 28, 4401 EA Yerseke, The Netherlands.

Bradshaw, A.D., Department of Environmental and Evolutionary Biology, University of Liverpool, Liverpool L69 3BX, U.K.

Clijsters, H., Limburgs Universitair Centrum, Universitaire Campus, B-3610 Diepenbeek, Belgium.

De Vos, C.H.R., Department of Ecology and Ecotoxicology, Vrije Universiteit, De Boelelaan 1087, 1081 HV Amsterdam, The Netherlands.

Diggelen, J. van, Pandata BV, Postbus 1923, 2280 DX Rijswijk, The Netherlands.

Dueck, Th. A., Department of Ecology & Soil Ecology, Research Institute for Plant Protection, P.O. Box 9060, 6700 GW Wageningen, The Netherlands.

Duin, W.E. van, Department of Ecology and Ecotoxicology, Vrije Universiteit, De Boelelaan 1087, 1081 HV Amsterdam, The Netherlands.

Gora, L., Limburgs Universitair Centrum, Universitaire Campus, B-3610 Diepenbeek, Belgium.

Griffioen, W.A.J., Department of Ecology and Ecotoxicology, Vrije Universiteit, De Boelelaan 1087, 1081 HV Amsterdam, The Netherlands.

Harmens, H., Department of Ecology and Ecotoxicology, Vrije Universiteit, De Boelelaan 1087, 1081 HV Amsterdam, The Netherlands.

Huiskes, A.H.L., Delta Institute for Hydrobiological Research, Vierstraat 28, 4401 EA Yerseke, The Netherlands.

Ietswaart, J.H., Department of Ecology and Ecotoxicology, Vrije Universiteit, De Boelelaan 1087, 1081 HV Amsterdam, The Netherlands.

Janiesch, P., Dept. Physiologische Ökologie, Universität Oldenburg, F.B. 7, Postfach 2503, D-2900 Oldenburg, FRG.

Joosse, E.N.G., Department of Ecology and Ecotoxicology, Vrije Universiteit, De Boelelaan 1087, 1081 HV Amsterdam, The Netherlands.

Kakes, P., Department of Ecology and Ecotoxicology, Vrije Universiteit, De Boelelaan 1087, 1081 HV Amsterdam, The Netherlands.

Koe, T. de, Dep^{to} de Biologia, Universidade de Tras os Montes e Alta Douro, 5000 Vila Real, Portugal.

Kuiters, A.T., Department of Ecology and Ecotoxicology, Vrije Universiteit, De Boelelaan 1087, 1081 HV, Amsterdam, The Netherlands.

Leendertse, P.C., Department of Ecology and Ecotoxicology, Vrije Universiteit, De Boelelaan 1087, 1081 HV Amsterdam, The Netherlands.

Lenssen, G.M., Department of Ecology and Ecotoxicology, Vrije Universiteit, De Boelelaan 1087, 1081 HV Amsterdam, The Netherlands.

Lolkema, P.C., Department of Ecology and Ecotoxicology, Vrije Universiteit, De Boelelaan 1087, 1081 HV Amsterdam, The Netherlands.

McNeilly, T., Department of Environmental and Evolutionary Biology, University of Liverpool, Liverpool L69 3BX, U.K.

Neeling, A.L. de, Rijksinstituut voor de Volksgezondheid en milieuhygiëne, P.O. Box 1, 3720 BA Bilthoven, The Netherlands.

Otte, M.L., Department of Ecology and Ecotoxicology, Vrije Universiteit, De Boelelaan 1087, 1081 HV Amsterdam, The Netherlands.

Pieterse, A.H., Royal Tropical Institute, Rural Development Programme, Mauritskade 63, 1092 AD Amsterdam, The Netherlands.

Posthumus, A.C., Department of Ecology and Ecotoxicology, Vrije Universiteit, De Boelelaan 1087, 1081 HV Amsterdam, The Netherlands.

Rozema, J., Department of Ecology and Ecotoxicology, Vrije Universiteit, De Boelelaan 1087, 1081 HV Amsterdam, The Netherlands.

Rozijn, N.A.M.G., Vrije Universiteit, De Boelelaan 1105, 1081 HV Amsterdam, The Netherlands.

Schat, H., Department of Ecology and Ecotoxicology, Vrije Universiteit, De Boelelaan 1087, 1081 HV Amsterdam, The Netherlands.

Scholten, M.C.T., M.T. – TNO, Department of Biology, Laboratory for Applied Marine Research, P.O. Box 57, 1740 AB Den Helder, The Netherlands.

Schroten, J., Department of Ecology and Ecotoxicology, Vrije Universiteit, De Boelelaan 1087, 1081 HV Amsterdam, The Netherlands.

Simons, J., Department of Ecology and Ecotoxicology, Vrije Universiteit, De Boelelaan 1087, 1081 HV Amsterdam, The Netherlands.

Staaij, J.W.M. van de, Department of Ecology and Ecotoxicology, Vrije Universiteit, De Boelelaan 1087, 1081 HV Amsterdam, The Netherlands.

Straalen, N.M. van, Department of Ecology and Ecotoxicology, Vrije Universiteit, De Boelelaan 1087, 1081 HV Amsterdam, The Netherlands.

Tietema, T., National Institute of Development, Research and Documentation, University of Botswana, Private Bag 0022, Gaborone, Botswana.

Tolsma, D.J., Department of Ecology and Ecotoxicology, Vrije Universiteit, De Boelelaan 1087, 1081 HV Amsterdam, The Netherlands.

Veenendaal, E.M., National Institute of Development, Research and Documentation, University of Botswana, Private Bag 0022, Gaborone, Botswana.

Verkleij, J.A.C., Department of Ecology and Ecotoxicology, Vrije Universiteit, De Boelelaan 1087, 1081 HV Amsterdam, The Netherlands.

Vries, P.J.R. de, Unie van Waterschappen, Joh. van Oldebarneveldlaan 5, Postbus 80200, 2508 GE 's-Gravenhage, The Netherlands.

Werff, M. van der, Department of Ecology and Ecotoxicology, Vrije Universiteit, De Boelelaan 1087, 1081 HV Amsterdam, The Netherlands.

Preface

September 1987, the Faculty Biology of the Vrije Universiteit, Amsterdam commemorated the fact that Prof. Dr. Wilfried Hans Otto Ernst had been active as a scientist for 25 years. This period of 25 years of scientific research started at the Institut für Angewandte Botanik (Institute of Applied Botany) of the University of Münster, FRG. In 1965 he completed his Ph.D. thesis, entitled "Untersuchungen der Schwermetallpflanzengesellschaften Mitteleuropas unter Einschluss der Alpen."

He was appointed full Professor at the Department of Ecology of the Vrije Universiteit, Amsterdam in 1973. On the occasion of his 25th anniversary as a scientist, a promise was made, though in covert terms, which we could not redeem at that time. The promise held to offer Prof. Ernst a book, in which his former and present staffmembers, Ph.D. students and colleages should write a review about their specialism concerning a central theme.

Now, at the beginning of 1990 we consider the chapters of "Ecological Responses to Environmental Stresses" to be completed. The book reflects the wide range of research approaches that has been initiated and organized by Wilfried Ernst.

The editors hope to have attained the primary aim of the production of the *book of friends,* that is to gather relevant papers of staff-members and colleagues of Wilfried Ernst. The title of the book "Ecological Responses to Environmental Stresses" covers the majority of the chapters included. The title still covers the research projects of the present Department of Ecology and Ecotoxicology, of the Vrije Universiteit.

The study of environmental stresses as occurring in ecosystems high in heavy metals, salt or organic contaminants, and the response of plants and animals is highly important for the understanding of the recent environmental problems. The chapters in the book are concise, reflect the vision of the authors but do not always intend to be thorough and lengthy reviews.

The writing of the texts and the editorial work has taken considerable time, but it has been a pleasure to compose this *book of friends*.

We hope that both the fundamental and applied ecological research of environmental stresses as presented here may for a long time be stimulated and guided by Wilfried Ernst.

J. ROZEMA
J. VERKLEIJ

W.H.O. Ernst

About Wilfried H.O. Ernst

Here we (the editors) intend to present a short history of the personal and scientific life of Wilfried Ernst. This short note therefore does not cover the complete curriculum vitae with all memberships and editorial activities of Wilfried Ernst. A list of publications is given in the second chapter of this book. The reader must realize that this list needs to be updated on the time of appearance of this book.

"About Wilfried Ernst", is nothing more than a rough and personal draft on the way that Wilfried Ernst went.

The period in Silesia and Westfalia

Wilfried Hans Otto Ernst was born April 18th, 1937 in Ottmachau, Silesia, at that time a part of Germany. His father was employed as a manager in the dairy industry. As a young boy Wilfried Ernst went to Grammar School in Gellenau in Silesia from 1943–1944. After the Second World War Silesia, as part of pre-war Germany, was under the control of the Sovjet Republic and Poland. Millions of citizens of Silesia were forced to move away from the border area that Silesia formed. For this reason, in 1946 Wilfried Ernst was transported with his father, mother and sisters to Brochterbeck, Westfalia, after a stay in the transit camp of the British-French forces in Magdeburg. In Westfalia Wilfried Ernst continued his education at the nearby Grammar School in Brochterbeck.

The secondary School "Goethe Gymnasium" was attended in Ibbenbüren in Westfalia from 1949–1958. At the end of this period Wilfried Ernst made a choice to academically develop two of the talents he was gifted with: skil for music, and physics and biology. At the age of 21 years Wilfried Ernst started his university education at the Willhelms Universität in Münster in the study of physics and biology. He took his degree of Candidate in Philosophy in 1961. His musical talent appeared to full advantage by playing the violin for many years in a regional orchestra. The foundation of his further scientific career was completed with the dissertation "Untersuchungen der Schwermetallpflanzengesellschaften Mitteleuropas unter Einschluss der Alpen". The supervisors for this Ph.D. study were Prof. Dr W. Baumeister and Prof. Dr E. Burrichter.

From 1964 until 1969 Wilfried Ernst was employed as a scientific assistant in the Institute of Applied Botany, with Professor Baumeister as director.

During this period he visited a number of countries to see as many interesting heavy metal contaminated sites as possible. In 1967 he met Anthony Bradshaw, at that time working in Bangor. The following year Anthony Bradshaw was appointed as Professor in Liverpool. With members of the Liverpool group, Janis Antonovics and Tom McNeilly, he visited famous mine sites like Halkyn and Dolfrwynog in North Wales. For the preparation of the Habilitationsschrift it was necessary to visit waste tips in the Eastern part of Germany for example in Jena. To perform this Wilfried Ernst, disguised himself as a worker in the sugar beet industry, joined the morning transport to Mansfeld and returned in the evening after having sampled the vegetation of mine waste tips. In 1969, July 9th the "Habilitationsarbeit für Botanik" was completed. Professor Dr. Hiram Wild from former Salisbury, Rhodesia (now Harare (Zimbabwe)) invited him to come to Salisbury for a study of the ecology of heavy metal vegetation in Southern Africa. The support by a Schimper fellowship H. & E. Walter Foundation enabled Wilfried Ernst to go to Rhodesia and soon after his marriage in spring 1969 to Ingela Niermann, he and Ingela travelled to Salisbury and stayed there for about six months.

Back in Germany their first son Fabian (1971) was born in Münster, and the second son Marius born in Amsterdam (1977). An appointment as lecturer in

Botany followed (1970–1971) and at the age of 34 Wilfried Ernst became Professor of "Plant Ecology and Environmental Protection", at the University of Münster. Peter Janiesch was the first Ph.D. student and he is now Professor of Plant Ecology at the University of Oldenburg, F.R.G.

Ecology in Amsterdam

In 1972 the head of the Plant Ecology group of the Department of Ecology at the Vrije Universiteit Amsterdam, Professor Kuilman resigned. A year later Wilfried Ernst got the position as full professor of Botany at the Vrije Universiteit, The Netherlands, at this Department (now the Department of Ecology and Ecotoxicology). He accepted the chair with an inaugural address entitled: Principes en concepten van een moderne plantenoecologie (1975).

As one of his earliest Dutch doctoral students (Jelte Rozema) and both as Ph.D. students of Wilfried Ernst, we still remember quite well the many excursions that were made with staffmembers (J. van Andel, F. W. van der Vegte, J. Rozema, J. Verkleij) technicians (T. F. Lugtenborg, H. J. M. Nelissen) and students to nearby (man-made) ecosystems such as the Amsterdamse Bos, Bijlmermeer or the North Holland Dune Reserve. Also the bark of a first tree on the way to a field site, could evoke lengthy eco-(physio)logical lectures so that we never reached the lovely dune reserve. Using a swiss army knife as a spade Wilfried Ernst excavated the root system of many plants including Rhizobium nodules.

There is no doubt that the coming of Wilfried Ernst to the Dutch and Amsterdam scene of plant ecologists has made great impact.

Overlooking the scientific career of Wilfried Ernst a remarkable transition becomes obvious. The early work starts with "ökologisch-soziologische Untersuchungen" (1964), but already in 1969 a more physiological study on "subzelluläre Speicherungsorte des Zinks" was published. In the line of research of the mechanism of heavy metal resistance, the physiological and biochemical investigations have been intensified and now focus on the structure of the plasma membrane and role of metal-binding compounds like phytochelatins and organic acids.

The ecological research not only went into more detail with physiological and biochemical studies of the mechanism of heavy metal tolerance, it also broadened, while settling in the Department of Ecology in Amsterdam.

His erudite and up to date knowledge of ecology and many related fields have encouraged his colleagues and staffmembers. He has always argued for causal analysis of ecological relationships. This experimental approach has, certainly at the time of his early work in the Netherlands broken a line of traditional descriptive ecological research. The main approach of his research is of empirical nature. Wilfried Ernst's message often implies suspicion for general ecological models and stresses the indispensable first hand, reliable ecological observations and data. As no one other he demonstrates his unconditional love for ecological field studies. His work on the population ecology of Phleum arenaria of the Dutch coastal dunes is of this kind. Also summer holidays in the Alps have always been a period for sampling and observations. The research of Wilfried Ernst's group has been diversified, branched out and intensified. This is illustrated by the broad range of ecological studies from the cellular level to the ecosystem approach. The line of heavy metal studies was continued. New research was started in the coastal environment with studies of the population ecology of dune species and ecological studies of salt marshes. In 1983 his co-worker Jelte van Andel was appointed as Professor of Plant Ecology at the State University of Groningen.

His way of ecological thinking has also found its way to nature management and environmental policy. Wilfried Ernst has contributed markedly to the development of alternatives for surface infiltration of pretreated river Rhine water into the North-Holland dune system.

In his first Amsterdam' years Wilfried Ernst set up new directions of ecological research. One of his motivations to come to the Vrije Universiteit was the presence of a Plant Ecology and an Animal Ecology Group and integration of these groups under a combined research programme has been his main aim. With some persistence he eventually succeeded to start a joint programme on "The effects of and adaptation to heavy metals on plant and animal organisms", which research has now led to the largest part in both groups.

Another new field what has his interest, is the population genetics. He holds the view that this discipline has to be integrated with the other disciplines, the population dynamics and ecophysiology. Ecological genetics as it is correctly called, is now an important part of the plant and animal ecology department.

Perhaps his greatest ambition has been to find scientific and financial means to carry out a tropical ecology programme. With the support of the Vrije Universiteit, NUFFIC (Netherlands University Foundation for International Cooperation) and WOTRO (Foundation of Tropical Research) he has led for the last ten years with great enthusiasm an ecolog-

ical research project in Botswana. He loves to visit this developmental country every year in the fall. Temperatures above 40—45° C are no problem for him and during his stay he debates about any scientific subject, gives lectures and participates in field trips and at the end after his departure from Botswana his Ph.D. students and coworkers are left exhausted.

Now that environmental sciences is a major issue the research in the Department of Ecology and Ecotoxicology, as guided by Wilfried Ernst, will face many new challenges and will be supported by his extended experience in the study of environmental problems.

List of publications of W.H.O. Ernst

Scientific publications

Ernst, W. 1964. Ökologisch-soziologische Untersuchungen der Schwermetallpflanzen – gesellschaften Mitteleuropas unter Einschluss der Alpen. — Dissertation der Math.-Nat. Fakultät der Universität Münster.

Ernst, W. 1965. Idem. Abhandl. Landesmus. Naturkunde Münster in Westfalen 27(1), 1—54.

Ernst, W. 1965. Über den Einfluss des Zinks auf die Keimung von Schwermetallpflanzen und auf die Entwicklung der Schwermetallpflanzengesellschaft. — Ber. Dtsch. Bot. Ges. 78, 205—212.

Ernst, W. 1966. Ökologisch-soziologische Untersuchungen an Schwermetallpflanzengesellschaften Südfrankreichs und des östlichen Harzvorlandes. — Flora B 156, 301—318.

Ernst, W. 1967. Bibliographie der Arbeiten über Pflanzengesellschaften auf schwermetallhaltigen Böden mit Ausnahme des Serpentins. — Excerpta Botanica B8, 50—61.

Baumeister, W., Ernst, W. & Rüther, F. 1967. Zur Soziologie und Ökologie europäischer Schwermetallpflanzengesellschaften. — Forschungsber. Land. Nordrhein-Westfalen 1803, 1—46.

Ernst, W. 1968. Der Einfluss der Phosphatversorgung sowie die Wirkung van ionogenem und chelatisiertem Zink auf die Zink- und Phosphataufnahme einiger Schwermetallpflanzen. — Physiol. Plant. 21, 323—333.

Ernst, W. 1968. Zur Kenntnis der Soziologie und Ökologie der Schwermetallvegetaton Grossbritanniens. — Ber. Dtsch. Bot. Ges. 81, 116—124.

Ernst, W. 1968. Das Violetum calaminariae westfalicum, eine Schwermetallpflanzengesellschaft bei Blankenrode in Westfalen. — Mitt. flor. soz. Arbgem. NF 13, 263—268.

Ernst, W. 1968. Ökologische Untersuchungen an Pflanzengesellschaften unterschiedlich stark gestörter schwermetallreicher Böden in Grossbritannien. — Flora B 158, 95—107.

Ernst, W. 1969. Pollenanalytischer Nachweis eines Schwermetallrasens in Wales. — Vegetatio (Tüxen-Festschrift) 18, 393—400.

Ernst, W. 1969. Ökologische Untersuchungen der Violetea calaminariae-Schwermetallpflanzengesellschaften. — In: R. Tüxen (ed.) "Experimentelle Pflanzensoziologie" pp. 146—155, Junk Publ. Den Haag.

Ernst, W. 1969. Zur Physiologie der Schwermetallpflanzen — Subzelluläre Speicherungsorte des Zinks. — Ber. Dtsch. Bot. Ges. 82, 161—164.

Ernst, W. 1969. Beitrag zur Kenntnis europäischer Spülsaumgesellschaften. I. Sand- und Kiesstrände. — Mitt. flor. soz. Arbgem. NF 14, 86—94.

Ernst, W. 1969. Die Schwermetallvegetation Europas. — Habilitationsschrift Math.-Nat. Fak. Universität Münster, 184 pp.

Ernst, W. 1971. Zur Ökologie der Miombo-Wälder. — Flora 160, 317—331.

Ernst, W. 1972. Ecophysiological studies on heavy metal plants in South Central Africa. — Kirkia 8, 125—145.

Ernst, W. 1972. Schwermetallresistenz und Mineralstoffhaushalt. — Forschungsber. Land. Nordrhein-Westfalen 2251, 1—38, Westdeutscher Verlag, Opladen.

Ernst, W. 1972. Zink- und Cadmium-Immissionen auf Böden und Pflanzen in der Umgebung einer Zinkhütte. — Ber. Dtsch. Bot. Ges. 85, 295—300.

Ernst, W. & Weinert, H. 1972. Lokalisation von Zink in den Blättern von Silene cucubalus Wib.— Z. Pflanzenphysiol. 66, 258—264.

Ernst, W. & Walker, B.H. 1973. Studies on the hydrature of trees in miombo woodland in South Central Africa. — J. Ecol. 61, 667—673.

Ernst, W. 1974. Mechanismen der Schwermetallresistenz. — Verhandl. Ges. Ökol. Erlangen 1974, 189—197.

Ernst, W. 1974. Schwermetallvegetation der Erde. — G. Fischer, Stuttgart, 192 pp.

Ernst, W., Mathys, W., Salaske, J. & Janiesch, P. 1974. Aspekte von Schwermetallbelastungen in Westfalen. — Abhandl. Landesmus. Naturkunde Münster in Westfalen 36(2), 1—30.

Ernst, W. & Feldermann, D. 1975. Auswirkungen der Wintersalzstreuung auf den Mineralstoffhaushalt von Linden. — Z. Pflanzenernähr. Bodenk. 1975, 629—640.

Ernst, W., Mathys, W. & Janiesch, P. 1975. Physiologische Grundlagen der Schwermetallresistenz — Enzymaktivitäten und organische Säuren. — Forschungsber. Land. Nordrhein-Westfalen 2496, 1—38. Westdeutscher Verlag, Opladen.

Ernst, W. 1975. Variation in the mineral contents of leaves of trees in miombo woodland in South Central Africa. — J. Ecol. 63, 801—808.

Ernst, W. 1975. Principes en concepten van een moderne plantenoecologie. — Oratie, Vrije Universiteit Amsterdam, 17 pp.

Kraal, H. & Ernst, W. 1976. Influence of copper high tension lines on plants and soils. — Environ. Pollut. 11, 131—135.

Ernst, W. 1976. Physiological and biochemical aspects of metal tolerance. In: Mansfield, T.A. (ed.), Effects of Air Pollutants on Plants. Soc. Exper. Biol. Seminar Ser. 1, 115—133.

Ernst, W. 1976. Ökologische Grenze zwischen Violetum calaminariae und Gentiano-Koelerietum. — Ber. Dtsch. Bot. Ges. 89, 381—390.

Ernst, W. 1976. Violetea calaminariae. In: Tüxen, R. (ed.), Prodrome of the European Plant Communities, Vol. 3, 1—133.

Ernst, W. 1977. Physiology of heavy metal resistance in plants. — Proc. Internat. Conf. Heavy Metals in the Environment, Toronto 1975. Vol. 2, 121—136.

Rozema, J., Nelissen, H.J.M., Kroft, M. van der & Ernst, W.H.O. 1977. Nitrogen mineralization in sandy salt marsh soils of the Netherlands. — Z. Pflanzenernaehr. Bodenk. 10, 707—717.

Ernst, W. 1978. Discrepancy between ecological and physiological optima of plant species: a re-interpretation. — Oecol. Plant. 13, 175—188.

Ernst, W. 1978. Chemical soil factors determining plant growth. In: Freijsen, A.H.J. & Woldendorp, J.W. (eds.). Structure and Functioning of Plant Populations. North Holland Publ. Comp., Amsterdam, Oxford, New York, 1978, 155—187.

Ernst, W. & Marquenie–Van der Werff, M. 1978. Aquatic angiosperms as indicators of copper contamination. — Arch. Hydrobiol. 83, 356—366.

Baumeister, W. & Ernst, W. 1978. Mineralstoffe und Pflanzenwachstum. — G. Fischer, Stuttgart, New York, 416 pp.

Van Andel, J., Bos, W. & Ernst, W. 1978. An experimental study on two populations of Chamaenerion angustifolium (L.) Scop. (= Epilobium angustifolium L.) occuring on contrasting soils, with particular reference to the response of bicarbonate. — New Phytol. 81, 763—772.

Ernst, W. & Nelissen, H.J.M. 1979. Growth and mineral nutrition of plant species from clearings on different horizons of an iron-humus podzol profile. — Oecologia (Berl.) 41, 175—182.

Ernst, W. 1979. Population biology of Allium ursinum in Northern Germany. — J. Ecol. 67, 347—362.

Ernst, W. 1979. Ökologische Aspekte eines Rumici-Alopecuretum geniculati in einem Feuchtegradienten von einem Typhetum latifoliae zu einem Lolio-Cynosuretum. — Phytocoenologia 6 (Festband Tüxen), 74—84.

Van Andel, J., Ernst, W., Nelissen, H. 1979. Mineralstoffkreislauf in Populationen von Kahlschlagarten in Beziehung zur Sukzession. — Verhandl. Ges. Ökol. Münster 1978, 7, 361—368.

Marquenie-Van der Werff, M. & Ernst, W.H.O. 1979. Kinetics of copper and zinc uptake by leaves and roots of an aquatic plant, Elodea nuttallii. — Z. Pflanzenphysiol. 92, 1—10.

Ernst, W.H.O. 1980. Biochemical aspects of cadmium in plants. — In: J.O. Nriagu (ed.), Cadmium in the Environment, J. Wiley & Sons, New York, 639—653.

Ernst, W.H.O. 1980. Problems of bioindication at the level of individuals. — In: R. Schubert & J. Schuh (eds.), Bioindikation, Vol. 3, Halle/Saale, 3—9.

Ernst, W.H.O. & Bast-Cramer, W.B. 1980. The effect of lead contamination of soils and air on its accumulation in pollen. — Plant Soil 57, 491—496.

Ernst, W.H.O. & Lugtenborg, T.F. 1980. Vergleichende Ökophysiologie von Juncus articulatus und Holcus lanatus. — Flora 169, 121—134.

Ernst, W.H.O. 1981. Ecological implication of fruit variability in Phleum arenarium L., an annual dune grass. — Flora 171, 387—398.

Ernst, W.H.O. 1981. Probleme bei der Begrünung und Aufforstung von Schwermetallhalden. — In: A. Schwabe-Braun (ed.), Vegetation als anthropo-ökologischer Gegenstand, J. Cramer, Vaduz, 237—248.

Marquenie-Van der Werff, M., Ernst, W.H.O. & Faber, J. 1981. Complexing agents in soil organic matter as factors in heavy metal toxicity in plants. — In: W.H.O. Ernst (ed.), Heavy Metals in the Environment, Intern. Conf. Amsterdam 1981, CEP Consultants Ltd., Edinburgh, 222—225.

Ernst, W.H.O., 1982. Schwermetallpflanzen. — In: H. Kinzel (ed.), Pflanzenökologie und Mineralstoffwechsel, E. Ulmer, Stuttgart, 507—519.

Ernst, W.H.O. 1982. Fluor- und Selenpflanzen. — In: H. Kinzel (ed.), Pflanzenökologie und Mineralstoffwechsel, E. Ulmer, Stuttgart, 507—519.

Ernst, W.H.O. 1982. Monitoring of particulate pollutants. — In: L. Steubing & H.-J. Jäger (eds.). Monitoring of Air Pollutants by Plants, MAB-Proceedings, W. Junk, Den Haag, 121—128.

Ernst, W.H.O. 1983. Population biology and mineral nutrition of Anemone nemorosa with emphasis on its parasitic fungi. — Flora 173, 335—348.

Ernst, W.H.O. 1983. Anpassungsstrategien einjähriger Dünenpflanzen. — Verhandl. Ges. Ökol. Mainz 1981, 10, 485—495.

Ernst, W.H.O. 1983. Element nutrition of two contrasted dune annuals. — J. Ecol. 71, 197—209.

Ernst, W.H.O. 1983. Ökologische Anpassungsstrategien an Bodenfaktoren. — Ber. Dtsch. Bot. Ges. 96, 49—71.

Ernst, W.H.O. 1983. Indicatoren van een overmaat aan zware metalen in terrestrische ecosystemen. In: E.D.H. Best & J. Haeck (eds.), Ecologische Indicatoren, Pudoc, Wageningen, 109—120.

Ernst, W.H.O., De Neeling, A.J. & Vooijs, R. 1983. Replacement of sodium and potassium in a natrophobe and a natrophile Senecio species. — Z. Pflanzenphysiol. 112, 147—154.

Ernst, W.H.O. & Joosse, E.N.G. 1983. Umweltbelastung durch Mineralstoffe. — VEB G. Fischer, Jena, und G. Fischer, Stuttgart, 234 pp.

Ernst, W.H.O., Verkleij, J.A.C. & Vooijs, R. 1983. Bioindication of a surplus of heavy metals in terrestrial ecosystems. — Environ. Monitor. Assessment 3, 297—305.

Ernst, W.H.O., Van Duin, W.E. & Oolbeeking, G.T. 1984. Vesicular-arbuscular mycorrhiza in dune vegetation. — Acta Bot. Neerl. 32, 151—160.

Van Baalen, J., Nelissen, A.J.M., Ernst, W.H.O., Wattel, J. & Vooijs, R. 1984. Reproductive processes of Senecio fuchsii (partly in comparison with Eupatorium cannabinum) as affected by temperature, irradiance and soil fertility. — Flora 175, 81—90.

Lolkema, P.C., Donker, M.H., Schouten, A.J. & Ernst, W.H.O. 1984. The possible role of metallothioneins in copper tolerance of Silene cucubalus. — Planta 162, 174—179.

Dueck, T.A., Ernst, W.H.O., Faber J. & Pasman, F. 1984. Heavy metal immission and genetic constitution of plant populations in the vicinity of two metal emission sources. — Angew. Bot. 58, 47—59.

Ernst, W.H.O. 1985. Bedeutung einer veränderten Mineralstoffverfügbarkeit (Schwermetalle, Al, Ti) für Wachstums- und Selektionsprozesse in Wäldern. — In: S.W. Breckle & H. Kahle (eds.), Schwermetalle und saure Depositionen. Bielefeld. Ökol. Beitr. 1, 143—158.

Ernst, W.H.O. 1985. Schwermetallimmissionen — ökophysiologische und populationsgenetische Aspekte. — Düsseldorf. Geobot. Kolloquien 2, 43—57.

Ernst, W.H.O. 1985. Some considerations of and perspectives in coastal ecology. — Vegetatio 62, 533—545.

Ernst, W.H.O. 1985. Impact of mycorrhiza on metal uptake and translocation by forest plants. — In: D. Lekkas (ed.), Heavy Metals in the Environment (Athens 1985), Vol. 1, 596—599.

Ernst, W.H.O. 1985. The effects of forest management of the genetic variability of plant species in the herb layer. — In: H.R. Gregorius (ed.), Population Genetics in Forestry. Springer Lecture Notes in Biomathematics 60, 200—212.

Ernst, W.H.O. & Van Andel, J. 1985. Autoecologie. Adaptaties, voornamelijk van oecofysiologische aard. — In: K. Bakker, Th. E. Cappenberg, N. Croin-Michielsen, A.H.J. Freijsen, P.H. Nienhuis, J.W. Woldendorp & J.J. Zijlstra (red.), Inleiding tot de oecologie. Bohn, Scheltema & Holkema, Utrecht, Antwerpen, pp. 69—100.

Ernst, W.H.O., Tonneijck, A.E.C. & Pasman, F.J.M. 1985. Ecotypic response of Silene cucubalus to air pollutants (SO₂, O₃). — J. Plant Physiol. 118, 439—450.

Dueck, Th.A., Ernst, W.H.O., Hulzebos, E. & Pasman, F. 1985. Growth and reproduction of Silene cucubalus exposed to heavy metals and air pollutants. — In: D. Lekkas (ed.), Heavy Metals in the Environment (Athens 1985), Vol. 2, 236—238.

Rozema, J., Laan, P., Broekman, R., Ernst, W.H.O. & Appelo, C.A.J. 1985. On the lime transition and decalcification in the coastal dunes of the province of North Holland and the island of Schiermonnikoog. — Acta Bot. Neerl. 34, 393—411.

Van Andel, J. & Ernst, W.H.O. 1985. Ecophysiological adaptation, plastic responses, and genetic variation of annuals, biennals and perennials in woodland clearings. — In: J. Haeck & J.W. Woldendorp (Eds.). Structure and Functioning of Plant Populations. II. Phenotypic and Genotypic Variation in Plant Populations. North Holland Publishing Comp., Amsterdam, Oxford, New York, pp. 27—49.

De Neeling, A.J. & Ernst, W.H.O. 1986. Manganese and aluminium tolerance of Senecio sylvaticus L. — Acta Oecol., Oecol. Plant. 7(21), 43—56.

De Neeling, A.J. & Ernst, W.H.O. 1986. Response of an acidic and a calcareous population of Chamaenerion angustifolium (L.) Scop. to iron, manganese and aluminium. — Flora 178, 85—92.

Dueck, Th.A., Ernst, W.H.O., Mooi, J. & Pasman, F.J.M. 1986. Effects of SO₂, NOₓ and O₃ in combination on the yield and reproduction of Silene cucubalus populations. — J. Plant Physiol. 122, 97—106.

Dueck, Th.A., Visser, P., Ernst, W.H.O. & Schat, H. 1986. Vesicular-arbuscular mycorrhiza decrease zinc toxicity to grasses growing in zinc-polluted soil. — Soil Biol. Biochem. 18, 331—333.

Dueck, Th. A., Visser, P., Ernst, W.H.O. & Schat, H. 1986. Relationship between VA-mycorrhiza and zinc-toxicity in Festuca rubra L. and Calamagrostis epigeijos (L.) Roth. — Mycorrhizae: Physiology and Genetics. INRA, Paris 1986, pp. 661—663.

Ernst, W.H.O. 1986. Mineral nutrition of Nicotiana tabacum cv. bursana during infection by Orobanche ramosa. In: J.S. ter Borg (Ed.), Biology and Control of Orobanche. LH/VPO, Wageningen, pp. 80—85.

Ernst, W.H.O., 1986. Die Wirkung chemischer Komponenten der Laubstreu auf das Wald-Greiskraut, Senecio sylvaticus. — Abhandl. Landesmuseum Naturkunde Münster (Burrichter-Festschrift) 48(2/3), 291—301.

Ernst, W.H.O. 1986. Longterm pollution and selection. — 2nd Int. Conf. Environmental Contamination, Amsterdam 1986. CEP Consultant Edinburgh, pp. 10—15.

Kuiters, A.T., Van Beckhoven, K. & Ernst, W.H.O. 1986. Chemical influences of tree litters on herbaceous vegetation. — In: J. Fanta (Ed.), Forest Dynamics Research in Western and Central Europe. IUFRO-workshop Wageningen 1985. Pudoc, Wageningen, pp. 103—111.

Lolkema, P.C., Doornhof, M. & Ernst, W.H.O. 1986. Interaction between a copper tolerant and a copper-sensitive population of Silene cucubalus. — Physiol. Plant. 67, 654—658.

Van Andel, J., Rozijn, N.A.M.G., Ernst, W.H.O. & Nelissen, H.J.M. 1986. Variability in growth and reproduction in F₁-families of an Erophila verna population. — Oecologia 69, 79—85.

Verkleij, J.A.C., Prast, J. & Ernst, W.H.O. 1986. Different effects of cadmium on biomass-production and metal-uptake in cadmium-tolerant, co-tolerant and sensitive populations of silene cucubalus. — 2nd Int. Conf. Environmental Contamination, Amsterdam 1986. CEP Consultants, Edinburgh, pp. 27—29.

Ernst, W.H.O. 1987. Scarcety of flower colour polymorphism in field populations of Digitalis purpurea. — Flora 179, 231—239.

Ernst, W.H.O. 1987. Population differentiation in grassland. In: J. van Andel, J.P. Bakker & R.W. Snaydon (eds.), Disturbance in Grasslands. Junk Publ., Dordrecht, pp. 213—228.

Ernst, W.H.O. 1987. Metal fluxes to coastal ecosystems and the response of coastal vegetation. In: A.H.L. Huiskes, C.W.P.M.

Blom & J. Rozema (eds.), Vegetation between Land and Sea. Junk Publ., Dordrecht, pp. 302—310.

Ernst, W.H.O. 1987. Impact of the aphid Aulacorthum solani Kltb. on growth and reproduction of winter and summer annual life forms of Senecio sylvaticus L. — Acta Oecol., Oecol. Gen. 8, 537—547.

Ernst, W.H.O., Kraak, M. & Stoots, L. 1987. Effects of humic and fulvic acid on growth and reproduction of Scrophularia nodosa. — J. Plant Physiol. 127, 171—175.

Ernst, W.H.O. & Leloup, S. 1987. Perennial herbs as monitor for moderate levels of metal fallout. — Chemosphere 16, 233—238.

Ernst, W.H.O., Nelissen, H.J.M. & De Hullu, E. 1987. Size hierarchy and mineral status of Rhinanthus angustifolius populations under different grassland management regimes. — Vegetatio 70, 93—103.

Ernst, W.H.O. & Sekhwela, M.B.M. 1987. The chemical composition of lerps from the mopane psyllid Arytaina mopane (Homoptera, Psyllidae). — Insect Biochem. 17, 905—909.

Ernst, W.H.O. & Van Rooij, L.F. 1987. $^{134/137}$Cs fallout from Chernobyl in Dutch forest. — 6th Int. Conf. "Heavy Metal in the Environment", New Orleans 1987, CEP Consultants Edinburgh, pp. 284—286.

Tolsma, D.J., Ernst, W.H.O., Verweij, R. & Vooijs, R. 1987. Seasonal variation of nutrient concentrations in a semi-arid savanna ecosystem in Botswana. — J. Ecol. 75, 755—770.

Tolsma, D.J., Ernst, W.H.O. & Verweij, R.A. 1987. Nutrients in soil and vegetation around two artificial waterpoints in Eastern Botswana. — J. Appl. Ecol. 24, 991—1000.

Van 't Riet, J., Van Rossenberg, M.C., Koevoets, P., Verkleij, J.A.C. & Ernst, W.H.O. 1987. Copper-binding compounds in metal-tolerant and non-tolerant Silene cucubalus. — 6th Int. Conf. "Heavy Metals in the Environment", CEP Consultants, Edinburgh, pp. 401—403.

Verkleij, J.A.C., Lolkema, P.C. & Ernst, W.H.O. 1987. The effect of heavy metals on isozyme gene expression in Silene cucubalus. — In: Isozymes: Current Topics in Biological and Medical Research. Vol. 16. Agriculture, Physiology and Medicine. A.R. Liss Inc., New York, pp. 209—221.

De Koe, T., Rozema, J., Broekman, R.A. & Ernst, W.H.O. 1988. Heavy metals in sediment and vegetation of the Aveiro lagoon, Portugal. In: M. Astruc & J.N. Lester (eds.), Chemicals (Heavy metals) in the Environment-Lissabon. Proc. Int. Conf. pp. 671—674.

De Koe, T., Rozema, J., Broekman, R.A., Otte, M.L., Ernst, W.H.O. 1988. Heavy metals and arsenicum in water sediment and plants near the Jales gold and silver mine in North-Portugal. In: A.A. Orio (ed.) Environmental Contamination, CEP Consultants Edinburgh pp. 152—155.

Dumon, J.C. & Ernst, W.H.O. 1988. Titanium in plants. J. Plant Physiol. 133, 203—209.

Ernst, W.H.O. 1988. Decontamination of mine sites by plants: An analysis of the efficiency. In: A.A. Orio (ed.) Environment Contamination, CEP Consultants Edinburgh, pp. 305—310.

Ernst, W.H.O. 1988. Metal fluxes to coastal ecosystems and the response of coastal vegetation. A review. In: A.H.L. Huiskes,

C.W.P.M. Blom & J. Rozema (eds.). Vegetation between Land and Sea pp. 302—310. Junk Publ. Dordrecht.

Ernst, W.H.O. 1988. Resonse of plants and vegetation to mine tailings and dredged materials. — In: W. Salomons & U. Förstner (eds.). Chemistry and Biology of Solid Waste, pp. 54—69. Springer Verlag, Berlin.

Ernst, W.H.O. 1988. Seed and seedling ecology of Brachystegia spiciformis, a predominant tree component in miombo woodlands in South Central Africa. — Forest Ecol. Management 25, 195—210.

Ernst, W.H.O., Tietema, T., Veenendaal, E. & Masene, R. 1988. Dormancy, germination and seedling growth of two Kalaharian perennials of the genus Harpagophytum (Pedaliaceae). J. Trop. Ecol. 4, 185—198.

Ernst, W.H.O. & Tolsma, D.J. 1988. Dormancy and germination of semi-arid annual plant species, Tragus berteronianus and Tribulus terrestris. Flora 181, 243—251.

Ernst, W.H.O. & Van der Ham, N.F. 1988. Population structure and rejuvination potential of Schoenus nigricans in coastal wet dune slacks. Acta Bot. Neerl. 37, 451—465.

Otte, M.L., Rozema, J., Beck, M.A., Ernst, W.H.O. & Broekman, R.A. 1988. Uptake of arsenic by vegetation of the former Rhine estuary. In: A.A. Orio (ed.) Environmental Contamination, CEP Consultants, Edinburgh, 529—531.

De Vos, C.H.R., Schat, H., Vooijs, R. & Ernst, W.H.O. 1989. Copper-induced damage to the permeability barrier in roots of Silene cucubalus. — J. Plant Physiol. 135, 164—169.

Ernst, W.H.O. 1989. Selection of winter and summer annual life form in population of Senecio sylvaticus L. — Flora 182, 221—231.

Ernst, W.H.O. 1990. Ecological aspects of sulfur metabolism. In: H. Rennenberg, Brunold, C., Kok, L.J. de & Stulen, I. (eds.). Sulfur Nutrition and Sulfur Assimilation in Higher Plants, pp. 131—144, SPB Academic Publishers, Den Haag.

Ernst, W.H.O. & Tolsma, D.J. 1989. Mineral nutrients in some Botswana savanna types. In: J. Proctor (ed.). Mineral Nutrients in Tropical Forest and Savanna Ecosystems, pp. 97—120. Blackwell Scientific Publ., Oxford.

Ernst, W.H.O., Tolsma, D.J. & Decelle, J. 1989. Predation of seeds of Acacia tortilis by insects. — Oikos 54, 294—300.

Harmens, H., Verkleij, J.A.C., Koevoets, P. & Ernst, W.H.O. 1989. The role of organic acids and phytochelatins in the mechanism of zinc tolerance in Silene vulgaris (= S. cucubalus). In: J.P. Vernet (ed.). Heavy Metals in the Environment (Genève 1989), Vol. 2, 178—181. CEP Consultants, Edinburgh.

Verkleij, J.A.C., Koevoets, P., Van 't Riet, J., Rossenberg, M.C., Bank, R. & Ernst, W.H.O. 1989. The role of metal-binding compounds in the copper tolerance mechanism of Silene cucubalus. In: D. Winge & D. Hamer (eds.), Metal Ion Homeostasis: Molecular Biology and Chemistry, pp. 347—357. Alan R. Liss Inc. New York.

Verkleij, J.A.C., Lugtenborg, T.F. & Ernst, W.H.O. 1989. The effect of geographic isolation on enzyme polymorphism of heavy metal tolerant populations of Minuartia verna (L.) Hiern. — Genetica 78, 133—143.

Verkleij, J.A.C., Koevoets, P.L.M., Van 't Riet, J., De Knecht,

J. & Ernst, W.H.O. 1990. The role of metal-binding compounds (phytochelatins) in the cadmiumtolerance mechanism of bladder campion (Silene cucubalus L.) (Wib.) In: H. Rennenberg, L.J. de Kok, C. Brunold & I. Stulen (eds.). Sulfur Nutrition and Sulfur Assimilation in higher Plants, pp. 255—260 SPB Academic publishing B.V., Den Haag.

Ernst, W.H.O. 1990. Mine vegetation in Europe. In: J. Shaw (ed.), Evolutionary Aspects of Heavy Metal Tolerance in Plants, pp. 211—237 CRC Press, Boca Raton/Fl.

Ernst, W.H.O. 1990. Element (re)translocation in plants and its impact on representative sampling. In: H. Lieth & B. Markert (ed.) Element Concentration Cadasters in Ecosystems (ECCE), pp. 17—40, Verlag Chemie, Weinheim.

Ernst, W.H.O. 1990. Ecophysiology of plants in waterlogged and flooded environments. — Aquat. Bot.

Ernst, W.H.O., Decelle, J.E. & Tolsma, D.J. 1990. Predispersal seed predation in native leguminous shrubs and trees in savannas in southern Botswana. — Afr. J. Ecol. 28, 45—54.

Ernst, W.H.O. & Tolsma, D.J. 1990. Dispersal of fruits and seeds in woody savanna plants in souther Botswana. — Beitr. Biol. Pflanz.

Rozijn, N.A.M.G., Ernst, W.H.O., Van Andel, J. & Nelissen, H.J.M. 1990. Growth response to different constant soil moisture levels in four winter annual species during the complete life cycle. — Flora 184, 303—312

Van Baalen, J., Ernst, W.H.O., Janssen, D.M. & Nelissen, H.J.M. 1990. Reproductive allocation of dry matter and nutrients in plants of Scrophularia nodosa grown at various levels of radiance and soil fertility. — Acta Bot. Neerl. 39, 183—196

Ernst, W.H.O., Schat, H. & Verkley, J.A.C. 1990. Evolutionary biology of the metal resistance in Silene vulgaris. — Evol. Trends in Plants 4, 45—51.

Griffioen, W.A.J. & Ernst, W.H.O. 1990. The role of VA mycorrhiza in the heavy metal tolerance of Agrostic capillaris. — Agric. Ecosyst. Environ. 29, 173–177.

Van Duin, W.E., Rozema, J. & Ernst, W.H.O. 1990. Seasonal and spatial variation in the occurrence of vesicular-arbuscular (VA) mycorrhiza in salt marsh plants. — Agric. Ecosyst. Environ. 29, 107—110.

Verkley, J.A.C., Koevoets, P.L.M., Van 't Riet, J., Bank, R., Nijdam, Y. & Ernst W.H.O. 1990. Poly (γ-glutamylcysteinyl) glycines or phytochelatins and their role in cadmium tolerance of Silene vulgaris. Plant, Cell and Environ. (in press).

General publications

Ernst, W. 1968. Bericht über die Tagung der Floristisch-soziologischen Arbeitsgemeinschaft in Münster (Westfalen) von 2 bis 4 Juni 1967. Mitt. Flor. soz. Arb. gem. 13, 258—260.

Ernst, W. 1976. Wieviel Schwermetall können Pflanzen "vertragen"? — Umschau Wiss. Technik 76, 355—356.

Ernst, W. 1978. Ökologische Risiken, verursacht durch emittierte Feinstäube aus Verbrennungsmotoren: Folgerungen für die analytische Chemie. — Chem. Rundschau 31(36), 32—35.

Ernst, W. 1980. De top 40 van de wetenschappelijke smaak of de scores in de Science Citation Index (SCI). — Vakbl. Biol. 17, 351—353.

Ernst, W. 1982. Populationsbiologie und Mineralstoffhaushalt von Anemone nemorosa. — Festschrift zum 70. Geburtstag von Prof. Dr. rer. nat. habil. Walter Baumeister. Univ. Münster. 34—50.

Ernst, W.H.O. 1982. Ecologische betekenis van fenol-, fulvine- en huminezuren. Vakbl. Biol. 62, 350—352.

Ernst, W. 1983. In memoram Hiram Wild. — Vegetatio 51, 125—128.

Ernst, W. 1983. Het fenomeen "impact factor". Vakbl. Biol. 63, 442—446.

Ernst, W. 1983. De regulering van ecosystemen of de fout van de evolutie. Vakbl. Biol. 63: 407—408.

Ernst, W.H.O., Riphagen, I. & Stoots, L. 1984. Plantengroei onder invloed van organisch bodemmateriaal. — Vakbl. Biol. 64, 392—396.

Ernst, W.H.O. 1985. De begroeiing als onderdeel van het duinecosysteem. Haalt het duin het jaar 2000? — Van Duingebruik naar Duinbeheer. Prov. Waterleidingbedrijf van Noord-Holland, Bloemendaal, pp. 30—39.

Ernst, W.H.O., Dueck, Th.A. & Lolkema, P.C. 1985. Genetische effekten van emissies van zware metalen op planten. — Lucht en Omgeving 2, 69—72.

Ernst, W.H.O. 1985. Gefährdete Pflanzen und Tiere — Sind sie noch zu retten? — Universitas 40, 1363—1374.

Ernst, W.H.O. 1986. Decompositie en een nieuwe synthese. — Vakbl. Biol. 66, 436—439.

Ernst, W.H.O. 1987. Stofgehalten in de bodem — opname door en effecten op de planten. — Symposium Bodemkwaliteit (1986), pp. 63—73. MB 86/44. Voorl. Techn. Commissie Bodem, Leidschendam.

Field study of heavy metal tolerance of plants growing in the localised area of zinc contamination under a zinc coated electricity pylon. (Photograph: T. McNeilly.)

CHAPTER 1

Stress tolerance in plants — the evolutionary framework

A. D. BRADSHAW & T. McNEILLY

Abstract. Stress environments, particularly those caused by man, provide a unique opportunity to observe evolution in action and the ways in which tolerance to stress originates. The basic mechanism is that proposed by Darwin, but the evolution is faster and more localised than we might expect. It is also limited by the availability of appropriate variation and by fitness in normal environments. Where stress conditions fluctuate, systems allowing facultative adaptation by phenotypic plasticity appear, but less is known about how they evolve.

1. The mechanism

The basic concept, proposed by Darwin, that evolution is driven by natural selection acting on heritable variation, is still, despite various proposals, the only acceptable explanation. It can be seen most elegantly in the evolution of tolerance to stress environments, since the stress occurring is obviously the factor providing the selective pressure, which in turn elicits the evolutionary change which can be observed.

It is of the essence of species that they vary, and that some of this variation is useful, some disadvantageous, and some neutral. Combining the possibility of useful — adaptive — variation, and selection favouring that variation, evolutionary change can be seen to be a very reasonable expectation. It is also reasonable to suggest that widespread species are widespread because of their ability to accommodate, by evolution, the breadth of ecological stresses they encounter.

2. The process in nature

How, then, do such changes come about in practice? To see evolution in action one needs to firstly go to species, then to populations within species, and ultimately to individuals within populations. Adaptive changes in species or populations are most easily appreciated when a species occurs already, or, better still, is in the process of spreading, across an environmental gradient. Changes in species structure and composition can then be directly related to the environmental change, and the different populations used to make critical comparisons about adaptation, uncomplicated by effects of different ancestry.

A particularly good example, for both morphological and physiological changes, is the genus *Achillea* along the transect from the coast to the high Sierra Nevada in California. Firstly there are two species which occupy differing climatic zones, and secondly each is subdivided into very distinct, locally adapted, populations. From the experiments where individuals were transplanted into

J. Rozema and J. A. C. Verkleij (eds.), Ecological Responses to Environmental Stresses, 2—5.

alien sites (Clausen *et al.* 1948), it is very clear that considerable adaptive changes have occurred within these species, and that without these changes the species would occupy a much narrower range of habitats.

Such evolution does not necessarily take place only over wide geographic distances. If the stress is localised then the evolution can be equally localised. Elegant evidence of this is shown by the extraordinarily localised (within 1 m) population differences in *Anthoxanthum odoratum* in relation to different lime and fertiliser treatments applied to meadow grassland in the Park Grass Experiment at Rothamsted Experimental Station (Snaydon & Davies 1976). An example from more natural environments is the very localised (within 5 m) occurrence of dwarf populations of *Agrostis stolonifera* on exposed cliffs at Abraham's Bosom, Anglesey (Aston & Bradshaw 1966).

Thus evolution is a crucial and interesting process leading to the occurrence of stress tolerant plants. Recently it has become clear that such evolution occurs not only in relation to naturally occurring stresses, but also to those brought about by man's activities, for example heavy metal tolerance, SO_2 resistance, and pesticide resistance (Bishop & Cook 1981). This again can be highly localised (Jain & Bradshaw 1966).

3. The process in man-made habitats

Tolerance to naturally occurring stresses may be of ancient origin and have evolved over a long period of time. The evolution to man-made stresses must be more recent. Indeed the evidence now accumulated (e.g., Ernst *et al.* 1982; Wu *et al.* 1975) shows us that such evolution can occur over very short time periods, not more than a few years. This is very different to what we would have expected 25 years ago. Furthermore, such adaptation can be very localised and also polyphyletic. Evolutionary adaptation to the same stress can occur independently in widely different geographical locations within the distribution of a species.

A perfect example of this is provided the replicated evolution of zinc tolerance in the populations of *Agrostis capillaris* beneath electricity pylons in Britain (Al-Hiyaly *et al.* 1988). Such evolution has occurred in numerous different situations since pylons were erected just 20–30 years ago. In each case it is restricted to the area immediately beneath the pylon, only 10 m square.

Ultimately evolution depends on the differential survival or growth of individuals. In certain metal contaminated habitats this can actually be seen. Not only can selective death and the survival of only metal tolerant individuals be shown, but also, among the survivors, those that are most tolerant growing the most successfully (Wu *et al.* 1975).

4. Limits to evolution

Electricity pylons provide a further insight into the evolutionary process. Each pylon causes at least some degree of zinc toxicity beneath it, sufficient to restrict plant growth severely, and there can be some areas completely devoid of vegetation. But although some degree of zinc tolerance has been found in five species, tolerant populations of four of these are restricted to a small proportion of the pylons examined, and only one shows a level of tolerance equal to that found on zinc mines. This implies that there are limits in the occurrence of zinc tolerance, which must be governed by the availability of genes conferring the appropriate adaptation. Furthermore, such genes as are present may be insufficient to allow full expression of the character in question.

The more familiar metal contaminated sites represented by abandoned metal mines show the same situation. Only certain species are found on them, despite the open habitats, again those that have the necessary variability for metal tolerance. But these sites combine metal toxicity with other stresses, such as low nutrient status, poor water-holding capacity, extremes of exposure, and erosion (Ernst 1974). The need for tolerance to these factors as well to metal toxicity clearly puts a further restraint on the frequency of occurrence of successful individuals from which tolerant population can develop. There are many deaths — life on a mine waste heap is very difficult (Farrow *et al.* 1981) — and the control over evolution exercised by availability of appropriate variability is probably commonplace (Bradshaw 1984).

5. Fitness and cost of evolution

Why should not tolerance, however, be common in all individuals and populations of a particular species? Why do populations from normal soils so rarely contain tolerant individuals? The evidence from heavy metal tolerance (Hickey & McNeilly 1976, Ernst 1982) is that the advantage enjoyed by tolerant individuals in toxic conditions is generally associated with disadvantages in normal conditions. Hence, in a normal population growing on normal soil, any individual showing tolerance is likely to be at a selective disadvantage. This itself provides a possible limit to evolution on contaminated soils, if the cost of tolerance is too high.

6. Other stress situations

Thus far, we have considered the evolution of metal tolerance. How generally, in terms of evolution in response to stress, can the model of metal tolerance be applied? What of other stress factors? Saline habitats are a good example. The ecological situation appears to be generally similar to metal contaminated habitats — there is the predominance of a single stress factor, but others may be involved. However, saline habitats have been in existence, firstly more generally, and secondly for a much longer period of time. Nevertheless, we again see the evolution of only a small number of species tolerant to salinity. Often these have both salt tolerant and normal populations in close proximity (Ashraf *et al*. 1986). But a considerable number of species are known which are endemic to saline habitats. While endemic metallophytes are known, they do not seem to occur in such numbers or be so well developed, perhaps because the habitats have not been so continuously available (Wild & Bradshaw 1977).

7. Facultative adaptation

Stress conditions due to salt and metal toxicity are more or less fixed, and stable with time, due to the magnitude of the toxicity, the relative immobility of the metal or salt in the soil, or its regular replacement by the sea. In such situations adaptations are likely to become fixed or constitutive, since changes in the environment which might reverse the advantage of the adaptation are unlikely. But there are many other stress environments, such as those associated with season, in which the stress can fluctuate considerably in time. Adaptations which can accommodate environmental changes during the lifetime of an individual organism cannot be fixed. They must be facultative, involving plasticity of the phenotype. We now realise that plants show a great wealth of such adaptations, both morphological and physiological (Bradshaw & Hardwick 1989).

Although facultative adaptations are different from constitutive adaptations in their modes of operation, their evolutionary significance is equivalent, in that they, too, can be genetically based. Hence their gene frequencies can also be altered by natural selection, and evolution of new and different systems of adaptation to stress is possible, subject to the constraints which we have already discussed. There is limited evidence for this, but, so far, more work has been done on the physiology of the mechanisms involved than on their evolution.

8. Conclusion

Many people still see stress environments as extreme and unusual, and therefore hardly worthy of study. We have aimed in this brief essay to show, from what can only be a limited viewpoint, how their study over the past 25 years has illuminated considerably our understanding of plant evolution and adaptation. The subsequent chapters will show just how broad the subject is now, and how much this owes to the enthusiasm, endeavour, and encouragement provided by Professor Dr W. H. O. Ernst, who has made stress environments an exciting and respectable subject.

References

Al-Hiyaly, S. A., McNeilly, T. & Bradshaw, A. D., 1988. The effects of zinc contamination from electricity pylons — evolution in a replicated situation. New Phytologist 110: 571—580.

Ashraf, M., McNeilly, T. & Bradshaw, A. D., 1986. Tolerance of sodium chloride and its genetic basis in natural populations of four grass species. New Phytologist 103: 725—734.

Aston, J. L. & Bradshaw, A. D., 1966. Evolution in closely adjacent plant populations II Agrostis stolonifera in maritime habitats. Heredity, London 21: 649—664.

Bishop, J. A. & Cook, L. M. (eds.), 1981. Genetic Consequences of Man Made Change. Academic Press, London.

Bradshaw, A. D. & Hardwick, K., 1989. Evolution and stress — genotypic and phenotypic components. In: P. Calow (ed) Evolution, Ecology and Environmental Stresss. Biological Journal of the Linnean Society 37: 137—155.

Bradshaw, A. D., 1984. The importance of evolutionary ideas in ecology — and vice versa. In: B. Shorrocks (ed.) Evolutionary Ecology. Blackwell, Oxford pp. 1—25.

Clausen, J., Keck, D. D. & Hiesey, W. M., 1948. Experimental studies on the nature of plant species 111 Environmental responses of climatic races of Achillea. Carnegie Institute of Washington Publication No. 581. Carnegie Institute, Washington.

Ernst, W. H. O., 1974. Schwermetallvegetation der Erde. Fischer, Stuttgart.

Ernst, W. H. O., 1982. Schwermetallpflanzen. In: H. Kinzel (ed.) Pflanzenökologie und Mineralstoffwechsel. Ulmer-Verlag, Stuttgart pp. 507—519.

Ernst, W. H. O., Verkleij, J. A. C. & Vooijs, R., 1982. Bioindication of surplus of heavy metals in terrestrial ecosystems. Environmental Monitoring and Assessment 3: 297—305.

Farrow, S., McNeilly, T. & Putwain, P. D., 1981. The dynamics of natural selection for tolerance in Agrostis canina L. subsp. montana Hartm. In: International Conference, Heavy Metals in the Environment, Amsterdam 1981. C.E.P. Consultants, Edinburgh pp. 289—295.

Hickey, D. A. & McNeilly, T., 1976. Competition between metal-tolerant and normal plant populations: a field experiment on normal soil. Evolution 29: 458—464.

Jain, S. K. & Bradshaw, A. D., 1966. Evolutionary divergence among adjacent plant populations I The evidence and its theoretical analysis. Heredity, London 21: 407—421.

Snaydon, R. W. & Davies, M. S., 1976. Rapid evolution in a mosaic environment IV Populations of Anthoxanthum odoratum at sharp boundaries. Heredity, London 37: 9—25.

Wild, H. & Bradshaw, A. D., 1977. The evolutionary effects of metalliferous and other anomalous soils in South Central Africa. Evolution 31: 282—293.

Wu, L., Bradshaw, A. D. & Thurman, D. A., 1975. The potential for evolution of heavy metal tolerance in plants 111 The rapid evolution of copper tolerance in Agrostis stolonifera. Heredity, London 34: 165—187.

Authors' Address
A.D. Bradshaw and T. McNeilly
Department of Environmental and Evolutionary Biology
University of Liverpool
Liverpool L69 3BX
U.K.

Effects of Cd on the root and shoot growth of metal-sensitive (a) and -tolerant (b) *Silene vulgaris* after three weeks exposure to 0 and 50 μM Cd. (Photograph: J. A. C. Verkley.)

CHAPTER 2

Heavy metal resistance in higher plants: biochemical and genetic aspects

J. A. C. VERKLEIJ, P. C. LOLKEMA, A. L. DE NEELING & H. HARMENS

Abstract. Copper, zinc, cadmium and aluminium resistance in higher plants is discussed with regard to genetic and biochemical aspects. In general, metal resistance is genetically determined and metal-specific. Due to the lack of a proper measure of metal resistance, it is still not known whether the resistance is controlled by major genes or whether it is polygenic. However, by applying a more accurate method evidence is brought forward that copper resistance is controlled by a major gene.

Resistance in higher plants is mainly based on tolerance, which implies uptake of heavy metals and the ability to "tolerate" excess internal metal concentrations. A complete exclusion of heavy metals is as yet not demonstrated.

Upon exposure to toxic concentrations of cadmium, zinc and copper metal tolerant and non tolerant plants synthesize metal-lothiopeptides or phytochelatins. However, there is no conclusive evidence that these compounds play an essential role in the tolerance mechanism. Other potential chelators such as organic acids could be of importance, but evidence for their role is lacking. Compartmentation of excess metals in subcellular bodies (vacuoles) or organs (leaves) seems an effective strategy to avoid toxic effects. For copper tolerance structural modifications at the level of the plasmamembrane could be of significance. Results of biochemical studies, carried out in cell suspension lines, have to be confirmed with studies at the cellular and higher integration levels in intact plants.

1. Introduction

The phenomenon of adaptation of plants to soils enriched with heavy metals is extensively documented (Ernst 1974, Antonovics *et al.* 1971, Baker 1987). A special vegetation is found on the various ore outcrops or near metal smelters and among these plant species, ecotypes are known with a specific resistance to different heavy metals. Plant species like *Agrostis capillaris* (= *A. tenuis*) and *Silene vulgaris* (= *S. cucubalus*) have been successful by changing their metabolism in such a way that they can grow and reproduce in metalliferous soils. Several heavy metals that may occur in metal-enriched soils are essential micronutrients such as Mn, Zn, Cu, Fe and Mo, whereas others such as Cd, Pb, Al and Ag are non-essential (Woolhouse 1983).

The nature and the degree of heavy metal resistance is determined by the soil conditions at a particular site. Metal resistance seems to be a quantitative characteristic (Urquart 1971), correlated with the metal concentration in the soil, which is available for the plant. Heavy metal resistance is largely metal-specific and only to those metals occurring at toxic concentrations in its environment. Multiple-metal resistance to two or more metals is often associated with the occurrence of high levels of those metals in the soil (Gregory & Bradshaw 1965). However, there are cases of cross-resistance or co-tolerance where resistance to one or more metals confers resistance to another metal present at a low non-toxic level (Cox & Hutchinson 1980, Walley *et al.* 1974, Lolkema 1985, Verkleij & Prast 1989). Moreover, low level resistance in plants to metals other than the primary contaminants has been suggested (Baker 1987) and observed in *Silene vulgaris* (Schat 1990).

J. Rozema and J. A. C. Verkleij (eds.), Ecological Responses to Environmental Stresses, 8—19.
© *1991 Kluwer Academic Publishers.*

Heavy metal resistant plants often show an increased demand for those metals the plants are resistant to as expressed by a maximal growth at elevated metal concentrations (Ernst 1983, Lolkema et al. 1984). Furthermore there is increasing evidence to suggest that species differ substantially in their sensitivities to heavy metals; some have high thresholds of resistance, others are very sensitive (Baker 1987). These phenomena can be interpreted as evidence for a constitutive rather than a plastic character of metal resistance (Ernst 1982). However, although the majority of metal resistance appears to be genetically determined, Baker et al. (1986) have described environmentally induced Cd-resistance in grass species. In cell suspensions of Datura the Cd-resistance can be increased by prior exposure to Cd (Jackson et al. 1987). These results at least suggest that heavy metal resistance is partly determined by metal induced changes in physiologically traits. However, we have to keep in mind that evolutionary acquired heavy metal resistance in intact plants might not be comparable with metal resistant cell suspension cultures of plant species, which normally do not show any adaptation at all in nature.

Although many genetic studies of heavy metal resistance have been carried out, it is still not known whether the resistance is controlled by major genes or whether it is polygenic (MacNair 1983). Because of the continuous spectrum of resistance when individual plants from resistant populations are sampled and the lack of clear segregation data in the F_1 between crosses of well-defined resistant and sensitive lines, the number of genes involved must be high (Bröker 1963, Gartside & McNeilly 1974). However, a number of genetic investigations have shown that resistance is probably controlled by one "major" gene (MacNair 1983, Schat & Van der Maarel 1990). In their genetic studies both authors did not use the frequently applied "Tolerance Index" (TI) as a measure of tolerance. As argued by MacNair (1983) this index lacks accuracy because of the inherently high level of statistical noise and the fact that root elongation is a complex process, determined by many genes unrelated to resistance. By applying

alternative methods (a qualitative index, MacNair 1983 or the EC_{100} for root growth, Schat & Van der Maarel 1990) their results indicate that Cu-resistance is governed by a single gene. This finding strongly supports the fact that the evolution of metal-resistant populations is extremely rapid (Bradshaw et al. 1965). However, even within a species the genetic basis of different Cu-resistant populations could be different indicating the complexity of the genetic system (Nicholls & McNeilly 1982, Schat & Van der Maarel 1990).

The physiological mechanisms of resistance to the various heavy metals are poorly understood (Ernst 1982, Woolhouse 1983). Resistance can be based on avoidance or tolerance sensu Levitt (1980). The former can be realized by metal exclusion or diminished metal uptake. A real exclusion will not be possible for essential metals like Cu, and Zn. Tolerance implies the uptake of heavy metals and the ability to "tolerate" elevated metal concentrations in the plant body.

Higher plants are in contrast to some microorganism (Silver et al. 1972) not capable of complete exclusion of heavy metals. Such an exclusion can be achieved by alteration of membrane permeability or increased exudation of metal chelating substances and have as yet not reported in higher plants. A third exclusion posibility namely by changing the metal binding capacity of the cell wall is claimed by Turner & Marshall (1972), who found a correlation between Zn-resistance and Zn-binding capacity of root cell walls of Agrostis tenuis. This cell wall hypothesis is extensively discussed in a recent review (Verkleij & Schat 1990) and we will briefly revert to it in discussing the individual elements.

Assuming that avoidance is a less important strategy for higher plants, this review will mainly concentrate on the tolerance mechanisms, particularly on the production of intracellular metal-binding compounds and storage of metals in cellular compartments (compartmentation). Because tolerance seems to be highly metal specific we will discuss the metal specific mechanisms at the cellular and tissue level. The main emphasis will be placed on the tolerance to the essential metals copper and zinc and the non-essential ones, cadmium and aluminium. Cadmium and copper have recently attracted much attention in the metal-detoxification research and the other two metals, especially zinc, have been investigated by the research group of Ernst since a long time.

1.1. Copper tolerance/genetic studies

Since the observation of Prat (1934) that plants of *Melandrium rubrum,* raised from seeds from a copper-mine, could survive on a Cu-enriched soil in contrast to plants from a normal soil, it is generally agreed that Cu-tolerance is genetically determined. However, the numerous studies of the genetic control of this phenomenon have produced controversial results. Using a polycross technique, Gartside & McNeilly (1974) interpreted the estimates of the heritability of Cu-tolerance in a number of grasses as evidence for polygenic control. Although they found that tolerance was dominant over non-tolerance, Allen & Sheppard (1971) observed in F_1 progeny of crosses between tolerant and non-tolerant *Mimulus guttatus* that this phenomenon of dominance was dependent on the Cu concentration (recessive at high Cu-concentrations).

As we previously mentioned in the introduction several investigators questioned the use of the so-called "root elongation method" as a proper measure for genetic studies (MacNair 1983, Baker 1987, Schat & Van der Maarel 1990). Instead of using the tolerance index, i.e.

$$TI = \frac{\text{root growth in heavy-metal solution}}{\text{root growth in control solution}}$$

MacNair screened cuttings of non-tolerant and tolerant *Mimulus guttatus* plants at two Cu-concentrations in such a way that non-tolerant plants failed to root in the highest concentration. The tolerant plants were still able to show normal rooting. By means of this qualitative index he cogently demonstrated that copper tolerance in the yellow monkey flower is governed by a major gene (MacNair 1983). However, Schat & Van de Maarel argued that MacNair's method is fundamentally unsuitable for an analysis of a genetic system that controls the level of tolerance because of its qualitative

character. In their study they introduced a quantitative tolerance index, the EC_{100} for root growth (Schat & Van der Maarel 1990). Based on screening for tolerance of Cu-tolerant and non-tolerant populations of *Silene cucubalus* and F_1 and F_2 families derived from crosses, they also came to the conclusion that tolerance is governed by a single gene. The level of tolerance seems to be controlled by a number of hypostatic "enhancers". Although the proposed model of the genetic control could link up with a physiological biochemical one, it is not testable because the biochemistry of Cu-tolerance is as yet virtually unknown.

1.2. Copper tolerance/biochemical studies

Tolerance to Cu has been recorded for a great number of higher plants (Woolhouse 1983). Diminished uptake of Cu is sometimes observed in tolerant plants, e.g., *Silene cucubalus* (Ernst 1972, Lolkema *et al.* 1984), and proposed as one of the strategies of Cu-tolerance. Although we have to keep in mind that differences in metal uptake between tolerant and sensitive plants are based on Cu-concentrations in root tissue after long treatment of high external concentrations (toxic for the sensitive plant, but not for the tolerant one, Ernst *et al.* 1990), short term exposure to Cu indicated similar results (Lolkema 1985, De Vos *et al.* 1991). However, a diminished capacity for Cu-uptake in Cu-tolerant plants will not prevent high internal concentrations. This means that an effective cellular detoxification mechanism must exist in the roots of tolerant plants.

Since the discovery of metallothioneins (MTs) in animals after exposure to certain heavy metals and their possible role in metal detoxification, MT-like compounds have been reported being important Cu- and other metal-complexing agents in higher plants (Rauser & Curvetto 1984, Lolkema *et al.* 1984, Robinson & Thurman 1986, Schulz & Hutchinson 1988). However there is increasing evidence that phytochelatins are the major metal-complexing compounds not only in higher plants or plant cell suspension cultures, but also in algae and some yeast species (Grill *et al.* 1987, Gekeler *et al.* 1988, Murasugi *et al.* 1983). These phytochelatins (PCs)

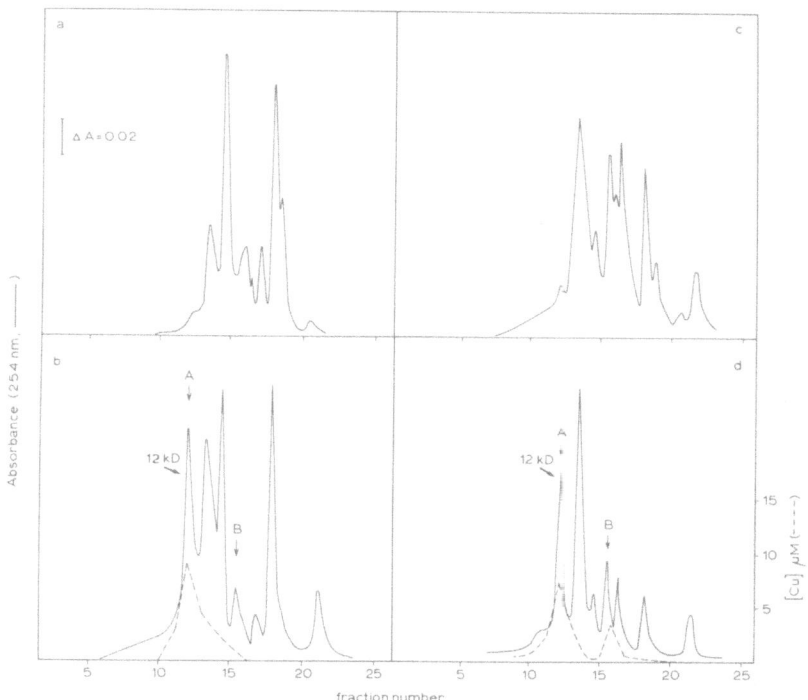

Fig. 1. FPLC-Superose 12 gel filtration profiles of root extracts of a copper-tolerant (a, b) and a copper-sensitive (c, d) plant of *Silene cucubalus* after three weeks growth on $1 \mu M$ (a, c) and $40 \mu M$ copper (b, d) (Verkleij *et al.* 1989).

which consist of poly (γ-glutamyl cysteinyl)glycine have been alternatively been called $(\gamma EC)_n G$ (Jackson *et al.* 1987), cadystins (Murasugi *et al.* 1983), γ-glutamyl peptides (Reese & Winge 1988) and metallothiopeptides (MTP, Verkleij *et al.* 1989). The occurrence of MT-like protein in early reports in plants containing apparent Mol. Wt. of

10 kD must now be interpreted in terms of PCs with n = 2–6 (Grill *et al.* 1987).

The synthesis of cytosolic metal-binding compounds after Cu-stress does not mean that these PCs play a major role in the tolerance mechanism. Upon growth in medium containing $40 \mu M$ Cu both Cu-tolerant and non-tolerant *Silene cucubalus* synthesize PCs having Mol. Wt. of 12.5 kD (Fig. 1, Verkleij *et al.* 1989). Also in sensitive and tolerant *Mimulus guttatus* these compounds are induced upon Cu-exposure (Robinson & Thurman 1986). In both plant species a small fraction of the total Cu-amount in the roots seems to be bound to the metal binding compounds. Moreover, after shortterm exposure to Cu, more PCs are synthesized in roots of Cu-sensitive *Silene vulgaris* compared with those of the tolerant one (Verkleij *et al.* 1990a). Apparently, induction of PCs depends on the intracellular toxic Cu-concentration, which is higher in the roots of sensitive plants (Table I, Verkleij *et al.* 1990a). Like metallothionein (MT), PCs may play a role in

Table I. Phytochelatin (PC) production (12.5 kD peak height in mm) and intracellular Cu-content (nmol/g dw.) in Cu-tolerant and non-tolerant plants of *Silene vulgaris* after exposure to $40 \mu M$ Cu during 4 day treatment.

Days after exposure	Cu-sensitive		Cu-tolerant	
	Cu	PC	Cu	PC
1/2	80	22	79	18
1	156	40	148	21
2	300	55	220	31
4	500	160	245	50

11

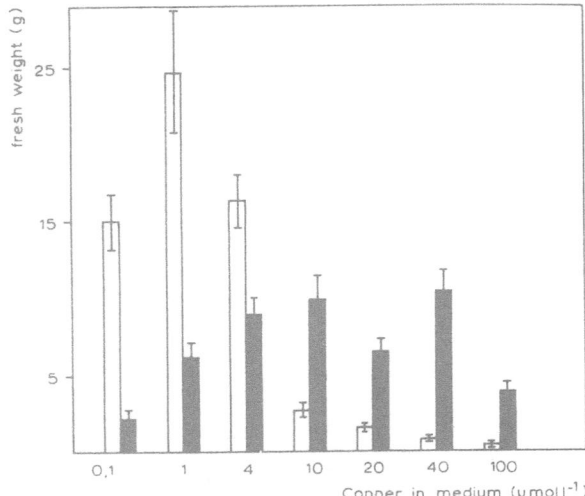

Fig. 2. Biomass production of a copper-sensitive (open columns) and a copper-tolerant (filled columns) population of *Silene cucubalus* expressed as fresh weight per plant after a growth period of 20 d on a range of copper concentrations in the nutrient solution (Lolkema *et al.* 1984).

the homeostatic control of cellular Cu-availability levels or in the Cu-detoxification. However, these compounds seem not be involved in Cu-tolerance.

Accumulation of excess Cu in organs or subcellular compartments is another strategy to achieve tolerance. In root cell walls of a number of dicotyledonous metallophytes only a few percent of the total root Cu content was found (Ernst 1972, 1974) and no difference was observed between tolerant and non-tolerant plants. Organelles such as chloroplasts contained similar amount of Cu "in vivo" in tolerant and sensitive *Silene cucubalus* (Lolkema & Vooijs 1986). The main storage organelle of metals is the vacuole and using electron microprobe technique Mullins *et al.* (1985) detected vacuolar Cu in both sensitive and tolerant plants. In contrast to the supposed role of the vacuole in the Zn- and Cd-tolerance mechanism, its function in Cu-tolerance is unclear. Especially the formation of soluble Cu-complexes in the Cu-hyperaccumulator species from Haut Shaba could be compartmented in the cell vacuoles (Brooks *et al.* 1985).

A lower root to shoot ratio for Cu-content was observed in Cu-tolerant *Silene cucubalus* indicating an increased root to shoot translocation (Lolkema *et al.* 1984). Although we have to be careful in interpreting these data because of the long term exposure (see Introduction) accumulation into the shoots could mean another possibility for removing excess Cu via leaf fall.

None of the above-mentioned strategies seems to be decisive for Cu-tolerance. The most promising results come from studies of Cu-induced damage to cell plasma membranes in roots of *Silene cucubalus* (De Vos *et al.* 1989). As was originally suggested by Wainwright & Woolhouse (1977) structural modifications at the level of the plasma membrane could be of significance for Cu-tolerance. Cu-tolerant plants of *S. cucubalus* showed a much lower K-leakage from their roots upon Cu-exposure than sensitive plants. Excessive Cu-levels cause lipid peroxidation and Cu-mediated thiol oxidation of the plasmamembrane (see De Vos & Schat, Chapter 3). In Cu-tolerant plants of *S. cucubalus* these damaging effects occur only at much higher concentrations than in sensitive plants, probably due to constitutional alteration of the plasmamembrane structure (De Vos *et al.* 1991). More detailed studies are needed to verify this hypothesis.

One of the consequences of being Cu- or Zn-tolerant is the diminished maximum growth rate (or biomass production) of tolerant plants compared with the sensitive ones (Fig. 2, Lolkema *et al.* 1984). The cost of tolerance, which could come to about 40% in terms of growth capacity, might go in energy dependent processes such as membrane transport (Ernst 1982). However, in order to clarify the costs of tolerance a better understanding of the physiological and biochemical mechanisms is required.

2. Zinc tolerance/genetic and biochemical studies

Zn-tolerance has been reported for a great number of plant species (Antonovics *et al.* 1971, Ernst 1974). It usually coincides with increased levels of Cd- and Pb-tolerance, which is not surprising in view of the frequent co-occurrence of increased levels of these metals in metalliferous soils (Cox

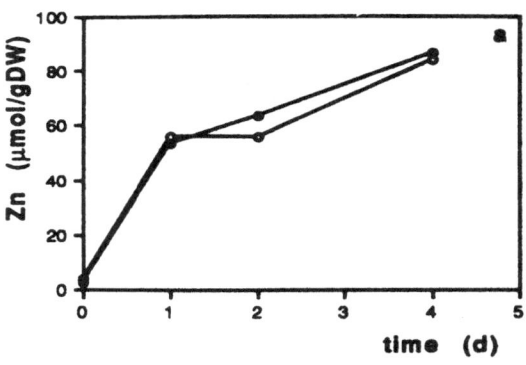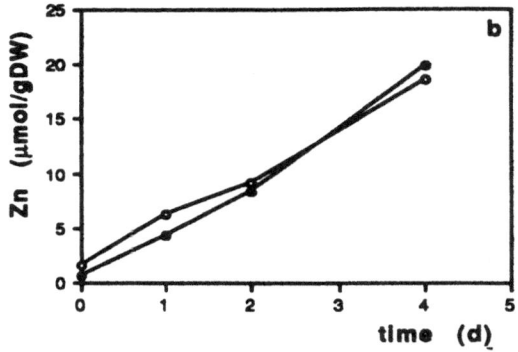

Fig. 3. Zinc content (μmol/g DW) of the roots (a) and the leaves (b) of zinc-sensitive (○) and tolerant (●) *Silene vulgaris* after 0–4 days of growth on a nutrient solution with 400 μM zinc (Harmens *et al.* 1989).

1986). Though Zn-tolerance has often developed in response to increased availability levels of Zn, a certain degree of Zn-tolerance has been observed in Cu-tolerant populations, which are not exposed to excessive Zn in the soil (Symeonides *et al.* 1985, Allen & Sheppard 1971, Schat & Van der Maarel 1990).

The high Zn-content of Zn-resistant plants shows that resistance is not achieved by an exclusion mechanism (See Introduction). Although there is a strong accumulation of Zn in the cell wall of root cells, no difference exists in the uptake of Zn and the Zn-binding capacity of the root cell wall from tolerant and non-tolerant populations (Fig. 3, Harmens *et al.* 1989, Thurman & Collins 1983). Only in British populations of *Agrostis tenuis* Turner and Marshall (1972) found a correlation between Zn-tolerance and the Zn-binding capacity of the root cell wall.

The mechanism of Zn-tolerance must be inside the cell. No difference is found in the sensitivity of enzymes to Zn (Mathys 1975, Ernst *et al.* 1975). Zn may induce increased PC levels, at least in suspension-cultured cells (Grill *et al.* 1987), though not always (Krotz *et al.* 1989). However, Buthionine Sulfoximine (BSO), an inhibitor of the glutathion synthesis, does not increase Zn-sensitivity suggesting that PCs are not decisive for Zn resistance (cf. Reese & Wagner 1987). After exposure to 400 μM Zn Harmens *et al.* (1989) could isolate in roots of Zn-sensitive and -tolerant plants of *Silene cucubalus* a low Mol. Wt. compound, which probably

consists of PCs. Grill and coworkers identified PCs in the roots of Zn-sensitive *Acer speudoplatanus* and zinc-tolerant *S. cucubalus* collected from the field (Grill *et al.* 1988). The very low PC/Zn-concentration ratio indicates that PCs probably do not play a primary role in the mechanism of Zn-tolerance.

Ernst and Mathys explained Zn-tolerance as resulting from an increased ability to transport Zn into the vacuole (Ernst 1975, Mathys 1977). Malate, which is usually present in higher contents in Zn-tolerant plants would serve as a carrier for Zn during cytoplasmic and tonoplast transport. After releasing Zn in the vacuole, malate is transported back into the cytoplasm, where it can pick up other Zn-ions (Fig. 4). In callus cultures of Zn- and Pb-tolerant and non-tolerant *Anthoxantum odoratum* higher levels of malate were found in both types of calli after exposure to Zn (Quereshi *et al.* 1986).

Unlike malate, the citrate contents of Zn-tolerant plants is not always higher in all investigated species (Thurman *et al.* 1982). There is no direct correlation between Zn-tolerance and oxalate contents (Mathys 1977). In Zn-tolerant plants of *S. cucubalus* a higher malate and citrate content was observed in the roots and leaves compared with those of the sensitive ones (Harmens *et al.* 1989). An induction of the malate and citrate synthesis was found in the roots of tolerant plants in the first two days after exposure to Zn. Both acids may play a role in the detoxification of Zn although it is difficult to explain why the malate and citrate con-

13

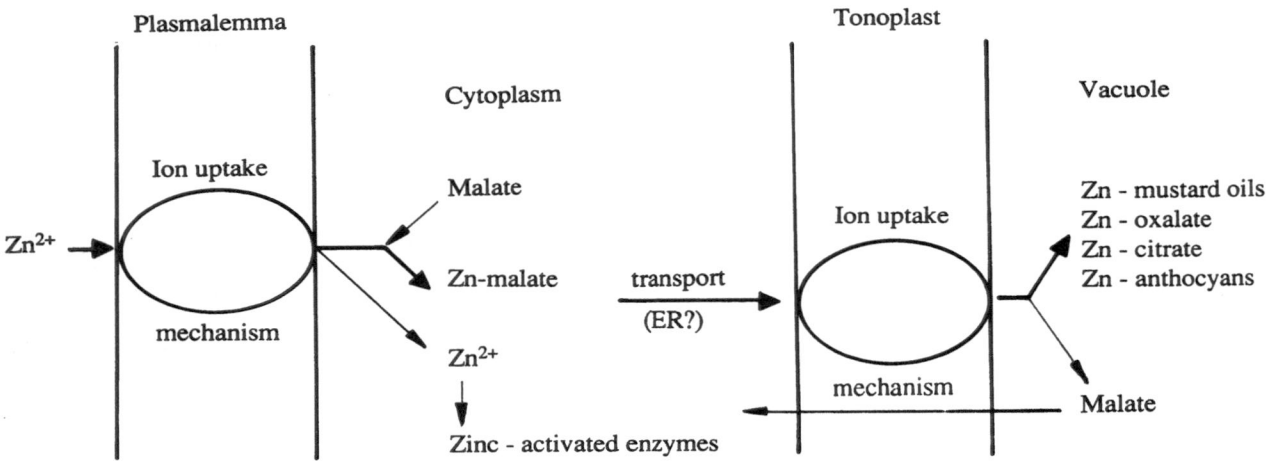

Fig. 4. Proposed model of the zinc-malate shuttle (Ernst 1975).

tent of the roots decreased between day two and four, while the Zn-content was still increasing. A primary role of organic acids in the mechanism of Zn tolerance has not been confirmed by any direct evidence. Van Steveninck suggested that Zn-tolerance in *Deschampsia caespitosa* might depend on the ability to detoxify Zn by storing it as Zn-phytate (Van Steveninck *et al.* 1987).

3.1. *Cadmium tolerance/genetic studies*

Cd-tolerance has been recorded in a number of plant species, primarily in grasses (Simon 1977, Ernst 1980), but also in dicotyledonous species (Verkleij & Prast 1989). Because Cd-enrichment inevitably co-occurs with that of Zn and Pb in metalliferous soils, these plant species show multiple-tolerance to these metals. It is even suggested that the mere presence of Zn- and Pb-tolerance automatically implies cross-resistance (co-tolerance) to Cd (Cox 1979). The data Cox showed for *Deschampsia caespitosa* to underline his thesis were recently supported by studies on *Silene cucubalus* (Verkleij *et al.* 1990b). However, not only Zn-Pb-tolerant, but also an extremely Cu-tolerant population of *S. cucubalus,* only exposed to Cu, showed a high co-tolerance to Cd (Verkleij & Prast 1989).

As a consequence Cd-tolerance might be just a phenomenon of cross-resistance and could be achieved by different mechanisms. The remarkable partial loss of Cd-tolerance of *Holcus lanatus,* which occurred after cultivation on an uncontaminated soil after 6 years could be explained by some metal-induced changes in physiological traits (Baker *et al.* 1986).

In addition to the evolutionary acquired metal tolerance in higher plants, metal-tolerant cell tissue and cell suspension cultures have been developed *in vitro* using plant species, normally not exposed to elevated concentrations in nature. Especially for the study of Cd-detoxification mechanisms at the cellular level, high Cd-tolerant cell lines are artificially selected in tomato, tobacco and *Datura innoxia* (Steffens *et al.* 1986, Reese & Wagner 1987, Jackson *et al.* 1987). Some of these Cd-tolerant cell lines were very sensitive to Zn (Steffens, personal communication). However, none of these cell lines have been successfully regenerated to whole plants, exhibiting the same high level of tolerance.

In contrast to Zn- and Cu-tolerant plants at increasing Zn- and Cu-concentrations (Ernst 1982, Lolkema *et al.* 1984) Cd-tolerant ones did not show a distinct growth optimum at any of the Cd-concentrations applied (Fig. 5, Verkleij *et al.* 1990c, Baker *et al.* 1986). Being a non-essential metal, Cd is not

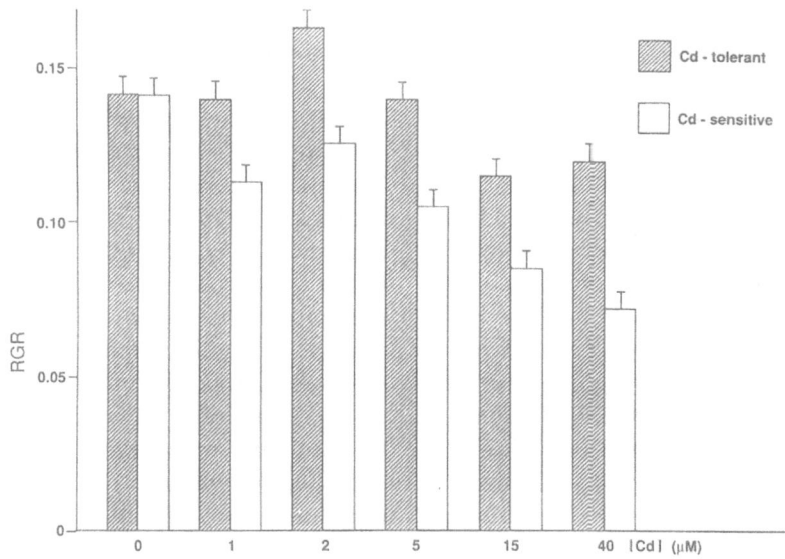

Fig. 5. Effects of cadmium on the relative growth rate (RGR) of Cd-sensitive and Cd-tolerant plants of *Silene vulgaris* (Verkleij *et al.* 1990c).

required in metabolic processes which could explain this phenomenon.

3.2. *Cadmium tolerance/biochemical aspects*

The observed higher root/shoot ratio of the Cd-content in Cd-tolerant plants of *Holcus lanatus* (Coughtry & Martin 1978) and *Silene vulgaris* (Verkleij & Prast 1989) after longterm exposure does not give conclusive evidence for a restricted translocation (because of secondary effects, see Introduction). However, the Cd-concentration in the roots by far exceeds that of the shoots. Cd-accumulation in the root cell wall is very low and no difference between tolerant and sensitive plants was observed (Verkleij *et al.* 1990c), which is in accordance with the non-specificity of the cell wall system. The excess of intracellular Cd has therefore to be detoxified to prevent harmful effects on the internal metabolic processes.

As already mentioned PCs are synthesized in higher plants upon Cd exposure. The high apparent Mol. Wt. estimates of these Cd-binding compounds, observed in various plant species, could either be explained by differences in chromatographic separation techniques and/or in the num-

ber of the γ-glu-cys moiety (n varies 2—6, Grill *et al.* 1987). Upon Cd-exposure both Cd-sensitive and Cd-tolerant plants or cell lines synthesized the same amounts of PCs (in *Datura innoxia*, Delhaize *et al.* 1989). However in tolerant tomato cell suspensions higher amounts of PCs were produced in combination with a higher γ-glutamyl cysteine synthetase activity (Steffens *et al.* 1986), suggesting that an increased capacity for PC-synthesis might be involved in Cd-tolerance. Also upon Cd-exposure glutathion depletion was less in tomato tolerant cell lines compared with sensitive ones (Scheller *et al.* 1987). Furthermore, addition of BSO to cell suspensions increased the Cd-sensitivity of the cell line (Reese & Wagner 1987). No difference in PC production and molecular size was found between Cd-sensitive and Cd-tolerant *Silene vulgaris* upon Cd-stress. However, the PCs in the Cd-tolerant ones contained twice as much Cd and acid labile sulfide (Verkleij *et al.* 1990c). Differences in labile sulfide content were already reported in fission yeast (Reese & Winge 1988) and tomato cell lines (Steffens *et al.* 1986) suggesting a crucial role of sulfide in the Cd-tolerance mechanism. In Cu-PC complexes no labile sulfide could be detected (Reese & Winge 1988, Verkleij *et al.* 1990a). An

15

essential role of sulfide could involve stimulation of the assimilatory SO_4^{2-}-reduction pathway (Robinson 1989) and recently elevated enzyme activities of this pathway were found in Cd-sensitive *Zea mays* upon Cd-treatment (Nussbaum *et al.* 1988).

The essentiality of Cd-PC in the Cd-detoxification with or without labile sulfide, is challenged by recent studies on tobacco suspension cells (Krotz *et al.* 1989). Their results suggest that tobacco cells accomodate the presence of toxic Cd and Zn by sequestration in the vacuole as complexes with endogenous organic acids. This strategy may be a principal means for accomodation of Cd as well as Zn in the presence and absence of Cd-PCs. Some evidence of this idea comes from vacuolar deposits of Zn and Cd, which were detected both in various sensitive and tolerant plant species (Ernst & Weinert 1972, Mullens *et al.* 1985, Rauser & Ackerly 1987). More profound studies of the mechanism of Zn and Cd transport systems at the tonoplast of sensitive and tolerant plants are necessary to prove this hypothesis.

4. Aluminium/biochemical and genetic studies

Aluminium (Al) toxicity occurs almost exclusively on acid soils due to the insolubility of Al at pH values higher than 4.5 (Woolhouse 1983). The major toxicity effects are inhibition of root elongation, cell division and disturbed uptake and transport of calcium, magnesium, potassium and phosphorus (Foy 1974).

Plant species and genotypes within species differ widely in tolerance to Al including a number of wild plants (e.g., De Neeling & Ernst 1986) and crop cultivars (e.g., Foy 1974) and some of these differences in tolerance are heritable. An avoidance strategy to limit entry of Al into the symplasm would be effective and several exclusion mechanisms are proposed and have received experimental support (Taylor & Foy 1985). Exclusion could be achieved by means of a pH barrier at the rhizosphere. However, avoidance of Al-uptake by a limited proton extrusion is not universal and could not represent a specific adaptation to Al-toxicity (Verkleij & Schat 1990). Other avoidance mecha-

nisms proposed are selective permeability of plasma membrane, exudation of chelates or immobilization of Al in the cell wall (Foy *et al.* 1978, Hecht-Buchholz & Schuster 1987).

Nonetheless, experimental support for exclusion is incomplete. Some species contain high concentrations of Al in their tissues and an internal tolerance mechanism must therefore be involved to detoxify the excess of Al. A number of Al-chelating compounds have been reported such as citrate, malate, phenolic compounds and proteins. It is unclear whether Al in these metal-ligand complexes will be stored in the vacuoles or will be transferred to alternative ligands already present in the vacuoles (Woolhouse 1983). Based on the numerous genetic and physiological studies it can be concluded that different mechanisms are involved with a varying degree of complexity, which are species-dependent (Woolhouse 1983, Verkleij & Schat 1990).

5. Conclusions

The genetics of heavy metal resistance in higher plants is hardly known. This is mainly due to the fact that the method applied for measuring resistance (i.e., root elongation) lacks accuracy. Moreover, this method could be very useful to assess Cu-resistance, but seems less appropriate for other metals (like Zn and Cd). In contrast to these two metals toxic external concentrations of Cu directly inhibit root growth and induce potassium leakage (De Vos *et al.* 1989). In order to solve the genetics of heavy metal resistance, more precise methods measuring the direct toxic effects of metals have to be developed.

In the recent research on the mechanism of heavy metal resistance main emphasis is placed on the metal binding compounds especially the metallothiopeptides or phytochelatins. However, this mechanism, as for the cell wall binding, fails to explain the metal specificity of the metal resistance. For, phytochelatin production is observed in both metal-sensitive and tolerant plants, phytochelatins are induced by a large number of heavy metals and phytochelatins are even found in plant species, not capable to evolve resistance to any metal in nature. Therefore it is likely that these metal binding peptides play a role in the homeostatic control of metal ions; in the mechanism of metal tolerance their role is limited and cannot be generalized.

The plasma membrane seems to be more important in the metal resistance process. Comparative studies onto the composition and structure of membranes isolated from sensitive and resistant plants should be carried out. Also the role of the cell wall in the metal uptake process must be evaluated. Compartmentation of metals in the various subcellular bodies could be another important process, which should be studied.

Furthermore it is worthwhile to study the metal transport via the xylem and phloeem vessels in the intact plants.

To evaluate the proposed mechanisms of heavy metal tolerance we really need to know the mechanisms of metal toxicity. However, the primary receptor molecules and target processes for the different metals are largely unknown. So a fundamental research on the molecular toxicology of heavy metals has to be started.

References

Antonovics, J., Bradshaw, A. D. & Turner, R. G., 1971. Heavy metal tolerance in plants. Adv. Ecol. Res. 7: 1—85.

Baker, A. J. M., 1978. Metal tolerance. New Phytol. 80: 635—642.

Baker, A. J. M., Grant, C. J., Martin, M. H., Shaw, S. C. & Whitebrook, J., 1986. Induction and loss of cadmium tolerance in *Holcus lanatus* L. and other grasses. New Phytol. 102: 575—587.

Bradshaw, A. D., McNeilly, T. & Gregory, R. P. G., 1965. Industrialization, evolution and the development of heavy metal tolerance in plants. pp. 327—344. In: G.T. Goodman, R.W. Edwards & J.M. Lambert (eds.), Ecology and the Industrial Society, The British Ecol. Soc. Sym. No 5. Blackwell, Oxford.

Bröker, W., 1963. Genetisch-physiologische Untersuchungen über die zinkverträglichkeit von *Silene inflata* Sm. Flora (Jena) 153: 122—156.

Brooks, R. R., Malaisse, F. & Empain, A., 1985. The heavy metal-tolerant flora of South Central Africa. A. A. Balkema, Rotterdam/Boston.

Coughtrey, P. J. & Martin, M. H., 1978. Cadmium uptake and distribution in tolerant and non-tolerant populations of *Holcus lanatus* grown in solution culture. Oikos 30: 555—560.

Cox, R. M., 1986. Contamination and effects of cadmium in native plants. pp. 101—109. In: H. Mislin & O. Ravera (eds.), Cadmium in the Environment, Experientia Supplementum Vol 50, Birkhauser Verlag, Basel.

Cox, R. M. & Hutchinson, T. C., 1980. Multiple metal tolerances in the grass *Deschampsia caespitosa* (L.) Beauv. from the Sudbury smelting area. New Phytol. 84: 631—647.

De Neeling, A. J. & Ernst, W. H. O., 1986. Manganese and aluminium tolerance of *Senecio sylvaticus* L. Acta Oecol./ Oecol. Plant. 7: 43—56.

De Vos, C. H. R., Schat, H., De Waal, M. A. M., Vooys, R. & Ernst, W. H. O., 1991. Increased resistance of the root cell plasmalemma to Cu^{2+}-induced permeability damage in copper tolerant *Silene cucubalus*. Physiol. Plant. (submitted).

De Vos, C. H. R., Schat, H., Vooijs, R. & Ernst, W. H. O., 1989. Copper-induced damage to the permeability barrier in roots of *Silene cucubalus*. J. Plant Physiol., 135: 164—169.

Delhaize, E., Jackson, P. J., Lujan, L. D. & Robinson, N. J., 1989. Poly(γ-glutamylcysteinyl)glycine synthesis in *Datura innoxia* and binding with cadmium. Plant Physiol. 89: 700—706.

Ernst, W., 1972. Schwermetallresistenz und Mineralstoffhaushalt. Westdeutscher Verlag: Opladen.

Ernst, W., 1974. Schwermetallvegetation der Erde. Gustav Fischer Verlag: Stuttgart.

Ernst, W. H. O., 1975. Physiology of heavy metal resistance in plants. pp 121—136. In: T.C. Hutchinson, S. Epstein, A.L. Page, J. Van Loon & T. Davey (eds.), Proc. Int. Conf. on Heavy Metals in the Environment. CEP Consultants, Toronto.

Ernst, W. H. O., 1980. Biochemical aspects of cadmium in plants. pp. 639—653. In: J.O. Nriagu (ed.), Cadmium in the Environment. J. Wiley and Sons, New York.

Ernst, W. H. O., 1982. Schwermetallpflanzen. pp. 472—505. In: H. Kinzel (ed.), Pflanzenökologie und Mineralstoffwechsel, Eugen Ulmer Verlag, Stuttgart.

Ernst, W. H. O., 1983. Ökologische Anpassungsstrategien an Bodenfaktoren. Ber. Deutsch. Bot. Ges., 96: 49—71.

Ernst, W. H. O., 1985. Schwermetallimmisionen-ökophysiologische und populations-genetische Aspekte. Düsseldorfer Geobotanische Kolloquien, 2: 43—57.

Ernst, W. & Weinert, H., 1972. Lokalisation von Zink in den Blättern von *Silene cucubalus* Wib. Z. Pflanzenphysiol. 66: 258—264.

Ernst, W. H. O., Schat, H. & Verkleij, J. A. C., 1990. Evolutionary biology of metal resistance in *Silene vulgaris*. Evol. Trends Plants 4: 45—51.

Foy, C. D., 1974. Effects of aluminium on plant growth. In: E.W. Carson (ed.) The plant root and its environment, Charlotte University Press, Charlotterville 601.

Foy, C. D., Chaney, R. L., & White, M. C., 1978. The physiology of metal toxicity in plants. Ann. Rev. Plant Physiol. 28: 511—566.

Gartside, D. W. & McNeilly, T., 1974. The potential for evolution of heavy metal tolerance in plants. II. Copper tolerance in normal populations of different plant species. Heredity 32: 335—348.

Gekeler, W., Grill, E., Winnacker, E.-L. & Zenk, M. H., 1988. Algae sequester heavy metals via synthesis of phytochelatin complexes. Archives of Microbiology 150: 197—202.

Gregory, R. P. G. & Bradshaw, A. D., 1965. Heavy metal tolerance in populations of *Agrostis tenuis* Sibth. and other grasses. New Phytol. 64: 131—143.

Gries, B., 1966. Zellphysiologische Untersuchungen über die Zinkresistenz bei Galmeiökotypen und Normalformen von *Silene cucubalus*. Flora, B156: 271—290.

Grill, E., Winnacker, E.-L. & Zenk, H. H., 1987. Phytochelatins, a class of heavy-metal binding peptides from plants, are functionally analogous to metallothioneins. Proceed. Nat. Acad. Sc. USA 84: 439—443.

Grill, E., Winnacker, E. L. & Zenk, M. H., 1988. Occurrence of heavy metal binding phytochelatins in plants growing in a mining refuse area. Experientia 44: 539—540.

Harmens, H., Verkleij, J. A. C., Koevoets, P. & Ernst, W. H. O., 1989. The role of organic acids and phytochelatins in the mechanism of zinc tolerance in *Silene vulgaris* (= *Silene cucubalus*). pp. 178—181. In: J.P. Vernet (ed.) Heavy Metals in the Environment, Geneva 1989, Vol. 2. CEP Consultants: Edinburgh.

Hecht-Buchholz, Ch. & Schuster, J., 1987. Responses of Al-tolerant Dayont and Al-sensitive Kearney barley cultivars to calcium and magnesium during Al-stress. Plant and Soil 99: 47—61.

Jackson, P. J., Unkefer, C. J., Doolen, J. A., Watt, K. & Robinson, N. J., 1987. Poly(γ-glutamylcysteinyl)glycine; its role in cadmium resistance in plant cells. Proceed. Nat. Acad. Sci. USA 84: 6619—6623.

Krotz, R. M., Evangelou, B. P. & Wagner, G. J., 1989. Relationships between cadmium, zinc, Cd-peptide, and organic acid in tobacco suspension cells. Plant Physiol. 91: 780—787.

Levitt, J., 1980. Responses of Plants to Environmental Stresses. Vol. 1. Academic Press: New York.

Lolkema, P. C., 1985. Copper resistance in higher plants. Thesis Vrije Universiteit, Amsterdam.

Lolkema, P. C., Donker, M. H., Schouten, A. J. & Ernst, W. H. O., 1984. The possible role of metallothioneins in copper tolerance of *Silene cucubalus*. Planta 162: 174—179.

Lolkema, P. C. & Vooijs, R., 1986. Copper tolerance in *Silene cucubalus*. Subcellular distribution of copper and its effects on chloroplasts and plastocyanin synthesis. Planta 167: 30—36.

MacNair, R. M., 1983. The genetic control of copper tolerance in the yellow monkey flower, *Mimulus guttatus*. Heredity 50: 283—293.

Mathys, W., 1975. Enzymes of heavy-metal resistant and non-resistant populations of *Silene cucubalus* and their interaction with some heavy metals in vitro and in vivo. Physiol. Plant. 33: 161.

Mathys, W., 1977. The role of malate, oxalate and mustard oil glycosides in the evolution of zinc-resistance in herbage plants. Physiol. Plant., 33: 161—165.

Mullins, M., Hardwick, K. & Thurman, D. A., 1985. Heavy metal location by analytical electron microscopy in conventionally fixed and freeze-substituted roots of metal tolerant and non tolerant ecotypes. pp. 43—46. In: Proceed. Inter. Conf. on Heavy Metals in the Environment, Athens. CEP Consultants: Edingburgh.

Murasugi, A., Wada, C. & Hayashi, Y., 1983. Occurrence of acid-labile sulfide in cadmium-binding peptide 1 from fission yeast. J. Biochem. 93: 661—664.

Nicholls, M. K. & McNeilly, T., 1979. Sensitivity of rooting and tolerance to copper in *Agrostis tenuis* Sibth. New Phytol. 83: 653—664.

Nussbaum, S., Schmutz, D. & Brunold, C., 1988. Regulation of assimilatory sulfate reduction by cadmium in *Zea mays* L. Plant Physiol. 88: 1407—1410.

Prat, S., 1934. Die Erblichkeit des Resistanz gegen Kupfer. Ber. Dt. Bot. Ges. 52: 65—67.

Quereshi, J. A., Hardwick, K. & Collin, H. A., 1986. Malic acid production in callus cultures of zinc and lead tolerant and non-tolerant *Anthoxanthum odoratum*. J. Plant. Physiol. 112: 477—479.

Rauser, W. E. & Ackerley, C. A., 1987. Localization of cadmium in granules within differentiating and mature root cells. Can. J. Bot. 65: 643—646.

Reese, R. N. & Wagner, G. J., 1987. Effects of buthionine sulfoximine on Cd-binding peptide levels in suspension-cultured cells treated with Cd, Zn or Cu. Plant Physiol. 84: 574—578.

Reese, R. N. & Winge, D. R., 1988. Sulfide stabilization of the cadmium-γ-glutamyl peptide complex of *Schizosaccharomyces pombe*. J. Biol. Chem. 114: 12832—12835.

Robinson, N. J., 1989. Algal metallothioneins: secondary metabolites and proteins. J. Appl. Phycol. 1: 5—18.

Robinson, N. J. & Jackson, P. J., 1986. "Metallothionein-like" metal complexes in angiosperms; their structure and function. Physiol. Plant. 67: 499—506.

Robinson, N. J. & Thurman, D. A., 1986. Isolation of a copper complex and its rate of appearance in roots of *Mimulus guttatus*. Planta 169: 192—197.

Schat, H. & Van der Maarel, W. M., 1990. Genetic control of copper tolerance in *Silene cucubalus* Wib. Funct. Ecol. (in press).

Schat, H., 1990. Manuscript in preparation.

Scheller, H. V., Huang, B., Hatch, E. & Goldsbrough, P. B., 1987. Phytochelatin synthesis and glutathion levels in response to heavy metals in tomato cells. Plant Physiol. 85: 1031—1035.

Schulz, C. L. & Hutchinson, T., 1988. Evidence against a key role for metallothionein-like protein in the copper tolerance mechanism of *Deschampsia cespitosa* (L.) Beauv. New Phytol. 110: 163—171.

Silver, S., Johnseine, P., Whitney, E. L. & Clark, D., 1972. Manganese-resistant mutants of *Escherichia coli*, physiological and genetical studies. J. Bacteriol. 110: 186—192.

Simon, E., 1977. Cadmium tolerance in populations of *Agrostis tenuis* and *Festuca ovina*. Nature 265: 328—330.

Steffens, H. V., Hunt, D. F. & Williams, B. G., 1986. Accumulation of non-protein-metal-binding polypeptides (γ-glutamyl-cysteinyl)$_n$ glycine in selected cadmium-resistant tomato cells. J. Biol. Chem. 261: 13879—13882.

Symeonides, L., McNeilly, T. & Bradshaw, A. D., 1985. Differential tolerance of three cultivars of *Agrostis capillaris* L. to cadmium, copper, lead, nickel and zinc. New Phytol. 101: 309—000.

Taylor, G. J. & Foy, C. D., 1985. Mechanisms of aluminium tolerance in *Triticum aestivum* L. (wheat). I. Differential pH induced by winter cultivars in nutrient solutions. Am. J. Bot. 72: 695—701.

Taylor, G. J., 1988. Mechanism of aluminium tolerance in *Triticum aestivum* (wheat). V. Nitrogen, nutrition, plant-induced pH and tolerance to aluminium; correlation without causality? Can. J. Bot. 66: 694—699.

Thurman, D. & Collins, J. C., 1983. Metal tolerance mechanisms in higher plants — A review. pp. 298—304. In: Proc. Inter. Conf. on Heavy Metals in the Environment, Heidelberg. CEP Consultants: Edinburgh.

Turner, R. G. & Marshall, C., 1972. The accumulation of zinc by subcellular fractions of roots of *Agrostis tenuis* Sibth. in relation to zinc tolerance. New Phytol., 71: 671—676.

Urquhart, C., 1971. Genetics of lead tolerance in *Festuca ovina*. Heredity 26: 19—33.

Van Steveninck, R. F. M., Van Steveninck, M. E., Fernando, D. R., Horst, W. J. & Marshner, H., 1987. Deposition of zinc phytate in globular bodies in roots of *Deschampsia caespitosa* ecotypes: a detoxification mechanism? J. Plant Physiol. 131: 247—257.

Verkleij, J. A. C., Bast-Cramer, W. B. & Levering, H., 1985. Effects of heavy-metal stress on the genetic structure of populations of *Silene cucubalus*. pp. 355—365. In: J. Haeck & J.W. Woldendorp (eds.) Structure and Functioning of Plant Populations. 2. Phenotypic and Genotypic Variation in Plant Populations. North-Holland Publishing Company: Amsterdam.

Verkleij, J. A. C. & Prast, J. E., 1989. Cadmium tolerance and cotolerance in *Silene vulgaris* (Moench.) Garcke [= *S. cucubalus* L. Wib.]. New Phytol. 111: 637—645.

Verkleij, J. A. C. & Schat, H., 1990. Mechanisms of metal resistance in higher plants. pp. 179—193. In: J. Shaw (ed.) Evolutionary Aspects of Heavy Metal Tolerance in Plants, CRC Press, New York.

Verkleij, J. A. C., De Nobel, W. T. & Nydam, Y. (1990a). Effects of Cd on germination, growth and reproduction in *Silene vulgaris* (manuscript in preparation).

Verkleij, J. A. C., Koevoets, P., Van 't Riet, J., Van Rossenberg, M. C., Bank, R. & Ernst, W. H. O., 1989. The role of metal-binding compounds in the copper tolerance mechanism of *Silene cucubalus*. pp. 347—357. In: D. Winge & D. E. Hamer (eds.) Metal Ion Homeostasis: Molecular Biology and Chemistry. Alan R. Liss Inc., New York.

Verkleij, J. A. C., Koevoets, P. L. M., Van 't Riet, J. & Ernst, W. H. O., 1990b. Phytochelatin production in tolerant and non-tolerant *Silene vulgaris* (manuscript in preparation).

Verkleij, J. A. C., Koevoets, P. L. M., Van 't Riet, J., Bank, R., Nijdam, Y. & Ernst, W. H. O., 1990c. Poly(γ-glutamyl-cysteinyl)glycines or phytochelatins and their role in cadmium tolerance of *Silene vulgaris*. Plant, Cell & Environ. (in press).

Wainwright, S. J. & Woolhouse, H. W., 1977. Some physiological aspects of copper and zinc tolerance in *Agrostis tenuis* Sibth.: Cell elongation and membrane damage. J. Exp. Bot., 25: 1025—1036.

Walley, K., Khan, M. S. I. & Bradshaw, A. D., 1974. The potential for evolution of heavy metal tolerance in plants. I. Copper and zinc tolerance in *Agrostis tenuis*. Heredity, 32: 309—319.

Woolhouse, H. W., 1983. Toxicity and tolerance in the responses of plants to metals. pp. 245—289. In: O.L. Lange, P.S. Nobel, C.B. Osmond & H. Ziegler (eds.) Encyclopedia of Plant Physiol., vol. 12, Physiol. Plant Ecol. III., Springer, Berlin.

Authors' Addresses
J. A. C. Verkleij, P. C. Lolkema & H. Harmens
Department of Ecology and Ecotoxicology
Vrije Universiteit,
De Boelelaan 1087
1081 HV Amsterdam
The Netherlands

A. L. De Neeling
Rijksinstituut voor de Volksgezondheid en milieuhygiëne
P.O. Box 1
3720 BA Bilthoven
The Netherlands

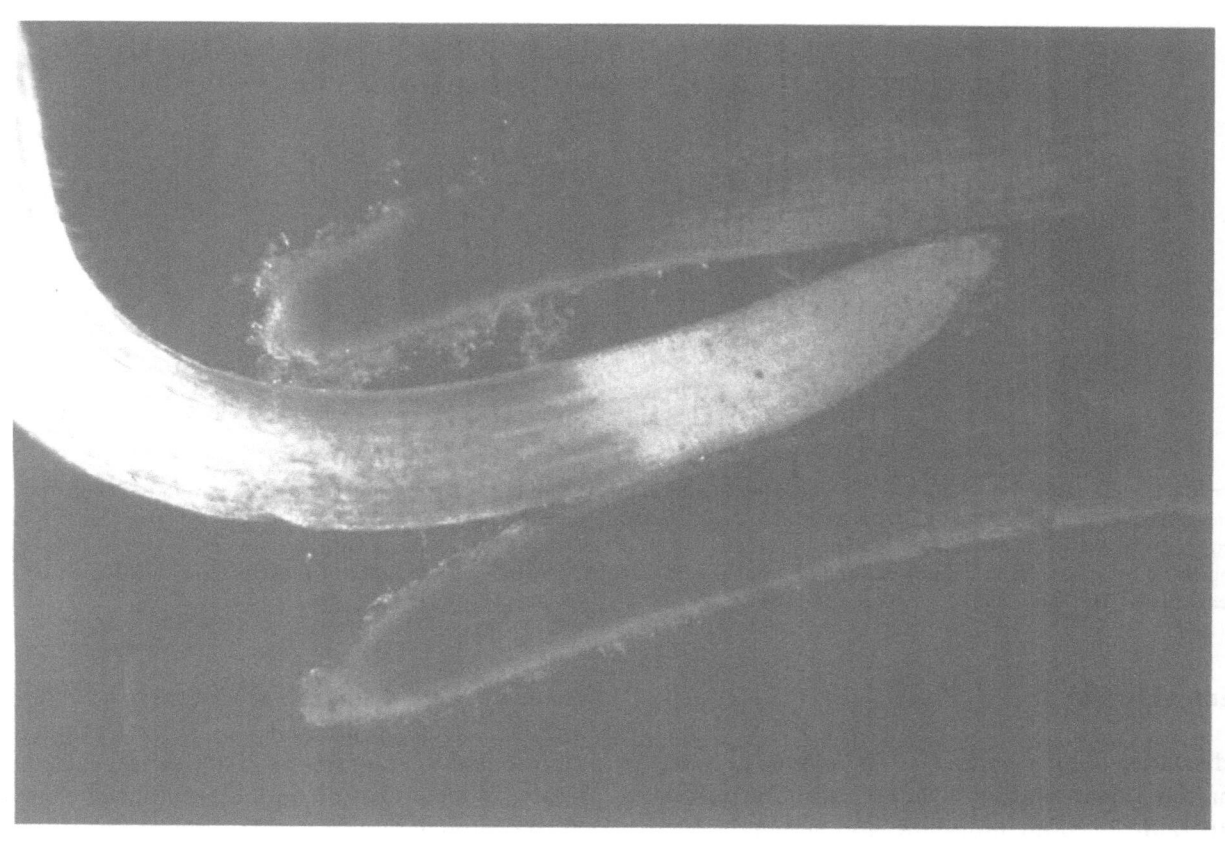

Trypan blue staining of cells, indicating loss of plasmalemma permeability and severe damage of cell viability, in root tips of *Silene cucubalus* after a 24 hour exposure to the free radical-producing compound cumene hydroperoxide (above), the sulfhydryl reagent N-ethylmaleimide (middle) or excess copper (below). (Photograph: C. H. R. De Vos.)

CHAPTER 3

Free radicals and heavy metal tolerance

C. H. R. DE VOS & H. SCHAT

Abstract. There are many possible ways in which heavy metals can affect the generation and the metabolization of cellular radical species. Due to a lack of *in vivo* studies, it is difficult to assess the significance of oxidative stress in metal toxicity in higher plants. Except for copper toxicity, which is associated with peroxidation of the root cell membranes, there are no indications that metal toxicity would primarily rely on a disturbance of the cellular radical balance. It is uncertain whether the evolution of metal tolerance involves changes in the capacity of the cellular defence against radicals.

1. Introduction

Radicals are defined as species having one or more unpaired electrons. Living cells continuously generate radicals, both inorganic and organic ones. Some of them are highly reactive and damaging. Heavy metals such as copper, iron and manganese are in many ways involved in the production and breakdown of cellular radicals. There are several reasons to suppose that shifts in the intracellular concentrations of these and other heavy metals, regardless of whether they are biologically essential or not, will have more or less pronounced effects on the production and/or the breakdown of radicals. It seems possible therefore, that radicals may be involved in the development of metal deficiency or metal toxicity symptoms. At present, it is difficult to estimate the precise role of radicals in metal toxicity phenomena; in general, the molecular toxicology of heavy metals is poorly understood as yet.

This paper aims to provide some ideas and evidence with respect to the role of radicals and cellular defence mechanisms in metal toxicity and metal tolerance phenomena in higher plants, especially with regard to copper. The questions that will be dealt with are the following. Firstly, in which ways can excessively accumulated metals affect the formation and degradation of reactive radicals? Secondly, to what extent can metal-imposed changes of the cellular radical balance be considered as a primarily toxic effect of the metal in question? Thirdly, to what extent does a metal's phytotoxicity depend on primary or secundary effects on the radical balance? Finally, are changes in the cellular defence against radicals involved in evolutionary acquired metal tolerance? The discussion of these problems is preceded by an introductory survey of some general radical reactions, radical damage and protective mechanisms in aerobic cells.

2. Oxygen radicals and lipid peroxidation

The partial oxygen pressure is an extremely important determinant of the radical production in biological systems. Although oxygen is essential for the great majority of organisms on earth, it is basically a toxic substance: organisms adapted to anoxic environments usually die upon exposure to oxygen and, moreover, increasing the oxygen pres-

J. Rozema and J. A. C. Verkleij (eds.), Ecological Responses to Environmental Stresses, 22—30.
© *1991 Kluwer Academic Publishers.*

sure above ambient levels is harmful to aerobic organisms too. There is no doubt that this is due to the formation of reactive oxygen species in their cells (Halliwell 1974).

In spite of its high oxidizing potential, molecular oxygen in the ground state is not extremely reactive, not even in the fairly reducing intracellular environment. This is due to the so-called spin restriction, imposed by the presence of two unpaired electrons with equal spin (molecular ground state oxygen is a di-radical). However, the cellular metabolism has oxygen activating properties: it converts molecular oxygen into other species of which the reactivity is not limited by the spin restriction. Most of them are the products of "incomplete" reduction of the oxygen molecule. A one electron reduction produces the superoxide anion (O_2^-). Superoxide is only moderately reactive: it can react with various macromolecules, though at a low rate. Its formation is mainly due to "leakage" of electrons from electron transport chains and to xanthine oxidase activity (Halliwell & Gutteridge 1985). A two electron reduction produces hydrogen peroxide (H_2O_2), which is a non-radical, in contrast to superoxide. H_2O_2 production mainly results from the reactions catalyzed by glycollate oxidase, urate oxidase and superoxide dismutase (SOD) (see below). H_2O_2 is not extremely reactive; it is known to inactivate SOD and several Calvin cycle enzymes (Halliwell & Gutteridge 1985). A three electron reduction of oxygen produces water (H_2O) and the hydroxyl radical ($OH^\bullet$), which is extremely reactive and damaging: it will oxidize any organic molecule that happens to be present at the site of its formation. It is formed by a one electron reduction of H_2O_2 by a reduced metal ion or metal chelate. This is the so-called Fenton reaction:

$$Me^{n+} + H_2O_2 \rightarrow Me^{(n+1)+} + OH^\bullet + OH^- \quad (1)$$

Superoxide may provide the reduced metal required:

$$M^{(n+1)+} + O_2^- \rightarrow Me^{n+} + O_2 \quad (2)$$

The net reaction is a metal-catalyzed H_2O_2 reduction by O_2^-

$$O_2^- + H_2O_2 \rightarrow O_2 + OH^\bullet + OH^- \quad (3)$$

Metals capable of the Fenton reaction are found in the d-block of the periodic table; many of them have unpaired electrons in the 3d and/or 4s orbitals and exhibit a variable valency ("transition metals"), which allows them to participate in one electron transfer reactions. Especially iron and copper are very effective catalysts of such reactions. Another damaging oxygen species is singlet oxygen ($^1O_2$), which is formed in so-called photosensibilization reactions. It is a highly reactive non-radical, which is formed when ground state oxygen absorbs energy from light-excited pigment molecules. It is responsible for "photobleaching", i.e., the oxidative destruction of pigments in light.

One of the most damaging effects of $OH^\bullet$ is the initiation of the peroxidative degradation of polyunsaturated fatty acid chains of membrane lipids. The vulnerability of polyunsaturated fatty acids to peroxidative breakdown relies on the fact that a double bond in a carbon chain weakens the C-H bond in the neighbouring methylene groups. Lipid peroxidation is a chain reaction, which can be subdivided into three phases. It starts with the abstraction of a hydrogen atom from a methylene group of a fatty acid chain by an initiator, which is usually $OH^\bullet$:

$$OH^\bullet + LH \rightarrow L^\bullet + H_2O \quad (4)$$

This is called the primary initiation. In the presence of oxygen, the lipid radical $L^\bullet$ will rapidly form a lipid peroxy radical:

$$L^\bullet + O_2 \rightarrow LOO^\bullet \quad (5)$$

The lipid peroxy radical is sufficiently reactive to abstract a hydrogen atom from another fatty acid chain:

$$LOO^\bullet + L'H \rightarrow LOOH + L'^\bullet \quad (6)$$

The lipid radical $L'^\bullet$ can enter reaction 5 and the result is a chain of the reactions 5 and 6. This is called the propagation phase, during which an increasing number of polyunsaturated fatty acids are transformed to lipid hydroperoxides (LOOH). The third phase is the termination phase, during which the peroxidation process is stopped by termination reactions in which radicals react with each other to form less reactive non-radical products, e.g.:

$$LOO^\bullet + L'OO^\bullet \rightarrow LOOL' + O_2 \qquad (7)$$

The lipid hydroxides formed during the propagation phase are fairly stable. However, in the presence of iron or copper ions they are transformed into more reactive species:

$$LOOH + Me^{n+} \rightarrow LO^\bullet +$$
$$Me^{(n+1)+} + OH^- \qquad (8)$$

$$LOOH + Me^{(n+1)+} \rightarrow LOO^\bullet +$$
$$Me^{n+} + H^+ \qquad (9)$$

Like peroxy radicals ($LOO^\bullet$), alkoxy radicals ($LO^\bullet$) are capable to abstract hydrogen from unsaturated fatty acids. Initiation by alkoxy radicals or peroxy radicals formed from hydroxyperoxides is called secundary initiation or hydroperoxide-dependent initiation.

The lipid peroxidation process generates several additional products: peroxy radicals may form cyclic endoperoxides which may hydrolyze to form malondialdehyde (MDA), low molecular fatty acids and various hydrocarbon radicals; alkoxy radicals may fall apart into alkane radicals and aldehydes. Aldehydes may attack proteins and alkane radicals abstract hydrogen to form volatile products such as ethane or pentane.

MDA and several endoperoxides react with thiobarbituric acid (TBA), which property is often used in quantitative measurements of lipid peroxidation. Also ethane or pentane production are used as a measure of lipid peroxidation. However, the products formed during peroxidation depend on the lipid composition of the membrane as well as on the availability of metal compounds capable of transforming lipid hydroperoxides.

3. Cellular defence mechanisms

As indicated above, free radicals and reactive oxygen species may exert a variety of damaging effects, which are often referred to as "oxidative stress". Apart from lipid peroxidation, oxidative stress may include direct damage to proteins, amino acids, nucleic acids, porphirines, phenolic substances, etc. Aerobic organisms have evolved very effective defence mechanisms against oxygen radicals and reactive intermediates of lipid peroxidation reactions.

The first line of defence includes factors that prevent the formation of, or scavenge oxygen species with a direct or indirect damaging effect. An important way to prevent the formation of reactive oxygen species is to increase the production of UV-absorbing substances, such as flavonoids. The protective effect of flavonoids and other phenolic compounds is not entirely based on UV-interception, but may as well rely on scavenging of oxygen radicals (Larson 1988). Other non-enzymic scavengers are ascorbate (O_2^-, $OH^\bullet$), glutathione ($OH^\bullet$, $^1O_2$), vitamin E (tocopherol) (O_2^-, $OH^\bullet$) and glucose ($OH^\bullet$). However, their importance *in vivo* in the first line of defence is uncertain, or at least not entirely based on direct scavenging of oxygen radicals (see below). Beside these and other non-enzymic scavengers, aerobic cells contain extremely effective enzymic scavengers of O_2^- and H_2O_2. At first sight this may seem surprising, in view of the comparatively low reactivity of O_2^- and H_2O_2, but it merely confirms the idea that the toxic potential of these species relies on their potency to participate in hydroxyl generating reactions, rather than on their reactivity *per sé* (see above). Moreover, scavenging $OH^\bullet$ will be less effective, due to the extremely high and aspecific reactivity of this species. Superoxide dismutase (SOD) catalyzes the dismutation of O_2^-.

$$2O_2^- + 2H^+ \rightarrow H_2O_2 + O_2 \qquad (10)$$

In eukaryotic cells SOD occurs in two forms, usually, viz. CuZnSOD, which is located in the cytoplasm and the chloroplasts, and MnSOD, which is located in the mitochondria (Halliwell & Gutteridge 1985). The copper and the manganese atom, respectively, participate in the dismutation reaction. Catalase is a haem-protein, located in the peroxisomes, which catalyzes the breakdown of H_2O_2:

$$2 H_2O_2 \rightarrow 2 H_2O + O_2 \qquad (11)$$

The functional analogue of catalase within the chloroplast is the non-metallo-enzyme ascorbate peroxidase, which uses ascorbate as an electron donor:

$$AH^- + H^+ + H_2O_2 \rightarrow A + 2 H_2O \qquad (12)$$

The resulting dehydro ascorbate (A) is reduced by dehydro ascorbate reductase, which uses glutathione as a reducing substrate:

$$A + 2 GSH \rightarrow AH^- + H^+ + GSSG \qquad (13)$$

The oxidized glutathione (GSSG) is subsequently reduced by glutathione reductase:

$$GSSG + NADPH + H^+ \rightarrow 2 GSH + NADP^+ \qquad (14)$$

Plant cells contain a number of other enzymes with peroxidase activity and a rather non-specific substrate-preference (Halliwell & Gutteridge 1985); their role in H_2O_2 breakdown *in vivo* is uncertain. The selenium-dependent glutathione peroxidase, which is assumed to degrade organic hydroperoxides and H_2O_2 in animal cells, does not occur in plants. Finally, metal binding compounds may as well contribute to the first line of defence, not only by reducing the availability of transition metals for the Fenton reaction, but also by direct scavenging of oxygen radicals. Metallothioneins, for example, are effective radical scavengers (Thornalley & Vasák 1985) and possibly their functional analogues in higher plants, the so-called phytochelatins, will have this property too.

The second line of defence includes factors that can stop the chain reaction in lipid peroxidation. The most important one is tocopherol (vitamin E), a lipid soluble antioxidant which traps peroxy radicals:

$$ROO^\bullet + ArOH \rightarrow ROOH + ArO^\bullet \qquad (15)$$

The resulting tocopheryl radical ($ArO^\bullet$) may react with a second peroxy radical, upon which the antioxidant activity is lost. It may be as well reduced by ascorbate, which leads to the recovery of ArOH:

$$ArO^\bullet + AH^- + H^+ \rightarrow ArOH + AH^\bullet \qquad (16)$$

The ascorbyl radical can be reduced by $AH^\bullet$-reductase (semidehydroascorbate-reductase):

$$AH^\bullet + NADH \rightarrow AH^- + NAD^+ + H^+ \qquad (17)$$

It may also spontaneously disproportionate to AH^- and A (dehydroascorbate). Beside preventing the destruction of vitamin E, ascorbate may act directly as a chain breaker. Glutathione may also play a role in the second line of defence, as it might prevent irreversible two electron oxidation of vitamin E (Haenen & Bast 1983).

The third line of defence includes factors that prevent LOOH-dependent initiation of lipid peroxidation, usually by converting lipid hydroperoxides into their corresponding alcohols. This reaction can be catalyzed by several enzymes, such as glutathione peroxidase (GSH-Px), which exists both in a selenium-dependent and a selenium-independent form, glutathion transferases with peroxidase activity and cytochrome P-450, a haem protein. The former ones catalyze the reaction:

$$LOOH + 2 GSH \rightarrow LOH + GSSG + H_2O \qquad (18)$$

4. Metal toxicity and oxidative stress

As outlined above, heavy metals such as copper, iron, zinc and manganese are essential elements in the free radical defence mechanism of living cells. However, excessive concentrations of these and other metals might disturb the cellular radical balance, leading to oxidative stress. The way in which the radical balance will be affected depends on the metal species involved.

It seems that under conditions of excessive uptake of redox-active metals free radical-generating reactions are likely to increase. Firstly, copper and iron are involved in the production of several highly reactive oxygen species. Hydroxyl radicals might be formed in the metal-catalyzed Haber-Weiss cycle (reaction 3), which is probably the most important source of this radical species *in vivo*. The required reduction of the metal can be provided by cellular reductants, such as O_2^-, ascorbate or thiols (Girotti 1985). However, the significance of increased $OH^\bullet$ formation in metal-induced tissue injury *in vivo* is still uncertain. In cells, only a small minority of the copper and iron will be available for the Haber-Weiss reaction (Czapski *et al.* 1983). Free ionic copper and iron will not be present and functionally bound metals are usually less liable to oxidation by H_2O_2 or reduction by O_2^- or other cellular compounds. Still, it seems likely that cells maintain a small reserve of low molecular iron chelates, such as Fe-ADP, Fe-GDP and Fe-citrate, which are capable to catalyze the formation of $OH^\bullet$ (Halliwell & Gutteridge 1985). Superoxide may stimulate the formation of these compounds, as it is able to reduce ferritin bound iron, which promotes its liberation from the ferritin molecule (Thomas *et al.* 1985). Chelated copper has also been reported to mediate the production of $OH^\bullet$. However, in contrast to the iron-catalyzed cycle, the biological damage is assumed to take place directly at the site of copper-binding (Czapski *et al.* 1983, Rowley & Halliwell 1983).

Superoxide radicals might be formed by the metal-mediated one-electron reduction of dioxygen. Kumar *et al.* (1978) demonstrated that erythrocyte membranes produce O_2^- in the presence of copper ions. In their experiments, the production of these radicals was mediated by the membrane protein thiols:

$$R\text{-}SH + Cu^{2+} \rightarrow R\text{-}S + Cu^+ + H^+ \qquad (18)$$

$$Cu^+ + O_2 \rightarrow Cu^{2+} + O_2^- \qquad (19)$$

Following the spontaneous or enzyme-catalyzed dismutation of O_2^-, the production of H_2O_2 might increase too. Apart from its own damaging effect, H_2O_2 formation might give rise to the generation of $OH^\bullet$, through the metal-catalyzed Fenton-reaction. In addition, $^1O_2$ has been reported to be produced during copper-mediated hydroperoxide-dependent lipid peroxidation reactions (Ding & Chan 1984). Heavy metals are reported to block electron transport chains of cellular membranes, such as thylacoids (Shioi *et al.* 1978). This might lead to incomplete reduction of molecular oxygen and subsequently to increased formation of oxygen free radicals. In isolated chloroplasts and green algae, the toxic effect of copper apparently relies on the direct action of the copper ions on the thylacoid membranes, involving the blocking of the photosynthetic electron transport (Shioi *et al.* 1978) and subsequent free radical damage to membrane lipids (Sandmann & Böger 1980a, b). However, reports on copper-dependent inhibition of photosynthetic electron transport in higher plants *in vivo* are contradictory (Gora *et al.* 1985, Baszynski *et al.* 1988). In addition, Lolkema & Vooijs (1986) did not find any considerable increase in the copper content of chloroplasts in copper-stressed *Silene cucubalus*. As yet, there is no evidence that in higher plants copper toxicity would primarily rely on a blocked electron transport and subsequent peroxidative destruction of thylacoids or other membranes.

Secondly, excessive uptake of redox-active heavy metals might give rise to increased formation of lipid free radicals, such as alkoxy and peroxy radicals, leading to peroxidative damage to cellular membranes. The involvement of redox metals in the initiation of lipid peroxidative processes might rely on the Haber-Weiss generation of $OH^\bullet$ and on the conversion of hydroperoxides (Girotti 1985). With respect to the hydroperoxide-dependent initiation, Chan *et al.* (1982) demonstrated that cupric ions are more efficient than ferric or ferrous iron. In their *in vitro* experiments, the other metals tested (zinc, cobalt, cadmium, nickel, manganese, mercury, chromium and tin) were without detectable effect, suggesting a unique role of copper and iron in the conversion of hydroperoxides and subsequent lipid peroxidation. Recently, it has been demonstrated that the absolute ratio's of reduced versus oxidized metal might account for the initiation of hydroperoxide-independent lipid peroxidation, without the involvement of $OH^\bullet$ (Braughler *et al.* 1986, Minotti & Aust 1987). This might suggest that *in vivo* OH• formation through Haber-Weiss catalyzation is not a necessity. In contrast the redox-status of the metal ions, which depends on the chelator species, might be the determinant for lipid peroxidation to occur. Anyway, irrespective of the nature of initiation, the crucial role of redox-metals in the induction of lipid peroxidation reactions has been well established (Girotti 1985). In fact, in many medical and pharmacological laboratory's iron is routinely used to start the lipid peroxidative process, for instance in order to examine the sensitivity of biological membranes for oxidative stress.

Another possibility for heavy metal-induced oxidative stress might be a decreased defence against free radicals. Heavy metals in general have a more or less high affinity for cellular thiols, such as glutathione and protein thiols, causing oxidation and/or cross-linking of sulfhydryls. Thiols, especially glutathione, play an important role in the free radical defence system of cells, and decreasing the cellular thiol level may cause oxidative stress. Thiol depletion leading to free radical damage has been frequently reported (Younes & Siegers 1981, Ribanov *et al.* 1982, DiMonte *et al.* 1984, Mehlhorn *et al.* 1986). In addition, due to competition for membrane carriers or via other mechanisms, high extracellular levels of one metal species might lead to a decreased uptake of metal species involved in the radical defence mechanisms. For instance, high copper availability may provoke iron-deficiency symptoms, expressed in a decreased activity of iron-containing enzymes, such as catalase (Permual & Beattie 1965). In the same way, metal species acting upon the copper, zinc or manganese homeostasis might affect the SOD-activity. Cardinaels *et al.* (1984) suggested that the lowered Cu/Zn SOD-activity in cadmium-stressed bean plants was due to a decreased zinc uptake. It is obvious that such a disturbance will lead to a lowered defence against reactive oxygen species, resulting in free radical-mediated tissue injury. However, the significance of oxidative stress imposed by such a mechanism is doubtful: in most cases, the direct effect of the increased concentration of the toxic metal itself will have more important effects on cell functioning.

In view of the numerous reports on metal-mediated free radical damage *in vitro,* including experiments with cell cultures such as erythrocytes, there are many reasons to suppose that the phytotoxicity of heavy metals might rely on damage by free radical formation, especially in the case of redox-active metals. However, there are only a small number of reports dealing with metal-induced oxidative stress *in vivo,* especially in the case of higher plants. In one of these reports, it has been suggested that oxygen radicals are involved in the toxicity of high concentrations of ferrous iron to higher plants growing in submerged soils (Hendry & Brocklebank 1985). In this report, it was demonstrated that the amount of TBA-reactive material, as a measure of lipid peroxidation, in the plant tissues was increased with increasing iron in the anaerobic nutrient solution. However, the authors failed to check the effect of the elevated tissue iron levels on the assay procedure, as chromogen formation strongly depends on the redox-metal concentration in the extract (Girotti 1985). Recently, oxidative damage has been suggested to play a primary role in the mechanism of copper-intoxication of higher plants *in vivo* (De Vos *et al.* 1989). These authors report that the toxic effects of high nutrient copper levels on roots of *Silene cucubalus,* viz. damage to plasma membrane permeability, loss of cell viability and accumulation of lipid peroxidation products, were also obtained with supply of cumene hydroperoxide, an organic peroxide known to induce lipid peroxidation by free radical formation in membranes (Fig. 1).

5. Role of free radical defence in metal tolerance

Populations of some plant species are known to tolerate high intracellular levels of heavy metals, including copper (Ernst 1974). How do these metal tolerant plants, especially the copper-tolerant ones, manage to live with such a potential of free radical producing agents, without suffering from oxidative stress? For higher plants, it has been reported that the free radical defence system might play an important role in the detoxification of compounds, of which the phytotoxicity can be ascribed to free radical formation (Lee & Bennett 1982,

27

Pauls & Thompson 1982, Schmidt & Kunert 1986). It seems likely therefore, that the cellular defence mechanism might play a role in metal tolerance phenomena too. Indeed, it has been reported that *in vivo* exposure to high concentrations of some heavy metals, such as copper, manganese and zinc, leads to an increased activity of SOD in various organisms (Naiki 1980, Del Rio *et al.* 1985, Lee & Hassan 1985, Palma *et al.* 1987). Lee & Hassan (1985) suggested that the SOD-activity in yeast increases in order to prevent free radical damage by the copper ion-induced increase of the intracellular flux of O_2^-.

There are some reports suggesting that reactive oxygen-related enzymes might play a role in metal-tolerance phenomena, especially in the case of copper. Using copper-resistant yeast cells, Naiki (1980) demonstrated that the more SOD the cells had, the higher was their survival after exposure to high extracellular copper levels. Interestingly, when exposed to cadmium the survival of the copper-resistant cells was independent of their SOD-activity. Cadmium is not a transition metal and hence will not directly generate O_2^-. The Cu/ZnSOD-activity in leaves of copper-tolerant *Silene cucubalus* plants has been reported to increase upon high copper supply, whereas the activity in copper-sensitive plants remains unchanged (Lolkema 1985). MnSOD and catalase in leaf peroxisomes have been suggested to play a role in the mechanism of copper-tolerance in pea plants (Palma *et al.* 1987).

However, as yet there is no conclusive evidence that increased enzymic defence against reactive oxygen species can prevent the toxic effects of copper and hence might explain copper-resistance. In rats, injections of copper sulphate solutions lead to elevated levels of lipid hydroperoxides, indicating free radical damage to membranes, in spite of the increased SOD-activity (Ljutakova *et al.* 1984). According to Naiki (1980), increasing the SOD-activity itself can not support the growth of copper-resistant yeast cells in a copper-containing medium. Moreover, in the report on the increased SOD-activity in copper-tolerant *Silene cucubalus* (Lolkema 1985), the enzyme activity was expressed as units per gram fresh weight. When calculated on

a protein basis, however, the copper-tolerant plants as well as the copper-sensitive plants appear to increase their specific SOD-activity, and both to the same extent (Schat & De Vos, unpublished results), suggesting that the mechanism of copper-tolerance in this higher plant species does not primarily rely on the increased SOD-activity.

6. Epilogue

It is difficult to estimate the significance of oxidative stress in metal toxicity phenomena in higher plants as yet. Firstly, the majority of studies into oxidative stress reports experiments with animals or lower organisms. Secondly, *in vivo* studies, which allow a validation of information drawn from *in vitro* experiments, are relatively scarce. Moreover, a thorough evaluation of the role of radicals in metal toxicity is hindered by a lack of fundamental knowledge concerning the molecular phytotoxicology of heavy metals in general. The specific receptor molecules and target processes are often unidentified as yet, which makes it impossible to distinguish between primary and secundary effects (Woolhouse 1983).

As yet, only in the case of copper, there is some evidence that lipid peroxidation has to be considered as an important primary effect of toxic levels of this metal (section 4). Anyway, there are different opinions with respect to the mechanisms of copper toxicity in plants and micro-organisms. Rao & Venkateswerlu (1986) ascribe copper toxicity in *Neurospora crossa* in part to a direct non-competitive inhibition of glutamine synthetase, which was observed both *in vivo* and *in vitro*. However, supplying the cultures with glutamine only partly restored the growth under excessive copper levels. Moreover, addition of glutamine to culture media will decrease the free ionic copper concentration, due to the formation of a glutamine-copper complex, which might as such explain the growth restoration observed. Copper toxicity in green algae has been ascribed to a direct action of copper on thylacoid membranes *in vivo*, which involves lipid peroxidation (Sandmann & Böger 1980b). For higher plants, on the other hand, there is no evidence that copper toxicity would primarily rely on a blocked photosynthetic electron transport or peroxidative thylacoid destruction (section 4). In general, toxic copper levels in the root environment are more rapidly and more conspicuously apparent by reduced growth and functioning of the root system itself; the growth and vitality of the shoot is usually less responsive, and there are several indications that copper-imposed loss of shoot vigour may represent a secundary effect, rather than a primary one, at least at sub-lethal concentration levels (Schat *et al.,* in prep.).

In general, it may be expected that copper exerts its toxic effects primarily in the cells in which it is taken up from the external solution; in these cells it will be taken up in the free ionic form (Graham 1981), which is the most toxic one. This may explain the difference between algae and higher plants as to the nature of the primary effects.

With respect to the role of defence mechanisms against oxidative stress in metal tolerance, there is a similar uncertainty, mainly due to a lack of studies at this point, as well as to a general lack of knowledge concerning the physiology and biochemistry of metal tolerance phenomena (Verkleij & Schat 1990). Metal tolerance is often believed to be due to an increased capacity to synthesize metal chelating peptides, such as phytochelatins, although many good arguments can be raised against this point of view (Verkleij & Schat 1990). Anyway, there is no reason to believe that the binding of excessively accumulated metals to specific peptides would completely prevent oxidative stress. Firstly, increased phytochelatin production could lead to a decrease in free glutathione, which is a most important cellular anti-oxidant. Secondly, at least for some metals the binding to phytochelatins may be too weak to prevent binding to other receptor molecules, such as proteins with electron transport functions. Thirdly, it is unknown whether binding to phytochelatins will completely prevent the redox-activity of transition metal ions, such as copper. All these points need to be investigated in the near future.

References

Baszýnski, T., Tukendorf, A., Ruszkowska, M., Skórzýnska & Maksymiec, W., 1988. Characteristics of the photosynthetic apparatus of copper non-tolerant spinach exposed to excess copper. J. Plant Physiol. 132: 708—713.

Braughler, J. M., Duncan, L. A. & Chase, R. L., 1986. The involvement of iron in lipid peroxidation. J. Biol. Chem. 261: 10282—10239.

Cardinaels, C., Put, C., Van Assche, F. & Clijsters, H., 1984. The superoxide dismutase as a biochemical indicator, discriminating between zinc and cadmium toxicity. Arch. Int. Physiol. Biochim. 92: 2—3.

Chan, P. C., Peller, O. G. & Kesner, L., 1982. Copper (II)-catalyzed lipid peroxidation in liposomes and erythrocyte membranes. Lipids 17: 331—337.

Czapski, G., Aronovitch, J., Samuni, A. & Chevion, M., 1983. The sensitization of the toxicity of superoxide and vitamin C by copper and iron: a site specific mechanism. In: G. Cohen & R.A. Greenwald (eds.). Oxy radicals and their scavenger systems (Vol. 1). Elsevier Biomedical, Amsterdam pp. 111—115.

De Vos, C. H. R., Schat, H., Vooijs, R. & Ernst, W. H. O., 1989. Copper-induced damage to the permeability barrier in roots of *Silene cucubalus.* J. Plant Physiol. 135: 164–169.

Del Rio, L. A., Sandalio, L. M., Yánez, J. & Gómez, M., 1985. Induction of manganese-containing superoxide dismutase in leaves of *Pisum sativum* L. by high nutrient levels of zinc and manganese. J. Inorg. Biochem. 24: 25—34.

DiMonte, D., Ross, D., Bellomo, G., Eklöw, L. & Orrenius, S., 1984. Alterations in intracellular thiol homeostasis during the metabolism of menadione by isolated rat hepatocytes. Arch. Biochem. Biophys. 235: 334—342.

Ding, A-H. & Chan, P. C., 1984. Singlet oxygen in copper-catalyzed lipid peroxidation in erythrocyte membranes. Lipids 19: 278—284.

Ernst, W. H. O., 1974. Schwermetallvegetation der Erde. Fischer, Stuttgart.

Girotti, A. W., 1985. Mechanisms of lipid peroxidation. J. Free Rad. Biol. Med. 1: 87—95.

Gora, L., Van Assche, F. & Clijsters, H., 1985. Effects of toxic copper treatment on chloroplast properties of *Phaseolus vulgaris.* Arch. Int. Physiol. Biochim. 93: 8.

Graham, R. D., 1981. Absorption of copper by plant roots. In: J.F. Loneragan, A.D. Robson, R.D. Graham (eds.). Copper in soils and plants. Academic Press, Sydney New York London pp. 141—163.

Haenen, G. R. M. M. & Bast, A., 1983. Protection against lipid peroxidation by a microsomal glutathione-dependent labile factor. FEBS Lett. 159: 24—28.

Halliwell, B., 1974. Superoxide dismutase, catalase and glutathione peroxidase: solutions to the problems of living with oxygen. New Phytol. 73: 1075—1086.

Halliwell, B. & Gutteridge, J. M., 1985. Free radicals in biology and medicine. Clarendon Press, London.

Hendry, G. A. F. & Brocklebank, K. J., 1985. Iron-induced

oxygen radical metabolism in waterlogged plants. New Phytol. 101: 199—206.

Kumar, K. S., Rowse, C. & Hochstein, P., 1978. Copper-induced generation of superoxide in human red cell membrane. Biochem. Biophys. Res. Comm. 83: 587—592.

Larson, R. A., 1988. The antioxidants of higher plants. Phytochem. 27: 969—978.

Lee, E. H. & Bennett, J. H., 1982. Superoxide dismutase. A possible protective enzyme against ozon injury in snap beans (*Phaseolus vulgaris* L.). Plant Physiol. 69: 1444—1449.

Lee, F. J. & Hassan, H. M., 1985. Biosynthesis of superoxide dismutase in *Saccharomyces cerevisiae:* effects of paraquat and copper. J. Free Rad. Biol. Med. 1: 319—325.

Ljutakova, S. G., Russanov, E. M. & Liochev, S. I., 1984. Copper increases superoxide dismutase activity in rat liver. Arch. Biochem. Biophys. 235: 636—643.

Lolkema, P. C., 1985. Copper resistance in higher plants. Thesis, Free University Press, Amsterdam.

Lolkema, P. C. & Vooijs, R., 1986. Copper tolerance in *Silene cucubalus*. Subcellular distribution of copper and its effect on chloroplast and plastocyanin synthesis. Planta (Berl.) 167: 30—36.

Mehlhorn, H., Seufert, G., Schmidt, A. & Kunert, K. J., 1986. Effect of SO_2 and O_3 on production of antioxidants in conifers. Plant Physiol. 82: 336—338.

Minotti, G. & Aust, S. D., 1987. An investigation into the mechanism of citrate-Fe^{2+}-dependent lipid peroxidation. J. Free Rad. Biol. Med. 3: 379—387.

Naiki, N., 1980 Role of superoxide dismutase in a copper-resistant strain of yeast. Plant Cell Physiol. 21: 775—783.

Palma, J. M., Gómez, M., Yánez, J. & Del Rio, L. A., 1987. Increased levels of peroxisomal active oxygen-related enzymes in copper-tolerant pea plants. Plant Physiol. 85: 570—574.

Pauls, K. P. & Thompson, J. E., 1982. Effects of cytokinins and antioxidants on the susceptability of membranes to ozone damage. Plant Cell Physiol. 23: 821—832.

Permual, A. & Beattie, J. M., 1965. Effect of different levels of copper on the activity of certain enzymes in leaves of apple. Am. Soc. Hort. Sc. 88: 41—47.

Rao, S. & Venkateswerlu, G., 1986. Glutamine metabolism in *Neurospora crassa* under conditions of copper toxicity. J. Exp. Bot. 37: 947—955.

Ribanov, S., Benor, L., Benchov, I., Monovich, O. & Markova, V., 1982. Hemolysis and peroxidation in heavy metal-treated erythrocytes; GSH content and activities of some protecting enzymes. Experimentia 38: 1354—1355.

Rowley, D. A. & Halliwell, B., 1983. Superoxide-dependent and ascorbate-dependent formation of hydroxyl radicals in the presence of copper-salts: a physiologically significant reaction? Arch. Biochem. Biophys. 225: 279—284.

Sandmann, G. & Böger, P., 1980a. Copper-mediated lipid peroxidation processes in photosynthetic membranes. Plant Physiol. 66: 797—800.

Sandmann, G. & Böger, P., 1980b. Copper deficiency and toxicity in *Scenedesmus*. Z. Pflanzenphysiol. 98: 53—59.

Schmidt, A. & Kunert, K. J., 1986. Lipid peroxidation in higher plants. The role of glutathione reductase. Plant Physiol. 82: 700—702.

Shioi, Y., Tamai, H. & Sasa, T., 1978. Effects of copper on photosynthetic electron transport systems in spinach chloroplasts. Plant Cell Physiol. 19: 203—209.

Thomas, C. E., Morehouse, L. A. & Aust, S. D., 1985. Ferritin and superoxide-dependent lipid peroxidation. J. Biol. Chem. 260: 3275—3280.

Thornalley, P.J. & Vasák, M., 1985. Possible role for metallothionein in protection against radiation-induced oxidative stress-kinetics and mechanisms of its reaction with superoxide and hydroxyl radicals. Biochim. Biophys. Acta 827: 36—44.

Verkleij, J. A. C. & Schat, H., 1990. Mechanisms of metal tolerance in higher plants. In: A.J. Shaw (ed.) Heavy metal tolerance in plants: evolutionary aspects. CRC Press Inc., Boca Raton, Florida pp. 179—193.

Woolhouse, H. W., 1983. Toxicity and tolerance in the responses of plants to metals. In: A. Läuchli & R.L. Bieleski (eds.) Encyclopedia of plant physiology (Vol. 12c). Springer Verlag, Berlin pp. 245—300.

Younes, M. & Siegers, C. P., 1981. Mechanistic aspects of enhanced lipid peroxidation following glutathione depletion *in vivo*. Chem.-Biol. Interactions 34: 257—266.

Authors' Address
C.H.R. de Vos & H. Schat
Department of Ecology and Ecotoxicology
De Boelelaan 1087
1081 HV Amsterdam
The Netherlands

Minuartia verna, a plant frequently occurring on soils of mine waste tips, rich in heavy metals. (Photograph: J. A. C. Verkley.)

CHAPTER 4

Physiological responses of higher plants to soil contamination with metals

H. CLIJSTERS, F. VAN ASSCHE & L. GORA

Abstract. Net photosynthesis of bean (*Phaseolus vulgaris* L.) seedlings decreased after assimilation of toxic amounts of zinc by the roots. This effect was related to the substitution of essential bivalent cations by zinc in at least two important metalloproteins of the chloroplast: 1) replacement of Mn^{2+} by Zn^{2+} in the water splitting enzyme of the thylakoid membrane inhibited photosynthetic electron transport at high electron flow rates; and 2) partial substitution of Mg^{2+} by Zn^{2+} in the ternary ribulose 1,5 bisphosphate carboxylase-CO_2-metal^{2+} complex *in vivo* decreased the carboxylase/oxygenase capacity ratio. These effects, obtained after application of toxic metal doses to intact plants, were compared with data on isolated chloroplasts to which zinc was applied *in vitro*. The relevance of both experimental approaches for assessing the physiological effects of heavy metals *in vivo* is discussed. The importance of metal concentration measurements at the subcellular level is emphasized. Enhancement of malondialdehyde content and ethane production in plants, treated with toxic amounts of Cu and Zn indicates that these metals affected membrane integrity by lipid peroxidation; the capacity of the enzyme lipoxygenase, involved in this process, also increased. Cellular decompartmentalization is a consequence of this effect. Similar changes were also observed after several kinds of environmental stress, and even during plant senescence.

1. Introduction

Due to industrial pollution and agricultural practice, phytotoxic amounts of metals like Zn, Cu, Cd and Pb can contaminate the soil as well as the ground and surface water. Growth inhibition and reduction of biomass production are general responses of higher plants to metal toxicity (Lepp 1981).

These elements interfere with the most important physiological processes: transpiration, respiration, photosynthesis (a.o., Carlson *et al*. 1975; Lee *et al*. 1976a, b; Van Assche *et al*. 1980). They modify mineral nutrition, biomass partitioning, the hormonal balance and affect plant development in general (Wallnöfer & Engelhardt 1984). Several mechanisms of metal action at the physiological and biochemical level were described (Foy *et al*. 1978; Van Assche & Clijsters 1989). In this paper

attention will be focused on two of these mechanisms:
- inhibitory effects of zinc on chloroplast activity
- membrane destabilization by copper and zinc.

2. Experimental approach

The physiological response of plants to metals has been intensively studied during the last two decades. Generally, two main experimental approaches can be distinguished:
- *in vitro* application of metals to isolated systems (plant organs, subcellular fractions or organelles, enzyme preparations ...) in the assay medium. The systems were isolated from plants which were not exposed to the metals beforehand.
- *in vivo* application of the metal to intact plants,

J. Rozema and J. A. C. Verkleij (eds.), Ecological Responses to Environmental Stresses, 32—39.

followed by comparative measurements on control plants and on plants, treated with toxic concentrations of the metal.

Both approaches were also applied in comparative studies on metal-sensitive and metal-resistant plant ecotypes.

For the experiments in this paper, *in vivo* metal application was used. Dwarf beans (*Phaseolus vulgaris* L. cv. Limburgse Vroege) were cultivated on vermiculite, saturated with full strength Hoagland solution. Additional doses of metals were applied as sulphate salts to the nutrient solution before sowing. The plants were grown under controlled environmental conditions and watered with deionized water. All the measurements described were performed either on 14 days old intact plants or on systems isolated from primary leaves of these plants without any further metal addition to the assay media.

3. Effects of zinc on photosynthesis

3.1. Gas exchange measurements

Net CO_2 fixation of the primary leaves was significantly inhibited after toxic zinc treatment (10 mM Zn^{2+} in the nutrient solution). Leaf dark respiration and transpiration however were not affected. Zn toxicity further induced a significant increase of the CO_2 compensation point concentration (Table I).

Modelling of the CO_2 diffusion pathway from the ambient air to the carboxylating site in the chloroplast revealed that the internal mesophyll conductance was rate limiting (Van Assche *et al.* 1980). These introductory experiments demonstrated that mainly photosynthesis was affected at the Zn concentration used. Stomatal closure could be excluded as the cause of this inhibition; the attention was therefore focused on chloroplast activity.

3.2. Effects on photosynthetic electron transport

Several partial reactions of the photosynthetic electron transport were measured; a variety of artificial electron donors and acceptors and specific inhibitors were applied (Van Assche & Clijsters 1986a). Some relevant data are summarized in Table II. The electron transport involving both photosystems 1 and 2 was significantly inhibited by Zn at high electron flow rates (ADP-stimulated, uncoupled). The highly active photosystem 2 (PS 2) specific photoreduction of oxidized diaminodurene (DAD^+) was also significantly affected, but the reaction with reduced DAD and methylviologen (MV), involving PS 1 only, was less sensitive (Table II). From these data it was clear that Zn affects photosynthetic electron transport at the site of PS 2.

When semicarbazide was used as electron donor at the oxidizing side of PS 2, no inhibition but a two-fold increase in activity was measured (Table II). This indicated that Zn inhibited electron flow at the water splitting site. The high rate of the semicarbazide mediated reaction could be ex-

Table I. Gas exchange measurements on the primary leaves of intact *Phaseolus* plants, as a function of zinc nutrition.

	zinc concentration in the nutrient medium	
	optimal 2 mM	toxic 10 mM
net CO_2-fixation (mg $CO_2 \cdot m^{-2} \cdot s^{-1}$)	0.36 ± 0.05[a]	$0.26^{**} \pm 0.04$
Dark respiration (mg $CO_2 \cdot m^{-2} \cdot s^{-1}$)	0.016 ± 0.004	0.018 ± 0.007
Transpiration (mg $H_2O \cdot m^{-2} \cdot s^{-1}$)	21.3 ± 6.5	23.8 ± 2.4
CO_2-compensation point concentration (mg $CO_2 \cdot m^{-3}$)	79 ± 8	$115^{**} \pm 26$

[a] mean $\pm$ standard deviation (N = 12).
*, ** significantly different from control (optimal concentration) at p = 0.05 and p = 0.01, respectively (Student's t-test).

plained by the uncoupling effect of this artificial electron donor.

Manganese ions are essential for the photolysis of water: with 5 to 8 Mn ions per 400 molecules of chlorophyll the water splitting activity is optimal. Mn remains bound to the thylakoid membrane after ethylenediamine tetraacetate (EDTA)-washing. Thylakoid membranes, prepared from control plants, contained 6 ions of manganese per 400 molecules of chlorophyll after washing with EDTA (Table III). This result agrees with the range of values cited above. After toxic Zn treatment (cfr. 3.1.) however, only 2,4 Mn were found/400 chl (Table III). In these plants the amount of thylakoid-bound Zn/400 chl rose from 5,5 in controls to about 30 after toxic Zn treatment (Table III). These data strongly suggest that the inhibition of the electron transport at the water splitting site was due to Mn deficiency, caused by substitution of Zn for this element.

Zn only limited electron transport at high electron flow rates (Table II). This is clearly shown in Fig. 1, illustrating the inhibition of DAD^+-photoreduction at photon flux densities above 150 μmol ·

$m^{-2} \cdot s^{-1}$, corresponding to electron flow rates, exceeding 120 μeq · (mg · Chl)$^{-1}$ · h^{-1}. At lower photon flux density, the quantum yield of the reaction, estimated from the slope of the curves, was essentially the same after optimal and toxic Zn treatment (see insets of Fig. 1). This phenomenon might be due to the limited amount of Mn, still bound to the thylakoids after toxic Zn treatment. At low electron flow rates, this Mn could accumulate sufficient oxidizing equivalents to maintain the corresponding low rates of water splitting activity. At higher flow rates however, more redox capacity is necessary; the Zn ions, substituting for Mn, have no redox behaviour, and electron transport is inhibited.

Substitution between bivalent cations, as described here, is common in metalloproteins, when the relative concentration of the competing ions changes markedly in the tissue (Clarkson & Hanson 1980). In the present example, such competition is evident at the thylakoid membrane under conditions of toxic Zn treatment.

The results, described above, can be compared with the effects of metals, applied to chloroplasts, isolated from control plants *in vitro* (for a review: see Clijsters & Van Assche 1985).

In this experimental approach Zn also inhibited the electron transport associated with PS 2, but significant differences could be observed with the data obtained after Zn assimilation by intact plants:
- application of Zn to isolated chloroplasts *in vitro* inhibited electron transport not at the water

Table II. Photosynthetic electron transport [micro-electron-equivalents (meq.) · (mg chlorophyll)$^{-1}$ · h^{-1}] of broken chloroplasts as a function of zinc treatment.

	zinc treatment	
	optimal 2 mM	toxic 10 mM
Photosystem 2 + 1		
$H_2O \rightarrow$ ferricyanide		
basal	233 ± 45[a]	259 ± 51
ADP-stimulated	393 ± 85	311* ± 61
uncoupled	838 ± 156	400** ± 102
H_2O/semicarbazide → ferricyanide	502 ± 152	1012** ± 184
Photosystem 2		
$H_2O \rightarrow DAD^+$/ferricyanide	827 ± 156	555** ± 61
Photosystem 1		
Ascorbate/$DAD^- \rightarrow$ MV	729 ± 109	620 ± 304

[a] mean ± standard deviation (N = 10).
*, ** significantly different from control (optimal concentration) at p = 0.05 and p = 0.01, respectively (Student's t-test).

Table III. Effect of Zn treatment on Mn and Zn content (number of ions per 400 molecules of total chlorophyll) and on Zn/Mn ratio of chloroplast thylakoids after two-fold washing with 5 mM EDTA. For each value 4 different chloroplast preparations were grouped. The error of analysis was about 10%.

	zinc treatment	
	optimal 2 mM	toxic 10 mM
Mn	6.0	2.4
Zn	5.5	29.8
Zn/Mn ratio	0.92	12.4

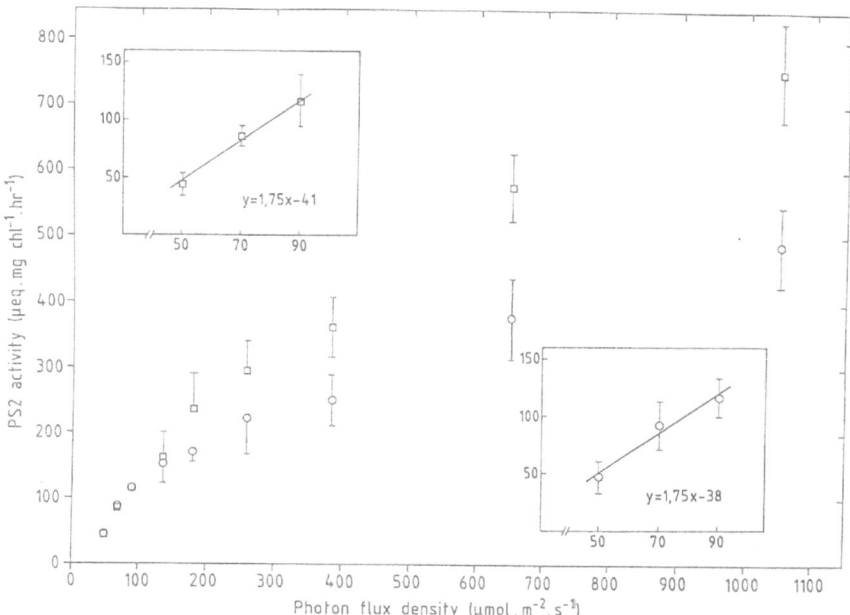

Fig. 1. PS 2 activity ($H_2O \rightarrow DAD^+$/ferricyanide) of chloroplasts prepared from plants which received control (□) and toxic (○) zinc treatment, as a function of the photon flux density. Data are mean values ± standard deviation of 5 extracts. Results obtained below $100\,\mu\text{mol} \cdot m^{-2} \cdot s^{-1}$ are given in insets.

splitting site, but more closely to the reaction centre of PS 2;
- the inhibitory effect was independent of photon flux density;
- there was evidence for an additional effect between PS 2 and PS 1, which was not observed after toxic Zn assimilation *in vivo*.

There may be several reasons for these discrepancies:
- the Zn concentration, applied to isolated chloroplasts *in vitro,* is usually much higher than the concentration that can be reached in the chloroplast *in vivo;*
- the incubation time of thylakoids with metal *in vitro* is extremely short (only a few minutes) as compared to the period of interaction *in vivo;*
- Zn-induced deficiencies of other (bivalent) cations (cf., the deficiency in thylakoid-bound Mn mentioned above) are excluded *in vitro.*

Considering these differences, it is our opinion that application of toxic metal doses to intact plants is the most relevant experimental approach to assess the physiological mechanisms of metal toxicity. *In vitro* experiments can suggest potential metabolic

sites of interaction, but do not provide evidence for the mechanism of action of metals, assimilated from a contaminated substrate. This is also illustrated by the effects described for Pb and Cu. Both elements are highly toxic when applied to isolated chloroplasts. Pb however is hardly assimilated by the roots and will therefore rarely affect the plant at the chloroplast level. Baszynski *et al.* (1982) were unable to show any inhibition on photosynthetic electron transport, when toxic concentrations of Cu were assimilated by intact spinach plants. Generally Cu accumulates merely in the root and is translocated to the leaf only to a limited extent. In algae and water plants however, living in contaminated surface water, Cu and Pb can accumulate more directly in the chloroplast.

These data as well as the results presented in the next section demonstrate the importance of allocation studies of metals at the subcellular level.

3.3. Zinc effects on ribulose 1,5 bisphosphate carboxylase/oxygenase

Bivalent cations play a major role in the activation

of ribulose 1,5 bisphosphate carboxylase, the key-enzyme of CO_2-fixation in the chloroplast. This enzyme has oxygenase activity as well. Activation of ribulose 1,5 bisphosphate carboxylase/oxygenase (RuBisCo) with CO_2 ($^ACO_2$) and a bivalent cation (Mg^{2+}) is necessary to form a functional ternary RuBisCo-$^ACO_2$-Mg^{2+} complex, which can bind a substrate CO_2 molecule. The cation is loosely bound to the protein; experimentally it can become free by precipitation of the protein with $(NH_4)_2SO_4$. The free RuBisCo protein can bind another bivalent cation after incubation *in vitro*, but this cation exchange causes a marked decrease of the carboxylase/oxygenase activity (C/O) ratio (for references see Van Assche & Clijsters 1986b).

Dwarf beans, treated with a toxic dosis of Zn (cf. 3.1.), showed significant decrease of the C/O ratio; the carboxylation capacity was significantly inhibited, but the oxygenase capacity was not affected (Table IV). The question therefore arose whether under these conditions Mg^{2+} was replaced by Zn^{2+} in the RuBisCo-$^ACO_2$-Me^{2+} complexes *in vivo*.

To check this hypothesis, a crude leaf protein extract was precipitated with $(NH_4)_2SO_4$. Consequently, the cations incorporated in the RuBisCo complexes *in vivo* were removed. The precipitate was resuspended in buffer, and desalted on a Sephadex G-25 column, equilibrated with 20 mM $MgCl_2$ and 10 mM $NaHCO_3$. This procedure optimally reactivated RuBisCo: the ternary enzyme complex was fully restored with Mg^{2+}.

This precipitation/reactivation process had no effect on the C/O ratio of the enzyme, isolated from control plants. The enzyme complex from plants, treated with toxic zinc concentration, however, showed a marked increase of the C/O ratio: the significant difference observed between control and zinc treated plants disappeared after the precipitation/reactivation procedure (Table V).

This experiment demonstrates that in plants, suffering from zinc toxicity, RuBisCo complexes did not only contain Mg^{2+}, but also another cation, which could be removed after enzyme extraction and replaced by Mg^{2+} *in vitro*.

Zn can compete with Mg in the ternary RuBisCo complex, when it is present in excess in the chloroplast of plants, receiving phytotoxic amounts of this metal. Table VI shows that the Mg/Zn ratio was lower in this organelle than in a crude leaf extract. After Zn treatment this ratio further decreased: in the chloroplast fraction the concentration of Mg and Zn became of the same order of magnitude. Since the affinity of Zn^{2+} for carboxyl groups is higher than that of Mg^{2+}, at least a partial substitution of Zn^{2+} for Mg^{2+} in the ternary RuBisCo-$^ACO_2$-Me^{2+}-complex is probable *in vivo*.

4. Cu and Zn affect membrane integrity

Table VII shows that the malondialdehyde content of the primary leaf as well as the ethane production

Table IV. RuBisCo capacity (mU · mg protein^{-1}) and carboxylase/oxygenase capacity ratio in crude leaf extracts of *Phaseolus vulgaris*, as a function of treatment with zinc.

	zinc treatment	
	optimal 2 mM	toxic 10 mM
RuBP carboxylase	235 ± 31[a]	180** ± 24
RuBP oxygenase	48 ± 6	49 ± 5
carboxylase/oxygenase	5.0 ± 0.7	3.7** ± 0.4

[a] mean ± standard deviation (N = 10).
*, ** significantly different from control (optimal concentration) at p = 0.05 and p = 0.01, respectively (Student's t-test).

Table V. RuBP carboxylase/oxygenase capacity ratio of fresh leaf extracts, and after an inactivation/reactivation procedure, as a function of zinc treatment to *Phaseolus vulgaris*.

	zinc treatment	
	optimal 2 mM	toxic 10 mM
fresh extract	5.5 ± 1.0[a]	3.1** ± 0.5
after precipitation with $(NH_4)_2SO_4$ and reactivation according to Lorimer *et al.* (1977)	5.3 ± 1.8	4.5 ± 1.2

[a] mean ± standard deviation (N = 9).
** significantly different from control (optimal concentration) at p = 0.01 (Student's t-test).

by intact plants were increased as a function of the Cu- or Zn-concentration applied. Both metabolites are products of membrane lipid peroxidation. (Kimmerer *et al.* 1982; Kunert *et al.* 1985).

High Cu-concentrations cause strong lipid peroxidation in isolated spinach chloroplast membranes (Sandmann & Böger 1980). The authors argue that this process is started by hydroxyl radicals, generated by Cu inhibition of the photosynthetic electron transport; the redox behaviour of this metal plays an important role in this process.

It is, however, not likely that this mechanism of action can explain the present effect of Cu on membrane lipid peroxidation. The photosynthetic electron transport of chloroplasts in primary bean leaves was not affected by Cu *in vivo* (Gora *et al.* 1985). Contrary to this metal Zn does not participate in one electron redox reactions; nevertheless it also produced intermediates of lipid peroxidation (Table VII).

Both metals induce the activity of a variety of enzymes (see Van Assche & Clijsters 1990 for a review). They also enhanced the capacity of lipoxygenase (Table VII), responsible for enzyme catalyzed peroxidation of polyunsaturated cis, cis 1–4 pentadiene lipic acids as linolenic acid, a typical component of plant cell membranes. This effect might explain the production of malondialdehyde and ethane by plants, intoxicated with Cu and Zn.

5. Discussion

Due to accumulation of Zn in the chloroplast, substitution for other essential bivalent cations in metalloproteins or metal protein complexes seems to be a major mechanism of zinc interaction with photosynthetic reactions. In tomato, treated with toxic amounts of Cd in hydroculture, a similar mechanism of action was observed; restoration of Cd-induced inhibition at the water splitting site was obtained by addition of extra Mn to the nutrient solution (Baszinsky *et al.* 1980).

For Cd, interaction with functional SH-groups on enzymes is often proposed as a possible mechanism of action, especially in experiments, where the metal is applied to isolated chloroplasts *in vitro*. The same mechanism is suggested to explain the inhibitory effects of Cu and Hg *in vitro* (see Clijsters & Van Assche 1985 for references). The latter two elements have very strong affinity for SH-groups indeed; consequently, this mechanism of interaction cannot be excluded *in vivo*, even at the relatively lower concentrations of these elements in the chloroplast.

Summing up, inhibition of net CO_2 fixation by Zn (Table I) can be the result of at least two mechanisms. At high photon flux density, both electron transport and carboxylase capacity are inhibited; at low photon flux density, only the effect on the C/O ratio is observed. This decrease in capacity ratio explains the enhanced CO_2 compensation point concentration, measured on intact plants (Table I).

At low photon flux density inhibition of the carboxylation capacity could lead to limited consumption of NADPH in the Calvin cycle and to overreduction of the NADP-pool. Under these conditions, the chloroplast might produce superoxide

Table VI. Total magnesium and zinc content of crude extracts and chloroplasts of *Phaseolus vulgaris,* as a function of zinc treatment. For each value 4 different preparations were grouped. The error of analysis was below 10%.

	zinc treatment					
	optimal (2 mM)			toxic (10 mM)		
	Mg	Zn	Mg/Zn	Mg	Zn	Mg/Zn
Crude extract [μg metal · (mg protein)$^{-1}$]	50	0.38	132	31	0.83	37
Chloroplast fraction:						
– thylakoids [μg metal · (mg chlorophyll)$^{-1}$]	20.7	1.7	12.2	34.4	12.0	2.9
– stroma [μg metal · (mg protein)$^{-1}$]	6.8	0.7	9.7	13.4	2.5	5.4

radicals at the reducing side of PS 1 and these radicals could be transformed into peroxide by superoxide dismutase. This peroxide can be eliminated by a reaction sequence, involving ascorbate, ascorbate peroxidase, glutathion and glutathion reductase (Halliwell 1978).

It is worth mentioning that we observed an increase of the peroxidase capacity without any inhibitory effect on superoxide dismutase in plants, treated with toxic Zn doses (Cardinaels et al. 1984). Research is in progress to examine whether the reaction sequence, mentioned above, operates in chloroplasts of Zn treated plants to eliminate the extra H_2O_2, possibly generated at low light intensities.

Zn as well as Cu also induced lipoxygenase capacity in primary leaves. This enzyme catalyzes membrane lipid peroxidation; metabolites of this process were detected (Table VII).

Membrane peroxidation is associated with lipid gel phase formation, modification of membrane permeability and conformational changes of membrane bound proteins (Chia et al. 1984). However other stress factors, e.g., ozone (Pauls & Thompson 1981), acid deposition (Chia et al. 1984) and freezing (Kacperska et al. 1984) all induce lipid peroxidation. Cu and Zn might therefore exhibit a non specific effect on metabolism by cell decompartmentalization and modification of membrane bound enzyme activity.

cesses in intact higher plants. Their effect varies as a function of the nature and concentration of the metal applied and of the plant species contaminated. Two different mechanisms of interaction were described at the subcellular level.

Interference of Zn with the process of photosynthesis appears to be a rather specific phenomenon: cation substitution in two metal-requiring enzymes of the chloroplast. A similar mode of action may be expected for several metals with other functional metalloproteins localized in the chloroplast or in other compartments.

The effect of Cu and Zn on peroxidative membrane desintegration is a more general reaction, not specific for metals, since other xenobiotics and several environmental stress factors induce similar changes. Some authors suggest that metals may cause premature leaf senescence, as membrane lipid peroxidation is characteristic for this developmental stage. Stimulation of ethylene production, described for several metals (for references see Gora et al. 1989), supports this hypothesis.

The data presented demonstrate the complexity of metal interaction at the subcellular level, which is reflected by the multitude of physiological effects in the intact plant. It is therefore obvious that the overall plant response is the result of this complex interference rather than a reaction on a clearly defined metal induced cellular effect.

6. Concluding remarks

Metals interfere with a range of physiological pro-

Acknowledgements

Skillful technical assistance of mrs. C. Bogaert-

Table VII. Malondialdehyde content (μmol · g fresh weight^{-1}), lipoxygenase activity (mU · μg protein^{-1}) and ethane production (n1 · h^{-1} · g fresh weight^{-1}) as a function of the Cu- or Zn-concentration (mM) in the nutrient solution.

metal concentration	malondialdehyde content	ethane production	lipoxygenase activity
Control	243 ± 7[a]	0.80 ± 0.37	0.44 ± 0.17
Cu 3.9	839** ± 326	1.33** ± 0.49	1.24** ± 0.50
7.9	1,367** ± 133	2.52** ± 1.23	3.00** ± 0.25
Zn 5.8	661* ± 286	1.57** ± 0.54	1.84** ± 0.45
15.4	1,032** ± 154	1.33** ± 0.49	2.11** ± 1.32

[a] mean ± standard deviation (N = 4).

*, ** significantly different from control (optimal concentration) at p = 0.05 and p = 0.01, respectively (Student's t-test).

Vanherle, G. Strijkers-Clerkx and B. Bamps-Vanacken is gratefully acknowledged. This research was supported by the "Instituut voor het Wetenschappelijk Onderzoek in Nijverheid en Landbouw" (IWONL) Brussels.

References

Baszinsky, T., Wajda, L., Krol, M., Wolinska, D., Krupa, Z. & Tukendorf, A., 1980. Photosynthetic activities of cadmium treated tomato plants. Physiol. Plant. 48: 365—370.

Baszinsky, T., Krol, M., Krupa, Z., Ruszowska, M., Wojcieska, U. & Wolinska, D., 1982. Photosynthetic apparatus of spinach exposed to excess copper. Z. Pflanzenphysiol. 108: 385—395.

Cardinaels, C., Put, C., Van Assche, F. & Clijsters, H., 1984. The superoxide dismutase as a biochemical indicator, discriminating between zinc and cadmium toxicity. Arch. Int. Physiol. Biochim. 92: PF 2—3.

Carlson, R. W., Bazzaz, F. A. & Rolfe, G. L., 1975. The effect of heavy metals on plants II. Net photosynthesis and transpiration of whole corn and sunflower plants treated with Pb, Cd, Ni, and Tl. Environm. Res. 10: 113—120.

Chia, L. S., Mayfield, C. I. & Thompson, J. E., 1984. Simulated acid rain induces lipid peroxidation and membrane damage in foliage. Plant, Cell and Environm. 7: 333—338.

Clarkson, D. T. & Hanson, J. B., 1980. The nutrition of higher plants. Annu. Rev. Plant Physiol. 31: 329—398.

Clijsters, H. & Van Assche, F., 1985. Inhibition of photosynthesis by heavy metals. Photosynth. Res. 7: 31—40.

Foy, C. D., Chaney, R. L. & White, M. C., 1978. The physiology of metal toxicity in plants. Annu. Rev. Plant Physiol. 29: 511—566.

Gora, L., Van Assche, F. & Clijsters, H., 1985. Effects of toxic copper treatment on chloroplast properties of Phaseolus vulgaris. Arch. Int. Physiol. Biochem. 93: pp 8.

Gora, L. & Clijsters, H., 1989. Effect of copper and zinc on the ethylene metabolism of Phaseolus vulgaris L. In: H. Clijsters, M. De Proft, R. Marcelle & M. Van Poucke (eds.) Biochemical and physiological aspects of ethylene production in lower and higher plants. Kluwer, Dordrecht pp 219—228.

Halliwell, B. & Foyer, C. H., 1978. Properties and physiological function of a glutathione reductase purified from spinach leaves by affinity chromatography. Planta 139: 9—17.

Kacperska, A. & Kubacka-Zebalska, M., 1985. Is lipoxygenase involved in the formation of ethylene from ACC? Physiol. Plant. 64: 333—338.

Kimmerer, T. W. & Kozlowski, T. T., 1982. Ethylene, ethane, acetaldehyde and ethanol production by plants under stress. Plant Physiol. 60: 840—847.

Kunert, K. J. & Ederer, M., 1985. Leaf aging and lipid peroxidation: The role of antioxidants vitamin C and E. Physiol. Plant. 65: 85—88.

Lee, K. C., Cunningham, B. A., Paulsen, G. M., Liang, G. H. & Moore, R. B., 1976a. Effects of cadmium on respiration rate and activities of several enzymes in soybean seedlings. Physiol. Plant. 36: 4—6.

Lee, K. C., Cunningham, B. A., Chung, K. H., Paulsen, G. M. & Liang, G. H., 1976b. Lead effects on several enzymes and nitrogenous compounds in soybean leaf. J. Environm. Qual. 5: 357—359.

Lepp, N. W. (ed.) 1981. The Effect of Heavy Metals on Plants. Applied Science Publishers, London.

Lorimer, G. H., Badger, M. R. & Andrews, T. J., 1977. D-ribulose-1,5-bisphosphate carboxylase-oxygenase. Improved methods for the activation and assay of catalytic activities. Anal. Biochem 78: 66—75.

Pauls, K. P. & Thompson, J. E., 1981. Effect of in vitro treatment with ozone on the physical and chemical properties of membranes. Physiol. Plant. 53: 255—262.

Sandmann, G. & Böger, P., 1980. Copper-mediated lipid peroxidation processes in photosynthetic membranes. Plant Physiol. 66: 797—800.

Van Assche, F., Ceulemans, R. & Clijsters, H., 1980. Zinc mediated effects on leaf CO_2 diffusion conductances and net photosynthesis in Phaseolus vulgaris. Photosynth. Res. 1: 171—180.

Van Assche, F. & Clijsters, H., 1986a. Inhibition of photosynthesis in Phaseolus vulgaris by treatment with toxic concentrations of zinc: effects on electron transport and photophosphorylation. Physiol. Plant. 66: 717—721.

Van Assche, F. & Clijsters, H., 1986b. Inhibition of photosynthesis in Phaseolus vulgaris by treatment with toxic concentrations of zinc: Effect on ribulose-1,5-bisphosphate carboxylase/oxygenase. J. Plant Physiol. 125: 355—360.

Van Assche, F. & Clijsters, H., 1990. Effects of metals on enzyme activity in plants. Plant, Cell & Environment 19: 195–206.

Wallnöfer, P.R. & Engelhardt, G., 1984. Schädstoffe, die aus dem Boden aufgenommen werden. In: B. Hock & E.F. Elstner (eds.), Pflanzentoxicologie. Bibliographisches Institut, Mannheim pp 95—117.

Authors' Address
H. Clijsters, F. van Assche & L. Gora
Limburgs Universitair Centrum
Universitaire Campus
B-3610 Diepenbeek
Belgium

Spoil heaps of the silver and gold mine of Jales, North Portugal. (Photograph: T. de Koe.)

CHAPTER 5

Arsenic in water, sediment and vegetation of the Jales gold mine, North Portugal

T. DE KOE

Abstract. The As content of water, sediment and vegetation in a gold mine area has been analysed by AAS methods. Highest levels have been found near and on the spoilheaps; 2550 mg · l^{-1} in water, 17000 mg · kg^{-1} dw in sediment and 3600 mg · kg^{-1} dw in roots of *Juncus acutiflorus*. A big difference has been found between As concentrations in green and senescent shoot tissue. Dead leaves of *Agrostis castellana* present an average of 200 mg · kg^{-1} dw As, against only 40 mg · kg^{-1} dw in green leaves, values for *Agrostis canina* are 1940 and 60 mg · kg^{-1} dw respectively. Possible translocation and tolerance mechanisms are discussed.

1. Introduction

The Jales gold and silver mines are situated in north Portugal (Fig. 1) in an area where mining activities can be dated back till the Roman times.

The spoilheaps have a function as sinkdeposits, they filter the process water in order to diminish the amount of suspended solids entering the surface waters. The, thus filtered, process water is gathered in a small canal (site 1, Fig. 1), that ends up in Jales brook (site 2 and 3). In Tinhela river, site 4 shows the influences of the mine waste, and site 5 the natural background values.

Research on aquatic invertebrates (Cortes *et al.* 1986; Cortes & Pereira in print 1987) revealed a drastic drop in both species diversity and abundance in Tinhela river, after receiving the mine waste bearing Jales brook.

The steep, barren and sandy spoilheaps are regularly sprayed with process water, during summer, to diminish eolic erosion. Only *Agrostis castella-na**, Agrostis delicatula* and *Holcus lanatus* have been found growing in sparse tufts, in places where these elsewhere barren spoils are able to retain some humidity.

The development of plant growth on goldmine tailings is generally difficult because of its sandy structure and poor water-holding capacity, its extremely low pH (2–4) due to the weathering of pyrites, and its low levels of organic matter and major plant nutrients in combination to high concentrations of heavy metals and arsenic (Porter & Peterson 1977; Bradshaw & Chadwick 1980; Williamson & Johnson 1981).

The availability of arsenic, that may be extremely toxic to plants, depends on the clay fraction of the tailings and the presence of iron and/or aluminium (Johnson & Hiltbold 1969; Jacobs *et al.* 1970; Woolson *et al.* 1971; Woolson *et al.* 1973; Wauchope 1975; Livesey & Huang 1981; Wauchope & McDowell 1984), the phosphorus content and pH (Benson 1953; Woolson *et al.* 1973; Carrow *et al.*

---

* The nomenclature of *Agrostis canina* L., *Agrostis castellana* Boiss. et Reut., *Agrostis delicatula* Pourret ex Lapeyr, *Holcus lanatus* L., *Juncus acutiflorus* Ehrh. ex Hoffm. and *Juncus effusus* L. is according to M.G. Rollan, Claves de la Flora de Espana, Vol. II, 2nd ed. Mundi-Prensa, Madrid, 1985.

J. Rozema and J. A. C. Verkleij (eds.), Ecological Responses to Environmental Stresses, 42—47.
© *1991 Kluwer Academic Publishers.*

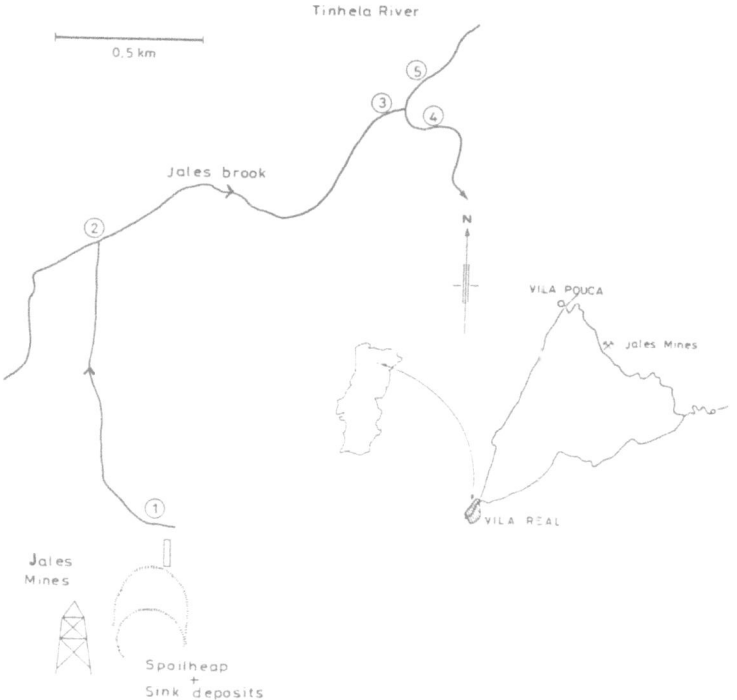

Fig. 1. Localization of the Jales gold and silver mine and the sampling sites. 1. canal near spoilheaps; 2. junction of canal and Jales brook; 3. Jales brook; 4. Tinhela river, polluted; 5. Tinhela river, unpolluted reference.

1975) and the activity of micro-organisms (Marcus-Wyner & Rains 1982; Craig 1986).

2. Arsenic in streams in the mining area

The arsenic content, analysed directly on a Perkin-Elmer 4000 AAS with a HMS-10 hydride system, of the water in the canal near the spoilheaps is extremely high (Table I): $2550 \, mg \cdot l^{-1}$ (compare e.g., the maximal $140 \, mg \cdot l^{-1}$ in a stream passing goldmine tailings in Nova Scotia, Brooks *et al.* 1982).

Table I. Arsenic content $(mg \cdot l^{-1})$ of water samples of streams in the mining area.

Site 1	canal	2550
Site 3	Jales brook	50
Site 4	Tinhela river, polluted	20
Site 5	Tinhela river, unpolluted	1

Dilution in the clean water of Jales brook and Tinhela river results in diminishing concentrations, but Tinhela river shows still $20 \, mg \cdot l^{-1}$ As downstreams the junction with Jales brook (site 4), a concentration that can cause growth inhibition in algae (Christensen & Zielski 1980; Bringman & Kuhn 1980) and is lethal for several aquatic invertebrates (Surber & Meehan 1930). Aquatic mosses and angiosperms, abundant at site 5, disappear completely from Tinhela river after receiving the mine waste (Cortes 1989). At site 4 the community of aquatic invertebrates presents a different composition, besides a drop in species diversity and abundance, compared to site 5 (Cortes *et al.* 1986). In the fish population (*Salmo trutta fario*), the mine waste affects mainly the youngest trouts (Cortes *et al.* 1988).

As the natural background value in this silicious mountain stream can be considered the value of $1 \, mg \cdot l^{-1}$ As at site 5.

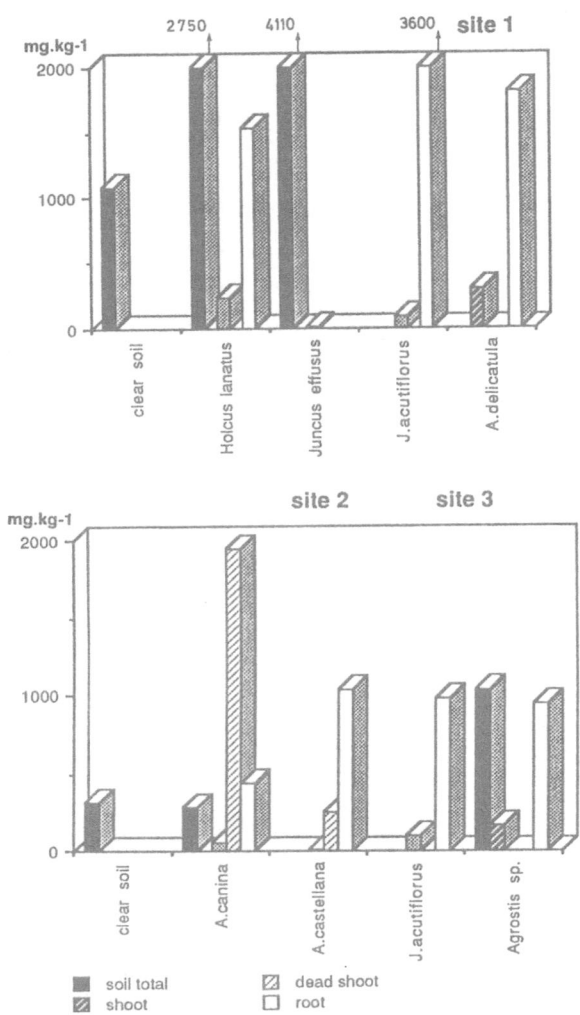

3. Arsenic in sediments

Sediment samples, taken either from the barren areas of the spoilheaps or from sediment adjacent to the roots of plants growing in contact with water containing the mine waste, have been analysed for total arsenic content (after the digestion of the organic matter with $HCl : HNO_3 = 3 : 1$ (v/v), after shaking during 2 hours with concentrated HCl or according to Jackson 1958), water soluble arsenic, ammonium acetate extractable As and calcium lactate extractable As, like described for the water samples.

Sediments show high total arsenic levels at site 1 ($17000 \, mg \cdot kg^{-1}$ dw), lower at site 2 ($300 \, mg \cdot kg^{-1}$ dw) and again higher values at site 3 ($1000 \, mg \cdot kg^{-1}$ dw), see Tables II and III, and Fig. 2. The higher values at site 3 are probably due to the accumulation of suspended materials that precipitate there, as a result of the sudden drop in current velocity of Jales brook.

The fraction of water soluble As is always higher than the ammonium acetate extractable fraction (Table II), opposite to data from mines in the United Kingdom (Porter & Peterson 1975), which possibly may lead to differences in As uptake by plants. More likely, however, is that the ammonium acetate extractable fraction does not represent the amount of As available to the plants. There is literature supporting the hypothesis that As is normally taken up by the phosphate uptake system due to the similarity of its ion arsenate to phosphate (Rothstein 1963; Borst Pauwels *et al.* 1965; Asher & Reay 1979; Macnair 1987). Ammonium acetate extracts positive ions bound to sediment particles and is thus not very suitable to give an indication

Fig. 2. Total arsenic contents of plant and sediment samples in the mining area. *Holcus lanatus, Juncus effusus, Juncus acutiflorus* and *Agrostis delicatula* at site 1; *Agrostis canina, Agrostis castellana* and *Juncus acutiflorus* at site 2; *Agrostis* sp. at site 3.

Table II. Arsenic content ($mg \cdot kg^{-1}$ dw) and pH of sediment samples adjacent to rootsystems of plants growing in contact with water containing the mine waste. Total As after digestion with a mixture of $HCl : HNO_3$ (3 : 1 v/v).

Site	Sample	Total	NH$_4$-acetate extractable	water soluble	pH
1	*Juncus effusus*	4100	2	7	6.7
1	*Holcus lanatus*	2800	3	10	7.1
2	*Agrostis canina*	300	2	10	6.2
2	*Agrostis castellana*	300	1	5	6.2
3	*Agrostis sp.*	1000	2	5	6.4

Table III. Mean, standard deviation and range (in brackets) of arsenic content (mg · kg^{-1} dw) and pH of sediment samples of barren areas and sites adjacent to rootsystems of plants growing on the spoilheaps. Total As analysis after Jackson (1958 #) or after shaking with concentrated HCl (*). n = number of samples.

Sample	Total	Ca-lactate extractable	Water soluble	pH
barren	# 1200 ± 1200	40 ± 40	7 ± 8	5.1 ± 1.4
(n = 5)	(100–2700)	(10–100)	(1–20)	(3.7–6.7)
Holcus lanatus	* 1200 ± 400	20 ± 10	3 ± 1	
(n = 5)	(600–1600)	(10–30)	(2–5)	
Agrostis castellana	# 4400 ± 5500	50 ± 20	10 ± 20	4.9 ± 1.2
(n = 10)	(300–17000)	(20–90)	(0.03–60)	(3.6–6.6)

about the bioavailability of the negative arsenates. Extraction with calcium lactate might be a better method, see Table III.

The spoils of the Jales mines do not have an extremely low pH, values around 5 in the rhizosphere (see Table III), maybe because its relatively recently mined materials did not develop acidity yet.

4. Arsenic in plants

Arsenic levels in plant tissue have been analysed after digestion of the organic matter by HNO$_3$/HClO$_4$ (10 : 1 v/v) or concentrated HNO$_3$. All plant parts have been carefully washed (3 times in tap water followed by 3 times in demineralised water) prior to analysis to avoid contamination of the samples by soil particles.

The striking difference in concentrations between green and senescent shoot tissue (see Table IV) stresses the necessity to analyse these parts separately. Values given for shoot tissue will largely depend on the proportion of dead leaves included in the sample.

In *Agrostis castellana* samples the As concentration is significantly higher (ANOVA p < 0.01) in the senescent leaves than in the green leaves (see Table IV). The senescent leaves might function as a sink and plants avoid in this way too high concentrations in more vital tissues. The As concentration in green shoot tissue is positively correlated (Pearson, p < 0.05) with the concentrations in the roots, the same tendency could be observed for Fe and Zn. The As concentrations in the roots and in the green leaves do not show significant correlations with the concentrations in the soil samples, in contrast to observations of Porter & Peterson (1975) for other *Agrostis* species. This indicates that either there is a barrier that avoids As entering the root system, or translocation of As in the plant maintains the concentrations on exceptable levels. Root tissue always shows higher concentrations of As than the calcium lactate extractable fraction of the soil samples (see Tables III and IV) which makes the barrier hypothesis not likely. The almost significant positive correlation between the calcium lactate extractable fraction of the soil samples and the As concentrations in the senescent leaves does not contradict the idea of translocation of As to the senescent leaves that function as a sink.

Also *Agrostis canina* presents higher concentra-

Table IV. Mean, standard deviation and range (in brackets) of arsenic content of several plant parts of *Agrostis castellana* growing on the spoilheaps. n = number of samples.

roots (n = 10)	senescent leaves (n = 10)	green leaves (n = 10)	stolon (n = 7)	inflorescence (n = 5)
300 ± 200	200 ± 100	40 ± 20	40 ± 10	30 ± 20
(100–600)	(100–300)	(20–80)	(30–60)	(10–70)

tions of As in the senescent leaves (1940 mg · kg⁻¹ dw) than in green shoot tissue (60 mg · kg⁻¹ dw) and roots (450 mg · kg⁻¹ dw).

Soil factors known to have influence on the As availability for plants, like % organic matter, pH, phosphorus and iron content, do not show significant correlations with the As concentration in the roots, senescent and green leaves of *Agrostis castellana*.

Micro-organisms may be of importance to the As availability for *Agrostis castellana* on the spoils of the Jales mines, VAM infection of its roots has been found.

Significant inter-element correlations (Pearson, $p < 0.05$) could be found only in the root samples between As and Pb. The *Agrostis castellana* samples did not show a significant correlation between As and Fe like Porter & Peterson (1977) reported for *Agrostis canina*, *A. tenuis* and *A. stolonifera*.

References

Asher, C. J. & Reay, P. F., 1979. Arsenic uptake by barley seedlings. Austr. J. Plant Phys. 6: 459—466.

Benson, N. R., 1953. Effect of season, phosphate, and acidity on plant growth in arsenic-toxic soils. Soil Sc. 76: 215—224.

Borst Pauwels, G. W. F. H., Peters, J. K., Jager, S. & Wijffels, C. C. B. M., 1965. A study of the arsenate uptake by yeast cells compared with phosphate uptake. Biochem. Biophys. Acta 94: 312—314.

Bradshaw, A. D. & Chadwick, M. J., 1980. The restoration of land. The ecology and reclamation of derelict and degraded land. Blackwell, Oxford.

Bringman, G. & Kuhn, R., 1980. Comparison of the toxicity thresholds of water pollutants to bacteria, algae and protozoa in the cell multiplication test. Wat. Res. 14: 231—241.

Brooks, R. R., Fergusson, J. E., Holzbecher, J., Ryan, D. E., Zhang, H. F., Dale, J. M. & Freedman, B., 1982. Pollution by arsenic in a gold mining district in Nova Scotia. Env. Poll. (B) 4: 109—117.

Carrow, R. N., Rieke, P. E. & Ellis, B. G., 1975. Growth of turfgrasses as affected by soils phosphorus and arsenic. Soil Sc. Soc. Am. Proc. 39: 1121—1124.

Christensen, E. R. & Zielski, P. A., 1980. Toxicity of arsenic and PCB to a green alga (Chlamydomonas). Bull. Env. Cont. Toxicol. 25: 43—48.

Cortes, R. M. V., 1989. Biotipologia de ecossistemas lóticos do nordeste de Portugal. Ph.D. Thesis Universidade de Trás-os-Montes e Alto Douro, Vila Real, Portugal.

Cortes, R. M. V., Carvalho, L. H. M. & Azevedo, J. C. 1988. Produtividade piscícola e estrutura das populaçoẽs de peixes na bacia do rio Tua. Actas Col. Luso-Esp. Ecol. Bacias Hidrog. e Rec. Zool., Porto pp. 65-72.

Cortes, R. M. V., De Koe, T. & Molles, M., 1986. Comparison of the effects of organic versus mineral pollution on the macro-invertebrate communities of two rivers in northern Portugal. Proc. Third Eur. Congr. Entom., Amsterdam pp. 79—82.

Cortes, R. M. V. & Pereira, J. O. B., 1987. Impacto de alguns poluentes na produtividade piscícola de alguns rios de Trás-os Montes. Actas Sem. Aquac. ICBAS, Porto. In press.

Craig, P. J., 1986. Organometallic compounds in the environment. Principles and Reactions. Longman, Harlow.

De Koe, T., Rozema, J., Broekman, R. A., Otte, M. L. & Ernst, W. H. O., 1988. Heavy metals and arsenic in water, sediment and plants near the Jales gold and silver mine in north Portugal. Third Int. Conf. Env. Contam., Venice pp. 152—154.

Jackson, M. L., 1958. Soil chemical analysis. Prentice-Hall, Inc. Englewood Cliffs, N.J.

Jacobs, L. W., Syers, J. K. & Keeney, D. R., 1970. Arsenic sorption by soils. Soil Sc. Soc. Am. Proc. 34: 750—754.

Johnson, L. R. & Hiltbold, A. E., 1969. Arsenic content of soil and crops following use of methanoarsenate herbicides. Soil Sc. Soc. Am. Proc. 33: 279—282.

Livesey, N. T. & Huang, P. M., 1981. Adsorption of arsenate by soils and its relation to selected chemical properties and anions. Soil Sc. 131: 88—94.

Macnair, M. R. & Cumbes, Q., 1987. Evidence that arsenic tolerance in *Holcus lanatus* L. is caused by an altered phosphate uptake system. New Phytol. 107: 387—394.

Marcus-Wyner, L. & Rains, D. W., 1982. Uptake, accumulation and translocation of arsenical compounds by cotton. J. Environ. Qual. 11: 715—719.

Porter, E. K. & Peterson, P. J., 1975. Arsenic accumulation by plants on mine waste (UK). Sc. Total Env. 4: 365—371.

Porter, E. K. & Peterson, P. J., 1977. Arsenic tolerance in grasses growing on mine waste. Env. Poll. 14: 255—265.

Rothstein, A., 1963. Interactions of arsenate with the phosphate-transporting system of yeast. J. Gen. Phys. 46: 1075—1085.

Surber, E. W. & Meehan, O. L., 1930. Lethal concentrations of arsenic for certain aquatic organisms. Am. Fish. Soc. Trans. 61: 225—239.

Wauchope, R. D., 1975. Fixation of arsenical herbicides, phosphate and arsenate in alluvial soils. J. Env. Qual. 4: 355—358.

Wauchope, R. D. & McDowell, L. L. 1984. Adsorption of phosphate, arsenate, methanearsonate and cacodylate by lake and stream sediments: comparisons with soils. J. Env. Qual. 13: 499—504.

Williamson, A. & Johnson, M. S., 1981. Reclamation of metalliferous mine wastes. In: N.W. Lepp (ed.) Effect of heavy metal pollution on plants. Vol. 2. Appl. Sc. Publ., London. pp. 185—212.

Woolson, E. A., Axley, J. H. & Kearney, P. C., 1971. The chemistry and phytotoxicity of arsenic in soils: I. Contaminated field soils. Soil Sc. Soc. Am. Proc. 35: 938—943.

Woolson, E. A., Axley, J. H. & Kearney, P. C., 1973. The chemistry and phytotoxicity of arsenic in soils: II. Effects of time and phosphorus. Soil Sc. Soc. Am. Proc. 37: 254—259.

Author's Address
T. de Koe
Dep[to]. de Biologia
Universidade de Trás-os-Montes e Alto Douro
5000 Vila Real
Portugal

Plant life in an amphibious environment. (Photograph: M. van der Werff.)

CHAPTER 6

Ecophysiological adaptations of higher plants in natural communities to waterlogging

P. JANIESCH

Abstract. Adaptations of plants to waterlogging are described as prerequisites of their occurrence in natural plant communities. Waterlogging in the soil results in oxygen deficiency and affects mineralization and solubility of mineral substances as well as the formation of phytotoxic compounds. Besides particularities in nitrogen supply, the changes in solubility of iron and manganese are discussed as the decisive factors in plant growth and distribution. Iron tolerance, oxygen transport capability and radial oxygen loss from the roots are considerably more important for plant survival than metabolic adaptations, as previously assumed.

1. Introduction

The soil conditions of a natural stand and the occurrence of plant communities are closely interrelated. These relationships can be seen in typical combinations of species on similar soils. Besides climatic and endogenous factors of the plants themselves the supply with mineral nutrients plays a major part for their distribution. Apart from mineral deficiency, an excess of minerals may result in toxic reactions and may thus be decisive for the occurrence of plants in natural communities (Ernst 1969, 1974, 1983; Albert & Kinzel 1973; Stewart *et al.* 1973; Janiesch 1980; Rozema *et al.* 1985).

Naturally occurring plant communities adapted to high levels of the water table are wide-spread. Among the forest communities these include alder swamps and flood plain forest. Other examples are wet meadows, peat bogs, river sides and salt marshes. The high water table and the special mineral conditions of these communities are so dominant that the composition of species is very uniform in Central Europe (Ellenberg 1978). Therefore these so-called "azonal communities" are especially suitable for research on soil — plant relationships.

The natural occurrence of these communities has become rare due to draining, intensive agriculture and industrialisation. The investigation of the special ecological conditions of these plant communities is therefore imperative in order to save or regenerate such biotopes with their special importance for plants and animals.

2. Effect of flooding on soil conditions

2.1. The redox potential

In aerobic soils 10% to 15% oxygen in the soil air is sufficient for plant growth. However, these conditions change as soon as the groundwater level rises. As a result of waterlogging water fills the air capillaries and reduces the normal processes of gaseous exchange between the air and soil. This alters the pressure gradients and mass flow of gases with the net effect of decreasing oxygen diffusion. In water the oxygen diffusion is 10,000 times lower than in the air and oxygen exchange between atmosphere and the soil is drastically reduced (Armstrong 1972). The remaining oxygen will be consumed by

J. Rozema and J. A. C. Verkleij (eds.), Ecological Responses to Environmental Stresses, 50—60.

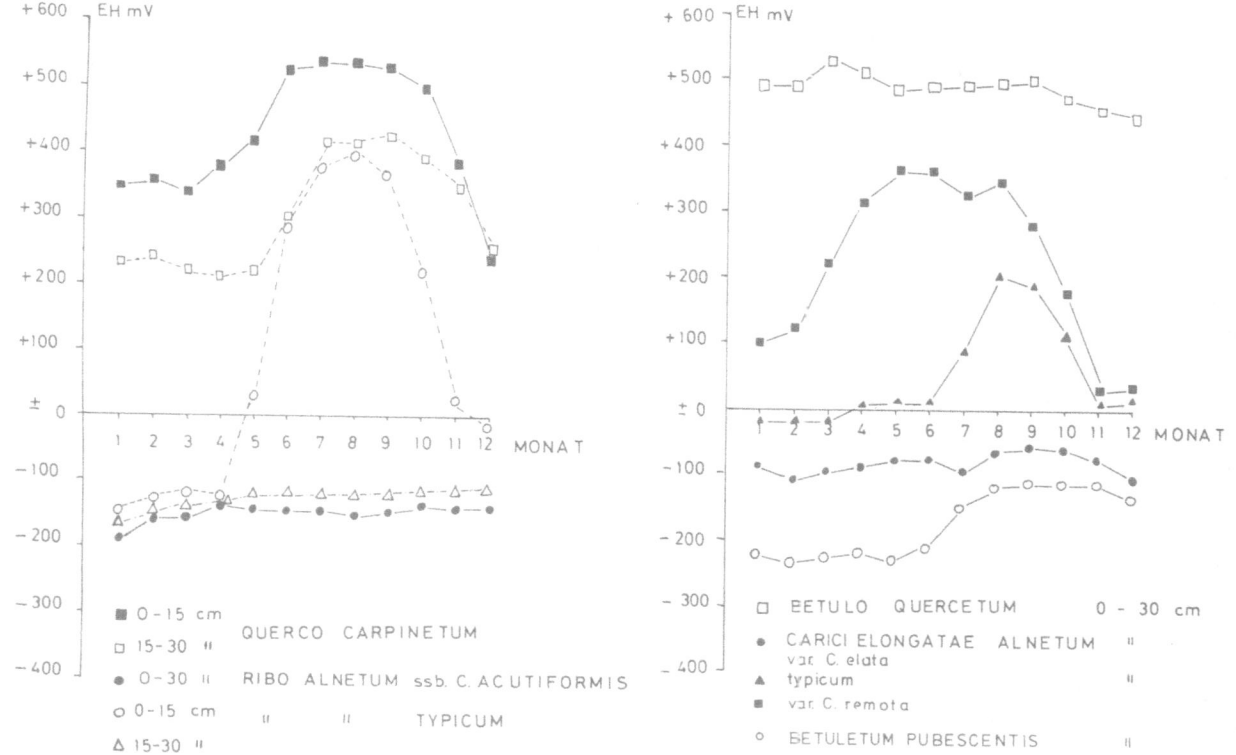

Fig. 1. Redox Potential of the soils of different woodland communities during the vegetation period (soil depth 0—30 cm, mVolt normal hydrogen electrode; methods see Janiesch 1981).

microbial fermentation. The immediate consequence is that anaerobic conditions occur. Only, when percolation of water in the soil exceeds a rate of 1.5 m per day, the oxygen demand of roots in flooded soils in some cases can be met (Grable 1966). The lowering of the redox potential is a first measurable event. Patrik & Mahaptra (1968) divided soils into four classes:

1. Oxidizing conditions Redox Potential > +0.4 Volt
2. Weak reducing conditions Redox Potential +0.4 to +0.1 Volt
3. Reducing conditions Redox Potential +0.1 to −0.1 Volt
4. Strongly reducing conditions Redox Potential −0.1 to −0.3 Volt

The redox potential of a flooded soil gives a fairly good indication of the intensity of oxidation or reduction in the soil. There are significant differences in the redox potentials of soils in natural plant communities in the course of a vegetation period. In alder swamps, for instance, measurement of the redox potential during the whole vegetation period (Janiesch 1981) shows reducing conditions over long periods, but conditions change when the water table is lowering during summer. On the other hand, oxidative conditions predominate in the soils of oak-hornbeam woods (Fig. 1). Similar reports are given for peat bogs, salt marshes (Armstrong *et al.* 1985) and rice soils. The pH of most soils tends to change towards the neutrality, with the pH of acid soils increasing and alkaline soils decreasing. Besides the direct influence of oxygen deficiency on plant growth, which will be discussed later, the redox potential indirectly influences plant growth by changing the mineral condition of the soils.

2.2. Fermentation processes and plant nutrients

A wide variety of toxic compounds accumulate in

waterlogged soils (Ponnamperuma 1972). Decomposition of organic matter in flooded soils differs from that in well drained soils in velocity, pathways and end products. The number of aerobic and facultatively anaerobic organisms, which exhaust the available soil oxygen supply, is reduced after flooding. When the oxygen and nitrate present in flooded soil are consumed, anaerobic bacteria use oxidized soil components such as hydrous oxides of Mn(IV), Fe(III) and sulfate, converting them to their reduced counterforms Mn(II), Fe(II) and sulfide, respectively. Besides these products of microbial fermentation increasing carbon dioxide, hydrogen, methane, hydrogen sulfide have an especially strong influence on plant growth because of their toxicity. These gases vary in concentration from 1 to 20% CO_2, 10 to 95% N_2, 15 to 75% CH_4, and 0 to 10% H_2 (see Patrick & Mikkelsen (1971)). The exact concentration depends upon the time after flooding as well as upon soil and other environmental conditions, such as microbial populations and the nature of the organic and inorganic substances.

Let us now focus upon changes in mineral conditions. The nitrogen relations and the high amounts of soluble iron and manganese are important factors for plant distribution (Janiesch 1979, 1981).

Nitrate and ammonium are the major sources of inorganic nitrogen taken up by the roots of higher plants and thus are decisive factors for plant growth in general (see Marschner 1986). The amount of nitrogen that can be taken up by plants is different owing to the fact that mineralization rates often change within small distances, but the form in which nitrogen is present (e.g., ammonium and/or nitrate) may also differ from site to site. The denitrification rates increase immediately upon flooding. The accumulation of ammonium indicates an inhibition of nitrification. The main nitrogen transformation processes in flooded soils are mineralization, nitrification – denitrification, volatilization and biological fixation.

Table I shows that the mineralization rate of nitrogen in different plant communities varies with water table. In addition, there are significant differences in nitrogen supply in various plant communities. It ranges from denitrification processes and a possible oxygen deficiency to very high mineralization rates of 350 kg N per year and ha. Mineralization rates, however, are clearly dependent on the groundwater level. The same is true when ammonium is converted into nitrate and vice versa.

In flooded soils ammonium is the stable form of nitrogen. Plants growing in these sites must therefore adapt themselves to a changing nitrogen supply in the form of ammonium and/or nitrate. *Carex* species in alder swamps show a significant dependence on the kind of nitrogen supply (Janiesch

Table I. Annual net nitrogen mineralization rate in different ecosystems.

	kg Nitrogen per year & hectare			
	NO_3-N	NH_4-N	N-min (total)	NH_4/NO_3 (ratio)
Ribo Alnetum typicum	87	4	91	0.05*
subass. C. acutiformis	8	10	18	1.25*
Carici elongatae – Alnetum typicum	74	86	160	1.16*
var. C. remota	225	117	342	0.52*
Betuletum pubescentis	2	16	18	8.00*
Querco – Carpinetum	205	7	212	0.03*
Phalaridetum			100	0.21#
Caricetum davallianae			5	1.70#
Caricetum elatae			30	19.0#
Sphagnetum magellanicum			5	19.0#

* Janiesch (1981).
# Ellenberg (1977).

52

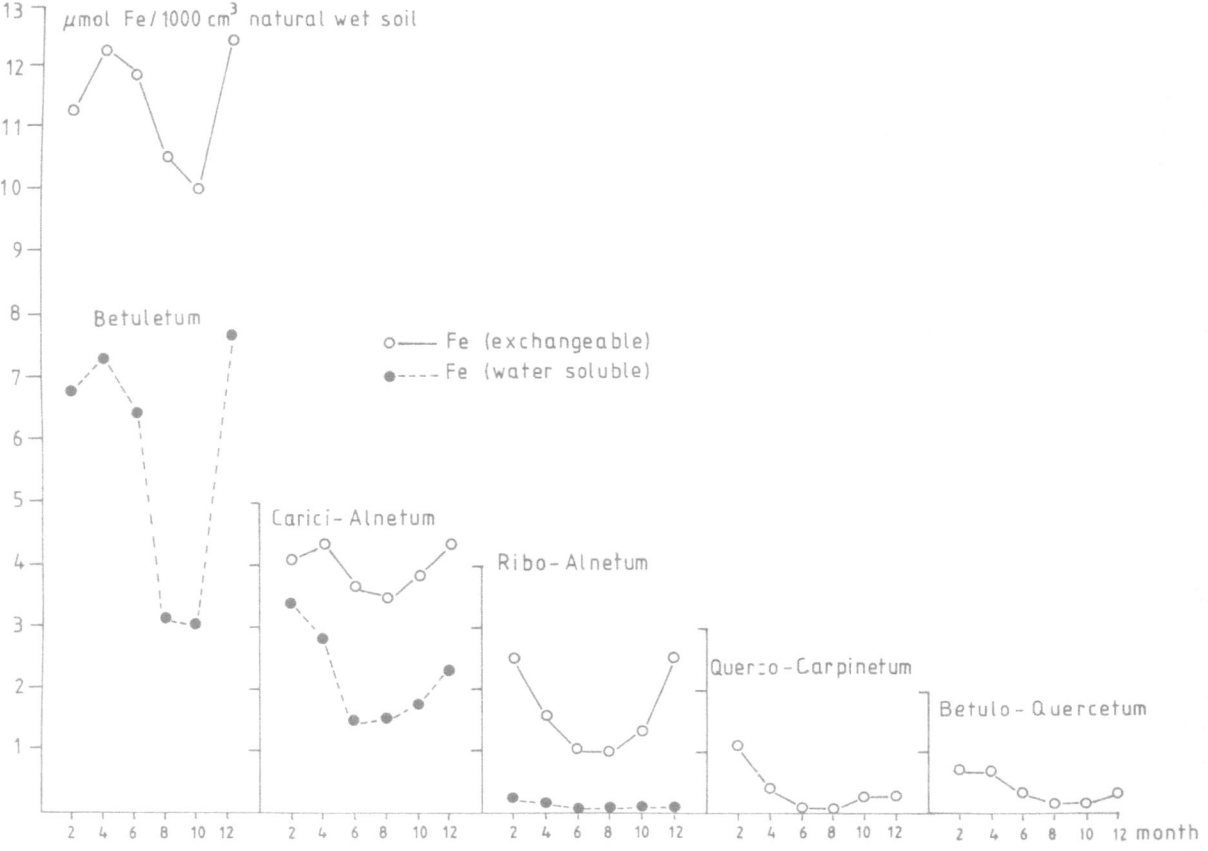

Fig. 2. Soluble and exchangeable iron in the soils of different woodland communities during the vegetation period. (soil depth 0–30 cm; methods see Janiesch 1981).

1986). Species in very damp sites, such as *Carex pseudocyperus,* prefer ammonium as nitrogen source, whereas species· in sites with changing moisture, such as *Carex remota,* prefer nitrate. One reason for this behaviour is the different ability of these species to produce the enzyme nitrate reductase which is responsible for nitrate assimilation. There are also significant differences in the degree of resistance to ammonium and nitrate (Janiesch 1986). Uptake of ammonium and nitrate also requires physiological adaptations in order to prevent acidification or alkalinization of the cell environment (Raven 1979).

There are numerous reports of increases in solubility of iron and manganese in flooded soils. Reports about the importance for plant growth and distribution in natural plant communities, how-

ever, are rare and contradictory. The transformation of insoluble to soluble iron compounds occurs already after a few days upon flooding (Mandel 1961). Five to fifty percent of active Fe(III) present in a soil may be reduced within a few weeks of submergence, depending on temperature, organic matter content, nitrate concentration, and chemical structure of the oxides (Yoshida 1978). Iron is thereby not totally water soluble but is exchangeable to soil colloids or bound to organic complexes (Ponnamperuma 1972; Turner & Patrick 1968). The soluble amounts of iron may often reach concentration toxic to plant roots (Baba *et al.* 1964; Martin 1968; Janiesch 1979).

There are significant differences in the solubility of different iron compounds in the soils of natural plant communities during the vegetation period

(Janiesch 1978, 1979, 1981). In Fig. 2, for example, the different soluble fractions of iron in the soils of natural plant communities are shown. The highest concentrations of soluble iron were found in fen birch woods, while they were relatively low in oak-hornbeam woods. Iron toxicity may be therefore an important growth-limiting factor for plants not adapted to flooded soils. At the moment there are few reports about plants from natural sites. In agricultural cultures this effect has been reported from rice soils (Ottow *et al.* 1982; Virmani 1977).

3. Waterlogging and plant growth

3.1. Introduction

Plants occurring in flooded sites show specific morphological, anatomical and physiological adaptations. All of these adaptations may ultimately be explained by physiological/biochemical processes. However, the connections and interrelations of metabolic pathways and their effects on morphological changes are not yet known in detail.

There are some differences in the reaction on tolerance of low oxygen conditions between herbaceous and woody plants (see Jackson & Drew 1984, Kozlowski 1984). In many plant communities herbaceous plants represent important character and differential species for diagnostic purposes. Therefore the adaptations of herbaceous plants to low oxygen concentrations and related changes in mineral contents of the soils will be discussed below.

Higher plants have an oxygen requirement for optimal growth. In well drained soils aerobic respiration of plant roots and soil microorganisms consume between 5–24 g oxygen per square meter of land surface (Russell 1973). Under flooded conditions, however, oxygen diffusion to the roots is restricted (see 2.2.).

Plants have evolved different strategies to overcome flooding stress: (1) obtaining energy by fermentation, (2) production of non-toxic end products and (3) production or transport of oxygen from the atmosphere to the roots. The period of resistance to anaerobic conditions ranges from few hours to several months (Crawford 1982, Kordan 1976, Webb & Armstrong 1983, Huck 1970). Plants at different sites differ in their ability to endure periods poor in oxygen. This suggests the existence of adaptative processes corresponding to the different soil conditions.

3.2. Morphological changes

Anaerobic conditions cause numerous morphological changes in higher plants. Plants change the direction of root growth (Guhman 1924, Webster & Eavis 1972). Especially under high carbon dioxide partial pressures of 200–500 mbar, roots display negative geotropism. Reduced water absorption and closure of stomata (Kozlowski 1982) as well as an increase in the ethylene content of root tissues, inhibition of root growth, and formation of adventitious roots (Tang & Kozlowski 1984) are the earliest plant responses to flooding. Several components like ethylene, carbohydrates, growth hormones, nitrogenous substances (Jackson & Drew 1984) are involved in the sequential events leading to morphological adaptation to flooding. Hypoxia and ethylene accumulation stimulate formation of interconnected gas-filled channels (aerenchyma within roots and shoots). As little as $0.1 \mu l$ per 1000 ml of exogenous ethylene in well-oxygenated nutrient solution stimulates the formation of aerenchyma (Drew *et al.* 1981).

3.3. Metabolic adaptation

The main influences of low oxygen concentration in plants at the cellular level are metabolic changes. Under fully aerobic conditions the oxidation of 1 mol sugar to carbon dioxide and water yields 38 mol ATP. During anaerobic conditions the transfer of electrons from cytochrome oxidase to oxygen cannot take place and synthesis of ATP via the electron-transport system is blocked. Furthermore the regeneration of NADH via the tricarboxylic acid cycle is impossible. Under anaerobic conditions only 2 ATP are available via glycolysis. This will generate an energy problem for the plant and lead to an unfavourable ATP/ADP ratio (Mendelssohn *et al.* 1981). Another problem is a mechanism

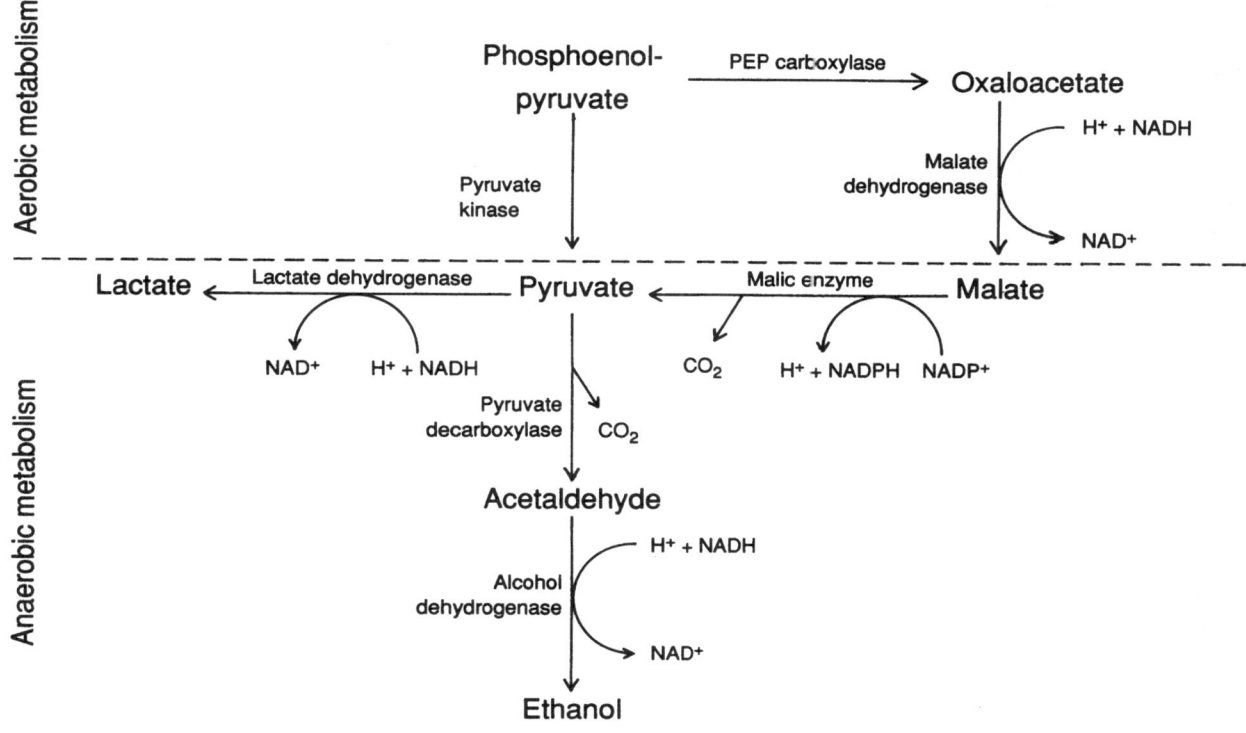

Fig. 3. Anaerobic metabolic pathway in plant tissue. (Adapted from Crawford, 1978).

to eliminate an excess of NADH. In the fermentation pathway NADH can be eliminated via pyruvate to produce lactate or ethanol. Ethanol can accumulate to toxic concentrations under these circumstances. Crawford & Tyler (1969) and Crawford (1978, 1982) reported an alternative pathway to prevent high ethanol accumulation in flooding tolerant plants. Non-toxic end products such as malate, aspartate or shikimate can be observed. (Fig. 3). This theory had widespread attention (Larcher 1980, Harbone 1982, Moore 1982).

It can be questioned whether anaerobic respiration can provide enough energy for uptake and transport of nutrients, growing processes and metabolism of plants in waterlogged soils. In the past years there have been a number of reports contrary to the metabolic theory of flooding tolerance, which holds that tolerance is due to the production of compounds other than ethanol. It is important to note that the accumulation of malate and other products is not related to the total amount of substrate metabolized. Ap Rees & Wilson (1984) were unsuccessful in confirming the metabolic theory in *Glyceria* roots. Earlier Jenkin & Ap Rees (1983) showed that malic enzyme is also present in *Glyceria*. Furthermore many *Carex* species, which can endure long periods of anaerobiosis like *C. elata* and *C. pseudocyperus,* produce ethanol under natural conditions and show no metabolic differences as compared to *Carex* species from aerobic locations (Moog & Janiesch 1988). Similar reports are given by Jackson *et al.* 1981).

3.4. Significance of aerenchyma

In addition to causing morphological and metabolical changes, anaerobic conditions also influence the mineral nutrition of plants. This relates to the uptake and translocation of nutrients as well as to the response to toxic substances, which may accumulate in the soil under anaerobic conditions.

In this context another question arises: what mechanisms are developed for adaptation to toxic products of microbial fermentation, to the changing solubility of toxic substances such as manganese and iron and sometimes high amounts of ammonium? Inhibition of aerobic root respiration may interfere with the uptake and transport of nutrients (Drew & Sisword 1977, 1979; Larkum & Loughman 1969). Blacquieres et al. (1988) reported that ATP requirement and root respiration were higher with ammonium than with nitrate. The response to root anaerobiosis is strongly dependent upon the supply with macro- and micronutrients, since optimal nutrient conditions for the roots — especially for nitrate — may prevent a large part of the growth rate alteration (Trought & Drew 1981; Priol & Guyot 1985; Moog & Janiesch 1989). The significance of stimulation of nitrate reductase in flood-tolerant species (Garcia-Novo & Crawford 1973, 1976) is also unsettled.

An alternative hypothesis may be substantiated by the observation of the oxygen transport from shoot to root in large gas-filled spaces, aerenchyma, of many species. The anatomical and biochemical adaptation to flooding is correlated with the development of aerenchyma in many plants (Armstrong et al. 1982). Studer & Brändle (1984) showed that after 20 minutes of anaerobiosis exchange of gas between roots and shoots had taken place in *Menyanthes, Typha* and *Glyceria*. From these investigations it seems clear that all of the oxygen for metabolism of roots may be transported exclusively via the shoots. The marked influence of oxygen transport in roots and shoots and the important Radial Oxygen Loss of roots in the rhizosphere for flooding tolerance has been reported in many publications (Coult & Vallance 1968; Armstrong 1964, 1967, 1979).

The oxidation of the rhizosphere seems to be one of the most important processes for survival of plants in these sites. A thin layer of oxidized iron hydrates around the root may be observed in almost all plants. This oxidation could have several functions. On the one hand, it serves as a barrier against toxic substances such as sulfide, on the other hand, high concentrations of iron(II) and manganese(II) compounds, respectively, are oxidized in the rhizosphere and thus become insoluble. In addition, oxygen is ventilated through the aerenchyma from the surface organs to the subterranean organs by means of oxygen release at the root surface.

How can sufficient oxygen be transported in aerenchyma? Diffusion alone cannot be sufficient for the oxygen demand of roots and does not explain the radial oxygen loss. Grosse & Schröder (1986) give an explanation for long-distance transport in aerenchyma by a physical way. Under anaerobic conditions an increase in carbon dioxide production occurs in plant root tissue. The solubility of carbon dioxide in water is higher than that of oxygen. This leads to a decrease in pressure in the intercellular system and initiates a downward air stream throughout the plant from the shoot to the root. Besides this the thermoosmosis of gas in aerenchyma seems to be widespread in flood-tolerant plants (Grosse & Schröder 1986). Depending on temperature differences between parts exposed to air and the above-ground parts a pressurization was observed in Nuphar. This creates a pressure difference strong enough to establish an air flow through the aerenchyma. This may also be the motor for the radial oxygen loss. In this care, if enough oxygen can be transported, no metabolic adaptation during anaerobiosis has to be established. The radial oxygen release avoids also a transport of toxic products to the roots via the rhizosphere.

3.5. *Iron and manganese toxicity*

One point of focus may be the tolerance of high amounts of soluble iron and manganese in plants. For many species 300–500 ppm soluble iron, which was measured in flooded soils, leads to toxic reactions (Foy et al. 1978; Janiesch 1979, 1981; Benckiser et al. 1984). Excessive uptake of these metals may lead to various physiological disorders (Foy et al. 1978) and this toxicity of iron and manganese may exclude certain species from waterlogged soil

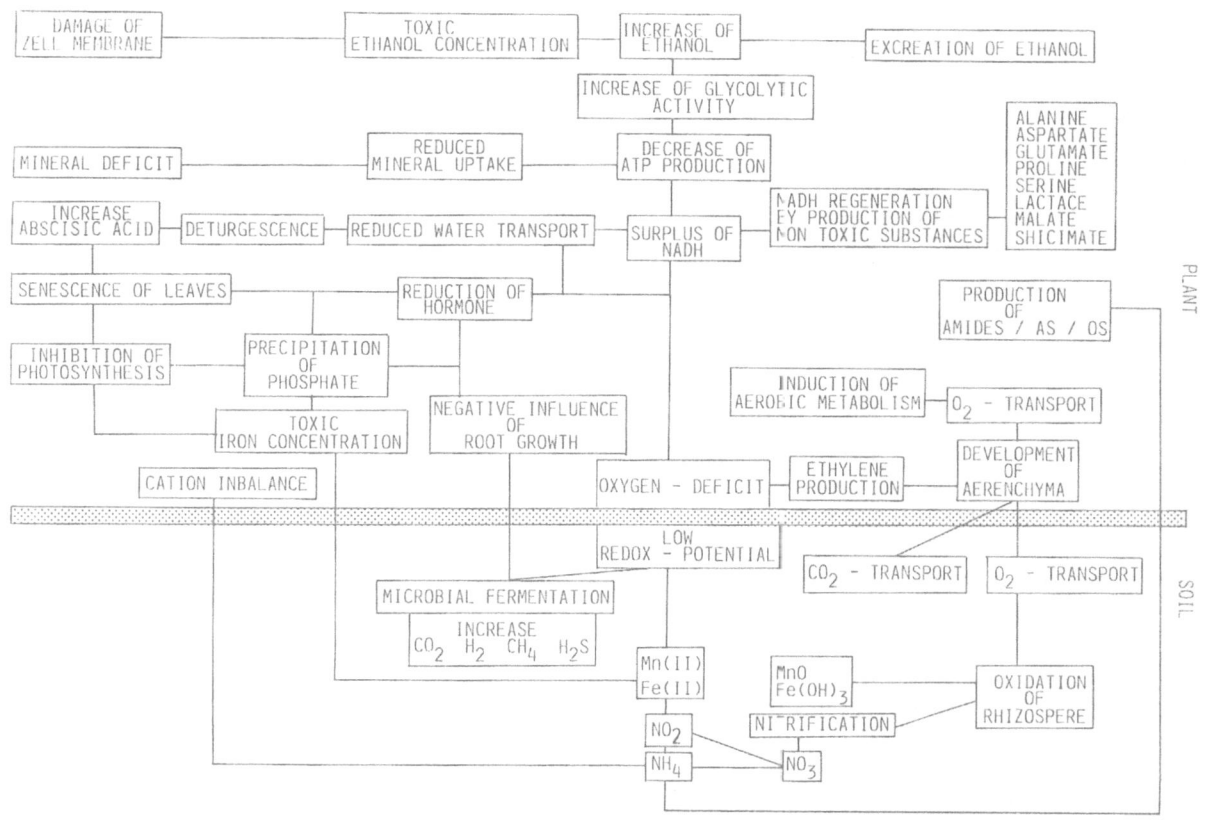

Fig. 4. Adaptations of flood-tolerant and flood-sensitive higher plants to anaerobic soil conditions.

Table II. Oxygen loss of roots of *Carex pseudocyperus* (eight weeks old) under aerobic and anaerobic conditions and increasing iron(II) concentrations in water nutrient solution (method see Janiesch 1981, Moog & Janiesch 1989).

μmol oxygen/cm² per hour

Fe (μMol)	aerobic	anaerobic
100	1.39 ± 0.21	20.38 ± 0.71
250	3.45 ± 0.73	31.53 ± 0.89
500	3.56 ± 0.69	63.06 ± 0.43
750	5.71 ± 0.47	90.47 ± 0.71
1000	6.78 ± 1.24	113.18 ± 1.03
1250	6.78 ± 1.21	115.17 ± 1.37
1500	5.84 ± 1.93	112.79 ± 3.57
1750	4.38 ± 2.03	101.27 ± 5.89
2000	2.24 ± 1.58	91.13 ± 7.38

(Martin 1968; Jones & Etherington 1970; Jones 1971, 1972; Wheeler *et al.* 1985). Iron toxicity is an important growth-limiting factor for agricultural lands, also in particular for flooded rice (Ottow & Benckiser 1982; Virmiani 1977).

Under natural conditions, however, iron concentration in plant tissue often does not vary in the manner expected from the iron content of the soil (Rozema *et al.* 1985; Ernst & Lugtenborg 1988). It is well known that roots are capable of oxidizing Fe(II) and Mn(II) to Fe(III) or Mn(IV) (Armstrong 1976, 1979, 1982; Janiesch 1981; Horiguchi, 1987). This indicates that iron is not taken up in the same amount as present in the soil because of the oxidation approximate the root.

Iron seems also to be necessary for optimal growth under anaerobic conditions, since iron and anaerobic conditions influence oxygen release by the roots (Janiesch 1981). With the polarographic method quantitative measurements are possible (Armstrong 1967). The amount of oxygen that can be released into the rhizosphere by the root, on the one hand, depends upon exterior conditions and on the other hand is different from species to species. (Janiesch 1981). Table II presents these conditions for *Carex pseudocyperus*. Anaerobic conditions prove to be a prerequisite of radial loss of oxygen. In addition, Fe^{2+} is able to stimulate the release of oxygen. Even though the physiological reasons for this behaviour are not yet clear, the reduced iron may act as a sink for oxygen diffusion into the soil and is a source for Fe^{2+} ions.

Several authors have reported differences in oxygen release among species adapted to anaerobic conditions; for example: *Menyanthes trifoliata* 6 μmol h^{-1} cm^{-2}, *Eriophorum angustifolium* 4.8 μ mol h^{-1} cm^{-2} and *Molinia coerulea* 0.3 μmol h^{-1} cm^{-2}, (Armstrong 1964). This may in part explain the distribution of plant species in natural communities. Anaerobic conditions, the occurrence of Fe (II), formation of aerenchyma and oxygen transport in the plant tissue seem to be some of the most important characteristics for colonization of flooded sites.

4. Conclusions

Adaptation to an anaerobic environment is not a single mechanism but a complex connection of morphological, biochemical and physiological adaptations, which leads to many complex and interrelated processes. In a summary table (Fig. 4) the response of both flooding-tolerant and flooding-sensitive species have been compiled as far as they are known to date. It has become apparent that laboratory studies are not sufficient to explain the manifold adaptations of plants to their environmental conditions. In particular, comparative studies in natural plant communities must be promoted if the results of ecological research are to be applied to the preservation of threatened ecosystems. An especially ecophysiological, multicausal

research seems to be necessary as shown by Rozema *et al.* (1985) for coastal halophytes.

References

Albert, R. & Kinzel, H., 1973. Unterscheidung von Physiotypen bei Halophyten des Neusiedlerseegebietes (Österreich). Z. Pflanzenphys. 70: 138—157.

Ap Rees, T. & Wilson, P. M., 1984. Effects of reduced supply of oxygen on the metabolism of roots of *Glyceria maxima* and *Pisum sativum*. Z. Pflanzenphys. 114: 493—503.

Armstrong, W., 1964. Oxygen diffusion from roots of some British bog plants. Nature 21: 801—802.

Armstrong, W., 1967. The use of polarography in the assay of oxygen diffusion from roots in anaerobic media. Physiol. Plant. 20: 540—553.

Armstrong, W., 1979. Aeration in higher plants. In: H.W. Woolhouse (ed.) Adv in Bot. Res. Acad. Press. pp. 226—332.

Armstrong, W., 1982. Waterlogged soils. In: J.R. Etherington (ed.) Environment and plant ecology. Wiley. New York. pp. 181—218.

Armstrong, W., Wright, E. J., Lythe, S. & Gaynard, T. J., 1985. Plant zonation and the effect of the spring-neap tidal cycle on soil aeration in a humber salt marsh. J. Ecol. 73: 323—339.

Baba, I., Inada, K. & Tayima, K., 1964. Mineral nutrition and the occurrence of physiological diseases. Proc. Symp. Rice Res. Inst.: 173—195.

Blacquieres, T., Voortman, E. & Stulen, I., 1988. Ammonium and nitrate nutrition in *Plantago lanceolata* L. and *Plantago major* L. ssp. Major. III. Nitrogen metabolism. Plant Soil 196: 23—34.

Coult, D. A. & Vallance, K. B., 1958. Observation on the gaseous exchange which take place between *Menyanthes trifoliata* L. and its environment. J. Exp. Bot. 27: 384—402.

Crawford, R. M. M., 1978. Metabolic adaptations to anoxia. In: D.D. Hook & R.M.M. Crawford (eds.) Anaerobic plant growth. Ann Arbor Science, Ann Arbor. pp. 119—136.

Crawford, R. M. M., 1982. The anaerobic retreat as a survival strategy for aerobic plants and animals. Trans. Bot. Soc. 44: 57—63.

Crawford, R. M. M. & Tyler, P. D., 1969. Organic acid metabolism in relation to flooding tolerance in roots. J. Ecol. 57: 235—244.

Drew, M. C. & Sisworo, E. J., 1977. Early effects of flooding on nitrogen deficiency and leaf chlorosis in barley. New Phytol. 79: 567—571.

Drew, M. C. & Sisworo, E. J., 1979. The development of waterlogging damage in young barley plants in relation to plant nutrient status and changes in soil properties. New Phytol. 82: 301—314.

Drew, M. C., Jackson, M. B., Giffard, S. C. & Campbell, R., 1981. Inhibition by silver ions of gas space (aerenchyma) formation in adventitious roots of *Zea mays* L. subjected to exogenous ethylene or to oxygen deficiency. Planta 153: 217—224.

Ellenberg, H., 1977. Stickstoff als Standortfaktor, insbesondere für mitteleuropäische Pflanzengesellschaften. Oecol. Plant. 12: 1—22.

Ellenberg, H., 1978. Vegetation Mitteleuropas mit den Alpen. E. Ulmer, Stuttgart.

Ernst, W. H. O., 1969. Beitrag zur Kenntnis der Ökologie europäischer Spülsaumgesellschaften. I. Mitteilung: Sand- und Kiesstrände. Mitt. Flor. soz. Arbeitsg. 14: 86—94.

Ernst, W. H. O., 1974. Schwermetallvegetation der Erde. G. Fischer. Stuttgart.

Ernst, W. H. O., 1983b. Element nutrition of two contrasted dune annuals. J. Ecol. 71: 213—263.

Ernst, W. H. O. & Lutgenborg, T. F., 1980. Vergleichende Ökophysiologie von Juncus articulatus und Holcus lanatus. Flora 169: 121—134.

Foy, C. D., Chaney, R. L. & White, M. C., 1978. The physiology of metal toxicity in plants. Ann. Rev. Plant Phys. 29: 511—566.

Garcia-Novo, F. & Crawford, R. M. M., 1973. Soil aeration, nitrate reduction and flooding tolerance in higher plants. New Phytol. 72: 1031—1038.

Grosse, W. & Schröder, P., 1986. Pflanzenleben unter anaeroben Umweltbedingungen, die physikalischen Grundlagen und anatomischen Voraussetzungen. Ber. Deutsch. Bot. Ges. 99: 367—381.

Grable, A. H., 1966. Soil aeration and plant growth. Adv. Agron. 18: 57—106.

Guhman, H., 1924. Variation in the root system of common everlast (Gnaphalium polycephalum). Ohio J. Sci. 24: 199—208.

Harborne, J. B., 1982. Introduction to ecological biochemistry. 2nd. ed., Acad. Press. London. New York.

Horiguchi, T., 1987. Mechanism of manganese toxicity and tolerance of plants. II. Deposition of oxidized manganese in plant tissue. Soil Sci. Plant Nutr. 33: 595—606.

Huck, M. G., 1970. Variation in taproot elongation rate as influenced by composition of soil air. Agr. J. 62: 815—818.

Jackson, M. B. & Drew, M. C., 1984. Effect of flooding on growth and metabolism of herbaceous plants. In: T.T. Kozlowski (ed.) Flooding and Plant Growth. Acad. Press. London. New York. pp. 47—128.

Jackson, M. B., Drew, M. C. & Giffard, S. C., 1981. Effect of applying ethylene to the root system of Zea mays on growth and nutrient concentration, in relation to flooding tolerance. Physiol. Plant. 52: 23—28.

Janiesch, P., 1978. Ökophysiologische Untersuchungen in Erlenbruchwäldern. I. Die edaphischen Faktoren. Oecol. Plant. 13: 43—57.

Janiesch, P., 1979. Eisen als Standortfaktor in Erlenbruchwäldern. Verh. Ges. Ökol. 7: 403—406.

Janiesch, P., 1980. Standortfaktoren in Quell-Erlenwäldern und pflanzensoziologische Gliederung. Ber. Inter. Symp. Epharmonie Rinteln 1979. Cramer. Lehre. 265—274.

Janiesch, P., 1981. Ökophysiologische Untersuchungen an Carex Arten aus Erlenbruchwäldern. Habilitationsschrift. FB Biologie. Münster.

Janiesch, P., 1986. Bedeutung einer Ernährung von Carex Arten mit Ammonium oder Nitrat für deren Vorkommen in Feuchtgesellschaften. Abhandl. Landesmus. Münster 48 (2/3): 341—354.

Jenkin, L. E. T. & Ap Rees, T., 1983. Effects of anoxia and flooding on alcohol dehydrogenase in roots of Glyceria maxima and Pisum sativum. Phytochem. 22: 2389—2393.

Jones, H. E., 1971. Comparative studies of plant growth and distribution in relation to waterlogging. III. The response of Erica cinerea L. to peats of different iron content. J. Ecol. 59: 583—591.

Jones, H. E., 1972. Comparative studies of plant growth and distribution in relation to waterlogging. V. The uptake of iron and manganese by dune and dune slack plants. J. Ecol. 60: 131—139.

Jones, H. E. & Etherington, J. R., 1971. Comparative studies of plant growth and distribution in relation to waterlogging. IV. The growth of dune slack plants. J. Ecol. 59: 793—801.

Kordan, H. A., 1976. Oxygen as an environmental factor in influencing normal morphogenetic development in germination rice seedlings. J. Exp. Bot. 27: 947—952.

Kozlowski, T. T., 1982. Water supply and tree growth. II. Flooding. For. Abstr. 43: 145—161.

Kozlowski, T. T., 1984. Responses of woody plants to flooding. In: T.T. Kozlowski (ed.) Flooding and plant growth. Acad. Press. London, New York. pp. 129—163.

Larcher, J. M., 1980. Physiological plant ecology. 2nd. Ed. Springer. Berlin, Heidelberg.

Larkum, A. W. D. & Loughman, B. C., 1969. Anaerobic phosphate uptake by barley plants. J. Exp. Bot. 62: 12—24.

Mandal, L. N., 1961. Transformation of iron and manganese in waterlogged rice soils. 2nd. ed., Soil Sci. 91: 121—126.

Marschner, H., 1986. Mineral nutrition of higher plants. pp. 674. Acad. Press. London. New York.

Martin, M. H., 1968. Conditions affecting the distribution of Mercurialis perennis L. in certain Cambridgshire woodlands. J. Ecol. 56: 775—793.

Mendelssohn, I. A., McKee, K. L. & Patrick, W. H., 1981. Oxygen deficiency in Spartina alterniflora roots: Metabolic adaptation to anoxia. Science 214: 439—440.

Moog, P. R. & Janiesch, P., 1988. Ethanol Biosynthese und Produktion organischer Säuren in Spross und Wurzeln von Carex remota unter aeroben und anaeroben Substratbedingungen. In: Tagungsb. Bot. Tagung Gießen. p. 39.

Moog, P. R. & Janiesch, P., 1989. Nutrient requirement of Carex species in solution culture. J. Plant. Nutr. 12: 497—508.

Moore, P., 1982. Survival mechanism in wetland plants. Nature 229: 581—582.

Ottow, J. C. G., Benckiser, G., Watanabe, I. & Santiago, S., 1982. Multiple nutrional soil stress as the prerequisite for iron toxicity of wetland rice (Oryza sativa L.). Trop. Agr. Trinid. 60: 102—106.

59

Patrick, W. H. & Mahaptra, I. C., 1968. Transformation and availability to rice of nitrogen and phosphorus in waterlogged soils. Adv. in Agron. 20: 323—359.

Patrick, W. H. & Mikkelsen, D. S., 1971. Plant nutrient behavior in flooded soil. In: Fertilizer technology and use, 2nd ed., Soil Sci. Soc. Amerc.

Ponnamperuma, F. N., 1972. The chemistry of submerged soils. Adv. Agron. 26: 29—96.

Prioul, J. L. & Guyot, C., 1985. Role of oxygen transport and nitrate metabolism in the adaptation of wheat plants to root anaerobiosis. Physiol. Veget. 23: 175—185.

Raven, J. A., 1979. Intracellular pH and its regulation. Ann. Rev. Plant. Phys. 30: 289—311.

Raven, J. A., 1988. Acquisition of nitrogen by the shoots of land plants: Its occurrence and implications for acid-base regulation. New Phytol. 109: 1—20.

Rozema, J., Luppes, E. & Broekman, R., 1985. Differential response of salt marsh species to variation of iron and manganese. In: W.G. Beeftink, J. Rozema & A.H.L. Huiskes (eds.) Ecology of coastal Vegetation. Vegetatio 61: 293—302.

Rozema, J., Bijwaard, P., Prast, G. & Broekman, R., 1985. Ecophysiological adaptations of coastal halophytes from foredunes and salt marshes. Vegetatio 62: 499—521.

Russell, E. W., 1973. Soil conditions and plant growth. 10th ed. Longmann, London.

Sachs, M. M., Freeling, M. & Okimoto, R., 1980. The anaerobic proteins of maize. Cell 20: 761—767.

Stewart, G. R., Lee, J. A. & Orebamjo, T. O., 1973. Nitrogen metabolism of halophytes. II. Nitrate availability and utilisation. New Phytol. 72: 539—546.

Studer, Chr. & Brändle, R., 1984. Sauerstoffkonsum und Versorgung der Rhizome von *Acorus calamus* L., *Glyceria maxima (Hartmann) Holmberg, Menyanthes trifoliata* L., *Phalaris communis Trin.* und *Typha latifolia* L. Bot. Helvet. 94: 23—31.

Tang, Z. C. & Kozlowski, T. T., 1984. Water relations, ethylene production, and morphological adaptation of *Fraxinus pennsylvanica* seedlings to flooding. Plant Soil 77: 183—192.

Trought, M. C. T. & Drew, M. C., 1981. Alleviation of injury to young wheat plants in anaerobic solution culture in relation to the supply of nitrate and other inorganic nutrients. J. Exp. Bot. 32: 509—522.

Turner, F. T. & Patrick, W. H., 1968. Chemical changes in waterlogged soils as a result of oxygen depletion. In: Int. Congr. Soil. Sci. Trans. 4: 53—65.

Virmani, S. S., 1977. Varietal tolerance of rice to iron toxicity in Liberia. Int. Rice Res. Newsl. 2: 4—10.

Webb, T. & Armstrong, W., 1983. The effect of anoxia and carbohydrates on the growth and viability of rice, pea and pumpkin roots. J. Exp. Bot. 34: 579—603.

Webster, P. L. & Eavis, B. W., 1972. Effects of flooding on sugarcane growth. I. Stage of growth and duration of flooding. Proc. Int. Sugar Cane Techn. pp. 708—714.

Wheeler, B. D., Al-Farraj, M. M. & Cook, R. E. D., 1985. Iron toxicity to plants in base-rich wetlands: Comparative effects on the distribution and growth of *Epilobium hirsutum* L. and *Juncus subnodulosus Schrank.* New Phytol. 100: 653—669.

Yoshida, T., 1978. Microbial metabolism in rice soils. In: Soil and Rice. pp. 445—463. Intern. Rice Res. Inst.

Author's Address
P. Janiesch
Dept. Physiologische Ökologie
Universität Oldenburg, FB 7
Postfach 2503
D 2900 Oldenburg
W-Germany

A beach plain salt marsh on the Frisian island of Schiermonnikoog. This type of salt marsh vegetation may be inundated for weeks or months in the winter period. (Photograph: J. Rozema.)

CHAPTER 7

Effects of inundation stress on salt marsh halophytes

J. VAN DIGGELEN

Abstract. In this contribution the effects of inundation on plant growth in the salt marsh habitat are discussed in relation to the experimental results of a Ph.D study on the ecophysiology of salt marsh halophytes.

In a study on the effects of flooding on 4 salt marsh species grown on clay or sand soil, it was found that all species grew better on clay soil. Flooding stimulated the growth of *Spartina anglica* and *Salicornia dolichostachya* and it inhibited growth of *Puccinellia maritima* and *Salicornia brachystachya*, which is in agreement with the distribution of these species in the field. Effects of high ferrous iron and sulphide concentrations, which may occur in reduced permanently flooded soils, were studied in nutrient solution experiments. In general, species responded to these potentially toxic substances as could be expected from their field distribution. The response of iron tolerant and sulphide tolerant species is discussed in relation to radial oxygen loss and sulphur metabolism, respectively.

1. Introduction

The literature on the ecology of salt marsh plants includes descriptive as well as experimental studies, on the importance of a great number of abiotic as well as biotic factors for the growth and distribution of species (Ernst 1985; Rozema *et al.* 1985). Inundation (seawater flooding), salinity and soil texture are regarded as the principal abiotic factors in salt marshes, determining spatial and temporal variation in physico-chemical soil properties, governing the performance and distribution of plant species in the salt marsh zonation. The performance and distribution limits, set by the physiological tolerance of species to these soil conditions, may be differentiated by competitive interactions and other biotic factors (Scholten *et al.* 1987). It was concluded that plant physiological tolerances, physical disturbance and interspecific competition interact to generate the salt marsh zonation (Bertness & Ellison 1987).

This contribution deals with the physiological tolerance and adaptation of salt marsh species to

inundation, with an accent on effects of reduced substances and submergence.

2. Effects of inundation on soils and plants

When a soil becomes saturated with water as a consequence of inundation, a complex sequence of interrelated physico-chemical and microbiological changes occurs. Except for a thin oxygenated zone at the surface, the soil becomes completely anaerobic within a few hours to several days, because the soil pore space is filled with water, and the remaining oxygen is depleted by respiration (plant roots, micro-organisms). Oxygen diffusion from the atmosphere is too slow to replenish oxygen at depths exceeding 5 to 10 millimeters.

The main chemical changes brought about by flooding of a well aerated soil are the disappearance of oxygen, accumulation of carbondioxide, anaerobic decomposition of organic matter, transformations of nitrogen, and reduction of Mn(IV), Fe(III) and sulphate (Ponnamperuma 1984). In salt

J. Rozema and J. A. C. Verkleij (eds.), Ecological Responses to Environmental Stresses, 62—73.
© *1991 Kluwer Academic Publishers.*

marshes, sulphate reduction is the terminal process of anaerobic mineralization of organic matter, while in fresh water marshes methane formation is the terminal process. The high concentration of sulphate in salt marsh soil (5–30 mM in interstitial water) prohibits depletion of the sulphate substrate and as long as sulphate reducing bacteria are active, the methanogenic bacteria are outcompeted (Wiebe et al. 1981; Carlson & Forrest 1982; Howard & Giblin 1983; Kristjansson & Schönheit 1983).

Plant growth is adversely affected by inundation induced changes in soil conditions in the following ways:
— inhibition of aerobic root respiration which may interfere with the uptake and transport of nutrients and also with the exclusion of sodium chloride in roots of salt marsh plants.
— self intoxication by accumulation of toxic fermentation products (ethanol, fatty acids).
— excessive uptake of reduced iron and manganese, which are accumulated in the soil solution.
— intoxication by sulphide which is accumulated in the soil.
— disturbance of hormonal metabolism and photosynthesis.

3. Adaptation to inundation stress

Two major types of adaptation to inundation stress can be distinguished: structural adaptations and metabolic adaptations. Structural adaptations are changes in root morphology (superficial rooting, increased branching of the roots, formation of adventitious roots) and the development of a continuous system of intercellular lacunae (aerenchyma), supplying the roots with oxygen via the shoot (Armstrong 1982).

Crawford (1967, 1978) developed a generalized theory of flooding tolerance based on metabolic adaptations, referred to as the metabolic theory. According to this theory, flooding-tolerant species would tend to accumulate non-toxic fermentation products (malate, glycerol, shikimate, lactate), whereas sensitive species would mainly accumulate toxic ethanol. The metabolic theory has been se-

verely criticised in recent years. In a review on effects of flooding on plant growth and metabolism, Jackson & Drew (1984) summarize a great deal of studies supporting their view that the evidence for the metabolic theory is incomplete.

Contrary to most non-saline wetlands where inundation is more or less permanent, coastal wetlands are much more dynamic due to the tidal cycle which leads to differences in frequency and duration of inundation along a height gradient in salt marshes. In a comparative study on the ecophysiology of salt marsh halophytes (Van Diggelen 1988) inundation was studied as a complex of factors by comparing plant growth on permanently flooded soils with growth on drained soils. In addition effects of chemical soil factors which may occur in permanently flooded, high reduced salt marsh soils, were studied by growing plants on nutrient solutions. In both the soil flooding studies and nutrient solution studies varying growth responses and degrees of tolerance were found between species, which correspond more or less to their field distribution.

4. Structural adaptations

The effects of flooding and salinity on the growth and mineral relations of seventeen salt marsh species were studied in a greenhouse. Plants were grown on sandy salt marsh soil under permanently flooded or drained conditions. With regard to structural adaptations, salt marsh species sensitive to inundation showed superficial rooting (*Atriplex littoralis, Halimione portulacoides*) on waterlogged soil, while tolerant species showed thin superficial roots and thick roots of high porosity penetrating deep into the soil (*Spartina anglica, Juncus maritimus, Juncus gerardii, Aster tripolium*).

In a comparative study of about 90 species Justin & Armstrong (1987) investigated root morphology, aerenchyma type and sensitivity or tolerance to waterlogging. In Table I their results obtained with salt marsh species are shown. In general these results are as anticipated: species from wet and waterlogged lower and middle salt marsh sites (*S. anglica, Puccinellia maritima, Aster tripolium* and

Limonium vulgare) show high root porosities, which tend to increase under flooded conditions, while species from the higher, more aerated soils (*Festuca rubra* and *Halimione portulacoides*) have low root porosities. An exception is formed by the typical wetland species from the genus *Salicornia* which have porosities similar to non-wetland species (Table I). According to Justin & Armstrong (1987), flooding tolerance in wetland species with non-wetland porosity characteristics depends on shallow rooting and a preference for more aerated wetland sites. The latter does not apply for *Salicornia* species, which may occur in the pioneer zone of the lower salt marsh. In the black anaerobic soil of these sites the root tips of *Salicornia* species die off due to anoxia and mechanical damage (Schat *et al.* 1987), resulting in shallow rooting. Despite this damage to the root system and impaired growth by anoxia (Schat *et al.* 1987), *Salicornia* survives. The fact that *Salicornia* species are obligate halophytes instead of salt excluders (preventing growth retardation due to excessive uptake of salt by damaged roots) together with their tolerance of ferrous iron up to 300 μM (Schat unpubl.) and their resistance to high sulphide concentrations (Ingold & Havill 1984; Van Diggelen *et al.* 1987) may explain their good performance in the pioneer zone. Flooding tolerance in *Salicornia* is probably based on radial oxygen loss (Schat *et al.* 1987), despite the low porosities, and metabolic tolerance to soil toxins (Havill *et al.* 1985).

5. Experimental studies on inundation stress

5.1. Soil flooding

In an experiment conducted under outdoor conditions in an experimental garden, growth and mineral relations of four species from the lower salt marsh (*Salicornia dolichostachya, Salicornia brachystachya, Puccinellia maritima,* and *Spartina anglica*) were studied in response to flooding and soil type under saline conditions. The experiment lasted three monhts. Flooding resulted in reduction of the soil: redox potentials were -122 and -42 mV for clay and sand respectively under flooded condi-

tions and 358 and 382 mV respectively under drained conditions. Biomass production of all species was lower on sand as compared to clay soil (Table II). The effect of soil type was significant in all species except for *Puccinellia maritima*. Lower nutrient availability in sand soil is likely to be the main cause of the effect of soil type on growth. This is substantiated by the significantly higher nitrogen concentrations found in shoots of *Puccinellia maritima* and *Spartina anglica* grown on clay (Table II).

Flooding enhanced growth of *Salicornia dolichostachya* and *Spartina anglica* and it inhibited the growth of *Puccinellia maritima* and *Salicornia brachystachya,* however, the effect of flooding was only significant in *Salicornia brachystachya.* The observed effects of flooding and soil type on both grass species fit in well with the competition experiments of Scholten & Rozema (1987), who found that increase of flooding frequency and clay fraction of the soil leads to a competitive advantage of *Spartina anglica* where lower flooding intensity and more sandy soils are more favourable to *Puccinellia maritima.* The response of both *Salicornia* species to flooding may explain their distribution in the field: the tolerant *Salicornia dolichostachya* occurring in the pioneer zone and the more sensitive *Salicornia brachystachya* growing at higher elevation (Schat *et al.* 1987).

The flooding treatment resulted in higher concentrations of iron and manganese in both roots and shoots of the four species studied (Fig. 1), corresponding with the reduced soil conditions. In *Puccinellia maritima* very high root concentrations of iron and manganese were found, probably most of which was precipitated on the root surface as a clearly visible iron plaque. Formation of a plaque by oxidation of ferrous and manganous ions outside the root by radial oxygen loss, is thought to be of adaptive value in preventing the uptake of large amounts of these ions into the plant (Chen *et al.* 1980; Armstrong 1982; Talbot *et al.* 1987). The relatively small increase of shoot iron as compared to root iron in plants on flooded soils found here, indicates that the immobilization of iron in the rhizosphere was quite effective, but cannot fully exclude iron (Talbot & Etherington 1987). Under flooded conditions, roots of *P. maritima* were covered with a thick layer of iron deposits, whereas *S. anglica* roots were white with some reddish-brown spots and traces. This may explain why the root concentrations of Fe and Mn in *P. maritima* were ten times those in *S. anglica*. Probably the oxides were deposited as a halo in the soil around *S. anglica* roots (Crowder & MacFie 1986), which was easily washed off, producing relatively white roots.

5.2. *Iron and Manganese stress*

Accumulated reduced substances such as ferrous iron, reduced manganese and sulphide are regarded as stress factors for the growth of salt marsh plants (Jeffrey 1987). There seems to be, however, very little experimental evidence to support this. Furthermore the determination of concentrations of these substances in salt marsh soils is complicated by technical problems. Ecophysiologists are interested in concentrations of plant available iron, manganese and sulphide. Recently it was shown that only pore water measurements provide reliable data with regard to the real composition of the soil solution (Luther & Church 1987). These authors proved that dissolved ferrous iron and free sulphide are mutually exclusive, giving way only

for high values of either of the two. In pore water of highly reduced lower salt marsh soils ferrous iron or sulphide concentrations up to a few millimolar may occur. Although seasonal and spatial variations in concentrations of reduced substances occur (depending on root density, bacterial activity and growing season), species growing on anaerobic lower salt marsh sites are confronted with high levels of these substances.

The effects of high ferrous iron concentrations

Table I. Root porosities of several salt marsh plant species grown on drained or flooded soil, according to Justin & Armstrong (1987). These authors distinguished between Wetland (W), Non-Wetland (N) and Intermediate (I) types of plants.

Species	Type	Root Porosity (%)	
		Drained	Flooded
Aster tripolium	W	26.5	25.6
Elymus pycnanthus	I	6.1	16.6
Festuca rubra	N	1.2	1.9
Halimione portulacoides	I	1.7	2.2
Limonium vulgare	W	13.2	18.0
Plantago maritima	W	7.7	21.7
Puccinellia maritima	W	17.4	25.5
Salicornia europaea	W	–	9.9
Spartina anglica	W	28.0	33.0
Suaeda maritima	I	5.4	3.1
Triglochin maritima	W	21.2	20.2
Salicornia brachystachya	W	5.8	7.1*
Salicornia dolichostachya	W	5.0	10.1*

* Data from Van Diggelen (1988).

Table II. Biomass production (g dry weight/plant) and shoot nitrogen concentrations (% of dry wt.) of four salt marsh species grown on sand or clay soil containing 350 mmol NaCl/l soil water, under drained and flooded conditions. Means of 4–12 plants. The effects of flooding and soil type were tested by analysis of variance (ANOVA).

Species	Soil type	Treatment	DW	N
Salicornia dolichostachya				
	Clay	drained	1.61	2.79
		flooded	1.95	3.28
	Sand	drained	0.45	3.21
		flooded	0.95	2.63
ANOVA		Flooding	ns	ns
		Soil type	***	ns
		Interaction	ns	ns
Salicornia brachystachya				
	Clay	drained	3.43	2.37
		flooded	2.42	2.29
	Sand	drained	2.42	2.50
		flooded	0.73	2.12
ANOVA		Flooding	***	ns
		Soil type	***	ns
		Interaction	ns	ns
Puccinellia maritima				
	Clay	drained	2.80	3.05
		flooded	1.14	3.85
	Sand	drained	1.94	2.04
		flooded	0.87	2.47
ANOVA		Flooding	ns	**
		Soil type	ns	***
		Interaction	ns	ns
Spartina anglica				
	Clay	drained	1.75	2.80
		flooded	3.20	3.49
	Sand	drained	0.41	1.68
		flooded	0.70	2.62
ANOVA		Flooding	ns	***
		Soil type	***	***
		Interaction	ns	ns

ns = not significant, ** = P < 0.01, *** = P < 0.001.

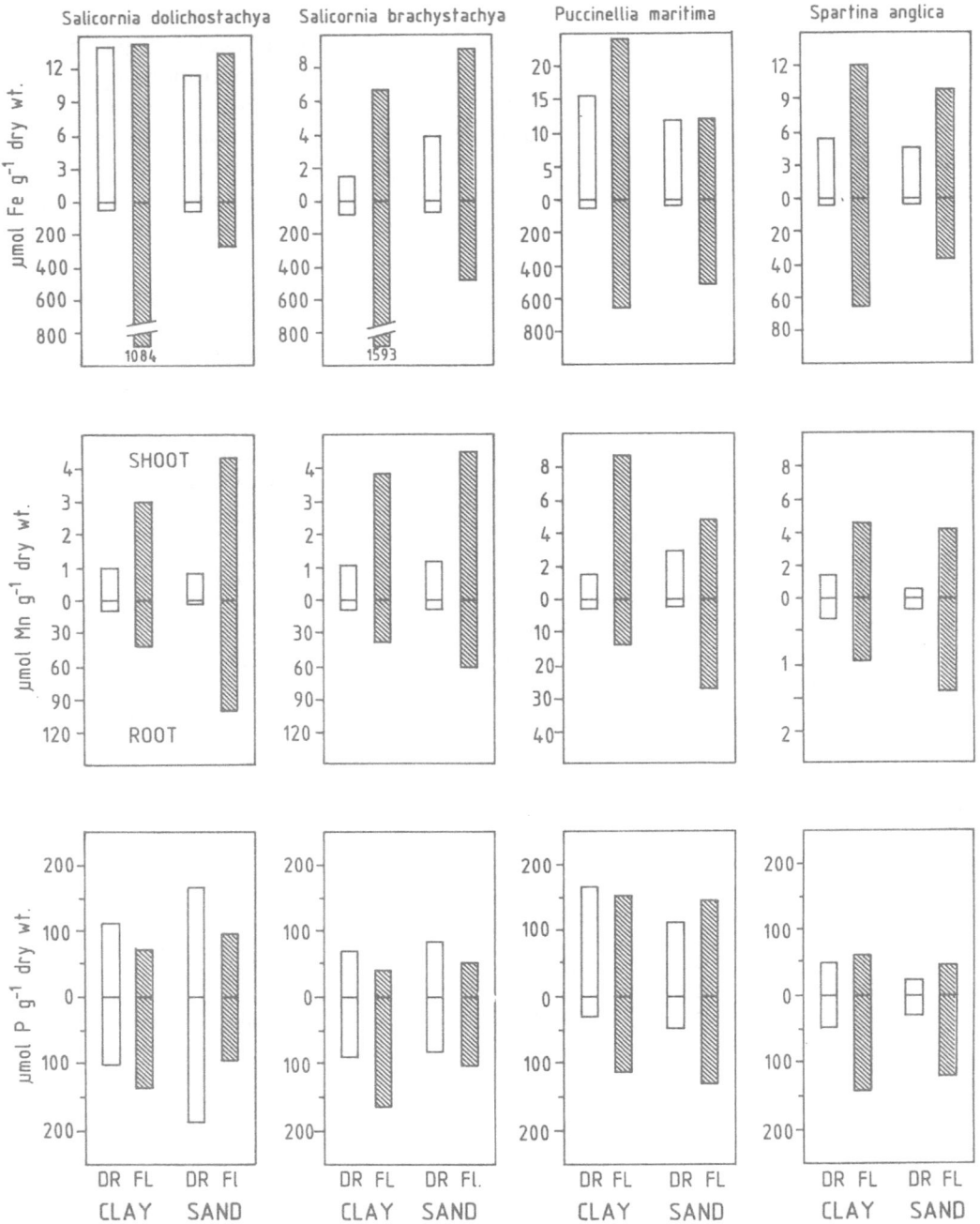

Fig. 1. Concentration of iron, manganese and phosphorus μmol g⁻¹ dry wt.) in dried root and shoot material of 4 salt marsh species grown on sand or clay soil, at 350 mM NaCl under drained or flooded conditions. Means of 4–10 and 7–15 plants from drained and flooded soil, respectively.

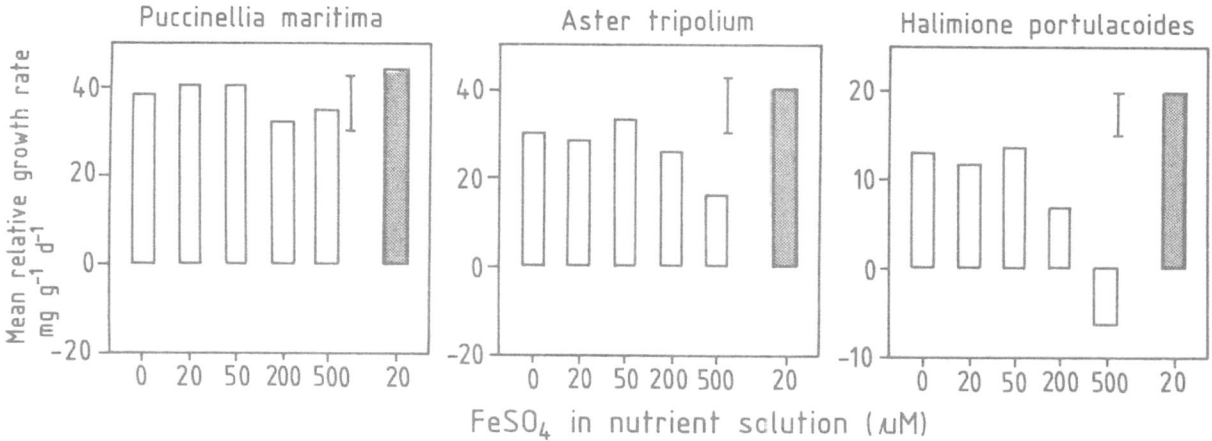

Fig. 2. Mean relative growth rate of three salt marsh species grown on anaerobic nutrient solutions containing 0–500 μM FeSO$_4$ (open columns), and on aerobic nutrient solutions containing 20 μM Fe-EDTA (shaded columns). Means of 8 replicates. Bars indicate LSD at P < 0.05 for the open columns.

on growth and physiology of three salt marsh species were investigated on anaerobic nutrient solutions at constant pH. Mean relative growth rates of *Halimione portulacoides* and *Aster tripolium* were considerably lowered at high ferrous iron concentrations (Fig. 2). Growth of *H. portulacoides* was significantly inhibited at 200 μM FeSO$_4$ and higher and *A. tripolium* was inhibited at 500 μM FeSO$_4$. These two species grew significantly better on the aerated control solution (shaded columns) as compared to plants grown on anaerobic solutions (open columns). The growth of *Puccinellia maritima* was not affected by lack of oxygen or increasing iron concentrations in the medium. From this and other experiments (Van Diggelen 1988) it was concluded that the growth response of salt marsh species to ferrous iron can be related to their distribution in the field: in general, lower salt marsh species are more tolerant than the ones from the higher salt marsh. Iron toxicity in *H. portulacoides* was not accompanied by increased concentrations of iron in the shoot, as usually found in plants under iron stress (Etherington & Thomas 1986; Talbot *et al.* 1987). Growth inhibition in this flooding sensitive species is probably caused by synergistic effects of anaerobiosis and iron toxicity, predominantly on the root system.

Crusts of precipitated ferric iron found on roots of plants grown on soil as well as on nutrient solu-

tion indicate the adaptive value of radial oxygen loss. Although most of the iron is precipitated as oxyhydroxides a considerable part may consist of insoluble ferric phosphates. This is evidenced by the linear relationship between root iron and phosphorus concentrations found in the nutrient solution experiment (Fig. 3). The environment of the root in a nutrient solution may be regarded as more or less artificial as compared to the rhizosphere in a solid substrate. Still precipitation of ferric phosphates also occurs in the rhizosphere of plants grown on soil as shown in Fig. 1: the increase of root iron under flooded conditions is correlated with an increase of root phosphorus in the flooding experiment discussed in paragraph 5.1. The proportion of iron precipitated as ferric phosphate depends on the phosphate concentration of the nutrient- or soil solution. Some authors suggested that immobilization of phosphate is the indirect cause of iron toxicity (Jones 1975; Waldren *et al.* 1987). This phosphate deficiency as the main cause of iron stress is not supported by our experimental results: with increasing iron concentrations in the nutrient solution growth was inhibited first while shoot phosphorus started to decrease at higher iron levels. Furthermore soil phosphate concentrations are increased by flooding.

In another experiment seven species were grown on nutrient solutions containing various ferrous

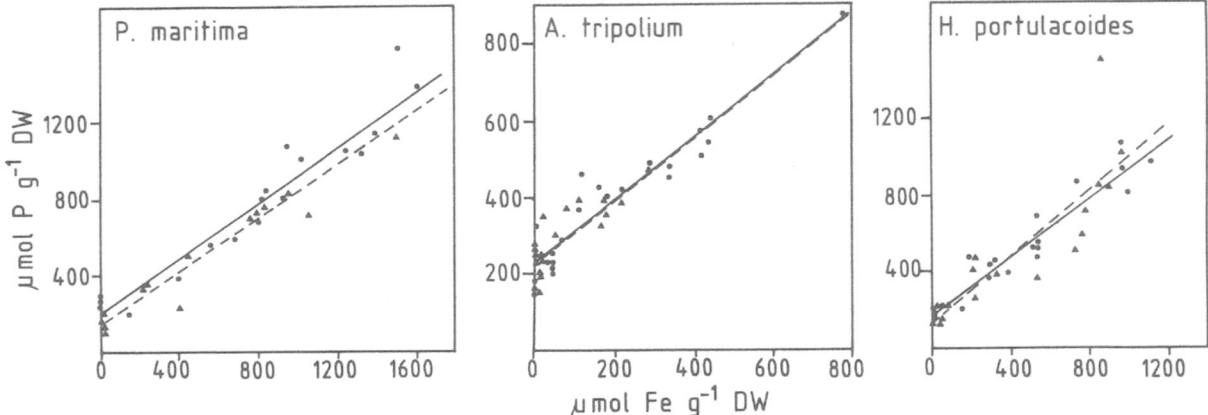

Fig. 3. Relationship between concentrations of phosphorus and iron in roots of plants grown on anaerobic nutrient solutions containing 0–500 μM ferrous sulphate. The solid line gives the relationship between total root P and Fe (●—●), and the dashed line (▲ --- ▲) gives the relationship between residual iron and phosphorus (after washing with dithionite). The relationship for *Puccinellia maritima*, *Aster tripolium* and *Halimione portulacoides*, respectively, is given by: solid lines P = 205 + 0.72 Fe, P = 225 + 0.81 Fe and P = 174 + 0.76 Fe; hatched lines: P = 140 + 0.70 Fe, P = 229 + 0.82 Fe and P = 138 + 0.86 Fe.

iron concentrations. The solutions were not aerated and were renewed every third or fourth day. Different sensitivities to ferrous iron were found in the dicotyledonous species with *Aster tripolium* being the most tolerant, *Atriplex littoralis* and *Plantago maritima* medium tolerant and *Spergularia maritima* as the most sensitive species. A remarkable higher tolerance to ferrous iron was found in grass species (*Puccinellia maritima*, *Festuca rubra*, *Elymus pycnanthus*) as compared to dicotyledonous species (Fig. 4). This may be related to the existence of an impermeable exodermis in grass roots, hampering the excessive uptake of iron (Smirnoff 1981).

Contrary to ferrous iron, it is improbable that manganese is a stress factor in salt marshes: concentrations of reduced manganese, comparable to those occurring in salt marsh soils (Singer & Havill 1985), did not affect the growth of salt marsh species in nutrient solution experiments (Cooper 1984).

5.2. Sulphide stress

Spartina and *Salicornia* are the only genera of terrestrial salt marsh halophytes known to be tolerant to a certain extent to high sulphide concentrations

(Carlson & Forrest 1982, Ingold & Havill 1984, Van Diggelen *et al.* 1987). Recently, Koch & Mendelssohn (1989) reported that the growth of *Spartina alterniflora* plants was stimulated by flooding as compared to plants grown on well drained soil. However, this growth stimulation was eliminated when sulphide was added to the soil up to a concentration of 1 mmol/l soil solution. These results prove that the growth of a flooding tolerant species may nevertheless be limited by high concentrations of soil toxins related to flooding.

In a nutrient solution experiment *Salicornia dolichostachya*, *S. brachystachya* and *Spartina anglica* were found to be tolerant up to 500 μM sodium sulphide (Table III), while the growth of *Halimione portulacoides* and *Agrostis stolonifera* was significantly inhibited at this concentration (Van Diggelen *et al.* 1987). The apparent growth stimulation by sulphide found in *Spartina anglica* is probably due to enhanced iron uptake: shoot and root iron concentrations increased significantly at 100 and 500 μM sulphide as compared to the control plants which showed visible iron deficiency, due to the high pH (6.5). Iron uptake may be facilitated by precipitation of FeS on the root, followed by subsequent oxidation to sulphate in the rhizosphere (Ponnamperuma 1978).

Oxidation of sulphide in the rhizosphere by radial oxygen loss (Yerly 1970; Tadano & Yoshida 1978; Howes *et al.* 1981) and bacteria (Joshi & Hollis 1977), oxidation in the plant (Carlson & Forrest 1982) and a higher sulphide resistance of enzymes and metabolic processes (Penhale & Wetzel 1983) may play a role in sulphide tolerance. An increase of sulphate concentration in *S. anglica* plants with increasing sulphide concentration in the growth medium may indicate the oxidation of sulphide in the plant (Fig. 5, Van Diggelen *et al.* 1986).

Havill *et al.* (1985) suggested that β-3-dimethyl-sulphoniopropionate (DMSP) might act as a sink for excess sulphur in salt marsh plants. Reduced sulphur may be detoxified by the synthesis of DMSP when sulphide oxidation in the plant is insufficient. This is substantiated by the increase of DMSP found in leaves of *S. anglica* grown at high sulphide concentrations (Fig. 5., Van Diggelen *et al.* 1986).

Dimethylsulphide (DMS) gas, formed by dissociation of DMSP, is emitted in large quantities in the marine environment and has attracted attention as a missing link in the global sulphur cycle (Lovelock *et al.* 1974). As the major source of cloud-condensation nuclei over the oceans, DMS seems to be involved in a biological regulation of the climate: production of DMS by planktonic algae in sea water influences formation and reflectance of clouds, which has an impact on the solar radiation intensity and temperature at sea level, which in turn affects the phytoplankton population and DMS emission. It was suggested that the system of phytoplankton, DMS cloud-condensation nuclei and clouds, might act as a planetary thermostat (Charlson *et al.* 1987). Though emission of DMS was measused in *Spartina* stands on American salt marshes (De Mello *et al.* 1987), coastal wetlands are thought to play a surprisingly small role in the global sulphur cycle, contributing only 2% to the total gaseous emissions (Charlson *et al.* 1987). We strongly doubt that DMS emissions in *Spartina* stands are related to environmental osmotic stress (De Mello *et al.* 1987): contrary to marine algae where DMSP was found to play a role as a compatible solute in osmoregulation, no evi-

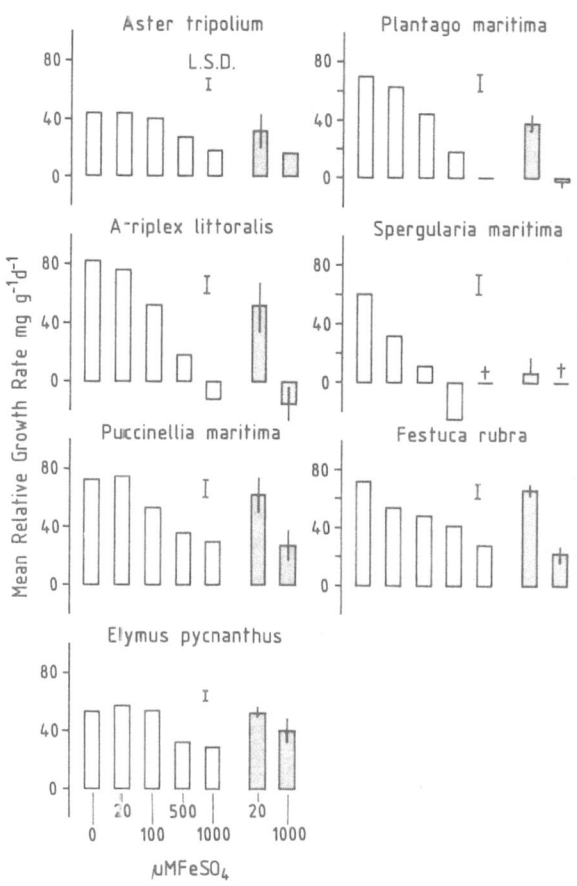

Fig. 4. Mean relative growth rate of 7 salt marsh species grown on nutrient solution containing 0–1000 μM FeSO$_4$ and 50 mM NaCl (open columns), or containing 20 or 1000 μM FeSO$_4$ without NaCl (shaded columns). Open columns: means of 3–8 replicates. Shaded columns: means and SD of 1–4 replicates.

dence for such a role was found in *S. anglica* (Van Diggelen 1986).

The observed increase of DMSP in *Spartina anglica* at high sulphide concentrations may indicate that DMSP is involved in the detoxification of excess reduced sulphur (Fig. 5). No relation was found, however, between soil sulphide concentrations and the DMSP concentration in *Spartina anglica* plants in the field. These experimental results and our measurements on plant material from the field may be explained by a temporary storage of excess sulphur in DMSP which is eliminated by emission of DMS gas, after dissociation of DMSP.

6. Submergence

Submergence, flooding of the whole shoot, may affect photosynthesis by reducing light intensity and gas exchange. Not only the uptake of carbon-dioxide is impaired, also the uptake of oxygen into the aerenchyma is prevented leading to oxygen deficiency in the root. Salt marsh plants are terrestrial plant species, not adapted to continuous submerged life. On the lower salt marsh, plants are subjected to submergence up to several hours a day. For *Salicornia* and *Spartina* species the seaward distribution limits may be governed by the duration of tidal immersion of the shoot (Hubbard 1969; Mahall & Park 1976).

In our experiments it was found that *Salicornia* species died within two weeks of submergence, whereas *Puccinellia maritima* and *Spartina anglica* survived submergence for two months. Softening of the tissue of *Salicornia* species may indicate rupture of the cells by excessive uptake of water due to the low osmotic potential of the shoots. It remains unclear to which extent photosynthesis is maintained in submerged shoots. Respiratory carbon dioxide from roots and soil bacteria might be used in submerged salt marsh halophytes to continue photosynthetic activity at a low level, as shown in submerged helophytes (Sondergaard & Wetzel 1980; Wetzel *et al.* 1983).

This may lead, however, to accumulation of starch which in turn reduces photosynthesis by feedback inhibition (Bradford 1983; Wample & Thornton 1984) resulting in early senescence of the shoot. Grass species such as *P. maritima* and *S. anglica* may benefit from storage of fructans in the vacuole as an alternative for starch (Hendry 1987), allowing photosynthesis to continue and delaying senescence and leaf death.

Grass species may also benefit from a thin air layer, trapped between the leaf surface and the surrounding water. This air layer supports leaf gas exchange of submerged grasses even if only a tip of a leaf is emerged (Raskin & Kende 1983; Jackson & Drew 1984).

Germination and growth of seedlings of *Aster tripolium*, *Spergularia maritima* and *Salicornia* species under submerged conditions was observed. This specific adaptation of young seedlings to submergence is important for species colonizing small open sites in depressions with a few centimeters of stagnant water. Submergence tolerance in these seedlings may be due to facilitated diffusion of nutrients and carbon sources through the whole small plant and development of physiological and structural adaptations starting with germination.

7. Conclusive remarks

It can be concluded that on lower salt marsh sites ferrous iron and sulphide stress and the effects of submergence play a role in the performance and distribution of salt marsh species. A well developed lacunar system throughout the plant constitutes the most important adaptation in these environments. In addition physiological tolerance and avoidance might play a role. In general species from the lower salt marsh (*Salicornia dolichostachya, Spartina an-*

Table III. Mean relative growth rate (mg/g/day) of 5 salt marsh species grown on nutrient solutions containing 0, 100 and 500 μM sodium sulphide. Means and standard deviations of 6–8 replicates. Means followed by the same letter are not significantly different at $P < 0.05$ (Student-Newman-Keuls test).

Species	Na$_2$S concentration (μM)		
	0	100	500
Salicornia dolichostachya	40.3a ± 7.6	46.1a ± 8.1	39.8a ± 11.3
Salicornia brachystachya	45.7b ± 15.5	48.1b ± 20.7	44.3b ± 11.6
Spartina anglica	7.5c ± 5.1	24.2d ± 4.0	21.4d ± 3.0
Agrostis stolonifera	53.2e ± 11.0	60.7e ± 4.8	37.4f ± 6.2
Halimione portulacoides	46.7g ± 10.0	44.3g ± 7.4	3.6h ± 8.2

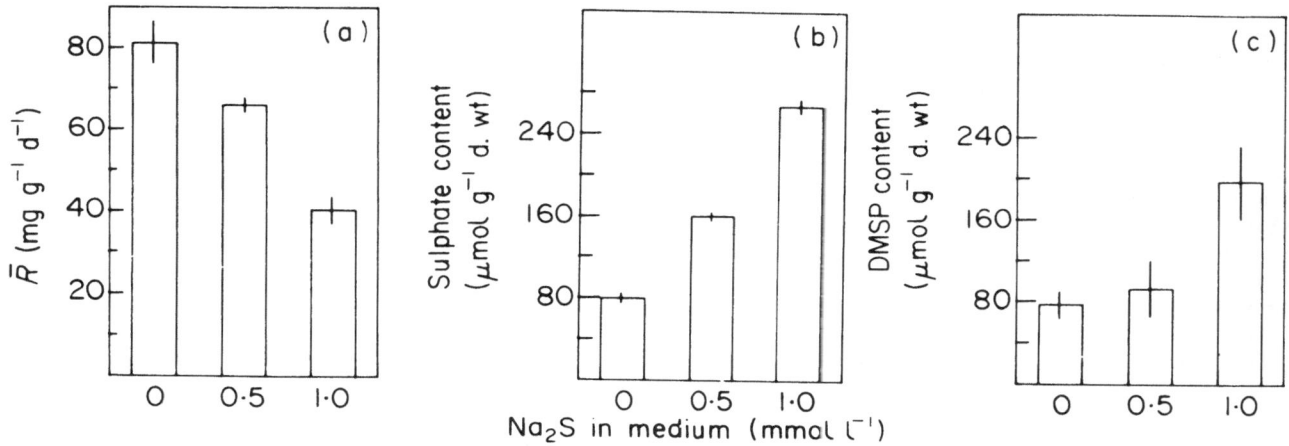

Fig. 5. Mean relative growth rate (a), sulphate content of shoot (b) and DMSP content of leaves (c) of *Spartina anglica,* grown on three different sodium sulphide concentrations. Means and SE of 4–6 replicates.

glica and *Puccinellia maritima*) are adapted to the conditions of their habitat, whereas species from higher elevations are less tolerant or sensitive to these conditions.

Yet, each of these species responds different to flooding and soil toxins and is adapted differently: growth of *Salicornia dolichostachya* is stimulated by flooding, it is not inhibited by high sulphide or ferrous iron concentrations but it is inhibited by anoxic conditions in nutrient solutions. Furthermore this species has low root porosities. Growth of *Spartina anglica* is stimulated by flooding but it is hampered by sulphide concentrations that may occur locally on lower salt marsh sites (1 mM), despite its very well developed lacunar system. *Puccinellia maritima* shows growth inhibition under flooded as compared to drained conditions while it was found to be tolerant to anoxia and high ferrous iron concentrations on nutrient solutions.

Salicornia brachystachya is less tolerant to flooding but does not respond differently to anoxia, sulphide or ferrous iron in comparison with *S. dolichostachya,* yet it occurs predominantly on middle and higher salt marsh sites. *Halimione portulacoides* is a typical example of a flooding sensitive species: it is sensitive to anoxia, to sulphide and to ferrous iron and it is restricted to well drained soils at higher elevations along creek banks.

Although it is clear that the zonated distribution of higher plants in salt marshes is governed by inundation, salinity and soil texture (of which salinity was not even taken into account in this paper), the responses of plant to the spatial and temporal variation in a great number of soil factors, is much more complicated than can be shown in experiments on single factors associated with flooding. Differences between species in response to flooding probably are more gradual than the prefixes "tolerant" or "sensitive" indicate. Studies on the causes of the zonated occurrence of species on middle and higher salt marsh sites where extreme flooding-related soil conditions are absent must be concentrated on differences between species in response to alternating periods of flooding and drainage and to hypoxic or anoxic nutrient solutions.

References

Armstrong, W., 1982. Waterlogged soils. In: J.R. Etherington (ed.) Environment and plant ecology. John Wiley, Chichester. pp. 290—330.

Bertness, M. D. & Ellison, A. M., 1987. Determinants of pattern in a New England salt marsh community. Ecol. Mon. 57: 129—147.

Bradford, K. J., 1983. Effects of soil flooding on leaf gas exchange of tomato plants. Plant Physiol. 73: 475—479.

Carlson, P. R. & Forrest, J., 1982. Uptake of dissolved sulphide by *Spartina alterniflora:* Evidence from natural sulphur isotope abundance ratios. Science 216: 633—635.

Charlson, R. J., Lovelock, J. E., Andrea, M. O. & Warren, G.,

1987. Oceanic phytoplankton, atmospheric sulphur, cloud albedo and climate. Nature 326: 655—661. .

Chen, C. C., Dixon, J. B. & Turner, F. T., 1980. Iron coatings on rice roots: mineralogy and quantity influencing factors. Soil Sci. Soc. Am. J. 44: 635—639.

Cooper, A., 1984. A comparative study of the tolerance of salt marsh plants to manganese. Plant Soil 81: 47—59.

Crawford, R. M. M., 1967. Alcohol dehydrogenase activity in relation to flooding tolerance in roots. J. Exp. Bot. 18: 458—464.

Crawford, R. M. M., 1978. Metabolic adaptations to anoxia. In: D.D. Hook & R.M.M. Crawford (eds.) Plant life in anaerobic environments. Ann Arbor Sci. Publ., Ann Arbor pp. 119—136.

Crowder, A. A. & MacFie, S. M., 1986. Seasonal deposition of ferric hydroxide plaque on roots of wetland plants. Can. J. Bot. 64: 2120—2124.

De Mello, W. Z., Cooper, D. J., Cooper, W. J., Saltzman, E. S., Zika, R. G., Savoie, D. G. & Prospero, J. M., 1987. Spatial and diel variability in the emissions of some biogenic sulfur compounds from a Florida *Spartina alterniflora* coastal zone. Atmos. Environ. 21: 987—990.

Ernst, W. H. O., 1985. Some considerations of and perspectives in coastal ecology. In: W.G. Beeftink, J. Rozema & A.H.L. Huiskes (eds.) Ecology of coastal vegetation. Junk, Dordrecht, pp. 533—545.

Etherington, J. R. & Thomas, A. M., 1986. Response to waterlogging and different sensitivity to divalent iron and manganese in clones of *Dactylis glomerata* L. derived from well drained and poorly drained soils. Ann. Bot. 58: 109—119.

Havill, D. C., Ingold, A. & Pearson, J., 1985. Sulphide tolerance in coastal halophytes. In: W.G. Beeftink, J. Rozema & A.H.L. Huiskes (eds.) Ecology of coastal vegetation. Junk, Dordrecht pp. 279—285.

Hendry, G., 1987. The ecological significance of fructan in a contemporary flora. New Phytol. 106: 201—216.

Howarth, R. W. & Giblin, A., 1983. Sulphate reduction in the salt marshes of Sapelo Island, Georgia. Limn. Oceanogr. 28: 70—82.

Howes, B. L., Howarth, R. W., Teal, J. M. & Valiela, I., 1981. Oxidation reduction potentials in a salt marsh: Spatial patterns and interaction with primary production. Limnol. Oceanogr. 26: 350—360.

Hubbard, J. C. E., 1969. Light in relation to tidal immersion and the growth of *Spartina townsendii* (s.l.) J. Ecol. 57: 795—804.

Ingold, A. & Havill, D. C., 1984. The influence of sulphide on the distribution of higher plants in salt marshes. J. Ecol. 72: 1043—1054.

Jackson, J. B. & Drew, M. C., 1984. Effects of flooding on growth and metabolism of herbaceous plants. In: T.T. Kozlowski (ed.) Flooding and plant growth. Academic Press, New York pp. 47—127.

Jeffrey, D. W., 1987. Soil-plant relationships: an ecological approach. Croomhelm, Beckenham pp. 235-256.

Jones, R., 1975. Comparative studies of plant growth and distribution in relation to waterlogging. VIII. The uptake of phosphorus by dune and dune-slack plants. J. Ecol. 63: 109—116.

Joshi, M. M. & Hollis, J. P., 1977. Interaction of *Beggiatoa* and Rice plant: Detoxification of hydrogen sulphide in the rice rhizosphere. Science 195: 179—180.

Justin, S. H. F. & Armstrong, W., 1987. The anatomical characteristics of roots and plant response to soil flooding. New Phytol. 106: 465—495.

Koch, M. S. & Mendelssohn, I. A., 1989. Sulphide as a soil phytotoxin: differential responses in two marsh species. J. Ecol. 77: 565—578.

Kristjansson, K. J. & Schönheit, P., 1983. Why do sulphate bacteria outcompete methanogenic bacteria for substrates? Oecologia 60: 264—266.

Lovelock, J. E., Maggs, R. J. & Rasmussen, R. A., 1972. Atmospheric dimethyl sulphide and the natural sulphur cycle. Nature 237: 452—453.

Luther, G. W. III & Church, T. M., 1988. Seasonal cycling of sulphur and iron in porewaters of a Delaware salt marsh. Mar. Chem. 23: 295—309.

Mahall, B. E. & Park, R. B., 1976. The ecotone between *Spartina foliosa* and *Salicornia virginica* L. in salt marshes of Northern San Francisco Bay. III. Soil aeration and tidal immersion. J. Ecol. 64: 811—819.

Penhale, P. A. & Wetzel, R. G., 1983. Structural and functional adaptations of eelgrass (*Zostera marina* L.) to the anaerobic sediment environment. Can. J. Bot. 61: 1421—1428.

Ponnamperuma, F. N., 1978. Electrochemical changes in submerged soils and the growth of rice. In: F. N. Ponnamperuma (ed.) Soils and Rice. International Rice Research Institute, Los Banos pp. 421—445.

Ponnamperuma, F. N., 1984. Effects of flooding on soils. In: T.T. Kozlowski (ed.) Flooding and plant growth. Academic Press, New York pp. 9—45.

Raskin, I. & Kende, H., 1983. How does deep water rice solve its aeration problem? Plant Physiol. 72: 447—454.

Rozema, J., Bijwaard, P., Prast, G. & Broekman, R., 1985. Ecophysiological adaptations of coastal halophytes from foredunes and salt marshes. In: W.G. Beeftink, J. Rozema & A.H.L. Huiskes (eds.) Ecology of coastal vegetation. Junk, Dordrecht pp. 499—521.

Schat, H., Van der List, J. C. & Rozema, J., 1987. Ecology and differentiation of the microspecies *Salicornia dolichostachya* Moss. and *Salicornia ramosissima* J. Woods: growth, mineral nutrition, carbon assimilation and development of the root system in anoxic and hypoxic culture solution. In: A.H.L. Huiskes, C.W.P.M. Blom & J. Rozema (eds.) Vegetation between land and sea. Junk, Dordrecht pp. 178—184.

Scholten, M., Blaauw, P., Stroetenga, M. & Rozema, J., 1987. The impact of competitive interactions on the growth and distribution of plant species in salt marshes. In: A.H.L. Huiskes, C.W.P.M. Blom & J. Rozema (eds.) Vegetation between land and sea. Junk, Dordrecht pp. 268—279.

Scholten, M. & Rozema, J., 1987. The competitive ability of *Spartina anglica* on Dutch salt marshes. In: A.J. Gray (ed.)

Spartina anglica: A review of current research. ITE, Furzebrook.

Singer, C. E. & Havill, D. C., 1985. Manganese as an ecological factor in salt marshes. In: W.G. Beeftink, J. Rozema & A.H.L. Huiskes (eds.) Ecology of coastal vegetation. Junk, Dordrecht pp. 287—292.

Smirnoff, N., 1981. Iron tolerance and the role of aerenchyma in wetland plants. Ph.D. Thesis, University of St. Andrews.

Sondergaard, M. & Wetzel, R. G., 1980. Photorespiration and internal recycling of CO_2 in the submersed angiosperm *Scirpus subterminalis.* Can. J. Bot. 58: 591—598.

Tadano, T. & Yoshida, S., 1978. Chemical changes in submerged soils and their effects on rice growth. In: F.N. Ponnamperuma (ed.) Soils and Rice. International Rice Research Institute, Los Banos pp. 399—421.

Talbot, R. J. & Etherington, J. R., 1987. Comparative studies of plant growth and distribution in relation to waterlogging. XIII. The effect of Fe^{2+} on photosynthesis and respiration of *Salix caprea* and *Salix cinerea* spp *oleifolia.* New Phytol. 105: 575—583.

Talbot, R. J., Etherington, J. R. & Bryant, J. A., 1987. Comparative studies of plant growth and distribution in relation to waterlogging. XII. Growth, photosynthetic capacity and metal ion uptake in *Salix caprea* and *Salix cinerea* ssp. *oleifolia.* New Phytol. 105: 563—574.

Van Diggelen, J., 1988. A comparative study on the ecophysiology of salt marsh halophytes. Ph.D. Thesis, Free University, Amsterdam 208 pp.

Van Diggelen, J., Rozema, J. & Broekman, R., 1987. Growth and mineral relations of salt marsh species on nutrient solutions containing various sodium sulphide concentrations. In: A.H.L. Huiskes, C.W.P.M. Blom, & J. Rozema (eds.) Vegetation between land and sea. Junk, Dordrecht pp. 258—266.

Van Diggelen, J., Rozema, J., Dickson, D. M. J. & Broekman, R., 1986. β-3-Dimethylsulphoniopropionate, proline and quaternary ammonium compounds in *Spartina anglica* in relation to sodium chloride, nitrogen and sulphur. New Phytol. 103: 573—586.

Waldren, S., Etherington, J. R. & Davies, M. S., 1987. Comparative studies of plant growth and distribution in relation to waterlogging. XIV. Iron, manganese, calcium and phosphorus concentrations in leaves and roots of *Geum rivale* L. and *G. urbanum* L. grown in waterlogged soil. New Phytol. 106: 689—696.

Wample, R. L. & Thornton, R. K., 1984. Differences in the response of sunflower (*Helianthus annuus*) subjected to flooding and drought stress. Physiol. Plant. 61: 611—616.

Wetzel, R. G., Brammer, E. S. & Forsberg, C., 1983. Photosynthesis of submerged macrophytes in acidified lakes. I. Carbon fluxes and recycling of CO_2 in *Juncus bulbosus* L. Aquat. Bot. 19: 329—342.

Wiebe, W. J., Christian, R. R., Hansen, J. A., King, G., Sherr, B. & Skyring, G., 1981. Anaerobic respiration and fermentation. In: L.R., Pomeroy & R.G. Wiegert (eds.) Springer, New York pp. 137—161.

Yerly, M., 1970. Ecologie comparée des prairies marécageuses dans les Préalpes de la Suisse occidentale. Veroff. Geobot. Inst. Zürich 55: 1—119.

Author's Address
J. van Diggelen
Department of Ecology and Ecotoxicology
Biological Laboratory
Free University of Amsterdam
P.O. Box 7161
1007 MC Amsterdam
The Netherlands

Present Address
Van Limburgstirumstraat 5
4286 BG Almkerk
The Netherlands

Parnassia palustris, growing in a wet dune scrub. (Photograph: H. Schat.)

CHAPTER 8

Water as a stress factor in the coastal dune system

H. SCHAT & K. VAN BECKHOVEN

Abstract. The depth of the groundwater table is an important determinant of the composition and structure of dune vegetation. The position of a species in a water availability gradient is strongly dependent on its life-history and its phenology. Vertical zonation patterns, imposed by variation in groundwater depth, can be significantly affected by interspecific competition or facilitation.

Throughout the first half of this century, the floristic richness of the Dutch coastal dune area has declined, partly due to groundwater exploitation for the production of drinking water. Since several years, the drop of the water table has been stopped or even reversed by the infiltration of eutrophic river water. The prospects for the return of characteristic slack communities are uncertain. The effects of groundwater raising on nutrient cycling in the top soil may be decisive. Unfortunately, nutrient conversion in dune soils is poorly studied.

1. Water as a determinant of zonation patterns in dune vegetation

1.1. Introduction

Due to its pronounced relief, which provides variation in exposition, inclination and groundwater depth, the coastal dune system exhibits an exceptional spatial heterogeneity as to the availability of water to plants. According to the depth of the groundwater table, one can distinguish extremely contrasting habitats, such as dune lakes, wet dune slacks and dry dunes, connected by various transitional habitats.

Species growing in wet slacks with an undisturbed groundwater regime, usually hygrophytes, must be able to cope with submergence in winter and waterlogging of the soil during some part of the growing season. In general, they will be able to exploit capillary groundwater throughout the summer, as the groundwater table, by definition, does not sink below a depth of one meter in summer (the capillary groundwater fringe is usually between 60

and 100 cm, depending on the grain size of the sand). Species growing on higher parts of dune slopes, usually xeromorphic on south-facing slopes or mesomorphic on north-facing ones, are permanently dependent on seapage water, which implies a risc of desiccation in the summer half year.

It is evident that the depth of the groundwater table is an extremely important determinant of the zonation patterns of dune vegetation. Considering the change in vegetation composition with increasing height above the bottom of a wet slack, it appears that the species turnover rate or β-diversity (*sensu* Whittaker 1972) depends strongly on the position in the gradient (Fig. 1). The majority of species reach the lower and/or upper limit of their vertical distribution range within a relatively narrow zone, roughly corresponding with the height trajectory between the modal water table level in the winter season and the mean maximal depth of capillary groundwater in summer (Fig. 1b, c). This is reflected by the low similarity of adjacent vegetation samples, taken at mutual height intervals of 5 cm only (Fig. 1a). Above this trajectory the simi-

J. Rozema and J. A. C. Verkleij (eds.), Ecological Responses to Environmental Stresses, 76—89.

larity increases with height until it reaches a constant level at about 35 cm above the mean maximal capillary groundwater depth in summer. The latter level probably marks the height above which continuous exploitation of groundwater becomes impossible for the majority of species (as soon as the groundwater depth falls below this level, its precise depth will be of little interest). Below the trajectory of maximal species turnover, the similarity between adjacent samples is also high, though not as high as on the higher parts of the slope (Fig. 1a), which may be taken to indicate that the very occurrence of a period submergence is more selective than its duration, at least beyond a certain minimal period of time.

In the following sections we will try to assess the importance of the possible selective factors, associated with the depth of the groundwater table, and to identify the specific properties of plant species, which determine their maximal and minimal height of occurrence, relative to the groundwater table. For this purpose we will compare a number of species found within the association *Centaurio-Saginetum moniliformis,* which typically occupies the transitional zone between the wet slack bottom and the higher parts of surrounding dune slopes (Diemont *et al.* 1940; Freijsen 1967a, b), i.e., the zone of maximal species turnover indicated in Fig. 1. The zonation patterns in the field that will be referred to, are those in the western part of the beach plain of the isle of Schiermonnikoog (The Netherlands).

1.2. Drought as a determinant of species zonation on a dune slope

1.2.1. Species zonation in relation to drought tolerance

The phreatic level in dune systems is characterized by a high degree of random temporal variation, imposed by the strongly unpredictable course of the precipitation/evaporation balance: the variation between years in the maximal depth and, in general, the depth at any particular date, frequently exceeds 50 cm, which corresponds with a much greater distance along the profile of a slope. Several species, especially therophytes, directly reflect these fluctuations by taking a different position on

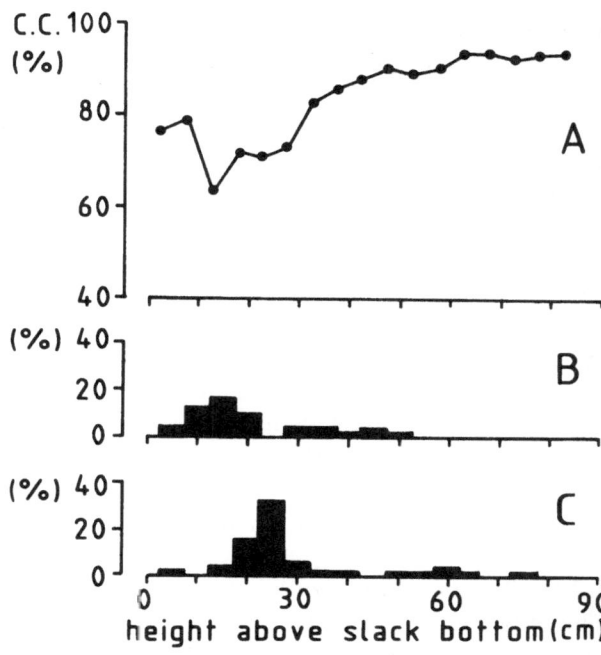

Fig. 1. Several characteristics of the height zonation of perennials in the western part of the beach plain at Schiermonnikoog. A) Similarity, measured by the coefficient of community (cf. Whittacker 1972), between adjacent vegetation samples taken at 5 cm height intervals along the profile of a slope. B) Frequency distribution of the position of the lower limits of distribution of adult representatives of the species present within the height trajectory depicted. C) The same for the upper limits.

the slope each year ("shuttling", cf. Beeftink, 1975), whereas others do not seem to respond at all. The sequence of species according to the position of their upper limit of occurrence on a slope is therefore subjected to variation between years (e.g., Schat 1982; Van Tooren *et al.* 1983). This could be one of the possible reasons for the apparently low degree of correlation between a species' position in this sequence and its ability to endure periods of water shortage, even when the comparison is limited to species with a similar rooting depth (Schat 1982; Van Tooren *et al.* 1983). This lack of correlation does not only apply to the sequence found in any particular year, but also to the sequence of the mean upper limits of distribution over a longer series of years (Table I). Of course, this may as well rely on interspecific differences in the ability to track random fluctuation in groundwater depth; however, even in exceptionally dry years the correlation, though much better (Van Tooren *et al.* 1983), still remains far from perfect. Even the latter does not necessarily imply that drought tolerance would be a less important determinant of the upper limit of distribution of the great majority of species. There are many indications that the general lack of precise correlation merely reflects interspecific differences in phenology, life history, growth form, reproductive strategy, dispersal and ontogenetic trends in the level of drought tolerance, which bring about corresponding differences in the way in which the risc of fatal desiccation depends on the time of the year and the phase of the life cycle, or in the ability to track seasonal and annual variation in water availability.

1.2.2. Phenology and zonation

The importance of a species' phenology is immediately clear from a comparison between winter annuals, which usually occur far above the groundwater table, in spite of their non-xeromorphic nature, and summer annuals, which are often confined to wet slacks and the lower parts of slopes (Van Tooren *et al.* 1983). The period of germina-

tion is an important phenological characteristic, as it determines the maximal height on the slope at which successful seedling establishment is still possible. After the winter season, as soon as a precipitation deficit starts to develop, the survival of young seedlings will become increasingly dependent on capillary groundwater reaching the soil surface. This implies that species which germinate in the winter half year, or in early spring (April) are usually capable to establish themselves higher on the slope than species which germinate not until May or June. Moreover, the former species are more likely to establish themselves at height levels, where the plants can not survive in summer. Their upper limit of distribution in summer will usually depend on the drought tolerance in a later phase of the life cycle. The latter species, on the other hand, will run a lower risc of desiccation in the adult phase of the life cycle, and their upper limit of distribution in summer will be more frequently determined by the drought tolerance of young seedlings, at least when there is no considerable lateral spread by rhizomes or stolons. The consequences of this for zonation patterns may be considerable. For example, the mean upper limit of distribution

Table I. Mean upper limit of distribution in August, period of germination, drought endurance in the adult phase and the frequency of fatal desiccation in the adult phase in the field of a number of short-lived, shallow-rooted dune slack species at the beach plain of Schiermonnikoog.

Species	Life history[1]	Upper limit in August[2]	Period of germination[3]	Drought endurance of adult plants in greenhouse[4]	Frequency of fatal desiccation of adults in the field[5]
Samolus valerandi	biennial	0	May	−	−
Centaurium pulchellum	summer annual	25	May	−	−
Parnassia palustris	pauciennial	30	April-May	− − −	+
Centaurium littorale	biennial	53	May	+	−
Linum catharticum	summer annual	59	April	−	+
Sagina nodosa	biennial	112	October-March	+ +	+
Plantago coronopus	pauciennial	140	September-October April-June	+ + +	+

[1] many species exhibit intraspecific variation; the life history given applies to the usual situation in the beach plain at Schiermonnikoog.
[2] the values given are expressed in cm above an arbitrary reference point.
[3] the period of germination depends on the height on the slope; the period given applies to higher levels on the slope. The period given for *Sagina* is the period of rooting of vegetative propagules (seedlings are hardly ever observed).
[4] the symbols denote the period of survival of three months old individuals, grown in mixed culture in a 10 cm sand layer, after arresting the water supply; the scale is semi-quantitative, ranging from very short (− − −) to fairly long (+ + +).
[5] the symbols refer to the frequency of mortality of adults due to desiccation, estimated from field observations over a series of years (6—10); −: (practically) nil, +: at least twice in 10 years.

in August of *Linum catharticum* lies above that of *Centaurium littorale*, the latter being much more drought tolerant than *Linum* (Table I). This is confirmed by the observation that in exceptionally dry summers (e.g., 1976) *Linum* wilts and dies off at heights on the slope where *Centaurium* does not even show any wilting at all (Van Tooren *et al.* 1983; Schat, unpublished). This is due to a difference in the time of germination of only three to four weeks. Next to the differential drought tolerance, this difference in germination period contributes to the marked difference in the risc of desiccation in the adult phase of the life cycle: the upper limit of *Linum* in summer is correlated with the cumulative precipitation deficit of the whole vegetation period, rather than that of the period of establishment, whereas the opposite is true for *Centaurium* (Van Tooren *et al.* 1983). The other data in Table I tend to confirm these relationships between the period of establishment, the mean upper limit of distribution in August, and the chance of desiccation in later phases of the life cycle.

1.2.3. Ontogenetic trends in drought tolerance

It is evident that the precise age-dependent pattern of the risc of desiccation will depend on ontogenetic trends in the ability to avoid or tolerate low levels of water availability. In general, the potency to cope with drought may be expected to increase throughout the juvenile phase, due to the increase in rooting depth and the decreasing trend in the area/volume ratio of leaves, as long as the full-grown leaf size continues to increase. Table II gives some information on the degree and sources of the interspecific variation as to the ontogenetic trends in the cuticular transpiration rate among several short-lived, shallow-rooted species. The ontogenetic pattern of the rate of cuticular transpiration per unit of leaf area varies strongly between species. In *Sagina nodosa* there is a two-fold increase with ageing of the rosette, which is due to the fact that the basal rosette leaves in fact represent the fairly waxy leaflets of the vegetative propagules, from which the rosettes were grown (Elliston Wright 1953); in *Centaurium littorale*, *Samolus valerandi* and *Parnassia palustris* there is a considerable decrease, associated with a visible change in

the appearance of the leaf surface (from dull to more or less shiny, which indicates a change in the structure or thickness of the epicuticular wax layer), whereas *Plantago coronopus* does not show any considerable change. This parameter is not simply correlated with a species' drought tolerance level (compare Table I). This is not surprising, as the rate of development of a tissue water deficit will directly depend on the fraction of tissue water lost per unit of time, which depends both on the amount of water lost per unit of time and per unit of leaf surface and the area/volume ratio of the leaf. The area/volume ratio is subjected to interspecific and ontogenetic variation. In each of the species, apart from *Plantago*, it decreases to about 50% of the initial value with increasing rosette age, in spite of the much greater interspecific differences in the relative increase in final leaf length (the area/volume ratio of a body increases linearly with its length, width or height, as long as its three-dimensional shape remains the same). Anyway, the sequence of the species according to their rates of cuticular transpiration, expressed as the percentage of water lost per unit of time, neither corresponds with the sequence of upper limits on the slope, mainly due to the position of *Plantago coronopus* (Table I). The reason for this lies in the exceptional capability to recover from desiccation of the latter species, relative to *Sagina nodosa* and *Centaurium littorale*. This phenomenon has been observed in the field as well as in greenhouse experiments, in which mixed cultures of the species combination given in Table II, grown in a 10 cm sand layer, were allowed to desiccate (Schat, unpublished). After arresting the water supply the species successively wilt in the sequence expected from their cuticular transpiration rates. However, when the water supply is restored as soon as the aerial parts of all the species are completely and irreversibly wilted, *Plantago* invariably re-sprouts, whereas the other species appear to have died, even *Sagina nodosa*, apart from its vegetative propagules. The ability to re-sprout after desiccation is apparently an important component of the drought tolerance syndrome. In the case of *Plantago coronopus*, it might be accomplished by the dense spiral of young leaves with long white hairs on the edges of their basal parts, which cover the terminal bud of the rosette axis, thereby delaying its desiccation. Moreover, the thick underground rosette axis and tap root might serve as a water reserve. It seems likely that a plant's architecture or "growth form" may greatly affect its potency to prevent or to recover from desiccation. It needs no comment that a plant's architecture changes with ageing. In agreement with this, the superior recovery of *Plantago*, relative to *Centaurium* and *Sagina*, is not or hardly apparent in the early phases of the life cycle.

1.2.4. Life history and zonation

The life history is another major determinant of a species' behaviour towards the water availability level. Firstly, the perennial architecture often includes underground perennating organs such as rhizomes, tap roots, corms, thickened or woody basal stem parts. These structures may be less vulnerable to desiccation than aerial parts and facilitate a plant's recovery from drought stress (see above): in exceptionally dry summers one may observe a complete aboveground die-back, followed by re-sprouting from surviving underground rhizomes in, for example, *Juncus* spec., *Carex* spec., *Parnassia*, *Plantago* (Van Tooren *et al.* 1983; Schat, unpublished observations). Furthermore, the life expectancy in the adult phase of the life cycle determines the extent to which a population's maintenance at a particular level on the slope depends on the drought tolerance level in the relatively vulnerable seedling phase. Moreover, vegetative reproduction and lateral spread by means of rhizomes or stolons, which is more or less confined to perennials, will further increase the ability to colonize higher levels on the slope, where successful seedling establishment may be not or only seldomly possible. A specific example of the importance of the mode of reproduction is provided by *Sagina nodosa*, which reproduces both sexually and vegetatively. Vegetative reproduction occurs by means of short densely-leaved branches ("buds"), which are developed from the axillary buds of the leaves of the flower-bearing stems and are shed in the autumn (Elliston Wright 1953). The latter allows the species to establish in the winter season (germination of seeds occurs in May), which guarantees that the upper limit on the slope is never restricted by desiccation in the early phase of the life cycle, in spite of the short-lived habit of this species (Schat 1982).

The life history is a major determinant of the extent to which the upper limit of distribution of adult, (potentially) reproductive plants in summer responds to random temporal variation in groundwater depth. In the case of annuals, this limit usually closely tracks the year to year variation in groundwater depth. The fairly strict biennial *Centaurium littorale* exhibits similar behaviour, but its response lags one year behind (Van Tooren *et al.* 1983). In the case of short-lived perennials, shifts in the upper limit of adult plants are less frequent or

Table II. Cuticular transpiration and area/volume ratios of full-grown leaves of young and old rosettes of several dune slack species.

Species	leaf length (mm)		area/volume ratio (mm^{-1})[1]		cuticular transpiration (g H$_2$O · m^{-2} · h^{-1})[2]		cuticular transpiration (% H$_2$O · h^{-1})	
	y[3]	o[3]	y	o	y	o	y	o
Samolus valerandi	2.0	42.0	16.2	7.6	11.6	6.2	23.4	7.7
Parnassia palustris	2.1	21.3	18.3	9.8	11.2	5.3	27.2	6.6
Centaurium littorale	1.8	13.8	16.8	8.6	1.8	1.2	3.7	1.3
Sagina nodosa	1.9	11.0	13.2	6.1	0.5	1.0	0.8	0.8
Plantago coronopus	8.1	46.1	12.9	4.2	4.8	5.1	7.6	2.7

[1] the data given apply to fully water-saturated leaves.
[2] transpiration rates were measured at 20° C, 65% R.H. at a constant wind speed; the data given represent the transpiration rate at a leaf water deficit of about 20%.
[3] y: the third leaf (cotyledons not included) of a young rosette; o: a leaf (usually the 15th) of an older rosette.

80

less conspicuous. The upper limit of adult long-lived perennials, at least of those species with a limited capacity for vegetative lateral spread, such as tussock-forming *Carex* species, *Juncus maritimus, J. alpino-articulatus* and *Schoenus nigricans,* seems to be completely inert. The individuals forming the upper limit are often comparatively large and equally sized, suggesting that they belong to a single seedling cohort, whereas at lower levels on the slope usually more cohorts seem to be present (Schat, unpublished). This indicates that the upper limit of adult representatives of these species tends to reflect highly exceptional conditions, which prevail only once or twice in a long sequence of years. In general, the chance to reach the reproductive phase at levels on the slope where complete mortality may frequently occur, will decrease with the duration of the pre-reproductive phase, which may be as long as three or four years in long-lived perennials. Moreover, the possibility of year to year survival in the adult phase precludes the necessity of a continuous rejuvenation by means of seedling establishment, which will further stabilize the upper limit of adult perennials on the slope. In the case of *Schoenus nigricans,* and probably many other long-lived perennials, the upper limit of adults seem to depend on exceptionally favourable conditions during the juvenile phase of the cohort in question, rather than on the drought tolerance level in the adult phase of the life cycle. This is not only suggested by the above described pattern of age distribution, but also by the fact that populations of this species are able to maintain themselves for longer periods in progressively desiccating slacks, notably without rejuvenating themselves by seedling establishment (Ernst & Van der Ham 1988; see below).

1.2.5. Dispersal and zonation

It is obvious that the ability to track year to year variation in groundwater depth, such as exhibited by short-lived species, requires a certain capacity for dispersal, either in space or in time. All the short-lived species mentioned thusfar, except one, viz. *Parnassia palustris,* have the ability to accumulate a persistent reserve of innately or enforcedly dormant seeds and exhibit a very limited capacity

of extensive dispersal in space (Schat 1982, 1983). *Parnassia palustris* exhibits extensive dispersal in space, which compensates for the short maximal life-span of its seeds, i.e. less than one year (Schat 1982, 1983). There are several indications that the possibilities to track year to year shifts in the maximal height on the slope at which successful establishment is possible, are usually not seriously restricted by insufficient dispersal (Van Tooren *et al.* 1983), although the rate of successful recruitment from more than one year old seed reserves is low. However, the latter seems to have more conspicuous consequences for the density than for the position of the upper limit of occurrence (Schat, in prep.). *Sagina nodosa* is an example of a species with a limited capacity to colonize higher levels on the slope over greater distances within one year, due to its peculiar mode of dispersal, viz. by means of vegetative propagules (see above). This implies a limited dispersal in space (Van Tooren *et al.* 1983), at least on a slope (compare Bruinenberg *et al.* 1980), and precludes any dispersal in time, since successful recruitment from seed banks is unlikely at higher levels on the slope — see above.

1.3. Waterlogging and submergence

It is evident that a species' lower limit of distribution, relative to the groundwater level, will be often determined by its ability to cope with waterlogging or submergence. Table III gives the mean lower limits of adult, reproductive individuals of a number of short-lived species, together with some experimentally established measures of tolerance of waterlogging and submergence and some field observations concerning the survival in a normal and an exceptionally wet winter season. There are interspecific differences in the capacity to penetrate into anoxic soil: only *Samolus* is capable to exploit the deeper soil layers, whereas root systems of the other species remain confined to a shallow upper layer. *Parnassia* and *Sagina* even show hardly any root penetration at all, but merely form a thin mat of roots lying on the soil surface. In all the species, apart from *Samolus,* the uptake of nitrogen, phosphorus and potassium, and, eventually, the growth of the shoot, is reduced (*Plantago*), or completely arrested (*Centaurium, Sagina, Parnassia*) under completely waterlogged conditions (Schat 1984; Schat unpublished). The ability to grow in waterlogged, anoxic soil is well correlated with the (potential) root porosity (Table III). The response to anoxia in hydroponic culture is essentially similar, (Schat unpublished), suggesting that anoxia *per se* sufficiently explains the effects of waterlogging in the sand culture experiments. It can be excluded that excessive accumulation of reduced iron or manganese have contributed to the effect of waterlogging under the experimental conditions chosen (Schat 1984). The idea that sensitivity to waterlogging would be entirely due to the incapability to tolerate or prevent excessive accumulation of iron or manganese (Etherington & Thomas 1986), is apparently not justified. Comparatively sensitive species, which are incapable to penetrate into anoxic soil layers, may suffer from waterlogging without showing any associated increase in the iron and manganese contents of their shoots (Schat 1984; Van Diggelen 1988); the fact that such species do not exhibit tolerance to increased levels of reduced iron or manganese is not surprising. It would not change their inability to cope with soil anoxia as such. Of course, this does not mean that iron or manganese will never act as a direct selective factor for dune slack species in general. Any-

way, such a selective role supposes some basic level of tolerance to soil anoxia *per se*, which is not present in the species dealt with here, except for *Samolus*. The response under experimental conditions correlates very well with the behaviour in the field. In spring, as long as the superficial soil layers remain reduced, which may last until the end of May or even later in exceptionally wet seasons, the growth of *Parnassia, Centaurium littorale, C. pulchellum, Sagina* and, though to a lesser extent, *Plantago* is extremely poor. New roots are exclusively developed in the upper 1 or 2 cm soil layer (3–4 cm in *Plantago*). The growth of *Samolus*, on the other hand, is not considerably affected and starts shortly after emergence of the plant. Anyway, waterlogging as such does not cause serious mortality; it merely shortens the period available for (normal) growth. For *Centaurium littorale* and *Sagina nodosa*, for example, the period of growth near the lower limit of distribution is 1–2 months shorter than near the upper limit.

The most important determinant of the lower limit of adult plants on the slope is the period of submergence in winter (Schat 1982). Under experimental conditions there seems to be a high degree of correlation between the ability to grow in waterlogged soil and the period of survival in completely submerged condition. *Samolus valerandi* is the

Table III. Mean lower limit of distribution (cm above slack bottom) of adult, reproductive plants; root porosity and maximal rooting depth in waterlogged sand, survival under submergence, and mortality during submergence in winter in the field of several dune slack species at the beach plain of Schiermonnikoog.

Species	lower limit of adults	root porosity (% air space volume)[1]	maximal rooting depth in water-logged sand (cm)[1]	survival under submergence (weeks)[2]	mortality in winter 1978/1979 (%)[3]	mortality in winter 1980/ 1981 (%)[3]
Samolus valerandi	0	18	> 10.0	>25	100	55
Centaurium pulchellum	0	7	2.5	5—10	–	–
Parnassia palustris	8	4	0.2	5—10	0	0
Plantago coronopus	8	15	5.0	10—15	100	80
Centaurium littorale	12	7	2.5	5—10	100	100
Sagina nodosa	12	3	0.5	5—10	100	100

[1] the values apply to plants grown in waterlogged sand with 10% peat.

[2] the values indicate the time of survival of 3 months old plants after imposing complete submergence in greenhouse experiments (20° C; $250\,\mu E \cdot m^{-2} \cdot sec \cdot ^{-1}$).

[3] the mortality rates apply to a level of 8 cm above the slack bottom in the case of *Parnassia* and 4 cm in the case of the other species; the corresponding periods of submergence were 10 and 12 weeks in 1978/1979 and 6 and 8 weeks in 1980/1981, respectively; *C. pulchellum* is a summer annual.

only species which keeps growing, though at a slow rate (Schat 1984). The other species die back and are eventually killed within 10 weeks, apart from *Plantago* (Table III). The response to submergence in the greenhouse does not correlate with the effects of submergence in winter in the field situation (Table III). *Samolus,* which survives under experimental conditions, exhibits even 100% mortality after 12 weeks of submergence in winter in the field. *Parnassia,* on the other hand, does not suffer at all from submergence in winter in the field; its inability to invade the lower parts of the slack (Table III) is due to its extreme sensitivity to salinity in this particular case (Schat 1982).

These inconsistencies may be due to differential seasonal patterns of metabolic activity, related to species-specific phenological characteristics. *Parnassia palustris* sprouts relatively late in spring and dies off before the onset of winter, including its root system; the rhizome passes the winter in a state of dormancy. It is well known that hibernating corms or rhizomes of some wetland plants are capable of surviving prolonged periods of complete anoxia, probably due to a low metabolic energy consumption and a specific mode of fermentation (Monk *et al.* 1984), which could also apply to *Parnassia*. The other species are potentially evergreen; they keep growing as long as the external conditions allow it and do not seem to show any degree of innate dormancy. Their energy requirements for survival under submergence in winter may be much higher. Therefore, they may more rapidly suffer from cellular ATP-depletion, which seems to be the primary cause of death under anoxia (Saglio *et al.* 1980; Webb & Armstrong 1983). *Samolus valerandi* seems to be the only species capable of maintaining some photosynthetic activity in submerged condition (Schat 1984). The unfavourable conditions for photosynthesis prevailing in the winter season might further reduce the cellular energy charge, which could explain the striking difference between its tolerance under summer conditions and that under winter conditions. Another example of the role of phenological characteristics, which affect a species' behaviour vis-à-vis high water levels in winter, is provided by *Centaurium pulchellum:* in spite of its inability to tolerate waterlogging or submergence, it reaches the bottom of the slack, due to its summer annual habit.

In general, the lower limit of adult plants on the slope will depend on phenology, life history, ontogenetic shifts in flooding tolerance, the reproductive strategy and the growth form.

1.4. Modifying effects of interspecific interactions

The examples given indicate that a species' vertical distribution range on a dune slope is determined by its ability to survive drought or submergence, in general. However, these examples apply to a relatively early successional stage, during which the plant cover was sufficiently open to allow establishment of whichever species at any level on the slope. In later successional stages the selective effect of the groundwater level will be increasingly indirect and mediated or overruled by competitive interactions between species. The impact of competitive interactions on the vertical distribution of a species is strongly dependent on the soil fertility level. This is clearly shown by the successional developments in the beach plain of Schiermonnikoog. This area provides a mosaic pattern of soil fertility levels, ranging from oligotrophic to eutrophic, due to the presence of a sea gull colony. At oligotrophic sites the vegetation is still open at present, i.e., about 30 years after the origin of the beach plain. There is an intense competition for nutrients (Schat 1982), which profoundly affects the growth of individual plants, but fails to restrict the vertical distribution range of single species to a significant extent. At the eutrophic sites the perennial vegetation cover has completely closed, apart from higher levels on the slope, where the destructive effect of trampling by the sea gulls overrules the fertilization effect, probably as a result of the low soil water content in summer. The base of the dune slopes, potentially the most species-rich zone, is densely vegetated with several grasses (mainly *Festuca rubra*) and *Potentilla anserina,* which excludes the establishment of seedlings. The distribution of short-lived species is confined to higher levels on the slope, where their persistence is at risk due to summer drought, except for the winter annuals.

Interspecific interactions with a modifying effect

on zonation patterns are not necessarily of a competitive nature, but may as well rely on "facilitation" (*sensu* Connell & Slatyer 1977). The importance of facilitation in dune slack vegetation is not necessarily increasing during succession, as indicated by the following example, which describes the effect of *Juncus maritimus* on the lower limit of distribution of *Plantago coronopus*. Considering the survival of *Plantago* during the winter season at various distances from large tussocks of *Juncus maritimus*, it is obvious that it is increased in the immediate surroundings of a tussock. This profoundly affects the density and persistence in the lower part of the slack (Table V). In 1978 the density was high and the plants were not markedly aggregated around the *Juncus*-tussocks, probably due to an accidental sequence of relatively dry winters. In the winter 1978/1979, characterized by an exceptionally long period of submergence, the population was completely killed, up to a level of 10 cm above the lowest point of the slack, apart from the ungerminated seeds (in this particular case, this height trajectory corresponds with a horizontal distance of about 20 m). The survival of the subpopulation of the lower part of the slack (more than 98% of the total population in 1979) from 1979 to 1980 depended completely on the survival of the seedling cohort of 1979. Firstly, because flowering does not occur until the second or third season after germination in this area and secondly because recruitment from more than one year old seed reserves never occurs in the lower parts of the slack. Moreover, the dispersal in space is very limited (less than 15 cm, in general), which precludes a rapid re-invasion of lost area (Schat 1982). In 1980

there was a complete extinction below a level of 5 cm above the lowest point of the slack (about 90% of the area of 1978), mainly due to heavy mortality in the winter 1979/1980 (Table IV), apart from a few individuals in the vicinity of scattered *Juncus* tussocks. Immediately above this level, there was some survival at greater distances from the tussocks, but the majority of individuals were concentrated around *Juncus*-tussocks (Fig. 2). During the next years the density increased again, due to seed production *in situ*. Up to 1986, when complete extinction occurred up to 12 cm above the lowest point of the slack (about 95% of the area of 1978), the density increased, but the strong aggregation around *Juncus*-tussocks remained. Since the winter 1985/1986, which was the second in a series of three winters with a long period of submergence, *Plantago coronopus* is virtually absent from the lower parts of the slack.

The above described facilitation of the maintenance in the lower parts of the slack by *Juncus maritimus* does not only apply to *Plantago coronopus*, but also to *Samolus valerandi*. The precise mechanism seems to be rather complex. The zone with increased survival, around the tussocks which extends to about 25 cm from the edges (Fig. 2), roughly corresponds with the surface of the soil volume which remains permanently oxidized throughout the winter season, due to radial oxygen loss from the dense mass of *Juncus* roots. This permanent oxygenation of the rhizosphere (sulphide precipitation occurs only at greater distances from the tussocks) relies on the fact that the tall shoots (60 cm) remain partly emerged during win-

Table IV. Survival (%) of *Plantago coronopus* in relation to distance from tussocks of *Juncus maritimus* during three successive winters (November till March) at 4 cm (I) and at 8 cm (II) above the lowest point of the slack.

distance from tussock (cm)	1978/1979		1979/1980		1980/1981	
	I	II	I	II	I	II
0—30	0	4	13	15	9	18
>30	0	0	1	7	—[1]	3

[1] no individuals present at the onset of the winter season.

Table V. Density of *Plantago coronopus* (number per dm^2) in the vicinity of and at greater distances from tussocks of *Juncus maritimus* in August of four successive years at 4 cm (I) and at 8 cm (II) above the lowest point of the slack.

distance from tussock (cm)	1978		1979[1]		1980[1]		1981	
	I	II	I	II	I	II	I	II
0—30	15	60	48	17	3.2	1.2	8.9	4.8
30	9	40	50	8	0.0	0.2	0.0	0.8

[1] the individuals present in 1979 and 1980 all belong to a single cohort, established in 1979.

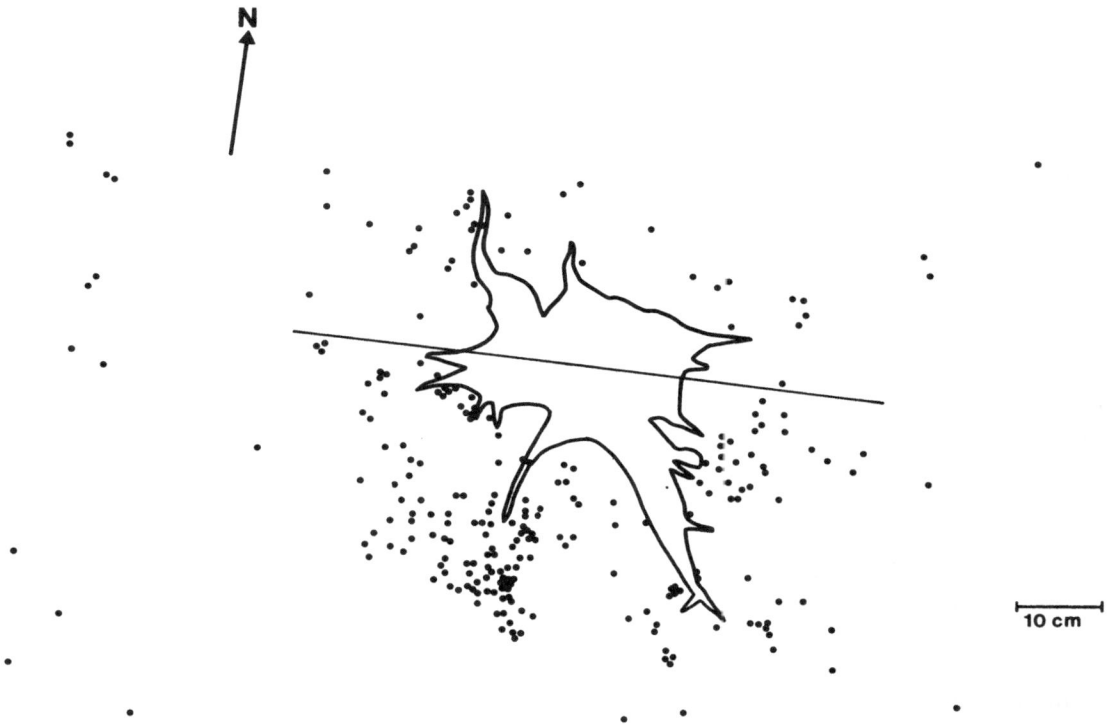

Fig. 2. Spatial pattern of distribution of individuals of *Plantago coronopus* (●) near the bottom of the beach plain at Schiermonnikoog in August 1980; the contours in the centre of the graph are those of a tussock of *Juncus maritimus;* the arrow points at the north.

ter flooding; less tall-growing species with a comparable potential for soil oxygenation, such as *Juncus alpino-articulatus* and *Juncus articulatus* do not prevent soil reduction throughout the whole winter, because they will be often subjected to complete submergence. This blocks the supply of atmospheric oxygen and, consequently, the oxygenation of the rhizosphere (Steinmann & Brändle 1981). The latter species do not significantly contribute to the survival of *Plantago* in winter. These observations suggest that soil oxygenation *per se* or the associated oxidation of iron, manganese, or sulphide could explain the facilitation. Greenhouse experiments (Schat 1984) suggest that oxygenation *per se* may account for increased survival under submerged conditions. However, the situation in the field may be more complex. Firstly, in the field the mortality is not strictly confined to the period of submergence. Many individuals still produce one or two leaves, immediately after emergence, be-

fore dying shortly afterwards. These are individuals of which the root system has almost completely died away throughout the preceding period of submergence. This "post-emergence" mortality is lower in the vicinity of *Juncus*-tussocks, which partly explains the decreased total mortality in the winter season. The degree of damage to the root system tends to be lower in the vicinity of a tussock. However, it can not be ruled out that the conditions during the period of sprouting are decisive in this respect. Also in this case, the oxygen status of the soil may play a major role. The growth rate of the plants in early spring, is much higher in the vicinity of a tussock. The plants produce large, dentate, green leaves, often with a hairy surface, whereas at greater distances from the tussocks the leaves remain small, linear and glabrous, apart from the leaf edge, and maintain a purplish colour till late in spring. The better growth near the tussocks is associated with an increased seed production in summer and a high P-content of the leaves (80 μmoles per g dry weight). At greater distances from the tussocks the P-content is only 40 μmoles g^{-1}, which indicates severe P-limitation. It is unlikely that the improved P-nutrition of *Plantago* (and other species as well, such as *Centaurium pulchellum, Samolus valerandi, Agrostis stolonifera*) contributes to the increased survival in the winter season. It is also observed around and within tussocks of other *Juncus* species and may be due to a stimulating effect of soil aeration on the P-mineralization rate, or the extension of the root system under waterlogged conditions (cf. Schat 1984). This may be considered as an additional facilitation mechanism, because it stimulates the seed production *in situ,* which is occasionally decisive for the year to year survival of the population (see above).

Another remarkable phenomenon is the orientation of the *Plantago* individuals around the tussocks: they are invariably concentrated on the south-exposed side of the tussock (Fig. 2). This suggests that the survival of *Plantago* in the winter season is sensitive to fairly subtle differences in temperature and/or light. Jerling (1981) observed a strongly negative effect of darkness on the tolerance to submergence in *Plantago maritima*. Temperature differences between the north-facing and south-facing side of the tussock are unlikely to occur during the period of submergence itself, but might arise after emergence, which could affect the "post-emergence" mortality (see above). A detailed explanation of these interesting phenomenon requires more detailed field observations and further experimentation.

Finally, the above example illustrates the importance of life history characteristics and dispersal strategies. The dependency of *Plantago* on the presence of *Juncus maritumus* is due to its limited dispersal in space and in time and to its inability to produce seeds within the first year after germination at this particular site. However, the latter is not an inherent property of the species, but a mere response to the low soil fertility level *in situ* (Schat et al. 1984). Obviously, the modifying effects of competitive and facilitating interactions are both dependent on the soil fertility level.

1.5. Conclusions

Though it is evident that the groundwater regime is of major importance for the zonation of dune vegetation, it is often not easy to disentangle the complex relationships involved. A detailed understanding of the behaviour of a species in relation to the groundwater regime requires a thorough knowledge of its population biology, which can only be achieved by long-term monitoring of natural populations and ecophysiological experimentation.

2. Effects of groundwater manipulation on soil processes and vegetation developments in dune slacks

2.1. Introduction

Since the beginning of this century, large parts of the dune system have been subject to a considerable drop of the groundwater table, due to various human activities (Bakker *et al.* 1979). Especially the increasing production of drinking water has contributed to the progressive desiccation of the dune system, which has lead to an enormous decline of the area of characteristic wet dune slack vegetation types (Bakker *et al.* 1979). Since 1950, the drop of the groundwater table has been stopped, or even reversed, at least locally, through the infiltration of surface water, mainly from the rivers Rhine and Meuse, in order to increase the capacity for the production of drinking water. However, the associated input of nutrients has lead to serious eutrophication of the infiltration areas (Londo 1975), which prevented a re-appearance of the original wet slack vegetation. Ruderal vegetations with a much lower species-diversity appeared instead (Van Dijk 1984). Moreover, the temporal pattern of the water table fluctuations still differs from that under undisturbed conditions, which might further restrict the possibilities for the recovery of the original vegetation.

Plans are made for a further rise of the water table in the near future. The conditions minimally required for the return of a more or less "natural" wet slack vegetation are not exactly known. The effects of wetting will depend on the present situation, i.e. the type of vegetation, the standing crop, the amount of soil organic matter, the soil pH, the calcium carbonate content and the mineralization rate, as well as on the way in which the wetting will be accomplished and on the accessibility of the slack to new invading species.

2.2. Soil factors affecting the development of the vegetation after raising the water table

The species-richness and characteristic floristic composition of wet slack vegetations are strongly dependent on the fertility and the $CaCO_3$-content of the soil. Especially the combination of a low level of phosphorus availability and a relatively high $CaCO_3$-content may represent favourable conditions for the return of many original slack communities (the productivity of dune vegetation is usually phosphorus-limited (Van Dijk 1984)).

A rise of the water table will have considerable effects, either direct or indirect, on the nutrient and carbon cycling in the vegetation/soil system. It will strongly affect the balance between biomass production, nutrient assimilation, litter production, decomposition, mineralization and the conversion of inorganic nitrogen in the soil (ammonification, nitrification, denitrification). Moreover, since many of these and other chemical conversions which depend on the water-status of the soil (e.g., sulphate reduction) involve the production or the consumption of protons, a water table rise may affect the soil pH, or at least the acid-neutralizing capacity, and the decalcification rate.

The chemistry and biology of dune soils is poorly studied as yet, which makes it difficult to predict the precise consequences of changing the water table level. In general, the mineralization rate will depend on the content and the type of organic matter and on the water content of the top soil (Grootjans 1985). Under either extremely dry or extremely wet conditions, the mineralization rate will be low, which may lead, in the latter case, to the formation of peat (Bakker *et al.* 1979). When the original conditions are dry, one might expect that wetting will cause a general increase in micro-

bial activity (Lund & Goksøyr 1980), associated with an increase in ammonium-nitrogen and nitrification (Birch 1958) and an increased mineralization rate as long as aerobic conditions are maintained. As soon as anaerobic conditions occur, the situation may be different; the mineralization and nitrification rate will decrease again, and the denitrification rate is expected to rise. Just as in marine sediments, the change in the nitrogen-status of the soil upon waterlogging may depend on the rate of reduction of nitrate to ammonium, relative to the reduction to molecular nitrogen (N_2) by denitrification (Sørensen 1978). It must be stressed, however, that the soil pH may play a major role in this respect (the optimal pH ranges for ammonification, nitrification and denitrification are 4–5, 6–7 and >5, respectively (Jeffrey 1987, Richards 1987)), and that these processes themselves affect the proton balance of the soil (see above).

Van Beckhoven (in prep.) compared the mineralization rate in dune soil along a moisture gradient ranging from a *Molinio-Arrhenatheretea*-community (moist) to a *Festuco-Galietum maritima* (dry) and found a strong positive correlation between soil water content, mineralization, microbial activity, measured as dehydrogenase and urease-activity, and the productivity of the vegetation, whereas the total soil nitrogen content was negatively correlated with soil moisture, which indicates an increased N-mineralization with increasing soil moisture. Moreover, increasing the water content of soil samples from the field under greenhouse conditions, also resulted in an increased microbial activity, which indicates a more or less direct effect of the soil water content *per se*. It remains to be investigated, however, to what extent changes in the nitrogen status of the soil will affect future vegetation developments, as phosphorus seems to be the main limiting factor (see above). However, this may be different under eutrophic conditions (Schat, unpublished).

2.3. The importance of the type of vegetation present at the start of the wetting process

2.3.1. Vegetation composition
The presence of relic dune slack species, such as

Schoenus nigricans, which often persists for long periods in desiccating slacks (section 1.2) may facilitate the colonization of other characteristic dune slack plants, such as *Centaurium littorale* (Schat 1984), *Samolus valerandi* and many other species after a rise of the water table (section 1.4).

2.3.2. Standing crop

The standing crop, i.e., the amount of living biomass per unit of soil surface, may greatly affect the developments after wetting. As soon as anaerobic conditions start to prevail, a large part of the standing crop might die away, resulting in a more or less sudden accumulation of dead organic matter, which will greatly affect microbial activity, soil pH and redox-status of the soil. Under drained conditions, which will usually prevail in summer, this may lead to an explosive increase in plant available nutrients. The latter might facilitate the spread of perennial, more or less waterlogging-tolerant ruderals, such as *Calamagrostis epigejos,* which will prevent the establishment of many other species by competitive exclusion.

2.4. The mode of groundwater manipulation

The way in which a rise of the groundwater table will be accomplished may to some extent determine the future vegetational developments. Firstly, increased surface infiltration with river water will lead to a further eutrophication; of course, it would be better to decrease the drinking water production. Moreover, desirable developments may be stimulated by controlling the rate and the degree of groundwater rising. In general, a sudden rise of greater magnitude will strongly disturb the carbon and nutrient cycling, which may lead to an undesirable sudden increase in nutrient availability (see above); therefore, a gradual rise over a longer series of years is preferable.

It may be that the developments can be influenced, at least to some extent, by restricting artificial groundwater manipulation to certain periods of the year, in view of the seasonality of soil processes and biomass production. However, as yet it is difficult to provide specific recommendations at this point. It will be important to maintain a more or less natural seasonal fluctuation in groundwater depth; the survival and growth of many dune slack plants depends on seasonal variation (section 1.2.; 1.3.).

2.5. Accessibility

The return of the characteristic wet slack vegetation after raising the groundwater table will partly depend on the species pool of the area in question. Many species have become extinct from large parts of the dune area and their return will depend on the accessibility of the area for them, at least when seed reserves are no longer present.

2.6. Conclusions and recommendations

The possibilities for regeneration of characteristic wet dune slack vegetation by means of groundwater level manipulation are difficult to estimate, partly due to a lack of knowledge on soil processes in dune systems; further monitoring and experimental studies, mainly concerning the mineralization rate in relation to the groundwater regime, will provide valuable information. Recently, experiments have been started to quantify the net mineralization rate, microbial activity, phosphate fixation, and nitrogen conversions in systems with and without plants, in relation the soil water content, both in the field and in the greenhouse.

It seems likely that additional management tools, such as mowing or sod cutting, may be indispensible, especially under eutrophic conditions.

References

Bakker, T. W. M., Klijn, J. A. & Van Zadelhoff, F. J., 1979. Duinen en duinvalleien. Een landschapsoecologische studie van het Nederlandse duingebied. Pudoc, Wageningen.

Beeftink, W. G., 1975. The ecological significance of embankment and drainage with respect to the vegetation of the southwest Netherlands. J. Ecol. 63: 423—458.

Birch, H. F., 1958. The effect of soil drying on humus decomposition and nitrogen availability. Pl. and Soil 10: 9—32.

Bruinenberg, J., Joenje, W. & Wieringa, T., 1980. Hapaxant species of coastal beach plains colonizing embanked sand flats. Acta Bot. Neerl. 29: 497—508.

Connell, J. H. & Slatyer, R. O., 1977. Mechanisms of succession in natural communities and their role in community stability and organization. Am. Nat. 111: 1119—1144.

Diemont, W. H., Sissingh, G. & Westhoff, V., 1940. Het dwerg-biezenverbond (*Nanocyperion flavescentis*) in Nederland. Ned. Kruidk. Arch. 50: 214—284.

Elliston Wright, F. R., 1953. Note on the dispersal of *Sagina nodosa* var. *moniliformis* Lange. Watsonia 2: 369—370.

Ernst, W. H. O. & Van der Ham, N. F., 1988. Population structure and rejuvenation potential of *Schoenus nigricans* in coastal wet dune slacks. Acta Bot. Neerl. 37: 451—465.

Etherington, J. R. & Thomas, A. M., 1986. Response to water-logging and different sensitivity to divalent iron and manganese in clones of *Dactylis glomerata* L. derived from well drained and poorly drained soils. Ann. Bot. 58: 109—119.

Freijsen, A. H. J., 1967a. Some observations on the transition zone between the xeroseve and the halosere on the Boschplaat (Terschelling, The Netherlands) with special attention to *Centaurium vulgare* Rafn. Acta Bot. Neerl. 15: 668—682.

Freijsen, A. H. J., 1967b. A field study on the ecology of *Centaurium vulgare*. Thesis, University of Utrecht.

Grootjans, A. P., Schipper, P. C. & Van der Windt, H. J., 1986. The influence of drainage on N-mineralization and vegetation response in wet meadows. II. *Cirsio-Molinietum* stands. Acta Oecol., Oecol. Plant. 7: 3—14.

Jeffrey, D. W., 1987. Soil-Plant Relationships. An ecological approach. Croom Helm, Timber Press, Great Britain.

Jerling, L., 1981. Effects of microtopography on the summer survival of *Plantago maritima* seedlings. Holarctic Ecology 4: 120—126.

Londo, G., 1975. Infiltreren is nivelleren. De Levende Natuur 78: 74—79.

Lund, V. & Goksøyr, J., 1980. Effects of water fluctuations on microbial mass and activity in soil. Microb. Ecol. 6: 115—123.

Monk, L. S., Crawford, R. M. M. & Brändle, R., 1984. Fermentation rates and ethanol accumulation in relation to flooding tolerance in rhizomes of monocotyledonous species. J. Exp. Bot. 35: 738—745.

Richards, B. N., 1987. The microbiology of terrestrial ecosystems. Longman Scientific & Technical, United Kingdom.

Saglio, P. H., Raymond, P. & Pradet, A., 1980. Metabolic activity and energy charge of excised maize root tips under anoxia. Plant Physiol. 66: 1053—1057.

Schat, H., 1982. On the ecology of some dutch dune slack plants. Thesis, Vrije Universiteit, Amsterdam.

Schat, H., 1983. Germination ecology of some dune slack pioneers. Acta Bot. Neerl. 32: 203—212.

Schat, H., 1984. A comparative ecophysiological study on the effects of waterlogging and submergence on dune slack plants: growth, survival and mineral nutrition in sand culture experiments. Oecologia (Berl.) 62: 279—286.

Schat, H., Bos, A. H. & Scholten, M., 1984. The mineral nutrition of some therophytes from oligotrophic dune slack soils. Acta Oecol., Oecol. Plant. 5: 119—131.

Sørensen, J., 1978. Capacity for denitrification and reduction of nitrate to ammonia in a coastal marine sediment. Appl. Environm. Microbiol. 35: 301—305.

Steinmann, F. & Brändle, R., 1981. Flooding tolerance in Bulrush (*Schoenus lacustris* (L.) Palla): relations between oxygen supply and "adenylate energy charge" of the rhizomes in dependence on the environmental oxygen concentration. Flora 171: 307—314.

Van Diggelen, J., 1988. A comparative study on the ecophysiology of salt marsh halophytes. Thesis, Vrije Universiteit. Amsterdam.

Van Dijk, H. W. J., 1984. Invloeden van oppervlakte-infiltratie ten behoeve van duinwater-winning op kruidachtige oevervegetaties. Thesis, Wageningen.

Van Tooren, B. F., Schat, H. & Ter Borg, S. J., 1983. Succession and fluctuation in the vegetation of a Dutch beach plain. Vegetatio 53: 139—151.

Webb, T. & Armstrong, W., 1983. The effects of anoxia and carbohydrates on the growth and viability of rice, pea and pumpkin roots. J. Exp. Bot. 34: 579—603.

Whittaker, R. H., 1972. Evolution and measurement of species diversity. Taxon 21: 213—251.

Authors' Address
H. Schat and K. van Beckhoven
Biological Laboratory
Vrije Universiteit Amsterdam
P.O. Box 7161
1007 MC Amsterdam
The Netherlands

The salt marsh Oosterkwelder at the Frisian island Schiermonnikoog with the mixed vegetation of *Juncus maritimus* (dark) and *Elymus pycnanthus*. The white vegetation at the front refers to *Artemisia maritima*. (Photograph: P. Leendertse.)

CHAPTER 9

Natural and man-made environmental stresses in coastal wetlands

J. ROZEMA & P.C. LEENDERTSE

Abstract. Various kinds of environmental stresses occur in coastal wetlands. The tidal floodings with seawater of sandy and muddy areas lead to a spatial and temporal variation of salinity, redox potential and many other derived factors. Growth of all terrestrial halophytes is significantly reduced at seawater salinity (-2.0 MPa). Salt marsh plants however, are morphologically and physiologically adapted to increased salinity and waterlogged conditions. In addition to salinity and anaerobiosis harsh physical conditions such as storms, and high rates of tidal currents may uproot *Salicornia* and erode the environment of *Spartina anglica*. Organic and inorganic contaminants transported by the large rivers are discharged in the open sea. The extent and consequences of eutrophication and contamination are discussed for the terrestric environment of estuarine and coastal salt marshes.

Increased concentrations of nitrogen and phosphorus have led to excessive growth of macroalgae. The flooding of salt marshes with nutrient enriched seawater seems to favour dominant growth of nitrophytic halophytes such as the grass species *Elymus pycnanthus* in ungrazed salt marshes. As a result the biodiversity of the salt marsh environment tends to decrease with eutrophication. The sediment of coastal and estuarine salt marshes may be highly contaminated with heavy metals, applied by polluted river water. Heavy metals are taken up by the root system and translocated to the shoot. Effects of heavy metal pollution on the growth of salt marsh plants in the field have not been reported as yet.

1. Coastal ecosystems

In coastal areas with a medium tidal range and with sufficient shelter against storms, salt marshes may develop. Salt marshes consist mainly of a grassland vegetation that is frequently inundated with seawater or estuarine water. Extensive areas of relatively species-rich salt marsh vegetation have developed on the Dutch-German-Danish Wadden islands and in estuaries. More special types of salt marshes are represented by the beach plain salt marsh (Beeftink 1977; Rozema 1978) and the more or less man-made foreland and "inlaag" type salt marshes in some parts of the Netherlands (Fig. 1). The sediment of salt marshes usually consists of silt and clay accreting in sheltered places to above the mean high waterline. A major part of the sediment that contributes to the development of salt marshes is of marine origin (about 80% on average, see Beeftink

& Rozema 1988) and the other part refers to fluvial supply.

The periodical flooding with seawater or estuarine water of most salt marshes governs the development of plant and animal life. The groundwater level of the lower salt marsh is mostly near the soil surface. In flooded salt marsh soils, the limited gaseous diffusion in the soil water leads to accumulation of nitrogen, carbon dioxide, methane, hydrogensulfide and dimethyl-sulfide in the reduced zone of the soil profile. Dimethyl-sulfide is also released by plants of *Spartina anglica*. Dimethylsulphonio-propionate (DMSP) (Fig. 2) is found in high concentrations in the tissue of *Spartina anglica* and is thought to function as a storage pool for excess sulphur (sulphide) (Van Diggelen *et al.* 1986). Sulfide compounds of these types are toxic to most higher plants (Van Diggelen *et al.* 1986). In addition to the accumulation of DMSP, there is an

J. Rozema and J. A. C. Verkleij (eds.), Ecological Responses to Environmental Stresses, 92—101.

Fig. 1. Map of the Netherlands with the estuarine area in the South-Western part (Western Scheldt, Eastern Scheldt, Haringvliet). In the Wadden area some of the Wadden Islands are indicated: Texel and Schiermonnikoog, as well as the Eems Dollard estuary.

glycine betaine
Chenopodiaceae
Gramineae
Solanaceae

proline
Gramineae
(Micro) algae

β-3-dimethyl

sulphoniopropionate

Chlorophyceae

higher plants

(*Spartina anglica*)

cf. Van Diggelen *et al.* 1986

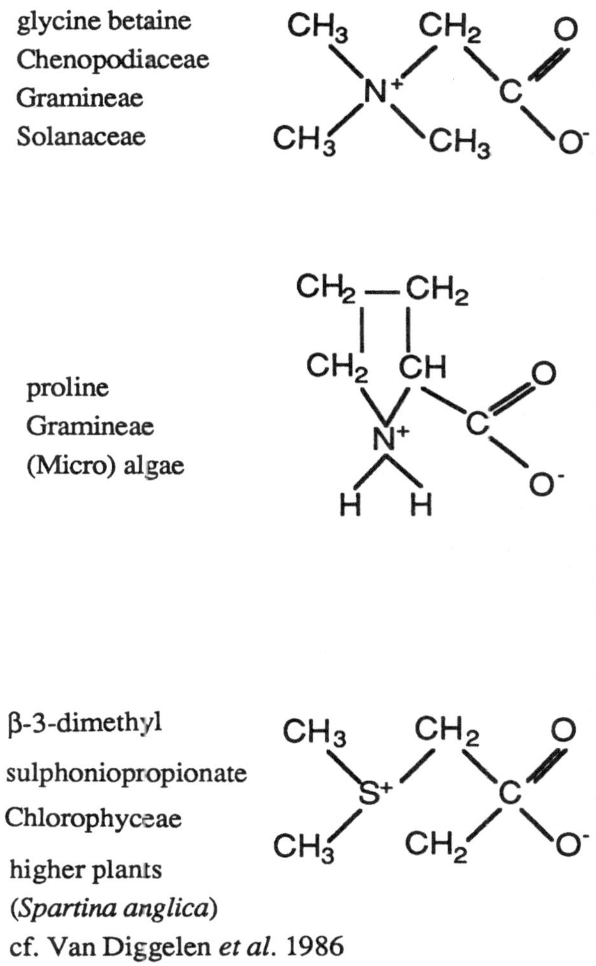

Fig. 2. Chemical structure and occurrence in plant groups of β-3-dimethylsulphoniopropionate (DMSP), proline and glycine betaine. In *Spartina anglica* the function of DMSP may be storage of excess sulphur. Proline and glycine betaine act as compatible osmotic solutes.

increase of the N-dipoles proline and glycine betaine in *Spartina anglica* (Fig. 2). Both proline and glycine betaine are thought to act as a compatible solute in *Spartina anglica,* contrary to DMSP. Air tissue in the cortex of plant roots (aerenchyma) and the occurrence of radial oxygen loss is a functional development in many plant species of the coastal wetlands, leading to avoidance of anaerobic conditions in the rhizosphere (Van Diggelen 1988).

2. Zonation of salt marsh vegetation: physiological tolerance and competitive relations

The vegetation of salt marshes is well known for its zonation, that is the more or less constant sequence of occurrence of plant species along a height gradient (Rozema 1978; Rozema *et al.* 1988). When variation exists in the zonation patterns of different salt marshes, this often relates to differences of frequency and duration of sea water flooding, and geomorphological differences between the salt marshes (Beeftink & Rozema 1988). In multi-species experiments in greenhouses it has been demonstrated that lower marsh plant species are significantly better adapted to sea water salinity and flooding than are species of the upper marsh (Van Diggelen 1988). This also implies that upper marsh species are sensitive to the kind of natural stress of the lower salt marsh environment: sea water salinity, flooding and anaerobic soil conditions. Differences in the physiological growth response to seawater salinity and flooding, together with differences in the time of emergence in the spring and resource exploitation (light and mineral nutrients) determine the outcome of competitive interactions between salt marsh plant species. In the field it has been shown that the distribution of salt marsh species is determined to a large extent by competitive interactions between salt marsh species (Scholten *et al.* 1987). The example of *Spartina anglica* and *Puccinellia maritima,* both species of the lower parts of salt marshes illustrates well the importance of physiological tolerance and competitive interactions in the distribution of plant species. The C4-species *Spartina anglica,* occurring at atlantic coastal sites of Europe, reaches its northern boundary of the geographical occurrence at the Skallingen salt marsh in Denmark. Sprouts of *Spartina anglica* develop earlier in the season in the salt marshes of the South-Western part of the Netherlands, than on the salt marshes of the Waddenarea. The tillers of the C3-grass species *Puccinellia maritima,* with a more northern Scandinavian and atlantic geographical distribution, develop earlier in the season in the Waddentype salt marshes, than sprouts of *Spartina anglica* do. This early seasonal emergence of *Puccinellia* provides *Puccinellia* with a competitive advantage relative to *Spartina.* This further leads to an advantage in the competition for light when this factor becomes more limiting later in the season when the canopy of the salt marsh grass vegetation closes. These competitive interactions explain the dominance of *Puccinellia* and the relative poor performance of *Spartina* in the Wadden salt marshes and the reverse situation in the estuarine salt marshes of the Eastern and Western Scheldt. However, the position of these two species in the salt marsh vegetation zonation does not differ in salt marshes in North and South-Western Netherlands (Rozema *et al.* 1988). Regarding the occurrence of the C4-species *Spartina anglica* along European coasts, its behaviour may be explained

Table I. Natural and Man-made environmental stresses of the salt marsh environment.

Natural Environmental stresses

Seawaterflooding

- reduced period of photosynthesis
- wave action
  uprooting
- seawater salinity
  osmotic and ionic stress
  fluctuations of soil moisture and salinity in terrestrial environments
- inundation
  anaerobiosis, low redox potential, phytotoxins $Fe(II)$, $Mn(II)$, sulfide
- salinity and flooding induced nutrient excess and deficiency

Man-made environmental stresses

Disturbance by various land developmental projects coastal defence, harbour development, land reclamation

Eutrophication (phosphorus and nitrogen) by sewage, agricultural use of fertilizers directly or indirectly by input of rivers Rhine, Meuse and Scheldt into coastal waters

Contamination

- heavy metals and various organic pollutants
- oil pollution

also by the fact that it is a hybrid species. *Spartina anglica* is the allotetraploid hybrid between *Spartina maritima,* a relatively small C4-grass of mediterranean and atlantic coasts of Europe, and *Spartina alterniflora,* a tall salt tolerant C4-grass species from the eastern coast of the United States.

Spartina maritima has always been a scarce species in Britain and the Netherlands, hardly setting seed. The low fitness of the C4-species *Spartina maritima* may relate to its high light requirement compared to the taller C3-plants dominating European salt marshes. Photosynthesis of *Spartina maritima* seems to be well adapted to the relative low temperatures of the atlantic European coast (Long & Incoll 1979). Temperatures of the summer season of the east coast of North America exceed those of atlantic European coasts (26–35° C) (Curtis *et al.* 1989). Maybe the temperature adaptations of the hybrid *Spartina anglica* are between that of the two parent species, explaining the better ecological performance of *Spartina anglica* in the South-Western parts of the Netherlands compared to the salt marshes of the Waddenarea. The competitive success of *Spartina anglica* in British and Dutch salt marshes in the South-Western parts of the Netherlands probably also relates to the tall shoots meeting the high light requirement of this C4 species relative to C3 neighbour plant species.

3. Natural environmental stresses in coastal wetlands

Only plant species with special adaptations are capable of inhabiting coastal wetlands. The physiological disturbance due to seawater salinity and submergence reduces the growth rate of many plant species substantially as studied in detail by Van Diggelen (1988). Furthermore, the physical consequences of wave action may limit growth of plants that are otherwise well adapted to seawater salinity and waterlogging conditions by uprooting the plants (Groenendijk 1986). The primary damaging effects of the rise of the sea level in the northern and southern parts of the Netherlands on the lower salt marsh vegetation results in erosion and uprooting rather than reduced growth due to

prolonged and more frequent inundation (Van Eerdt 1985; Rozema *et al.* 1985a). Also, successful establishment and maintenance in the lower parts of salt marshes is determined in part by a strong, branched root system that is not easily ruptured or uprooted, which prevents erosion and promotes the accretion of silt (Van Eerdt 1985). Knowledge of the physical and spatial properties of the root systems of salt marsh plants is scanty and should be extended.

The ecological differentiation between the species *Salicornia dolichostachya* Moss and *Salicornia ramosissima* J. Woods occurring in the lower and upper parts of salt marshes respectively, does not relate to differences with respect to adaptations to anaerobic conditions. The relative growth rate, photosynthetic, transpiration and respiration rate and uptake of mineral nutrients were the same for the two species. From field observations it was concluded that the more thinner and fragile root system of *Salicornia ramosissima* is more susceptible to wave action than the root system of *Salicornia dolichostachya*. Rupture of the roots, intrusion of water into the aerenchyma and development of iron sulphide coatings on the roots of *Salicornia ramosissima,* leads to growth reduction or the death of this species in the lower parts of the marsh system. Again, physical properties of the root system of these salt marsh species appear to determine the primary adaptation to wave action, rather than physiological adaptations to anaerobic conditions relating to flooding with seawater (Schat *et al.* 1987; Rozema *et al.* 1987).

4. Man-made environmental stresses in coastal wetlands

Intensive industrial activities along the mouth of the large rivers and the coast of, for example the North Sea, as well as industrial activities in the upstream hinterland and the development of harbours have led to the discharge of large amounts of organic and inorganic contaminants (Salomons *et al.* 1988). There is a tremendous eutrophication of the coastal water due to discharge of sewage of industrial domestic and agricultural origin. A much smaller part of contamination of coastal water is due to atmospheric deposition. A spectacular symptom of the consequences of eutrophication of the coastal seawater is the bloom of (toxic) algae such as in the lagoon of Venice (Sfriso 1987) and the massive development of macro algal species like *Ulva lactuca,* the development and spreading of bad smell due to sulphurous gases released from decomposing organic algal material in the Waddensea in the summer of 1989. High respiration rates of phytoplankton and other algae may cause anaerobic conditions in the coastal water and the death of large populations of fish species (Cadée 1984; Beukema & Cadée 1986).

Since the pollution of coastal areas by the discharge of industrial and domestic waste is via the rivers and estuarine water, a lot of research has focussed on the direct effects on plants and animals of aquatic estuarine and marine ecosystems (Wolff 1988). Far less research attention has been paid to the terrestrial coastal wetlands with dominance of higher plants.

Different aspects of contamination of the coastal waters will be considered here: i. eutrophication due to the discharge of domestic and agricultural waste water, ii. inorganic microcontaminants, that is heavy metals. Oil pollution is discussed in the chapter by Scholten & Leendertse (this volume). Huiskes & Beeftink (this volume) discuss the stress caused by economic uses of the salt marsh environment, while Otte (this volume) considers the contamination with heavy metals and arsenic in more detail.

A major part of the microcontaminants is adsorbed to small particles that are transported in the river, estuarine and sea water. Because of sedimentation of particulate matter with trace metals and organic micropollutants adsorbed, estuaries and coastal mudflats are considered to be a sink for micro-contaminants. Rozema *et al.* (1985b, 1986) describe the pollution with heavy metals of coastal salt marshes of the Netherlands. In the Western Scheldt area salt marsh sediment from seaward sites is more contaminated with heavy metals than more inland situated estuarine salt marshes. A major part of heavy metals is located in and on the root

systems. A relatively small fraction remains in the shoot. In those parts of the Dutch delta region where estuarine parts of the river or branches of the river are screened off from the sea by a dam, discharge of polluted river and estuarine water is no longer directly possible into the open North Sea. The closure of the Hollandsch Diep/Haringvliet estuary by the construction of the Haringvlietdam in 1970 has led to increased rates of sedimentation of mud and accumulation of heavy metals and organic micropollutants in this estuary (Fig. 1). The Hollandsch Diep/Haringvliet system is therefore called the "settling-tank of Western-Europe" (Saris & Duel 1987). Similarly in the Biesbosch, a former freshwater tidal system area, (Fig. 1), also screened off from the tidal action of the North Sea in 1970, enormous accretion of mud and accumulation of heavy metals and organic micro pollutants have taken place (Rozema & Otte 1989; Rozema *et al.* 1986).

5. Eutrophication of coastal waters and impact on the terrestrial ecosystem of coastal wetlands

Eutrophication refers to enrichment of the aquatic or terrestric ecosystem with mineral nutrients to such an extent, that primary production of the ecosystem is enhanced. The so-called macronutrients such as nitrogen and phosphorus are the most general inorganic compounds that limit the primary production of many ecosystems. Therefore the increase of the content of nitrogen and phosphorus in rivers, lakes, soil water and the soil often plays a major part in problems of eutrophication (Valiela & Teal 1974). The total concentration of inorganic nitrogen in the river Rhine in 1986 exceeded the content of the same river in 1930 by a factor of 6, for the Western Scheldt estuary this factor of enrichment for total inorganic nitrogen amounts to 8 (Van Buuren 1988). The discharge of sewage sludge has led to concentrations of ammonium in the Long Island Sound near New York of 45–100 μgat per liter, while in other sites of this coastal water, the ammonium concentration amounts to 0–5 μgat per liter (Valiela 1984). In coastal seawater ammonium concentrations are in the range of 1–8 μgat per liter, with much lower values in the open sea (0–1 μgat per liter) (Table II).

In the coastal wetlands, the fate of nitrogen enrichment as a result of the supply of eutrophicated estuarine and coastal waters is still incompletely known. Ammonium may be transformed to nitrate by nitrifying bacteria. Under anaerobic conditions denitrification of nitrate to N_2 and N_2O gas will occur. Plants will take up nitrate or ammonium and assimilate this inorganic nitrogen to organic nitrogen. In the process of decomposition of dead animal and plant material, mineralisation of organic material leads to the transfer of organic nitrogen into inorganic nitrogen. Supply and discharge of nitrogen in coastal wetlands consists mainly of the tidal transport of water. Although only a few detailed research projects have been carried out in a limited number of coastal wetlands, the following preliminary results are obtained.

There seems to be a net discharge of reduced compounds of nitrogen (ammonium and dissolved organic nitrogen) from salt marsh ecosystems with tidal currents (Leendertse 1989), when the nitrogen budget is considered for a year. At the same time there seems to be a net intake of nitrate into salt marshes on the basis of a year's budget (Jordan *et al.* 1983 (Table III)).

Removal of organic material by grazing or mowing, may represent an important part of the nitrogen budget for the salt marsh ecosystem. Fixation of nitrogen by cyanobacteria, blue green algae and free living bacteria may amount to about 10% of the total input of nitrogen into salt marshes dominated by *Spartina alterniflora*. Fixation of atmospheric nitrogen may take place by bacteria operating in the rhizosphere of *Spartina* (cf., Aziz & Nedwell 1979). Denitrification under anaerobic conditions in salt marshes of the eastern coastline of the United States may vary from 17% to 100% of the total discharge of nitrogen. For these salt marshes the output of nitrogen is 11% higher than the input of nitrogen (Valiela & Teal 1979). Taking the variation into account of the estimated values of the various fluxes, it is concluded from the above studies, that input of nitrogen roughly equals output. Much more detailed studies are required to make general conclusions possible. Also, the nitrogen budget of a salt marsh ecosystem may become different both qualitatively and quantitatively when instead of a year's period smaller time intervals are considered.

The content of phosphorus in coastal seawater has greatly increased as a result of discharge of phosphates from the rivers. It has been estimated that the phosphate content increased twofold to threefold in the Waddensea water during the period 1950—1970 (Van Buuren 1988). For the Scheldt estuary it has been assessed that the discharge of phosphates by urban sewage, industrial effluents and agricultural drainage is ten times higher than the highest and hundred times higher than the lowest phosphorus contents found in non-polluted estuaries (Beeftink & Rozema 1988).

Since 1977, the content of phosphorus in the river Rhine water tends to decrease near Lobith, but this is not found near Maassluis, where the river Rhine enters the North Sea (Van Buuren 1988). Budget studies conducted indicate that there is a net output of phosphate from marshes and a net input of organic phosphorus. In the spring, the phosphate content of coastal water is relatively low by uptake and assimilation by algae (Woodwell & Whitney 1977). The phosphate content of the water of the tidal zone of the Waddenzee is much higher in the summertime. For example in this period, relatively high temperatures will stimulate mineralisation of organic material. The rate of mineralisation of nitrogen was found to be extremely low in the winter period in the Danish Skallingen salt marsh (Jensen *et al.* 1985; see Rozema *et al.* 1977).

6. Consequences of increased levels of nitrogen and phosphorus in coastal and estuarine water

Massive development of mats of macro-algae like *Ulva lactuca* occur in the shallow Waddensea and in the lagoon of Venice in the Adriatic Sea (Sfriso 1987), as clear symptoms of increased primary production. Also, the production of phytoplankton biomass has greatly increased in the Waddensea.

Table II. Mean monthly concentrations (μmol/l) of nitrogen in tidal and coastal waters of Western Europe and eastern coast of North America.

Authors	Period	Location	Month	NO$_3^-$	NH$_4^+$	DON	PN	total
Western Europe								
Aziz & Nedwell (1986)*	1977/1978	Colne Point salt marsh (UK) (flood water)	March	100	6	50	28	184
			May	36	7	91	24	158
			July	12	8	56	19	95
			October	22	6	54	19	101
Dankers et al. (1984)	1979/1980	Eems-Dollard salt marsh (NL) (tidal water)	spring	100—250				
			June—Nov.	40— 80				
			winter	100—250				
			May—August		15— 50			
			Autumn—Winter		80—200			
Jensen et al. (1985)	1981/1982	Skallingen salt marsh (DK) (tidal water)	March	60	7			
			May	1	2			
			July	3	2			
			October	16	5			
			December	41	5			
Helder (1974)	1971/1972	Eastern Waddensea (NL) (coastal water)	March	25	15			
			May	16	5			
			July	2	5			
			October	4	15			
			December	10	17			
Helder (1974)	1971/1972	North Sea (Borndiep, NL) (coastal water)	March	25	2			
			May	15	4			
			July	1	2			
			October	2	4			
			December	10	10			
North America								
Woodwell et al. (1979)	1972—1974	Long Island Sound (tidal water)	March	7	1			
			May	2	3			
			July	1	2			
			October	3	3			
			December	9	7			
Heinle & Flemer (1976)	1972—1974	Patuxent River salt marsh (tidalwater)	March	45	29	36	31	141
			May	15	7	45	51	118
			July	33	18	20	31	102
			October	6	5	21	42	74
			December	30	27	16	32	105
Valiela et al. (1978)	1974	Great Sippewisset Marsh (tidalwater)	March	0.6	1.0		5	
			May	0.2	1.2		2	
			July	0.1	0.7		2	
			October	0.4	3.0		6	
			December	0.5	2.0		–	
Daly & Mathieson (1981)	1976/1977	Crommet Creek salt marsh (tidalwater)	March	4	4			
			May	4	4			
			July	2	5			
			October	7	3			
			December	14	3			
Whiting et al. (1987)*	1979	North Inlet, South Jones (tidalwater)	February	6	3			36
			May	4	2			55
			July	3	3			–
			October	4	6			62

The doubling of biomass and annual production of macrobenthic animals has also been related to eutrophication (Beukema & Cadée 1986).

In an experimental study application of sewage sludge to the Great Sippe Wisset salt marsh in Northern America, plant biomass increased significantly in the upper and lower parts of the salt marsh (Valiela & Teal 1974). The tall form of *Spartina alterniflora* in the lower marsh increased strongly with nutrient enrichment. In the upper marsh, biomass of both *Spartina patens* and *Distichlis spicata* increased in the second year. Thereafter competitive interference between the species led to a reduction of *Distichlis spicata*.

The general conclusion of the field study was: i) increased primary production of grasses and algae; ii) increased production of herbivores, detrivores and iii) increased rate of mineralisation of plant material. In addition, changes in the species composition of salt marsh vegetation occurred. Effects of eutrophication on the salt marsh ecosystem have hardly been investigated, with the exception of a few American studies of nutrient enrichment experiments in the field (Valiela & Teal 1974).

In a recent study Asjes (1988) recorded significant increase of biomass of salt marsh plant species after application of NPK fertilizer. The response of the Spiekeroog salt marsh vegetation with a more nutrient-rich status was less than the Schiermonnikoog salt marsh where the soil is more poor in nutrients. The number of plant species in the salt marshes of the Wadden islands is significantly higher than that of the salt marshes in the South Western Netherlands. This relates to a more muddy, nutrient-rich sediment in the south and more sandy, nutrient-poor soil in the Wadden area. Under conditions of low nutrient levels in the soil probably more species will compete and coexist with low relative growth rates, while in nutrient enriched systems there will be a shift to competition for light, with only a few species becoming dominant. There is some evidence that in eutrophicated salt marshes with no grazing and mowing, species diversity will decrease. In those salt marshes dominance of the "nitrophilous", perennial grass species like *Elymus pycnanthus* has been noted (Bakker 1986, Bakker 1989, Leendertse 1989, Leendertse *et al.* 1989).

A secondary effect of eutrophication on the salt marsh vegetation refers to interactive effects with micro-contaminants such as heavy metals. Teal (1986) suggested that eutrophicated salt marshes, polluted with lead showed increased biomass production and increased uptake of lead from the sedi-

Table III. Nitrogen budget (tidal exchange) (kg N ha^{-1} yr^{-1}) + = import; − = export.

Authors	area	Type of salt marsh		NO$_3^-$ + NO$_2^-$	NH$_4^+$	DON	PN	total
Dankers *et al.* (1984)	Eems-Dollard (NL)	lower salt marsh (Puccinellietum maritimae)		+ 39.5	− 12.6			
Aziz & Nedwell (1986)	Colne Point (UK)	salt marsh with Puccinellia and Halimione		+ 31.0	− 2.4	− 43.0	+ 12.4	− 2.2
Woodwell *et al.* (1979)	Flax Pond (USA)	Spartina alterniflora		+ 12.0	− 21.0			
Valiela & Teal (1979)	GSM (USA)	Spartina alterniflora		− 17.4	− 18.9	− 44.1	− 30.2	− 110.6
Whiting *et al.* (1987)	North Inlet (USA)	Spartina alterniflora		− 5.8	− 47.0			
Daly & Mathieson (1981)	Crommet Creek	Spartina alterniflora		+ 3.2	+ 21.0			
Wolaver *et al.* (1983)	Carter Creek	Spartina alterniflora (mesohaline)		+ 8.4	+ 19.0	+ 48.0	+ 241	+ 316.4
Jordan *et al.* (1983)	Rhode River (USA)	brackish salt marsh	upper	− 0.7	+ 0.9	− 49.0	+ 25.0	− 28.2
			lower	+ 3.4	− 14.0	− 16.0	+ 35.4	+ 8.7
		(Typha angustifolia, Spartina patens)						
Heinle & Flemer (1976)	Gott's marsh (USA)	brackish salt marsh (Spartina cynusiroides, Scirpus olneyi)		− 9.5	− 3.9	− 2.12	− 2.5	− 37.1

ment. Removal of dead plant material with the tidal currents could thus represent a significant input of heavy metals to the coastal waters. A perhaps more important interactive effect of eutrophication relates to increased growth of the root system. Increased root growth was associated with an increase of the redoxpotential, a reduction of insoluble heavy metal-sulphide compounds and increased levels of heavy metals in salt marsh plants (Giblin *et al.* 1986).

Acknowledgements

The research of Drs. P. C. Leendertse is financed by the Dutch Water Ways authorities (RWS-DGW) under the contract POLWAD which is gratefully acknowledged. The authors are much indebted to Ms. D. Hoonhout for typing of the manuscript and to Mr M. B. Meyer for correction of the English text.

References

Asjes, J., 1988. The effect of fertilizer application on interacting *Puccinellia maritima* and *Spartina anglica*. In: M.C.Th. Scholten & H. Schat (eds.) The competitive ability of Puccinellia. TVC-TNO-Delft. pp. 19—21.

Aziz, A. S. A. & Nedwell, D. B., 1979. Microbial nitrogen transformations in the salt marsh environment. In: R.L. Jefferies & A.J. Davy (eds.) Ecological processes in coastal environments. Blackwell, Oxford pp. 385—398.

Aziz, A. S. A. & Nedwell, D. B., 1986. The nitrogen cycle of an East Coast, U.K., saltmarsh. II. Nitrogen fixation, nitrification, denitrification, tidal exchange. Estuarine Coastal Shelf Sci. 22: 689—704.

Bakker, J. P., 1986. Beheersvormen en veranderingen in kweldervegetatie. In: J. Rozema (ed.) Oecologie van estuariene vegetatie. Vrije Universiteit. Amsterdam pp. 87—105.

Bakker, J. P., 1989. Nature management by grazing and cutting. Ph.D. Thesis. Kluwer, Dordrecht, 400 pp.

Beeftink, W. G., 1977. Salt marshes. In: V.J. Chapman (ed.) Wet coastal ecosystems. Elsevier, Amsterdam, pp. 109—155.

Beeftink, W. G. & Rozema, J., 1988. The Nature and functioning of salt marshes. In: W. Salomons, B.L. Bayne, E.K. Duursma & U. Förstner (eds.) Pollution of the North Sea: an assessment. Springer, Berlin, pp. 59—87.

Beukema, J. J. & Cadée, G. C., 1986. Zoobenthos responses to eutrophication of the Dutch Waddensea. Ophelia 26: 55—64.

Buuren, J. T., van, 1988. De Noordzee en eutrofiëring. H$_2$O 21: 591—595.

Cadée, G. C., 1984. Has input of organic matter into the western part of the Dutch Wadden Sea increased during the last decades? Neth. Inst. Sea Res. Publ. Ser. 10: 71—82.

Curtis, P. S., Drake, B. G. & Whigham, D. F., 1989. Nitrogen and carbon dynamics in C3 and C4 estuarine plants grown under elevated CO$_2$ *in situ*. Oecologia 78: 297—301.

Daly, M. A. & Mathieson, A. C., 1981. Nutrient fluxes within a small north temperate salt marsh. Mar. Biol. 61: 337—344.

Danker, N., Binsbergen, M., Zegers, K., Laane, R. & Rutgers van der Loeff, M., 1984. Transportation of water, particulate and dissolved organic and inorganic matter between a salt marsh and the Eems-Dollard estuary, The Netherlands. Estuarine Coastal Shelf Sci. 19: 143—165.

Diggelen, J. van, 1988. A comparative study on the ecophysiology of salt marsh halophytes. Ph.D. Thesis Vrije Universiteit Amsterdam. 208 pp.

Diggelen, J. van, Rozema, J., Dickson, D. M. J. & Broekman, R. A., 1986. β-3-Dimethylsulphoniopropionate, proline and quaternary ammonium compounds in *Spartina anglica* in relation to sodium chloride, nitrogen and sulphur. New Phytol. 103: 573—586.

Eerdt, M. van, 1985. The influence of vegetation on erosion and accretion in salt marshes of the Oosterschelde, The Netherlands. In: W.G. Beeftink, J. Rozema & A.H.L. Huiskes (eds.) Ecology of coastal vegetation. Junk, Dordrecht, pp. 367—373.

Giblin, A. E., Luther III, G. W. & Valiela, I., 1986. Trace metal solubility in salt marsh sediments contaminated with sewage sludge. Estuarine Coastal Shelf Sci. 23: 477—498.

Groenendijk, A. M., 1986. Ecological consequences of a storm-surge barrier in the Oosterschelde: the salt marshes. Ph.D. Thesis pp. 177.

Helder, W., 1974. The cycle of dissolved inorganic nitrogen compounds in the Dutch Waddensea. Neth. J. Sea Res. 8: 154—174.

Heinle, D. R. & Flemer, D. A., 1976. Flows of materials between poorly flooded tidal marshes and an estuary. Mar. Biol. 35: 359—373.

Huiskes, A. H. L. & Rozema, J., 1988. The impact of anthropogenic activities on the coastal wetlands of the North Sea. In: W. Salomons, B.L. Bayne, E.K. Duursma & U. Förstner (eds.) Pollution of the North Sea: an assessment. Springer, Berlin. pp. 455—473.

Jensen, A., Henriksen, K. & Rasmussen, M. B., 1985. The distribution and interconversion of ammonium and nitrate in the Skallingen salt marsh (Denmark) and their exchange with the adjacent coastal water. In: W.G. Beeftink, J. Rozema & A.H.L. Huiskes (eds.) Ecology of coastal vegetation. Junk, Dordrecht, pp. 367—366.

Jordan, T. E., Correll, D. L. & Whigham, D. F., 1983. Nutrient flux in the Rhode River: tidal exchange of nutrients by brackish marshes. Estuarine Coastal Shelf Sci. 17: 651—667.

Leendertse, P. C., 1989. De effecten van verontreiniging (eutrofiëring, microverontreiniging en olie) van het wadden-zeewater op kweldervegetaties. RWS/DGW-Vakgroep Oecologie & Oecotoxicologie VUA. 64 pp.

Leendertse, P. C., Rozema, J. & Janssen, G. M., 1989. Kweldervegetaties door vervuiling in de knel? Waddenbulletin 24: 187—189.

Long, S. P. & Incoll, L. D., 1979. The prediction and measurement of photosynthetic rate of *Spartina × townsendii (sensu lato)* in the field. J. appl. Ecol. 16: 879—891.

Rozema, J., Nelissen, H. J. M., Van der Kroft, M. & Ernst, W. H. O., 1977. Nitrogen mineralization in sandy salt marsh soil of the Netherlands. Z. Pflanzenernähr. Bodenkd. 140: 707—717.

Rozema, J., 1978. On the ecology of some halophytes from a beach plain in the Netherlands. Ph.D. Thesis. Free Univeristy, Amsterdam, 191 pp.

Rozema, J., Bijwaard, P., Prast, G. & Broekman, R. 1985a. Ecophysiological adaptations of coastal halophytes from foredunes and salt marshes. Vegetatio 62: 499—521.

Rozema, J., Otte, M. L., Broekman, R. A. & Punte, H., 1985b. Accumulation of heavy metals in estuarine salt marsh sediment and uptake of heavy metals by salt marsh halophytes. In: Proc. Int. Conf. Heavy metals in the environment. Athens, Greece, pp. 545—547.

Rozema, J., Otte, M. L., Broekman, R. A. & Wezenbeek, J. M., 1986. The uptake and translocation of heavy metals by salt marsh plants from contaminated estuarine salt marsh sediment: possibilities of bioindication. Proc. Int. Conf. Environmental Contamination, Amsterdam pp. 123—126.

Rozema, J., Scholten, M. C. Th., Blaauw, P. A. & Van Diggelen, J., 1988. Distribution limits and physiological tolerances with particular reference to the salt marsh environment. In: A.J. Davy, M.J. Hutchings & A.R. Watkinson (eds.) Plant population ecology. Blackwell, Oxford, pp. 137—164.

Rozema, J. & Otte, M. L., 1989. Zware metalen in sediment en vegetatie van uiterwaarden en andere buitendijkse gebieden, met bijzondere aandacht voor het Noordelijk Deltabekken. Bijdrage studiedag Onderwaterbodems, 11 mei 1988, Ede.

Salomons, W., Bayne, B. L., Duursma, E. K. & Förstner, U. (eds.), 1988. Pollution of the North Sea: an assessment, Springer, Berlin. 687 pp.

Saris, F. J. A. & Duel, H., 1987. Het Haringvliet, het bezinkputje van West-Europa. Landschap 3: 216—232.

Schat, H., Van der List, J. C. & Rozema, J., 1987. Ecological differentiation of the microspecies *Salicornia dolichostachya*
Moss and *Salicornia ramosissima* J. Woods. In: A.H.L. Huiskes, C.W.P.M. Blom & J. Rozema (eds.) Vegetation between land and sea. Junk, Dordrecht pp. 164—178.

Scholten, M., Blaauw, P.A., Stroetenga, M. & Rozema, J. 1987. The impact of competitive interactions on the growth and distribution of plant species in salt marshes. In: A.H.L. Huiskes, C.W.P.M. Blom & J. Rozema (eds.) Vegetation between land and sea. Junk, Dordrecht pp. 270—281.

Sfriso, A., 1987. Flora and vertical distribution of macroalgae in the lagoon of Venice: a comparison with previous studies. Estratto da Giornale Botanico Italiano 121: 69—85.

Teal, J. M., 1986. The ecology of regularly flooded salt marshes of New England: a community profile. U.S. Fish Wildl. Serv. Biol. Rep. 85: 61 pp.

Valiela, I. & Teal, J. H., 1974. Nutrient limitation in salt marsh vegetation. In: R.J. Reimold & W.H. Queen (eds.) Ecology of halophytes. Academic Press, London, pp. 547—563.

Valiela, I., Teal, J. M., Volkmann, S., Shafer, D. & Carpenter, E. J., 1978 Nutrients and particulate fluxes in a salt marsh ecosystem: tidal exchanges and inputs by precipitation and ground water. Limnol. Oceanogr. 23: 798—812.

Valiela, I. & Teal, J. M., 1979. The nitrogen budget of a salt marsh ecosystem. Nature 280: 652—656.

Valiela, I., 1984. Marine ecological processes. Springer, New York. 546 pp.

Whiting, G. J., McKellar, Jr., H.N., Kjerve, B. & Spurrier, J. D., 1987. Nitrogen exchange between a southeastern USA salt marsh ecosystem and the coastal ocean. Mar. Biol. 95: 173—182.

Woodwell, G. M. & Whitney, D. E., 1977. Flaxpond ecosystem study: exchanges of phosphorus between a salt marsh and the coastal waters of Long Island Sound. Mar. Biol. 41: 1—6.

Woodwell, G. M., Hall, C. A. S., Whitney, D. E. & Houghton, R. A., 1979. The Flax Pond Ecosystem: Exchanges of inorganic nitrogen between an estuarine marsh and Long Island Sound. Ecology 60: 695—702.

Wolaver, T. G., Zieman, J. C., Wetzel, R. & Webb, U. L., 1983. Tidal exchange of nitrogen and phosphorus between a mesohaline vegetated marsh and the surrounding estuary in the lower Chesapeake Bay. Estuarine Coastal Shelf Sci. 16: 321—332.

Wolff, W. J., 1988. Impact of pollution on the Wadden Sea. In: W. Salomons, B.L. Bayne, E.K. Duursma & U. Förstner (eds.) Pollution of the North Sea: an assessment, Springer, Berlin. pp. 441—454.

Authors' address
J. Rozema and P. C. Leendertse
Department of Ecology and Ecotoxicology
Vrije Universiteit
De Boelelaan 1087
1081 HV Amsterdam
The Netherlands

Salicornia dolichostachya being collected as a vegetable in France, Belgium, The Netherlands and the United Kingdom.
(Photograph: J. Rozema.)

CHAPTER 10

Economic uses of salt marshes*

A. H. L. HUISKES & W. G. BEEFTINK

1. Introduction

Since prehistoric times, man populates the deltas and estuarine areas of rivers. Ancient cultures often developed in areas such as the Nile delta, the deltas of the Indus and Ganges, of the Hoangho and Yangtsekiang, the deltas of the Song Khoi and the Mekong, the Rhine-Meuse-Scheldt system, the Chesapeake Bay-area.

Settlers through the ages often choose river mouths as a point of access and as a site of their first base. For instance: harbour facilities and religious culture (e.g., the remains of a Nehalennia Temple in the mouth of the Oosterschelde) connected with the salt trade with the British Isles were found in the Rhine-Meuse-Scheldt estuarine area (Stuart 1971).

We do not know if the presence of salt marshes was a prerequisite for the choice of the site for a settlement. It may be assumed that other geographic characteristics played a more important role, like a safe harbouring area and the presence of fresh water. Archeological, paleobotanical and historical research indicates however that man has made use of the salt marshes since time immemorial (Waterbolk 1976).

The extent and intensity of human influence on the salt marshes changed and increased markedly during the ages and especially in recent times adverse effects tend to prevail, resulting in changes in the ecosystem or even physical disappearance of the salt marsh (Beeftink 1975).

2. Gathering of plants and animals

As in many other types of landscape, the earliest known human influence was rather insignificant and was restricted to collecting plants and catching animals. It exists till the present day in a rather unchanged way. Gathering of certain plant species was and is done for a variety of reasons.

The longest tradition perhaps has the reaping of certain plant species for human consumption. In Western Europe species like *Salicornia spec.*, *Aster tripolium*, *Plantago maritima*, *Triglochin maritima* and *Atriplex prostrata* are known as vegetables even at present. *Salicornia* species are eaten as a vegetable in France, Belgium and the south-west Netherlands, strangely enough not in the northern part of the country or in northern Germany. Calkoen (1897) mentions this difference and states that the species is used in the northern provinces as animal fodder only. One may assume that *Salicornia* has been collected as a vegetable for many centuries. Today with the crave for new vegetables, species like *Salicornia* gain in interest and attempts are made to grow them commercially in greenhouses, so far not in economically interesting quantities.

The same situation exists for *Aster tripolium*.

* Communication nr. 444, Delta Institute for Hydrobiological Research.

J. Rozema and J. A. C. Verkleij (eds.), Ecological Responses to Environmental Stresses, 104—111.

This species is also eaten in France, Belgium and the south-west Netherlands. There is even some export from the Netherlands to Belgium and France (Arts & Bouma, 1982). Hondius (1621) mentions the species already as a vegetable.

Triglochin maritima is eaten in northern Germany (Warburg 1921). *Atriplex* species were used as a seed crop in prehistoric times (Van Zeist 1974).

Gathering the species mentioned has not changed over the centuries. The plants are collected by local people at low tide for their own use or to be sold in the market or door-to-door. The impact on the salt-marsh vegetation is small. Paalvast and Beeftink (unpubl.) found that collecting *Salicornia* for consumption did not have an effect on the population in the salt marsh when assessed on a year to year basis. The remaining plants apparently produce enough seeds to maintain a relatively constant population size. They estimate that 40% of the total seed production is more than sufficient.

For *Aster tripolium* the danger of local extinction is even more remote as this perennial species shows a vigorous regrowth after cutting (Braber 1983).

In Egypt *Juncus*-species, especially *J. rigidus* and *J. acutus,* have been used for centuries. In pharaonic times the species, mainly *J. rigidus,* were used as brushes for painting hieroglyphs (Täckholm 1976). They were and are used for making mats and baskets (cf., the story of Mozes in the Bible book Exodus; records by Abu Hanifa (875) and Ibn El-Beitar (1248) cited in Zahran & Wahid (1982)). Recently investigations are made about the use of the species in paper production (Zahran & Wahid 1982). *J. rigidus* is also used in oriental medicine as diuretic and as a remedy for diarrhoea. In western Europe *Aster tripolium* was gathered for the same purpose (Dodonaeus 1554).

Other salt-marsh plants have been used for medical purposes: *Salsola tetrandra* (Zahran & Negm 1973), *Cochlearia officinalis* and *Althaea officinalis*. *Spartina alterniflora* was used for roof-thatching (Teal & Teal 1969). *Scirpus lacustris subsp. tabernaemontani* has been used for mats, seats of chairs and other matting; the species has been deliberately planted for this purpose in the estuarine area of the Rhine (Haringvliet). Various species are reported to be used for packing material (Queen 1977). *Artemisia maritima* was used in the province of Zeeland for its smell in linen closets and as a flea repellent. *Limonium vulgare* is presently collected for ornamental purposes in bouquets; in order to prevent local extinction certain municipalities have special bylaws for the protection of the species (e.g., the municipality of Zierikzee).

Animal species are numerous in salt marshes both in vegetated parts as well as in the tidal creeks. Few species of terrestrial origin inhabit the marsh though, and of these even less are used by man.

There are no mammal species that inhabit the marsh permanently. They are there mainly at low tide and some of them were and are interesting to man. Racoons, Muskrats, Mink and Nutria's have been caught for their furs in American salt marshes (Queen 1977). Muskrat, Stoat, European polecat and Mole have been caught for the same purpose in European salt marshes (Ranwell 1972). Rabbit and hare are hunted or snared as a food source. Hunting in salt marshes is still a widespread activity but nowadays mainly for ducks and geese (Long & Mason 1983).

Much more abundant are the animal species of marine origin. Fish, shrimps, bivalves, lobsters and crabs are caught for food; worms for bait; snails (*Littorina* and *Nucella*) for dyes (purple) (Brühl 1929). For some of these species, e.g., Sole and Shrimp the salt-marsh creeks play an important role as a nursery (De Veen et al. 1979; Queen 1977). Estuaries bordered by marshes and coastal waters adjacent to estuarine-marsh systems are among the worlds most productive waters (Queen 1977).

Fishing has been one of the oldest activities in the salt marshes. Quite often techniques are used that did not change for many centuries. Sponselee & Buise (1979) describe the use of the, now illegal, "elger", a seven-tined pike for spearing flounder. This tool resembles closely Neptunes trident. Another way of catching flounder, which does not seem to be very advanced, is the socalled "flounder-treading", practised in the salt marshes of Saeftinge, S.W. Netherlands. Wading barefoot through the creeks, the fisherman carefully touches the creekbed. As soon as he feels something, a quick reaction of fixing the flounder under the foot and grabbing it with the hands may, according to Sponselee & Buise (1979), result in catches of 10–20 kilogrammes per tide.

Collecting bivalve species has evolved nowadays into certain forms of aquaculture (e.g., Mussels and Carpet shells).

2. Grazing and hay-making

By far the most widespread use of salt marshes has been open range grazing with domestic animals (Ranwell 1972). Salt marshes in Europe are mainly grazed by sheep, but also by cattle and horses and domestic geese (Schleswig-Holstein). In the USA also pigs are sometimes grazed in the salt marsh.

This activity is after gathering the oldest use man has made of the salt marshes. Behre (1979) and

Table I. Influence of grazing by cattle and sheep on salt-marsh plant species in the SW-Netherlands (Adapted from Beeftink 1977).

	Influence of grazing	
	moderate grazing	intensive grazing
Salicornia spp.	+	+
Puccinellia maritima	+	+
Puccinellia distans	+	0
Spergularia marina	+	+
Juncus gerardii	+	0
Suaeda maritima	0	0
Aster tripolium	0	−
Triglochin maritima	0	0
Limonium vulgare	0	−
Plantago maritima	0	0
Spergularia media	0	0
Artemisia maritima	0	−
Halimione portulacoides	0	−
Spartina anglica	0	−
Phragmites australis	0	−
Elymus pycnanthus	0	0
Agrostis stolonifera	0	0
Festuca rubra	0	0
Glaux maritima	0	0
Armeria maritima	0	+
Scirpus maritimus	−	−
Atriplex spp.	−	−

Körber-Grohne (1967) proved prehistoric grazing and hay-making in northern Germany. In The Netherlands records of agricultural use are scarce, but there have been findings of Neolithic settlements in the higher salt-marsh areas (Waterbolk 1976). From Roman times many settlements in the higher marsh are known.

Unlike gathering, grazing and hay-making have a profound effect on the salt-marsh vegetation and also on the animal community in the marsh. In an ungrazed marsh a few plant species dominate the vegetation zones arranged in accordance with the flooding frequency (Beeftink 1965). In the United States salt marshes have a rather uniform vegetation generally of few species, e.g., *Distichlis spicata, Spartina alterniflora, S. patens* and *Salicornia virginica. Distichlis* and *Salicornia* under ungrazed circumstances are found in the higher marsh areas, while under grazed circumstances these species occur also in the lower areas of the marsh (Shanholtzer 1974). Mean plant size however is smaller in grazed areas (Shanholtzer 1974; Turner 1987).

In west-European salt marshes plant species diversity is greater in general than in American salt marshes. Although the vegetation is also arranged according to the flooding frequency, the diversity of the geomorphology of the marsh induces a more diverse vegetation pattern. Ungrazed marhes have a less intricate pattern than grazed marshes (Bakker & Ruyter 1981; Jensen 1985). Bakker (1985) showed for the salt marsh of the Dutch Wadden island of Schiermonnikoog an increase in the number of species per unit area of more than 100 per cent due to grazing. Table I summarizes effects of grazing on salt marsh species. In general it can be said that grazing depresses species that form large monocultures and that it increases diversity. Because of this phenomenon grazing is nowadays often applied to salt marshes as a management measure, especially if one wants to host geese. Geese prefer a short turf resulting from summer grazing by cattle and sheep (Ebbinge & Canters 1973; Ebbinge et al. 1981). In turn, the geese graze the short turf thus increasing the diversity even further.

Grazing is not just the removal of aboveground parts. Beeftink (1977a) shows that consumption is just one factor of the grazing process. Another is the destruction of the turf in certain areas which will favour a secondary succession. Bakker and Ruyter (1981) report the increase of species like *Sagina maritima* and *Parapholis strigosa* which may indicate this process. Bakker (1987) indicates that grazing has a greater diversifying effect on the vegetation than mowing. He also showed an increase in cover of species from the lower marsh in higher marsh areas, indicating the recolonisation of temporary open areas in a formerly closed vegetation and also indicating higher soil salinities in relation to the effect of trampling of the soil by cattle and sheep. Differences in grazing pressure have different effects on the species composition in the salt marsh. Dijkema (1984) reported that moderate grazing favoured, e.g., *Limonium vulgare* and *Artemisia maritima*, while heavy grazing has the opposite effect.

Although grazing is favoured as a management measure, it has its drawbacks. It has to be strictly employed since terminating or altering the grazing will have immediate effects, mostly unfavourable effects in the sense of species diversity. Jensen (1985) showed an increase in vegetation cover, especially of species like *Festuca rubra* and *Halimione portulacoides*, while species like *Salicornia spec.*, *Suaeda maritima* and *Glaux maritima* disappeared. This indicates that the process described by Bakker (1987) is reversed when grazing ceases.

Dijkema (1984) and Bakker (1987) give figures on the amount of salt marshes in western Europe being grazed (Table II). The area of grazed salt marsh however is decreasing — about 10,000 ha of salt marsh has been abandoned over the last 50 years (Bakker 1987) —, especially in the southwest part of The Netherlands with the aforementioned result of decrease in species diversity (Bouwman, pers. comm.).

The intensity of grazing or grazing pressure may have an impact on the vegetation as well. Bakker (1987) states that a high grazing pressure may not only be detrimental to the vegetation but also to birds that use the vegetation in one way or another (e.g., for cover). Table I indicates the differences in moderate and heavy grazing on salt-marsh species. Beeftink (1977b) elaborates on this subject. Table IV gives figures indicating the optimum grazing pressure in order to maintain a maximum diversity. A negative effect of grazing is the decrease in diversity of invertebrate fauna (Table III, Dijkema 1984).

Hay-making is believed to be as old an activity as grazing. Bakker *et al.* (1983) show that it has not the same effect on the salt marsh vegetation as grazing. Although hay-making for agricultural purposes is not important anymore, it is increasingly used in salt marsh management as a substitute for grazing (Dijkema 1984).

In North America hay-making is still important to farmers and is common in *Spartina* marshes (Ranwell 1972).

3. Destructive use of salt marshes

The aforementioned human activities in the salt marshes are perturbations close to grazing by wild animals and disturbance by the tides and wave action in stead of destructions.

However man has used salt marshes in a destructive way and is still doing this despite of the threatened existence of these areas.

Table II. Salt marsh area in hectares in some countries in NW-Europe and its use. (adapted from Dijkema 1984 & Bakker 1987).

Country	Area (ha)	Protected (e.g. nature reserve)	Grazed or Mown	Not used	Unknown
Denmark	19,500	9,300	12,800	200	6,500
FRG	19,000	7,400	18,100	500	600
Netherlands	12,800	10,500	9,500	3,300	0
Belgium	400	200	200	200	0
U.K.	37,500	17,000	19,800	15,300	2,500
France	14,400	1,000	8,100	6,300	0

3.1. Aquaculture

Certain forms of aquaculture are not destructive to salt marshes. A logical extension of the gathering of shellfish, especially bivalves, is the cultivation of these species by improvement of the settling substrate or by collecting larvae or young animals and transporting them to sites which are favourable for growth.

Although in The Netherlands the shellfish culture is executed almost entirely on the tidal flats and not in the salt-marsh creeks, in other countries marsh creeks play an important role in this respect. Oyster beds in Louisiana (USA) or in the Sado estuary in Portugal lay adjacent to salt marshes or in the salt-marsh creeks. Carpet shells (*Tapes (Venerupis) spec.*) are cultivated in the salt-marsh creeks in the South of Portugal.

As soon as aquaculture is performed in a more intensive way the negative effects on the salt-marsh ecosystem may also increase, especially by the increase in area. Fish culture for instance needs rearing ponds, dikes, weirs, channels, etcetera. Although the culture itself may not harm the ecosystem of the marsh, the construction of the structures needed will occupy salt-marsh area. The culture is dependent on a natural process, that of the production of organic nutrients by marsh plants, which come available by a detrital or herbivoral food chain (Queen 1977). There are also cases known in which the fish eat the salt-marsh plants directly (Lewis & Peters 1984).

3.2. Salt-winning

Salt has played and still plays an important role in the history of mankind. The winning of sea salt has therefore been a very important industry for centuries. In the south-west part of The Netherlands salt was won by burning peat, formed in Neolithic times and soaked with sea water since periods of sea level transgression. In Roman times thriving communities existed in the coastal plain of the Low Countries deriving their wealth from the salt winning and the salt trade. This way of winning was most destructive as large quantities of peat were processed increasing the vulnerability of the land with respect to storm floods. The salt winning ceased in the 15th century when salt winning around the Mediterranean became profitable. This way of winning by means of salinas (saltpans, evaporation basins) is profitable only in those countries where evaporation is high. The method is still widely used in the South and South-West of Europe (up to South Brittany). In most cases the salinas are situated close to the sea and frequently in salt-marshes since Roman times already (cf., the Via Salaria (Salt Way) from Ostia to ancient Rome).

Although a number of European countries exploit rock salt nowadays, salt winning in salt pans is

Table III. Effects of grazing on the species diversity of the invertebrate fauna in a *Puccinellia*-vegetation in August. (modified after Dijkema 1984).

	number of species found	
	ungrazed	grazed
Spiders (Araneae)	8	3
Crickets (Cicadidae)	4	2
Lice (Aphididae)	1	2
Gall wasps (Cynipidae)	1	0
Chalcid wasps (Chalcidoidea, Proctotrupoidea)	9	2
Parasitic wasps (Ichneumonoidea)	2	0
Beetles (Coleoptera)	1	1
Other insects	6	2
Crane flies (Tipulidae)	4	1
Non-biting midges (Chironomidae)	2	0
Biting midges (Ceratopogonidae)	4	1
Gall midges (Cecidomyiidae)	2	0
Other Nematocera	1	0
Robber flies (Empididae)	0	0
Long-legged flies (Dolichopodidae)	10	4
Other Brachycera	3	2

Table IV. Recommended grazing pressures in order to maintain maximum vegetation diversity (in maximum numbers of animals per hectare) in salt marshes. (Data from Beeftink & Daane, unpubl.)

	Salt marsh	Brackish marsh	Beach plain
Sheep	2	3	0.5—1.0
Cattle	0.33	0.5	0.25

still an important industry in Mediterranean countries.

Excavating salt pans in salt marshes is harmful for most plant and animal species. The edges of the pans and the dikes between them are often the only salt-marsh habitats left.

3.3. Open cast mining

A very old mining activity in Dutch salt marshes was the winning of peat for salt (see above) but also for fuel. In the SW-Netherlands, where only the dunes and other higher areas were covered with woody species, peat was an important source of fuel for heating and cooking. Peat mining for fuel lasted till the 1950s (Sponselee & Buise 1979). Peat mining had very adverse effects on salt marshes and on the land behind it. The function of the salt marshes as a breakwater was weakened and some of the flood disasters in the 13th–15th century had an extra serious impact because of the peat mining. It was for safety reasons that in 1515 the emperor Charles V prohibited this activity in the Low Countries. People were allowed to plant trees as a substitute fuel source, which had a profound effect on the landscape of the area (Ovaa 1975).

Clay mining has been an activity in the salt marshes in north-western Europe for centuries, especially for reinforcement of the dikes (Beeftink 1977b). For the same reason turf cutting was practised (Kamps 1962). Effects of peat mining, clay winning and turf cutting on salt-marsh vegetation are largely unknown.

Experiments carried out by Beeftink and co-workers (unpubl.) revealed that turf cutting and clay winning throw back the vegetation to earlier developmental stages proportionate to the impact of the interference. Secondary succession towards the original state of the vegetation takes many years and is dependent on the regeneration rate of the substrate through sedimentation and the course of developmental processes in the new top soil.

3.4. Other uses

In many countries salt marshes have been reclaimed, mainly for agricultural purposes, al-

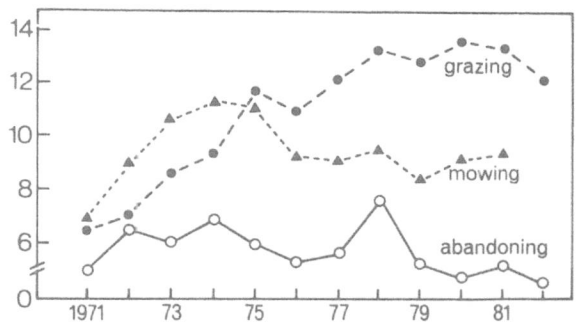

Fig. 1. Changes in mean species diversity with grazing, mowing and prolonged abandoning in various communities at the middle and high salt marsh on the island of Schiermonnikoog (mean of five permanent quadrats of $2 \times 2\,m^2$). (After Bakker 1987)

though they are also favourable locations for airports nowadays. This is not so much using the salt marsh, but rather land use. After embanking salt marshes for reclamation, only remnants of the original plant and animal life, added with some species characteristic of more brackish non-tidal environments might be found in former creek beds and other bottom land left (Beeftink 1977b). In The Netherlands the increase of salt-marsh area was stimulated, for reclamation purposes, by planting *Spartina anglica* imported from Poole Harbour (U.K.) (Verhoeven 1929) and by ditching. The introduction of this species and its subsequent spontaneous dispersal throughout coastal and estuarine areas, seriously disturbs the original zonation and succession, especially to the detriment of the pioneer stages with a vegetation of *Salicornia* species, *Spartina maritima*, *Puccinellia maritima*, *Scirpus maritimus* and specific salt-marsh algae (Beeftink 1977a, b).

Studies have been performed to assess the possible use of salt marshes in waste-water treatment. Various criticisms have been raised against this use. The scientific data for some of the marsh processes are so little that the monetary values that have been assigned to the marshes in this respect seem as yet rather unrealistic (Queen 1977).

In recent years the salt marshes gain interest as wildlife area (especially since the Wetlands convention of Ramsar in 1971).

Recreation pressures on the salt marshes are increasing. The marshes will never serve as areas

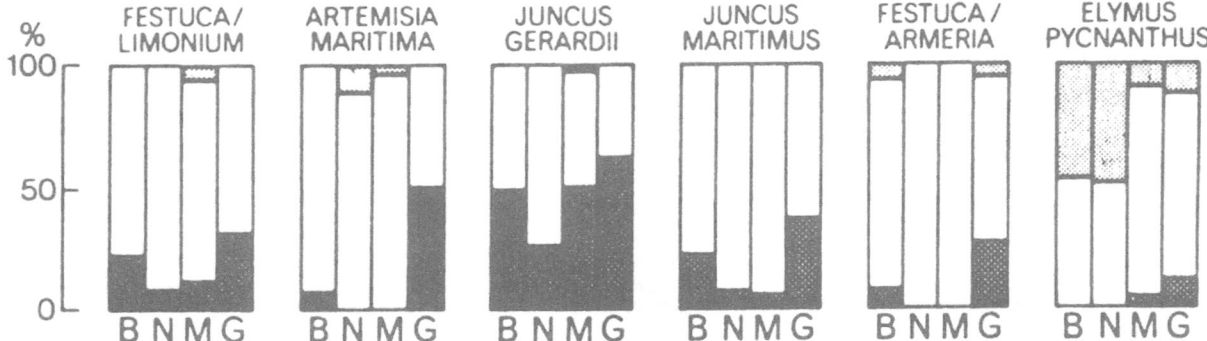

Fig. 2. The amount of low marsh, middle marsh and high marsh species and dune species in six vegetation types, 10 years after the start of the treatments (B = at the start; N = No treatment, M = Mowing, G = Grazing).

The cross hatched area represents the low-marsh species, the open area, the middle- and high marsh species and the shaded area, the dune species (after Bakker 1987).

for mass recreation. The salt-marsh ecosystem is too vulnerable. The salt marsh soil is rather weak and the winter buds of some species lie so close to the surface that trampling might be very harmful to the vegetation. The creeks may be used for sailing although this disturbs wildlife, especially birds in the nesting season.

4. Conclusion

Paul Gallico, in his novel *The Snow Goose,* describes salt marshes as "desolate, utterly lonely, and made lonelier by the calls and cries of the wild birds that make their homes on these marshes". This may suggest that these areas are generally avoided by human beings and deemed inhospitable. Yet man has made use of salt marshes for many centuries in a number of different ways as is shown in the present paper. It is however the influence of the tides that prevents our species — and many other species — from taking permanent possession of the salt marshes and preserves them in a more or less pristine state.

Acknowledgements

This paper is dedicated to Prof. Dr. W. H. O. Erst, whom the authors appreciate as a colleague and a friend.

We thank Prof. Dr. C. H. R. Heip for his comments on the draft text.

References

Arts, S. F. W. & Bouma W. J., 1982. Zeegroente. Student report, University of Groningen, 62 pp.

Bakker, J. P., 1985. The impact of grazing on plant communities, plant populations and soil conditions on salt marshes. Vegetatio 62: 391—398.

Bakker, J. P., 1987. Pflegeformen und Änderungen in der Salzwiesenvegetation. In: N. Kempf, J. Lamp & P. Prokosch (eds.) Salzwiesen: Geformt von Küstenschutz, Landwirtschaft oder Natur? pp. 215—241, Umweltstiftung WWF — Deutschland, Bonn.

Bakker, J.P. & Ruyter, J. C., 1981. Effects of five years of grazing on a salt-marsh vegetation. Vegetatio 44: 81—100.

Bakker, J.P., de Leeuw, J. & van Wieren, S. E., 1983. Micropattern in grassland vegetation created and sustained by sheep grazing. Vegetatio 55: 153—161.

Beeftink, W. G., 1965. De Zoutvegetatie van ZW-Nederland beschouwd in Europees verband. Meded. Landb. Hogeschool Wageningen 65: 1—167.

Beeftink, W. G., 1975. The ecological significance of embankment and drainage with respect to the vegetation of the South-West Netherlands. J. Ecol. 63: 423—458.

Beeftink, W. G., 1977a. The coastal salt marshes of western and northern Europe: An ecological and phytosociological approach. In: V.J. Chapman (ed.) Wet Coastal Ecosystems pp. 109—156, Elsevier, Amsterdam.

Beeftink, W. G., 1977b. Salt marshes. In: R.S.K. Barnes (ed.) The Coastline, pp. 83—121, Wiley and Sons, London.

Behre, K. E., 1979. Zur Rekonstruktion ehemaliger Pflanzengesellschaften an der deutschen Nordseekuste. In: O. Williams & R. Tuxen (eds.) Werden und Vergehen von Pflanzengesellschaften, pp. 181—214, Cramer, Vaduz.

Braber, L., 1983. Opzet en eerste resultaten van een proefveld voor bedrijfsmatige teelt van Zeeaster. Student Report, Higher Agricultural School, Dordrecht, 30 pp.

Brühl, L., 1929. Purpur und Sepia. Die Rohstoffe des Tierreichs. Vol 2: 358—379 and 380—396.

Calkoen, H. W., 1897. Plantenatlas, Sijthoff, Leiden, 223 pp.

De Veen, J. F., Boddeke, R. & Postuma, K. H., 1979. Tien jaar kinderkamer opnames in Nederland I. Het Zeeuwse Estuarium. Visserij 32: 3—23.

Dodonaeus, R., 1554. Cruydeboeck. Antwerpen.

Dijkema, K. S., 1984. Salt marshes in Europe. Council of Europe, Strasbourg, 178 pp.

Ebbinge, B. S. & Canters, K., 1973. The Brandgans en zijn overwinteringsgebied. Student report, State Univ. Groningen, 82 pp.

Ebbinge, B. S., Fog, M. & Prokosch, P. 1981. Brent Goose (Branta bernicla). In: C.J. Smit & W.J. Wolff (eds.) Birds of the Wadden Sea, pp. 28—37, Balkema, Rotterdam.

Hondius, P., 1621. Dapes inemptae oft de Mouffe schans dat is de soeticheydt des buytenlevens, vergeselschapt met de boucken. Leiden.

Jensen, A., 1985. The effect of cattle and sheep grazing on salt-marsh vegetation at Skallingen, Denmark. Vegetatio 60: 37—48.

Kamps, L. F., 1962. Mud distribution and land reclamation in the eastern Wadden shallows. Rijkswaterstaat comm. 4, The Hague, 73 pp.

Korber-Grohne, U., 1967. Geobotanische Untersuchungen auf der Feddersen Wierde 1. Steiner Verlag, Wiesbaden, 357 pp.

Lewis, V. P. & Peters, D. S., 1984. Mehaden — a single step from vascular plant to fishery harvest. J. Exp. Mar. Biol. Ecol. 84: 95—100.

Long, S. P. & Mason, C. F., 1983. Saltmarsh Ecology. Blackie, Glasgow, 160 pp.

Ovaa, I., 1975. De zoutwinning in het zuidwestelijk zeekleigebied en de invloed daarvan op het landschap. Boor en Spade 19: 54—68.

Queen, W., 1977. Human uses of salt marshes. In: V.J. Chapman (ed.), Wet coastal ecosystems, pp. 363—368, Elsevier, Amsterdam.

Ranwell, D. S., 1972. Ecology of salt marshes and sand dunes. Chapman and Hall, London, 258 pp.

Shanholtzer, G. F., 1974. Relationship of vertebrates to salt marsh plants. In: R.J. Reimold & W.H. Queen (eds.) Ecology of Halophytes, pp. 463—474, Academic Press, New York.

Sponselee, G. M. P. & Buise, M. A., 1979. Het verdronken land van Saeftinghe. Duerinck-Krachten, Kloosterzande, 135 pp.

Stuart, P., 1971. A new temple of Nehalennia. Oudheidk. Meded. Rijksmus. Oudh. Leiden 52: 76—78.

Täckholm, V., 1976. Ancient Egypt, landscape, flora and agriculture. In: J. Rzoska (ed.), The Nile, biology of an ancient river, pp. 51—54, Junk Publishers, The Hague.

Teal, J. & Teal, M., 1969. The life and death of a salt marsh. Little, Brown and Co., Boston, Mass, 274 pp.

Turner, M. G., 1987. Effects of grazing by feral horses, clipping, trampling, and burning on a Georgia salt marsh. Estuaries 10: 54—60.

Van Zeist, W., 1974. Palaeobotanical studies of settlement sites in the coastal area of The Netherlands. Palaeohistoria 16: 223—371.

Verhoeven, A. G., 1951. Bevordering van landaanwinning in en inpoldering van een gedeelte van het Zuider-sloe. Voordrachten Kon. Inst. voor Ingenieurs 36: 579—604.

Warburg, O., 1921. Die Pflanzenwelt, vol 2., Bibliographisches Institut, Leipzig, 544 pp.

Waterbolk, H. T., 1976. Oude bewoning in het Waddengebied. In: J. Abrahamse, W. Joenje & N. Van Leeuwen-Seelt (eds) Waddenzee, pp. 211—223, Land. Ver. Beh. Waddenzee, Harlingen en Natuurmonumenten, 's-Graveland.

Zahran, M. A. & Negm, S. A., 1973. Ecological and pharmacological studies on *Salsola tetrandra* Forsk. Bull. Fac. Sci. Mansoura Univ. Egypt 1: 67—75.

Zahran, M. A. & Wahid, A. A., 1982. Halophytes and human welfare. In: D.N. Sen & K.S. Rajpurohit (eds.) Contributions to the ecology of halophytes, pp. 235—257, Junk Publishers, The Hague.

Authors' Address
A.H.L. Huiskes & W.G. Beeftink
Delta Institute for Hydrobiological Research
Vierstraat 28
4401 EA Yerseke
The Netherlands

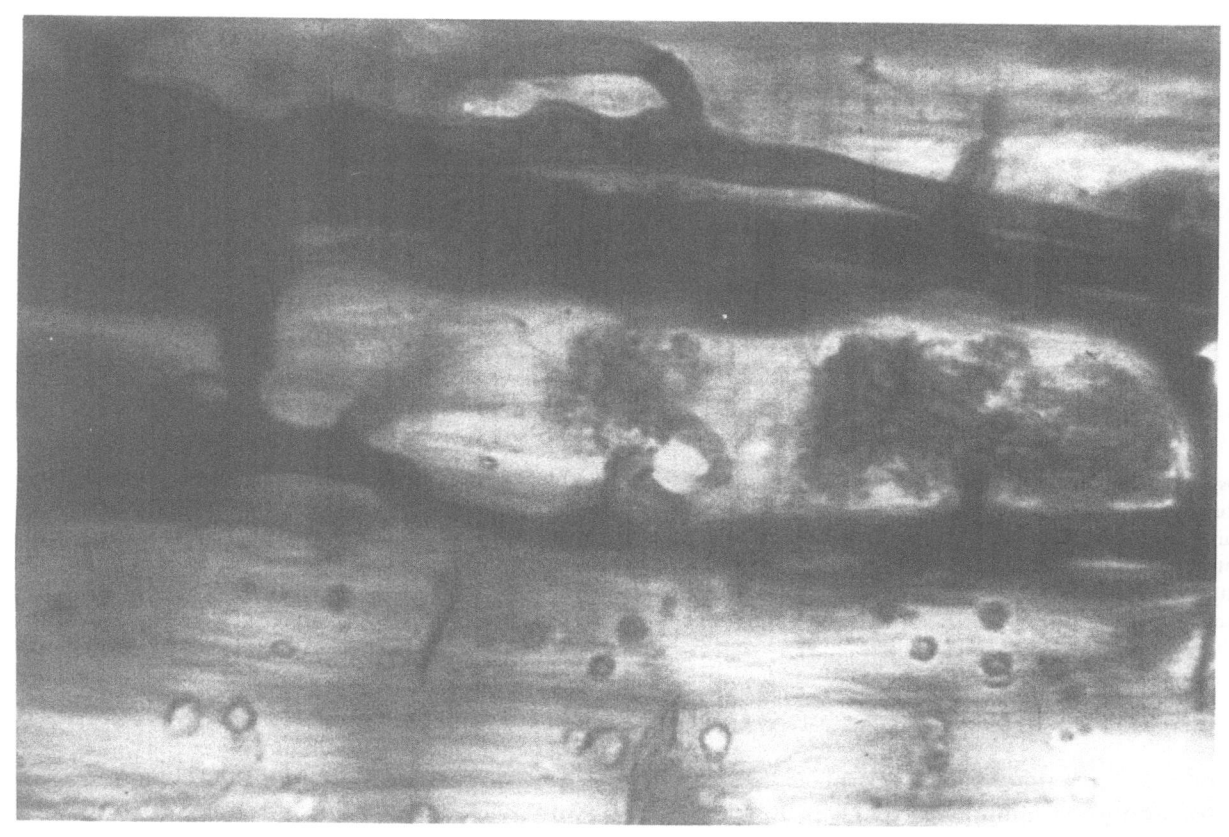

VA mycorrhizal infection in *Aster tripolium* ($\times 570$), with arbuscles and hyphae. (Photograph: W. van Duin.)

CHAPTER 11

Occurrence and function of mycorrhiza in environmentally stressed soils

W. E. VAN DUIN, W. A. J. GRIFFIOEN & J. H. IETSWAART

Abstract. Morphological characteristics are given for the different types of mycorrhiza, and their occurrence through the plant kingdom is surveyed, considering some geographical and ecological aspects. Attention is paid to the functioning of mycorrhizas, particulary in a number of stress situations, with an emphasis on salty and heavy metal contaminated soils. For natural populations it can be concluded that in moderately stressed situations a lot of seed plants grow profited by the presence of mycorrhizas, while in heavily stressed situations only a few species grow, having (nearly) no mycorrhizas. Finally it is concluded that fairly much is known on the function of mycorrhizas for crop plants, but very little for natural populations.

1. Introduction

Symbiotic fungi that form mycorrhizas, have existed for about 400 million of years, probably since the first occurrence of land plants in the Devonian (Stubblefield *et al.* 1987) and are found worldwide. They are well adapted to variations in several climatological factors, e.g., temperature and water, and soil conditions, e.g. aeration, pH, nutrient availability (deficiency and toxicity), and salinity.

Over the last thirty years research interest in mycorrhizas has increased enormously. In the 1950's a few workers were concerned with ectomycorrhizas, while very little was known about vesicular-arbuscular (VA) mycorrhizas, then still called "endomycorrhizas". Nowadays, the many studies on ectomycorrhizas and VA mycorrhizas, with approximately one half of both relating to agriculture, fruit-culture and forestry, show that mycorrhizology is an important part of both scientific and applied ecology, with its own methods and techniques (Schenck 1984).

2. Systematics and morphology

Mycorrhizas are symbiontic combinations of roots of vascular plants or mosses, the phytobiont, and hyphae of fungi, the mycobiont. The phytobiont will belong to the Anthophyta, Coniferophyta, Pteridophyta, Bryophyta, Anthocerotophyta or Hepatophyta, and commonly contains photosynthetic pigments. These pigments are absent in the mycobiont which will belong to one of the Amastigomycota groups: Zygomycotina, Ascomycotina, Basidiomycotina or Fungi Imperfecti. Of the Zygomycotina only a few genera are involved, each rich in species, while the Ascomycotina, Basidiomycotina and Fungi Imperfecti provide mycobionts from a range of genera. The representatives of these groups are often confined to one or a few phytobiont species, whereas the Zygomycotina taxa are not specifically bound. The Zygomycotina show very few cross-walls (aseptate), while the other groups possess septate hyphae. Under natural conditions the mycobiont representatives from the Ascomicotina and Basidiomycotina may produce above-ground fruit-bodies, while the Zygomycotina species produce spores below-ground, singly or

J. Rozema and J. A. C. Verkleij (eds.), Ecological Responses to Environmental Stresses, 114—123.

in simple sporangia. The function of the fungi involved in mycorrhizal infection varies from a symbiontic one to a saprophytic or even parasitic one.

A gradual transition in characteristics can be seen from ectomycorrhizas via ectendomycorrhizas to VA mycorrhizas. Ectomycorrhizas are characterized by the external mycelium forming a thickly interwoven layer of hyphae, called a sheath or mantle. The hyphae also grow in the epidermis and cortex of the phytobiont, in a net-like way known as Hartig-net, but remain intercellulary. The mycobiont will belong commonly to the Basidio- or Ascomycotina and the phytobiont to the Conifero- or the woody Anthophyta.

Ectendomycorrhizas are similar to ectomycorrhizas, but the sheath is less developed and the hyphae enter the phytobiont cells. They are nearly always found in young trees. In combination with Ericales four mycorrhizas types are discerned: arbutoid, in *Arbutus* and a few other Ericaceae genera, pyroloid, in Pyrolaceae, monotropoid, in Monotropaceae, and ericoid mycorrhizas, in most of the Ericales, particularly in Ericaceae and Empetraceae. In arbutoid mycorrhizas the mycobiont is a representative of the Basidiomycotina, and is similar to that in the ectomycorrhizas, with the exception that the hyphae penetrate the cells. In the pyroloid mycorrhizas Basidiomycotina as well as Ascomycotina are involved. With the exception of a Hartig-net of variable thickness, the situation is similar to the arbutoid type. In the monotropoid mycorrhiza type the well-developed sheath is connected by hyphal strands to the ectomycorrhiza of trees, thus the chlorophyll-free plant probably benefits from the photosynthetic products of the tree, as does the mycobiont. This symbiosis consisting of three types of organisms may also occur with other achlorophyllous plants or plant stages, e.g., fern prothallia. In ericoid mycorrhizas normally an Ascomycotina fungus is the mycobiont, though Basidiomycotina sometimes occur. No sheaths are formed, but the hyphae enter the cells of the phytobiont.

Orchidoid mycorrhizas are vitally important to the phytobiont in the achlorophyllous stage, just after the germination. It might be that the mycobiont behaves more or less as a parasite during the adult stage of the orchid plant. The Basidiomycotina fungi involved are otherwise saprophytes or parasites capable of digesting cellulose. In other respects the situation is similar as in ericoid mycorrhizas.

The group, formed by the VA mycorrhizas, is found mainly in the cryptogams mentioned and in the herbaceous angiosperms, but also occurs in trees and shrubs. The mycobiont usually belongs to the group of lower fungi called Zygomycotina, in which quite a number of parasites occur. A major part of the mycelium grows in the cortex of the roots of the phytobiont, while the outer hyphae do not form a sheath. Inside the phytobiont hyphae grow inter- as well as intracellulary, forming two structures from which the name of this mycorrhiza is derived: tree-like ramifications called arbuscules and bladder-like structures called vesicles. The first are formed only intracellulary and the latter intra- as well as intercellulary. Both arbuscules and vesicles play an important role in the transfer of nutrients.

3. Occurrence of mycorrhizas

The main body of knowledge on mycorrhizae originates from studies in temperate regions, while the (sub)tropical areas are nearly unexplored. This is particularly true of checklists, for one ecological biotope (Ernst *et al.* 1984), as well as ones for a particular geographical region. The comprehensive list covering the British Flora is of the latter type, though it refers to many publications from the European continent and other temperate regions of the world (Harley & Harley 1987).

Currently it is estimated that in circa 90% of the terrestrial plant families mycotropic species are present. In the Chenopodiaceae, Cruciferae and Cyperaceae, families previously thought to be nonmycorrhiza, mycorrhizal species have been found (Harley & Harley 1987). A survey of the occurrence of mycorrhizas in different plant groups is given in Table I.

The ectomycorrhizas in particular have been found in trees of the temperate region. VA mycorrhizas are much less confined to one plant group or

Table I. Presence of mycorrhizas in representatives of the British flora (compiled from Harley & Harley, 1987) va = vesicular-arbuscular mycorrhiza, ecto = ectomycorrhiza, ectendo = ectendomycorrhiza, endo = endomycorrhiza.

Group	Family name/ Number of families	Number of species			% of species		Mycorrhizal type	Note(s)
		Represented in British flora	Examined	Mycorrhizal	of number examined	of total nr. represented		
microphyllous ferns	4	23	21	14	67	61	va	no mycorrhiza in hydrophytic Isoetaceae
macrophyllous ferns	15	56	40	30	75	54	va	endo in Hymenophyllaceae, no mycorrhiza in ferns on rocks or walls and hydrophytic Azollaceae and Marsiliaceae
gymnosperms	3	14	14	14	100	100	ecto, extendo, va	
woody a- and choripetalous angiosperms	23	83	70	68	97	82	va, ecto	ectendo in Betulaceae
choripetalous herbaceous dicots (incl. Rosaceae)	10*	586	287	204	71	35	va (ecto)	ecto particularly in a number of woody representives
sympetalous herbaceous dicots	8**	400	227	200	88	50	va (ecto)	
Ericales	Monotropaceae	1	1	1	100	100	monotropoid	
	Pyrolaceae	5	5	5	100	100	pyroloid	seems to be different from other Ericales
	others: 3	27	23	23	100	85	ericoid	
terrestrial petaloid monocots	Orchidaceae	53	49	49	100	92	orchidoid	
	other large families: 2	56	34	34	100	61	va	
terrestrial nonpetaloid grass-like monocots	large families: 3	319	181	121	67	38	va	

*, **: only those represented by 20 or 10 species per family respectively.

116

climate. They are found in temperate, (sub)tropical and even antarctic regions, and in every type of soil stressed in any way or not. Because the hosts of VA mycorrhiza include a lot of economically important crops such as tea, coffee, rubber, citrus, fruit trees and shrubs and all grain crops, it is not surprising that much of the research on the effects of VA mycorrhizas has been done on crops.

VA mycorrhizal infection seems to be highest in soils that are moderately rich in nutrients and relatively low in water content. For the latter reason mycorrhizas seldom occur in waterlogged or flooded soils, although mycorrhizal infection has been found in some aquatic plants (Søndergaard & Laegaard 1977).

Since the bulk of infectious propagules, spores and hyphae, is found in the top 25 cm of the soil, representing the rooting range of most herbaceous plants, it is not surprising that non-infected plants are often found on sites where the soil has been disturbed, for example by deep ploughing, a vulcanic eruption or after severe erosion. Recolonization of these sites by mycorrhizal fungi sometimes takes several years. But once present they can play a role in the stabilization of the soil in these habitats by aggregating it (Rose 1988).

4. Ecophysiology of VA mycorrhiza in stress situations

In this section we will be concentrating on mycorrhizas, growing in stress situations, in particular on those found in salt marshes and in heavy metal contaminated soils. Many plants may be confronted with some form of stress: excess or deficiency of nutrients or water, irradiation, competition, infection by pathogens. In these natural or man maded stress situations, mycorrhizas can play an important role in the maintenance of populations of plant species and the yield of crops.

4.1. Macro-nutrients

After initial uptake of macro-nutrients by the roots, depletion zones may develop, because phosphate, and nitrogen ions are immobile in soil. By extending the root system of the phytobiont, the mycobiont enlarges the capacity for ion and water absorption by the host plant which may lead to improved growth. Mycorrhizal plants have another advantage over non-mycorrhizal plants: macro- and micro-nutrient transport may take place via hyphal links between mycorrhizal plants (intra- and interspecific) as is shown in Fig. 1. Carbon-compounds can be transported through mycelial links from one plant to another, not only as carbohydrates but probably as amino-compounds as well. Therefore, if the "donor" plant 2 ("source") is also infected with a N_2-fixating micro-organism, the "receiver" plant 1 ("sink") can be supplied with nitrogen-compounds through the mycelial links and would not have to compete with other plants and/or micro-organisms for a part of its nitrogen. The fungus might be able to explore nutrient sources which are not available to the plant, but the literature is conflicting, particularly concerning the uptake of ("rock") phosphate (Cooper 1984).

Enhanced nutrient uptake has been demonstrated for VA and ectomycorrhiza with regard to water and the major plant nutrients nitrogen, phosphor and potassium (Harley & Smith 1983; Powell & Bagyaraj 1984). It has been shown that mycobiont phosphatases stimulate the uptake of phosphorus by the phytobiont. In nutrient-poor soils deficiency symptoms can disappear after inoculation with mycorrhizas. In many studies a strong correlation is found between the concentration of phosphorus in the soil or the internal concentrations of phosphorus in plant tissues and the percentage mycorrhizal infection (Jasper et al. 1979; Menge et al. 1978). A theoretically representation of these relationships is given in Fig. 2.

In nutrient-poor soil plant growth conditions are sub-optimal, and carbohydrate transport towards the root is limited. Therefore the mycorrhizal infection can not fully develop although the plant is usually very susceptible to invading micro-organisms under these conditions. When the concentration of phosphorus in the soil increases, nutrient exchange between host and fungus reaches an optimum and infection is maximal (Schubert & Hayman 1986). In richer soils the roots are able to provide the plant with sufficient phosphorus with-

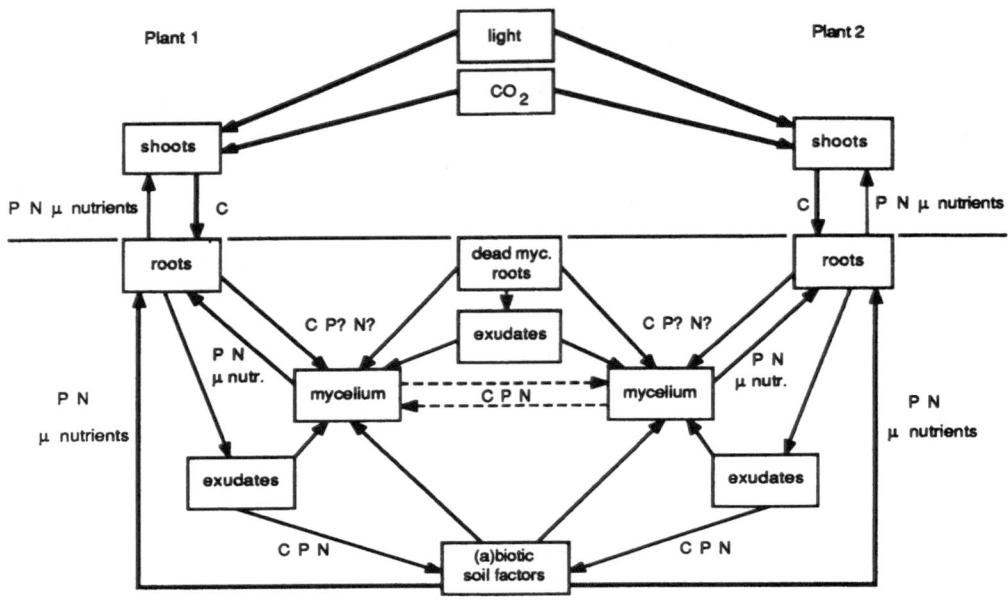

Fig. 1. Carbon, phosphorus, nitrogen and micro-nutrient pathways via hyphal links between two plants.

out the help of the external mycelium of the fungus. The susceptibility of the host for infection decreases (Thomson *et al.* 1986), however, because of the increased content of phosphorus of the plant tissue. In soils with a very high phosphorus content the infection may be low. The role of mycorrhizas in rich soils is not fully understood and it is not known whether the mutualistic relationship between host and fungus becomes more parasitic or not. Although most attention is still paid to the role of phosphorus in the mycorrhizal symbiosis, the enhanced uptake of nitrogen and other nutrients may be relevant as well.

Generally the macro-nutrient uptake in ericoid mycorrhiza is similar to that in VA mycorrhiza, but this is based on a limited number of experiments. In the case of an achlorophyllous host the mycobiont also supplies the host with carbohydrates derived from its own reserves, or from a chlorophyllous host, to which it is connected as well. As a rule, however, the carbohydrates pass from the phytobiont into the mycobiont, being the reverse of the VA mycorrhizal symbiosis.

4.2. Micro-nutrients and heavy metals

While there is consensus on the function of mycorrhizas concerning the uptake and transfer of macronutrients, there is much less clarity about their role regarding micronutrients and other elements. Generally it is assumed that mycorrhizas treat macro- and micronutrients in the same way, but there are only a few experiments supporting this hypothesis. Among the heavy metals there seems to be a difference. Some heavy metals, such as copper, manganese, molybdenium and zinc, are necessary in low quantities for plant growth and are mainly involved as cofactors in enzyme functions. If the supply of these micronutrients is restricted, mycorrhizas can be benificial to plants giving an enhanced uptake of these heavy metals. In soils enriched with phosphate, zinc and copper deficiency may evolve. With increasing phosphate availability VA mycorrhizal infection decreases and gives rise to a lower zinc and copper content in the plant. Levels at which deficiency symptoms become apparent may be reached then (Vander Zaag *et al.* 1979). Other metals which are not known to be necessary will not cause deficiency symptoms.

118

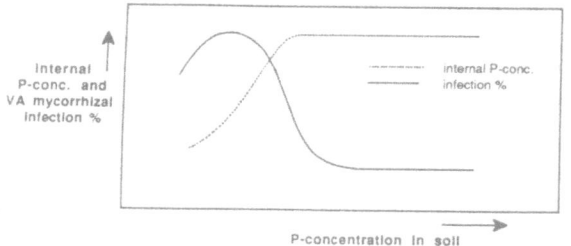

Fig. 2. Theoretical path of the percentage infection curve in relation to the phosphorus content of the soil and plant (after Amijee *et al.* 1989).

Heavy metals may be present in soils in high or even toxic concentrations naturally, or as a result of increasing acidity (acid rain), sewage sludge deposition or as a direct result of industrial activities, such as mining and metal smelting, or agriculture (e.g. use of pesticides and spreading of copper-enriched manure). If in these soils mycorrhizal infection also leads to an increased uptake of heavy metals it would easily lead to toxic concentrations, and one would expect only non-mycorrhizal plants to occur in these habitats. Indeed many disturbed environments are revegetated by plant species which are not known to be mycorrhizal (Reeves *et al.* 1979). Due to the lack of mycorrhizal inoculum these plants are at an advantage over the normally mycorrhizal plant species. This has also been found for two copper mining areas examined in Germany (Griffioen in prep.). Plant species growing on these sites are not known to be mycorrhizal or no infections have been observed. However, areas contaminated with other heavy metals are covered with many plant species forming several types of vegetation (Ernst 1974; Griffioen in prep.). Plant species growing in heavy metal contaminated environments have to evolve mechanisms to resist the high heavy metal content of the soil. Firstly, plant developed a restricted uptake or even an exclusion mechanism. Secondly, the heavy metals are translocated in a non-toxic form. This translocation of the heavy metals in a non-toxic form may be a result of specific metal-binding compounds or of complexation by organic acids. Instead of or after translocation the heavy metals may be bound to cell-walls (Nishizono *et al.* 1987) or compartimentation can take place.

Although a lot of research on heavy metal tolerance in seed plants has been done, the role of microorganisms, especially mycorrhizas, in this tolerance is often neglected. Bradley *et al.* (1982) found that plants with ericoid mycorrhizas grew better in copper and zinc enriched sand compared to non-mycorrhizal plants. The isolates used in this experiment were grown in pure culture to see the direct effects of heavy metals on fungal growth. The mycorrhizal plants showed a greater tolerance for heavy metals, which is a consequence of the reduced translocation to the shoots. Denny & Wilkins (1987) found the same for ectomycorrhizal birch seedlings. Both Bradly *et al.* (1982) and Denny & Wilkins (1987) attribute this exclusion to the binding of heavy metals to the cell-walls of the fungal hyphae, which have a greater affinity for metal ions than those of plant root cells. In this way ectomycorrhizal fungi may function as a filter by accumulating heavy metals in the external mycelium, mantle and carpophores. Well developed ectomycorrhizas are of vital interest to trees because of this retention, as can be concluded from studies on dry and wet deposits (acid rain) in conifer woods (e.g., Ernst 1985).

Though not all ectomycorrhizas and ericoid mycorrhizas can be cultured, it is possible to investigate the physiology of the fungus and heavy metal tolerance mechanisms, and produce large quantities of inoculum material. Axenic culture of VA mycorrhiza is, as yet, not possible. Experiments on germination and germ tube growth are the only two processes that can be studied on the fungus *in vitro*.

For plants with VA mycorrhiza, Gildon & Tinker (1983) found a decreased concentration of zinc in shoots compared to non-mycorrhizal plants, but Killham & Firestone (1983) and Griffioen & Ernst (1990) did not find this decrease in heavy metal content. With increasing acidity and heavy metal applications Killham & Firestone (1983) found a decrease of plant dry weight and they concluded that VAM infection could be not benificial to plants in these environments. Griffioen & Ernst (1990), however, reported no reduction in biomass production of VA mycorrhizal plants in a heavy metal contaminated soil, but they found higher heavy metal shoot: root ratios for the mycorrhizal plants. Therefore, if VA mycorrhizas play a role in the heavy metal tolerance of plants, it may be that they

119

translocate the heavy metal to the shoots in a detoxified form.

Morselt et al. (1986) reported the presence of heavy metal binding, metallothionein-like, proteins that may be responsible for detoxification, in the metal tolerant ectomycorrhizal fungus *Pisolithus tinctorius*. Production of these proteins is not restricted to the fungi. Rauser (1984) has isolated a similar copper-binding protein in *Agrostis gigantea* and Verkleij et al. (1989) reported the role of phytochelatins in the heavy metal tolerance of higher plants.

Binding to organic acids is another detoxification mechanism found in plants (Mathys 1977). There are no reports on the ability of mycorrhizal fungi to use organic acids as metal complexators, but it is known that oxalate can play a role in the copper binding of the fungus *Endothia parasitica* (Englander & Corden 1971).

Besides what may be a direct role in the detoxification of heavy metals, VA mycorrhizas can help the phytobiont in an indirect way by providing extra macro-nutrients. The benefit caused by the enhanced uptake of macro-nutrients may overrule a possible negative effect of higher heavy metal concentration. The heavy metal concentration in the plant tissues may not reach toxic levels, however, as a result of a dilution effect, caused by better plant growth. When interpreting experiments one should remember that tolerance for heavy metals may not only occur in strains of the phytobiont, but also in that of the mycobiont, which has been shown for *Glomus mosseae* by Gildon & Tinker (1981).

4.3. Salinity

The positive influence of VA mycorrhiza on the water economy of cultivated plants has been demonstrated, particularly in climatologically and physiologically dry (salty or brackish) soils. Little, however, is known about the effects of this symbiosis on natural populations of herbaceous plants in deserts or salt marshes (Rozema et al. 1985, 1986a, 1986b).

In the salt marsh a significant correlation between the percentage of infection and flooding frequency was found by Van Duin et al. (1990). Species growing on the upper and middle parts of the salt marsh, *Artemisia maritima*, *Aster tripolium*, *Atriplex prostrata*, *Elymus pycnanthus*, *Festuca rubra* ssp. *litoralis*, *Glaux maritima*, *Halimione portulacoides*, and *Plantago maritima*, were highly infected, while species from the middle to lower, more frequently flooded areas, *Limonium vulgare*, *Puccinellia maritima*, *Salicornia* spp. (a mixture of *S. dolichostachya* and *S. ramosissima*), *Spartina anglica*, *Spergularia media*, *Suaeda maritima*, and *Trichlochin maritima*, were hardly infected or not at all (Fig. 3). This may be caused by abiotic factors or by a combination of abiotic and biotic factors. For example *Aster tripolium* plants from the higher areas showed up to 60 percent infection, while plants from the inundated areas were not infected. This could be caused by a lower inoculum potential, and/or a lack of oxygen for the fungus, although normally sufficient O_2 is dissolved in water and aerenchym is present in *Aster* (as in most other infected species). Further causes could be the production of substances toxic for the mycorrhiza such as H_2S, Mn and some organic acids, a mechanical barrier on the roots formed by iron plaque, or a change in root morphology due to inundation. In a greenhouse experiment where monocultures and mixed cultures of *Triglochin maritima* and *Aster* were grown under conditions resembling those of the dryer, middle areas of the salt marsh, *Triglochin*, never highly infected in the field (maximum 5 percent), was not infected in the monoculture while infections of up to 40 percent were found in the mixed culture (Van Duin unpubl.). Fitter (1977) had similar results for mixed and monocultures of *Lolium perenne* and *Holcus lanatus*. It appears therefore that *Triglochin* may be more easily infected by external mycelium from already infected roots than from an inoculum containing spores only. A possible explanation for this phenomenon is given by Tommerup (1985), who found that living roots of some herbaceous species impose dormancy on VAM spores. As a result of this, seedlings are often infected by already existing external mycelium from neighbouring plants. The hyphal links between plants, intra- and interspecific, that are formed in this way (Haystead et al. 1988) may have an effect on the competitive interactions.

Safir et al. (1972) attribute the effects of mycor-

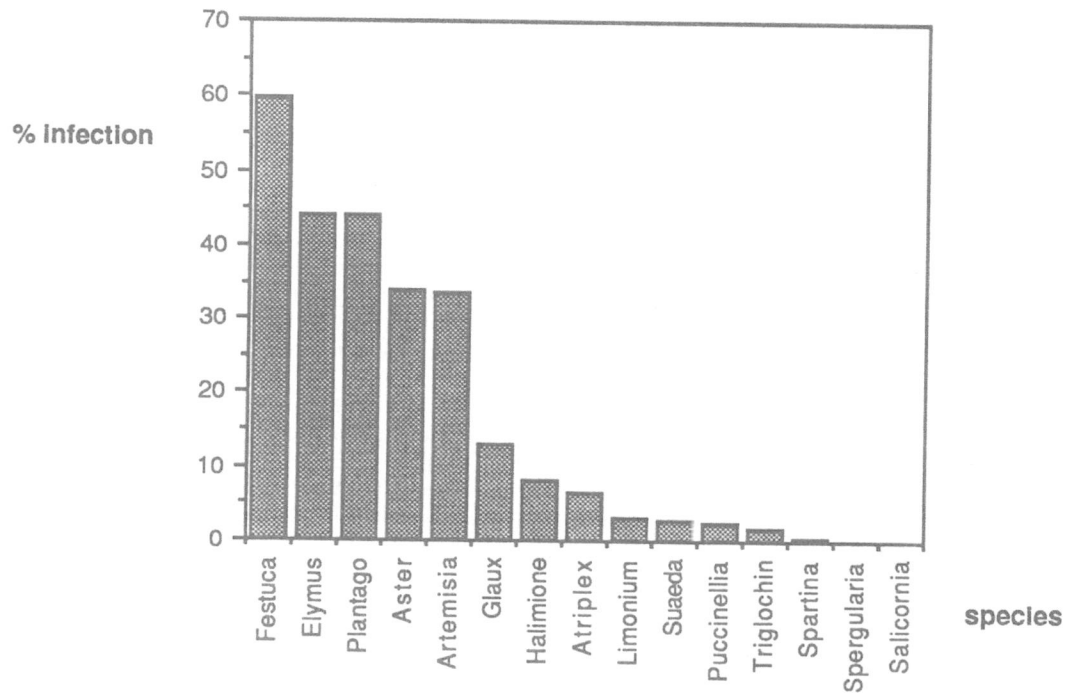

Fig. 3. Mean percentage VAM infection in 15 halophytes sampled monthly from May—November, 1987 and 1988, on a Dutch salt marsh.

rhizas on water relationships to a better nutritional status of the mycorrhizal plant. In wet habitats this could be true if phosphorus is a limiting factor. It may also be so in arid regions when the soil content of phosphorus is relatively high, because under very dry circumstances the rate of P-diffusion can be extremely low. Allen *et al.* (1981) found both an effect on water relationships and, also, an effect on the nutritional status of *Bouteloua gracilis.* Mycorrhizal infection increased the phosphorus concentration by 70 percent and transpiration rates in these plants were increased by more than 100 percent, while the stomatal and mesophyll resistances to CO_2 uptake were reduced. As a result of this, the chlorophyll concentration increased by 28 percent, and the photosynthetic rate increased by 68 percent. What factor(s) cause(s) the increased stomatal opening is not quite clear yet. The increased potassium concentrations in mycorrhizal plants may be involved, since the concentration of K^+-ions is high in the guard cells of open stomata. Changes in hormone levels (e.g., cytokinin) after

mycorrhizal infection as found by Allen *et al.* (1980) may also have an effect on stomatal opening (Incoll & Whitelam 1977).

4.4. *Pathogenic micro-organisms*

Infection with mycorrhizas often gives the plant a certain resistance or tolerance, but never a complete protection against fungal root pathogens, e.g., *Phytophtora, Pythium, Rhizoctonia* and *Fusarium,* and nematodes (Schönbeck 1979). The mechanisms causing this protection are amongst others: mechanical exclusion by the fungal sheath, especially in ectomycorrhizas (Barham *et al.* 1974), competition for infection sites (Dehne & Schönbeck 1979), for example *Pythium* species are capable of making vesicle and arbuscle like structures in the root cortex as well. Production of chemical compounds like antibiotics, phytoalexins or polyphenols, and an enhanced uptake of minerals resulting in a better nutritional status for the plant can play a role in this protection as well.

Pathogenic micro-organisms infect the roots much faster than VA mycorrhizas, because they are better competitors than the symbiotic fungi. In general, therefore, mycorrhizas will not be able to decrease the distribution of the pathogen unless they are well established when the pathogen attacks or when the inoculum density of the pathogen increases (Dehne & Schönbeck 1979). Mycorrhizal infection may increase the susceptibility of the plants to leaf pathogens, nematodes and viruses (Schönbeck 1979). Viruses appear to multiply best in cells containing arbuscules which may be caused by a better nutrient (phosphorus) status of these cells (Schönbeck & Spengler 1979).

5. Concluding remarks

The importance of mycorrhizas for the vascular plants has generally been demonstrated, particularly for crops and other plants under cultivation in situations of shortage of nutrients. Cultivated plants often react more strongly on mycorrhizal infection than their wild relatives, probably because the latter are better adapted to nutrient deficient soils with a high root: shoot ratio and a low growth rate (Koide *et al.* 1988). Examples have been given of the importance of mycorrhizas to wild plants under stress circumstances.

Contradictory results for the functioning of mycorrhizas, however, are frequently seen in literature. This is not surprising because physiologically different organisms are tested in various combinations and under differing experimental conditions.

From an ecological point of view more studies on the function of mycorrhizas in natural populations are highly desirable, not only in temperate, but also in tropical regions.

References

Allen, M. F., Moore, T. S. Jr. & Christensen, M., 1980. Phytohormone changes in *Bouteloua gracilis* infected by vesicular-arbuscular mycorrhizae. I. Cytokinin increases in the host plant. Can. J. Bot. 58: 371—374.

Allen, M. F., Smith, W. K., Moore, T. S. Jr. & Christensen, M., 1981. Comparative water relations and photosynthesis of mycorrhizal and non-mycorrhizal *Bouteloua gracilis*. New Phytol. 88: 683—693.

Amijee, F., Tinker, P. B. & Stribley, D. P., 1989. The development of endomycorrhizal root systems. VII. A detailed study of effects of soil phosphorus on colonization. New Phytol. 111: 435—446.

Barham, R. O., Marx, D. H. & Ruehle, J. L., 1974. Infections of ectomycorrhizal and nonmycorrhizal roots of shortleaf pine by nematodes and *Phytophthora cinnamoni*. Phytopathol. 64: 1260—1264.

Bradley, R., Burt, A. J. & Read, D. J., 1982. The biology of mycorrhiza in the Ericaceae. VIII. The role of mycorrhizal infection in heavy metal resistance. New Phytol. 91: 197—209.

Cooper, K. M., 1984. Physiology of VA mycorrhizal associations. In: Powell, C. L. & Bagyaraj, D. J. (eds), VA Mycorrhiza, CRC Press, Boca Raton pp. 155—186.

Dehne, H.-W. & Schönbeck, F., 1979. Untersuchungen zum Einfluß der endotrophen Mykorrhiza auf Pflanzenkrankheiten. I. Ausbreitung von *Fusarium oxysporum* f.sp. *lycopersici* in Tomaten. Phytopathol. Z. 95: 105—110.

Denny, H. J. & Wilkins, D. A., 1987. Zinc tolerance in *Betula* spp. IV. The mechanism of ectomycorrhizal amelioration of zinc toxicity. New Phytol. 106: 545—553.

Englander, C. M. & Corden, M. E., 1971. Stimulation of mycelial growth of *Endothia parasitica* by heavy metals. Appl. Microbiol. 22: 1012—1016.

Ernst, W. H. O., 1974. Schwermetallvegetation der Erde. Fischer Verlag, Stuttgart.

Ernst, W. H. O., Van Duin, W. E. & Oolbekkink, G. T., 1984. Vesicular-arbuscular mycorrhizae in dune vegetation. Acta Bot. Neerl. 35: 151—160.

Ernst, W. H. O., 1985. Impact of mycorrhiza on metal uptake and translocation by forest plants. Proc. Int. Conf. Heavy Metals Environ. Edinburgh pp. 596—599.

Fitter, A. H., 1977. Influence of mycorrhizal infection on competition for phosphorus and potassium by two grasses. New Phytol. 79: 119—125.

Gildon, A. & Tinker, P. B., 1981. A heavy metal-tolerant strain of a mycorrhizal fungus. Trans. Br. mycol. Soc. 77: 648—649.

Gildon, A. & Tinker, P. B., 1983. Interactions of vesicular-arbuscular mycorrhizal infection and heavy metals in plants. I. The effects of heavy metals on the development of vesicular-arbuscular mycorrhizas. New Phytol. 95: 247—261.

Griffioen, W. A. J. & Ernst, W. H. O., 1990. The role of VA mycorrhiza in the heavy metal tolerance of *Agrostis capillaris* L. Agric. Ecosyst. & Environ. 29: 173–177.

Harley, J. L. & Harley, E. L., 1987. A check-list of mycorrhiza in the British flora. New Phytol. suppl. to 105: 1—102.

Harley, J. L. & Smith, S. E., 1983. Mycorrhizal Symbiosis. Academic Press, London, New York.

Maystead, A., Malajczuk, N. & Grove, T. S., 1988. Underground transfer of nitrogen between pasture plants infected with vesicular-arbuscular mycorrhizal fungi. New Phytol. 108: 417—423.

Incoll, L. D. & Whitelam, G. C., 1977. The effect of kinetin on stomata of the grass *Anthephera pubescens* Nees. Planta 137: 243—245.

Jasper, D. A., Robson, A. D. & Abbott, L. K., 1979. Phosphorus and the formation of vesicular-arbuscular mycorrhizas. Soil Biol. Biochem. 11: 501—505.

Killham, K. and Firestone, M. K., 1983. Vesicular arbuscular mycorrhizal mediation of grass response to acidic and heavy metal depositions. Plant Soil 72: 39—48.

Koide, R., Li, M., Lewis, J. & Irby, C., 1988. Role of mycorrhizal infection in growth and reproduction of wild vs. cultivated plants. I. Wild vs. cultivated oats. Oecologia 77: 537—543.

Mathys, W., 1977. The role of malate, oxalate, and musterd oil glucosides in the evolution of zinc-resistance in herbage plants. Physiol. Plant. 40: 130—136.

Menge, J. A., Steirle, D., Bagyaraj, D. J., Johnson, E. L. & Leonard, R. T., 1978. Phosphorus concentrations in plants responsible for inhibition of mycorrhizal infection. New Phytol. 80: 575—578.

Morselt, A. F. W., Smits, W. T. M. & Limonard, T., 1986. Histochemical demonstration of heavy metal tolerance in ectomycorrhizal fungi. Plant Soil 96: 417—420.

Nishizono, H., Ichikawa, H. Suzuki, S. & Ishii, F., 1987. The role of the root cell wall in the heavy metal tolerance of *Athyrium yokoscense*. Plant Soil 101: 15—20.

Powell, C. L. & Bagyaraj, D. J. (eds.), 1984. VA Mycorrhiza. CRC Press, Boca Raton.

Rauser, W. E., 1984. Copper-binding protein and copper tolerance in *Agrostis gigantea*. Plant Sci. Lett. 33: 239—247.

Reeves, F. B., Wagner, D., Moorman, T. & Kiel, J., 1979. The role of endomycorrhizae in revegetation practices in the semi-arid west. I. A comparison of mycorrhizae in several disturbed vs. natural environments. Amer. J. Bot. 66: 6—13.

Rose, S. L., 1988. Above and below ground community development in a marine sand dune ecosystem. Plant Soil 109: 215—226.

Rozema, J., Arp, W., Van Esbroek, M. & Broekman, R., 1985. Relaties tussen autotrofe en heterotrofe planten op kwelders. Vakbl. Biol. 65: 465—468.

Rozema, J., Arp, W., Van Esbroek, M., Broekman, R., Punte, H. & Schat, H., 1986a. Vesicular-arbuscular mycorrhiza in salt marsh plants in response to soil salinity and flooding and the significance to the water relations. In: Physiological and Genetical Aspects of Mycorrhizae. INRA, Paris pp. 657—660.

Rozema, J., Arp, W., Van Diggelen, J., Van Esbroek, M.,

Broekman, R. & Punte, H., 1986b. Occurrence and ecological significance of vesicular-arbuscular mycorrhiza in the salt marsh environment. Acta Bot. Neerl. 35: 457—467.

Safir, G. R., Boyer, J. S. & Gerdemann, J. W., 1972. Nutrient status and mycorrhizal enhancement of water transport in soybean. Plant Physiol. 49: 700—703.

Schenck, N. C. (ed.), 1984. Methods and Principles of Mycorrhizal Research. American Phytopathology Society, St. Paul, Minnesota.

Schönbeck, F., 1979. Endomycorrhiza in relation to plant diseases. In: B. Schippers & W. Gams (eds.), Soil-borne Plant Pathogens, Academic Press, London pp. 271—280.

Schönbeck, F. & Spengler, G., 1979. Nachweis von TMV in Mykorrhiza-haltigen Zellen der Tomate mit Hilfe der Immunofluoreszenz. Phytopathol. Z. 94: 84—86.

Schubert, A. & Hayman, D. S., 1986. Plant growth responses to vesicular-arbuscular mycorrhiza. XVI. Effectiveness of different endophytes at different levels of soil phosphate. New Phytol. 103: 79—90.

Søndergaard, M. & Laegaard, S., 1977. Vesicular-arbuscular mycorrhiza in some aquatic plants. Nature 268: 232—233.

Stubblefield, S. P., Taylor, T. N. & Trappe, J. M., 1987. Vesicular-arbuscular mycorrhizae from the Triassic of Antarctica. Amer. J. Bot. 74: 1904—1911.

Thomson, B.D., Robson, A.D. & Abbott, L.K., 1986. Effects of phosphorus on the formation of mycorrhizas by *Gigaspora calospora* and *Glomus fasciculatum* in relation to root carbohydrates. New Phytol. 103: 751—765.

Tommerup, I. C., 1985. Inhibition of spore germination of vesicular-arbuscular mycorrhizal fungi in soil. Trans. Br. Mycol. Soc. 85: 267—278.

Vander Zaag, P., Fox, R. L., De La Pena, R. S. & Yost, R. S., 1979. P nutrition of Cassava including mycorrhizal effects on P, K, S, Zn and Ca uptake. Field Crops Res. 2: 253—263.

Van Duin, W. E., Rozema, J. & Ernst, W. H. O., 1990. Seasonal and spatial variation in the occurrence of vesicular arbuscular (VA) mycorrhiza in salt marsh plants. Agric. Ecosyst. & Environ. 29: 107–110.

Verkleij, J. A. C., Koevoets, P., Van 't Riet, J., Van Rossenberg, M., Bank, R. & Ernst, W. H. O., 1989. The role of metal-binding compounds in the copper tolerance mechanism of *Silene cucubalus*. In: D. H. Hamer & D. R. Winge (eds.), Metal Ion Homeostasis: Molecular Biology and Chemistry. Alan R. Liss, Inc., New York pp. 347—357.

Authors' address
W. E. van Duin, W. A. J. Griffioen and J. H. Ietswaart
Dep. of Ecology and Ecotoxicology
Vrije Universiteit
De Boelelaan 1087
1081 HV Amsterdam
The Netherlands

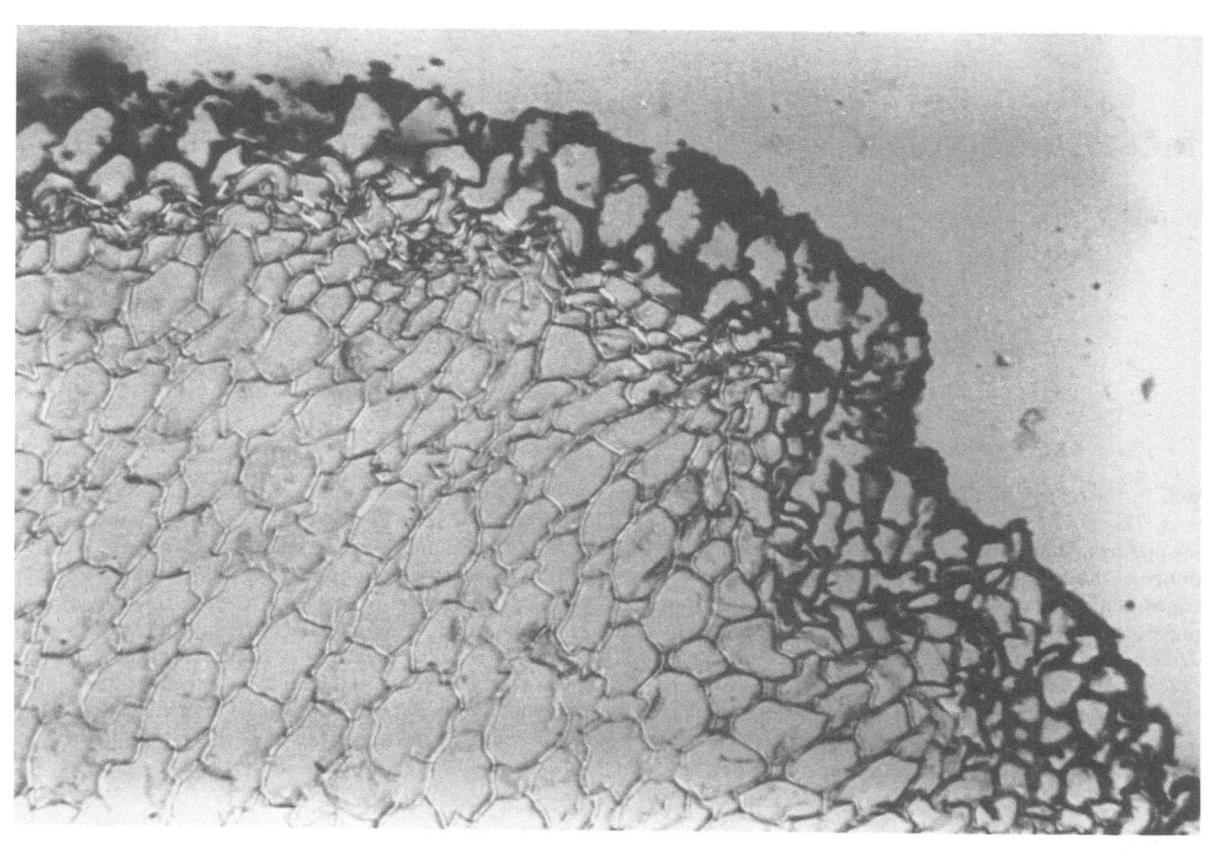

Cross section of root of *Spartina anglica,* showing the iron plaque at the root surface. To increase contrast iron plaque was stained using an Fe^{3+}-specific method with potassium ferrocyanide (Gray 1973).

CHAPTER 12

Contamination of coastal wetlands with heavy metals: factors affecting uptake of heavy metals by salt marsh plants

M. L. OTTE

Abstract. In this paper, the factors affecting uptake of heavy metals by salt marsh plants from contaminated salt marsh soils are discussed. Three routes of uptake can be distinguished: from the soil through the roots, from the water through the shoots during flooding and from atmospheric deposits through the shoots. The first route is considered most important. In the soil, the mobility and availability of the metals is determined by their chemical speciation. One of the most important factors determining the chemical speciation of the metals in salt marsh soils is the oxidation state of the soil. Due to the oxidative activity of the roots, plants are capable of changing the oxidation state of the soil in the rhizosphere. The resulting increase of the redox potential of the soil may strongly influence the mobility and availability of heavy metals to salt marsh plants. Apart from the chemical speciation of heavy metals in salt marsh soils, other factors, like interactions with micro-organisms, inter- and intra-specific differences in uptake systems of the plants and (micro-)climatic characteristics, may strongly determine uptake of heavy metals by salt marsh plants. It is discussed that in future research these factors should get more attention. The concept of biomonitoring may prove a helpful tool for this purpose.

1. Introduction

Salt marshes are often located near or along estuaries. River-borne pollutants, such as heavy metals, are transported through these estuaries. Due to the slow current in estuaries and mixing processes of riverine with marine water, large quantities of these pollutants are deposited in the sediment of salt marshes (Salomons & Förstner 1984). Another source of heavy metal input in salt marshes is the deposition of sea-borne material, such as algae and seaweeds (Ernst 1987). Salt marsh plants may thus be confronted with high concentrations of potentially toxic chemicals in their environment. This paper discusses the factors that are considered important in the uptake of heavy metals by salt marsh plants.

2. Routes of uptake of heavy metals in plants

The sediment may act as a main source of heavy metals for plants (Beeftink et al. 1982; Rozema et al. 1986b; Beeftink & Rozema 1988; Huiskes & Rozema 1988). During high tides plants growing on the lower parts of the salt marsh are submerged with contaminated water, that may act as another source of heavy metals. Furthermore atmospheric deposition may form a third source of heavy metals. Uptake of heavy metals by salt marsh plants may therefore not only occur through the roots, but also through the shoots.

Rozema et al. (1988) tried to quantify uptake of heavy metals by Elymus pycnanthus, Triglochin maritima and Atriplex littoralis through the shoots during submergence in sea water. By immersion of the shoots of the plants in water containing 15 μM Zn or 15 μM Cu and 50 mM NaCl during 2 hours at five subsequent days submergence was simulated. Uptake of heavy metals by the shoots was signif-

J. Rozema and J. A. C. Verkleij (eds.), *Ecological Responses to Environmental Stresses*, 126–133.

icant, but the concentrations used in the incubation solution were much higher compared to concentrations found in the water of the heavily polluted Western Scheldt river. It was concluded that uptake of heavy metals through the shoots of salt marsh plants can occur, but that this is of minor importance compared to uptake through the roots.

The same may be true for uptake of heavy metals from atmospheric deposits. Plants can be used as biomonitors of atmospheric deposition (Martin & Coughtrey 1982; Ernst & Leloup 1987). This illustrates that atmospheric deposition affects the heavy metals content of the (shoots of) plants. However the bulk of these atmospherically deposited metals is adsorbed to the surface of the leaves. It is not really taken up into the plant and concentrations are relatively low. It seems therefore likely that uptake of heavy metals from atmospheric deposits by salt marsh plants is negligible compared to uptake from the sediment through the roots. The amount of metals taken up by a plant is determined by the amount of metals available in the external medium (soil) and by the uptake activity of the plant.

3. Factors affecting the availability of heavy metals in salt marsh soil

The amount of metals in the soil that can be taken up by a plant (the bioavailable fraction of the total metal concentration of the soil) depends on its mobility in the soil and therefore on the speciation of the metal. Important factors determining speciation of metals in soil are oxidation state, pH, particle size, organic matter content, salinity, interactions between metals and the activity of micro-organisms and fungi.

3.1. Oxidation state of the soil

One of the most important factors determining the speciation of metals in soil is its oxidation state. In general, salt marsh soils are permanently waterlogged and oxygen supply in the soil is low (Armstrong 1978). Some micro-organisms use other oxidators than oxygen for their respiration. Along a step-wise "reduction chain" Fe^{3+}, Mn^{4+} and SO_4^{2-} are reduced to Fe^{2+}, Mn^{2+} and S^{2-}, respectively (Ponnamperuma 1984). Heavy metals like Zn, Cu and Cd form stable insoluble sulphides. In sulphide rich soils, with redox potential $E_h \approx -200\,mV$, mobility of heavy metals is low. But oxidation of sulphides in the upper layers of the soil may occur (Armstrong et al. 1985) resulting in mobilization of heavy metals. A further oxidation of iron and manganese to their insoluble (hydr-)oxides may again lead to immobilization of heavy metals due to adsorption to iron- and manganese (hydr-)oxides. Tidal fluctuations thus result in alternating mobilization and immobilization of heavy metals in the upper layer of salt marsh soils.

3.2. Alteration of the redox potential in the rhizosphere by salt marsh plants

The oxidation state of the soil is not only determined by tidal fluctuations. The plants themselves are capable of altering the redox potential in the rhizosphere (Trolldenier 1988). Due to this process the soil at vegetated sites, in most cases, has a much higher redox potential compared to the soil at nearby non-vegetated sites (Fig. 1).

Many (salt-) marsh plants have aerenchymous tissue to supply the roots with oxygen in anaerobic soils. The amount of oxygen transported to the roots exceeds the demand of the root tissue. The resulting radial oxygen loss (ROL) into the rhizosphere (Armstrong 1978) leads to oxidation of iron and manganese. This process results in formation of a so called iron plaque on the roots of the plants (see page 125). Iron plaque formation also occurs in plants that do not possess aerenchyma (e.g. conifers) indicating that other processes, like oxidation by exuded substances or by microbial activity in the rhizosphere, are of importance too (Levan & Riha 1986, Crowder et al. 1987, Trolldenier 1988).

Oxidation processes in the rhizosphere and iron plaque formation drastically change the speciation of heavy metals. Iron and manganese (hydr-)oxides are capable of adsorbing large quantities of heavy metals (Brümmer et al. 1988).

It has been suggested that the iron plaque could form a barrier to uptake of heavy metals by plants

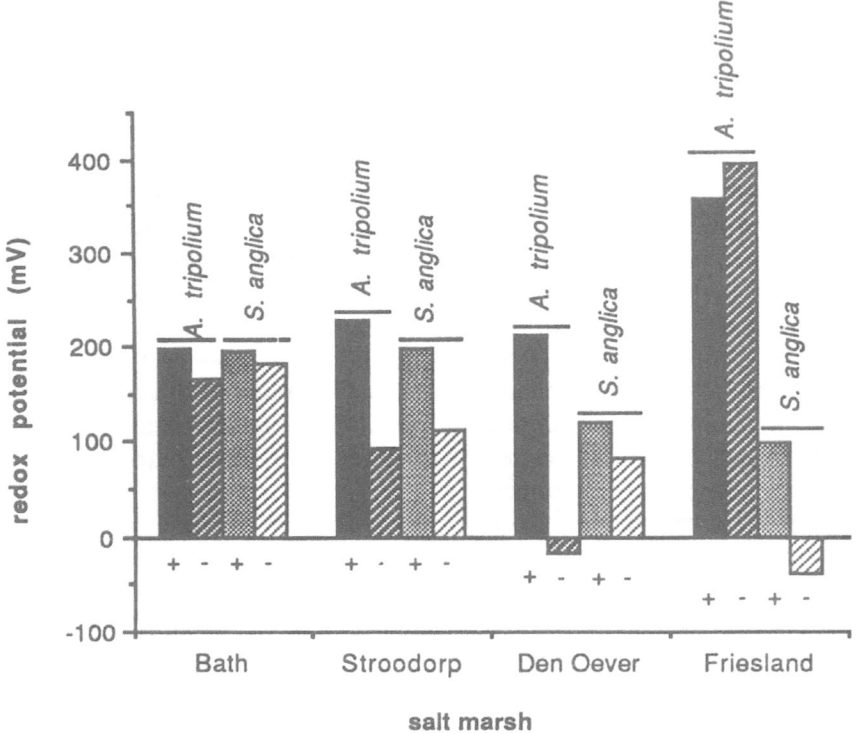

Fig. 1. Comparison of redox potential in soil between the roots (+) of *Aster tripolium* or *Spartina anglica* and in soil at non-vegetated places near the plants (−). The redox potential between the roots of plants is significantly higher compared to non-vegetated places. Furthermore differences between species and between salt marshes are significant. *Aster tripolium* grows at places with higher redox potential compared to *Spartina anglica*. At growing sites of *Aster tripolium* at the artificial salt marsh near Nieuwe Bildtdijk in the province of Friesland the redox potential is significantly higher compared to growing sites at other salt marshes (Otte, unpublished data).

(Taylor & Crowder 1983), but recent results of experiments performed by Otte *et al.* (1987, 1989) suggest that the iron plaque in fact enhances uptake of zinc by *Aster tripolium*. The reduced state of the bulk soil in salt marshes favours immobilization of heavy metals due to formation of metal sulphides, but oxidation processes in the rhizosphere may mobilize the metals. If due to oxidation of iron and manganese formation of an iron plaque occurs, zinc (and possibly other metals) adsorbs to the iron-/manganese-(oxo-)hydroxides. As a result zinc accumulates in the iron plaque to concentrations much higher than those found in the bulk soil. However zinc adsorbed to iron plaque still seems to be available to the plant. An explanation could be that the plant is capable of taking up zinc by complexating it with organic acids or other chelators

(Baumeister & Ernst 1978), that bind zinc stronger than iron-/manganese (oxo-)hydroxides. The results of Otte *et al.* (1989) also suggest that iron plaque can act as a barrier to uptake of zinc if the iron plaque is very thick (> 2000 nmol Fe cm^{-2} root surface). This may indicate that depletion of zinc at the root-iron plaque interface occurs under these circumstances.

Under uncontaminated conditions enhancement of the uptake of Zn due to formation of an iron plaque could be an advantage. Zn is an essential element for plants. The generally low availability of Zn in salt marsh sediment may lead to deficiency of zinc. Crowder & Macfie (1986) found that iron plaque formation is seasonal with an optimum during the summer. An increased uptake of zinc due to formation of an iron plaque during the growing season may therefore be of an advantage. But under contaminated conditions increased uptake of zinc may lead to toxic concentrations in the plants.

3.3. The pH of the soil

Another important factor determining speciation of heavy metals in soil is the pH. In general the mobility of heavy metals increasaes with decreasing pH. Little research is done on the pH changes in salt marsh soils. Redox reactions in the soil, such as the oxidation of sulphide, may lead to a decrease of pH. Together with redox potential pH may change drastically in the rhizosphere of salt marsh plants. Changes of two pH units in the rhizosphere of plants are theoretically possible (Nye 1981) and were actually measured by Marshner *et al.* (1986) for *Lupinus albus* under laboratory conditions.

3.4. Particle size and organic matter content of the soil

Other factors determining the mobility of heavy metals in salt marsh soils are particle size of the sediment and organic matter content. In general mobility of heavy metals decreases with increasing fraction of particles $< 63 \mu m$ in the soil due to adsorption of the metals to these particles (Salomons & Förstner 1984). Organic matter is capable of binding heavy metals by complexation. Although it is believed that plants are capable of taking up organo-metal complexes of low molecular weight, mobility of heavy metals in general decreases with increasing organic matter content of the soil (Schierup & Larssen 1981, Ernst *et al.* 1987).

3.5. Salinity

Yet another factor determining mobility of heavy metals in salt marsh soils is salinity. With increasing salinity heavy metals are mobilized. It still remains an open question whether this process increases availability of heavy metals to plants, or that formation of chloride -metal complexes (such as $CdCl_4^{2-}$) decreases availability (Luoma 1988). Furthermore chloride interferes with the physiological

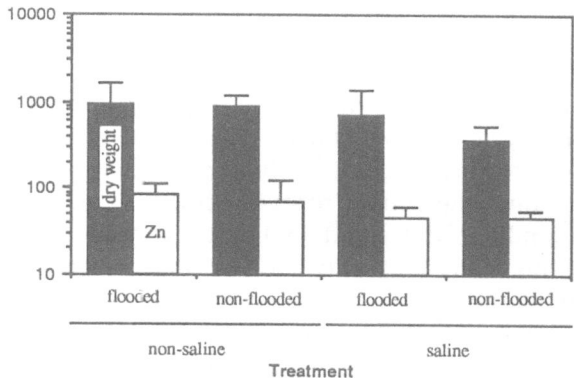

Fig. 2. Dry weight (mg) and Zn content (mg kg^{-1} dry weight) of shoots of *Aster tripolium* plants grown on uncontaminated salt marsh sediment for 7 weeks. Differences in dry weight were significant due to salinity. Zn content of the shoots differed significantly due to salinity, but there was also a significant interaction between salinity and flooding. The results show a relationship between zinc content and biomass production. This indicates that differences in zinc content of the shoots may not (only) be the result of differences in zinc mobility in the soil, but also to differences in physiological activity of the plants (Otte, unpublished data).

activity of the plant itself (Levitt 1980). A low concentration of heavy metals in plants under saline conditions may well be the result of a low uptake activity of the roots (Fig. 2).

3.6. Interactions between heavy metals in the soil

Contamination of salt marshes mostly occurs with several heavy metals simultaneously, such as Cd, Cr, Cu, Ni, Pb, and Zn. Interactions between the pollutants may therefore occur. In the sediment, competition between heavy metals for binding sites could influence the mobility of the metals. At the root-soil interface competition between metals for uptake sites could interfere with uptake of these metals (Christensen 1987, Elliot *et al.* 1986, Abdel-Sabour *et al.* 1988).

3.7. Micro-organisms and fungi

As mentioned before micro organisms in the rhizosphere could be of importance in the formation of an iron plaque on the roots of salt marsh plants by oxidation of Fe(II). Furthermore, micro organisms seem to play an important role in the mobilization of heavy metals. In the acquisition of iron by plants micro organisms may form organic substances (siderophores) that enhance availability of the metals (Crowley *et al.* 1988). Morel *et al.* (1986) showed that mucilages on the roots of *Zea mays* are capable of binding Cd, Cu and Pb, but it is not clear whether this leads to mobilization or immobilization of the metals.

Symbiosis of salt marsh plants with fungi, like vesicular arbuscular (VA) mycorrhiza, could also have an important influence on the uptake of heavy metals by (salt marsh) plants (Tinker & Gildon 1983). Like other micro-organisms fungi could affect the speciation of metals in the rhizosphere, but it is also theorized that VA-mycorrhiza could change the uptake activity of plant roots. Rozema *et al.* (1986a) and Van Duin *et al.* (1990) have shown that VA mycorrhiza occur in salt marsh vegetation, but the importance of these fungi to uptake of nutrients is not clear. Surely, the role of micro-organisms and fungi in the uptake of (heavy) metals by (salt marsh) plants needs much more attention.

4. Inter and intra specific differences in uptake systems of salt marsh plants

The speciation of metals in the soil is just one of the many factors determining the uptake of heavy metals by salt marsh plants. The uptake system of the plant itself is also of great importance. Large differences occur between different plant species and even between different populations of one species (Baumeister & Ernst 1978). Rozema *et al.* (1986b) suggested that, analogous to uptake of sodium and chloride, monocotyledonous salt marsh plants in general can be considered as excluders of heavy metals, whereas dicotyledons may act as accumulators. This is supported by the review of Baker (1981) on accumulator/excluder strategies in the uptake of heavy metals by plants tolerant to heavy metals. However within dicotyledon plant groups differentiation may exist into plant families with an accumulation or exclusion type of ion uptake system. Kuboi *et al.* (1986) showed that several dicoty-

ledonous families (Chenopodiaceae, Compositae) should be regarded as excluders of Cd.

Apart from differences between species and between populations, the uptake activity of heavy metals within an individual plant will vary with its development. Uptake of essential elements, such as zinc and copper, shows an optimum during the growing season. However, due to simultaneous biomass production and translocation of the elements between various plant parts, it may occur that the concentration of an element in a particular plant part remains constant or even decreases during the growing season (Baumeister & Ernst 1978).

4.1. Salt excretion

Some salt marsh species possess salt glands, e.g., *Glaux maritima*, *Limonium vulgare* and *Spartina anglica*. Through these salt glands plants are capable of excreting sodium and chloride, thus preventing the plant from accumulating too high concentrations of salt. Rozema & Roosenstein (1985) suggested that this mechanism could play a role in heavy metal metabolism of salt marsh plants. They showed that *Spartina anglica* is capable of excreting Zn and Cu through salt glands, but concluded that this is relatively ineffective to lower the internal heavy metal concentrations significantly.

5. Effects of microclimate on the uptake activity of (salt marsh) plants

The uptake of heavy metals by plants is not only determined by the availability of the metals in the soil, but also by microclimate. This is illustrated in Fig. 3. Although the experiment described here was not performed in a salt marsh, but in a fresh water marsh, the results show that in a situation where it may be assumed that the chemical speciation of the metals in the soil remains the same for all treatments during the experiment, a two-fold range in metal concentrations can be reached in the roots of plants due to (micro climatic) differences in vegetation type. In the field differences due to growing site will cause considerable variation in heavy metal content of plants.

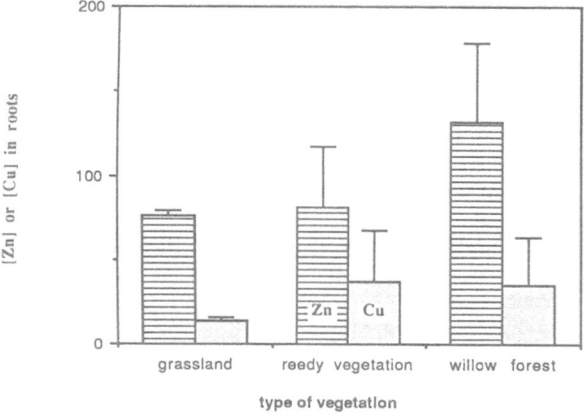

Fig. 3. Mean Zn and Cu concentrations (mg kg^{-1} dry weight) in roots of *Urtica dioica* (bars indicate standard deviations, n = 5) grown in pots, that were placed in three different types of vegetation (grassland, reedy vegetation, willow forest), for 7 weeks. The soil in the pots in each treatment contained 4 mg Cd kg^{-1} dry soil, 100 mg Cu kg^{-1} dry soil and 600 mg Zn kg^{-1} dry soil. The experiment was performed in the Dordtsche Biesbosch, Rhine estuary. The Zn concentration in the roots of plants grown in willow forest is significantly higher ($P < 0.05$) than in the other types of vegetation, whereas the Cu concentration in roots of plants grown in grassland is significantly lower ($P < 0.05$) than in reedy vegetation or willow forest. The results can not be explained by differences in growth of the plants, since growth (biomass production after 7 weeks) did not differ between types of vegetation. The results indicate that uptake of heavy metals is strongly affected by differences in the characteristics of the type of vegetation in which the plants grow. Such characteristics are radiation influx, temperature, humidity, etc. (Otte, unpublished data).

6. Discussion

The enormous variability that occurs in the salt marsh environment makes it very complicated to identify the main factors that determine uptake of heavy metals by salt marsh plants. Considerable research is done on the geophysical and geochemical factors determining the speciation of heavy metals in soil. Efforts to relate uptake of heavy metals by salt marsh plants under natural conditions to the speciation of the metals in the soil have seldom been successful. In most investigations attempts are made to correlate concentrations in the plants to concentrations in the soil, whereas actual uptake is not measured. Concentrations are the result of uptake and internal translocation pro-

cesses and as discussed above these processes are affected strongly by genetic differences between plants as well as by seasonal and (micro-)climatic changes. If one takes into account how many factors affect the uptake of heavy metals by salt marsh plants, it becomes clear that a simple chemical extraction procedure, to estimate "bioavailable" concentrations in the soil and to predict uptake of heavy metals by salt marsh plants is not likely to be found. Since large differences in uptake systems occur between species a "bioavailable fraction" should at least be defined per species. It is furthermore important to separate the factors that affect uptake of heavy metals by plants and determine their relative impact on heavy metal uptake, within the ecological range of the species under investigation.

To investigate uptake and effects of heavy metals in (salt marsh) plants the concept of biomonitoring seems most useful. In particular active biomonitoring, with plants pregrown under identical conditions, makes estimation of real uptake possible (Martin & Coughtrey 1982). It gives the researcher an opportunity to minimize intraspecific variability. And for the purpose of standardization of legislation of "acceptable" concentrations of heavy metals in (salt marsh) soil this approach could prove a powerful tool.

References

Abdel-Sabour, M. F., Mortvedt, J. J. & Kelsoe, J. J., 1988. Cadmium-zinc interactions in plants and extractable cadmium and zinc fractions in soil. Soil Sci. 145: 425—431.

Armstrong, W., 1978. Root aeration in the wetland condition. In: D. D. Hook & R. M. M. Crawford (eds.), Plant life in anaërobic environments. Ann Arbor inc. Mich.: 269—297.

Armstrong, W., Wright, E. J., Lythe, S. & Gaynard, T. J., 1985. Plant zonation and effects of the spring-neap tidal cycle on soil aeration in a Humber salt marsh. J. Ecol. 73: 323—339.

Baker, A. J. M., 1981. Accumulators and excluders — Strategies in the response of plants to heavy metals. J. Plant Nutr. 3: 643—654.

Baumeister, W. & Ernst, W. H. O., 1978. Mineralstoffe und Pflanzenwachstum. Gustav Fisher Verlag, Stuttgart, New York.

Beeftink, W. G., Nieuwenhuize, J., Stoeppler, M. & Mohl, C., 1982. Heavy metal accumulation in salt marshes from the

Eastern and Western Scheldt. Sci. Total Environ. 25: 199—223.

Beeftink, W. G. & Rozema, J., 1988. The nature and functioning of salt marshes. In: W. Salomons, B. L. Bayne, E. K. Duursma & U. Förstner (eds.). Pollution of the North Sea: an assessment. pp. 59—87.

Brümmer, G. W., Gerth, J. A. & Tiller, K. G., 1988. Reaction kinetics of the adsorption and desorption of nickel, zinc and cadmium by goethite. I. Adsorption and diffusion of metals. J. Soil Sci. 39: 37—52.

Christensen, T. H., 1987. Heavy metal competition for soil sorption sites at low concentrations. Proc. Int. Conf. Heavy metals in the environment, New Orleans. CEP consultants Ltd., Edinburgh, UK: 394—396.

Crowder, A. A. & Macfie, S. M., 1986. Seasonal deposition of ferric hydroxide plaque on roots of wetland plants. Can. J. Bot. 64: 2120—2124.

Crowder, A. A., Macfie, S., St.-Cyr, L., Conlin, T., Badgery, J. & Johnson-Green, P., 1987. Root iron plaques and metal uptake by wetland plants. Proc. Int. Conf. Wetlands/Peatlands Edmonton, Alberta, Canada: 503—508.

Crowley, D. E., Reid, C. P. P. & Szaniszlo, P. J., 1988. Utilization of microbial siderophores in iron acquisition by oat. Plant Physiol. 87: 680—685.

Elliot, H. A., Liberati, M. R. & Huang, C. P., 1986. Competitive adsorption of heavy metals by soils. J. Environ. Qual. 15: 214—219.

Ernst, W. H. O., 1987. Metal fluxes to coastal ecosystem and the response of coastal vegetation — a review. In: Vegetation between Land and Sea. A. H. L. Huiskes, C. W. P. M. Blom & J. Rozema, (eds.) pp. 302—310. Dr. W. Junk Publishers, Dordrecht.

Ernst, W. H. O., Kraak, M. H. S. & Stoots, L., 1987. Growth and mineral nutrition of *Scrophularia nodosa* with various combinations of fulvic and humic acids. J. Plant Physiol. 127: 171—175.

Ernst, W. H. O. & Leloup, S., 1987. Perennial herbs as monitors for moderate levels of metal fall-out. Chemosphere: 233—238.

Gray, P., 1973. Encyclopedia of microscopy and microtechnique. Van Nostrand Reinhold Publ., New York. pp. 234—238.

Huiskes, A. H. L. & Rozema, J., 1988. The impact of anthropogenic activities on the coastal wetlands of the North Sea. In: W. Salomons, B. L. Bayne, E. K. Duursma & U. Förstner (eds.). Pollution of the North Sea: an assessment. pp. 455—473.

Kuboi, T., Noguchi, A., Yazaki, J., 1986. Family-dependent cadmium accumulation characteristics in higher plants. Plant Soil 92: 405—415.

Levan, M. A. & Riha, S. J., 1986. The precipitation of black oxide coatings on flooded conifer roots of low internal porosity. Plant Soil 95: 33—42.

Levitt, J., 1980. Responses of plants to environmental stresses. Vol. II. Academic Press, New York.

Luoma, S. N., 1988. A comparison of field and bioassay approaches for assessing the bioavailability of sediment-bound trace metals. Presented at the International symposium on the fate and effects of toxic chemicals in large rivers and their estuaries, Québec, Canada, oct. 1988.

Marschner, H., Römheld, V., Horst, W. J. & Martin, P., 1986. Root-induced changes in the rhizosphere: importance for the mineral nutrition of plants. Z. Pflanzenernähr. Bodenk. 149: 441—456.

Martin, M. H. & Coughtrey, P. J., 1982. Biological monitoring of heavy metal pollution. Land and air. Applied Science Publ., London.

Morel, J. L., Mench, M. & Guckert, A., 1986. Measurement of Pb^{2+}, Cu^{2+} and Cd^{2+} binding with mucilage exudates from maize (*Zea mays* L) roots. Biol. Fertil. Soils 2: 29—34.

Nye, P. H., 1981. Changes of pH across the rhizosphere induced by roots. Plant Soil 61: 7—26.

Otte, M. L., Buijs, E. P., Riemer, L., Rozema, J. & Broekman, R. A., 1987. The iron plaque on roots of salt marsh plants: a barrier to heavy metal uptake? Proc. Int. Conf. Heavy metals in the Environment, New Orleans (USA). CEP consultants, Edinburgh (UK): 407—409.

Otte, M. L., Rozema, J., Koster, L., Haarsma, M. S. & Broekman, R. A., 1989. Iron plaque on roots of Aster tripolium L.: interaction with zinc uptake. New Phytol. 111: 309—317.

Ponnamperuma, F. N., 1984. Effects of flooding on soils. In: T. T. Kozlowski (ed.). Flooding and plant growth. Academic Press Inc., London.

Rozema, J., Arp, W., Van Diggelen, J., Van Esbroek, M., Broekman, R. A. & Punte, H., 1986a. Occurrence and ecological significance of vesicular arbuscular mycorrhiza in the salt marsh environment. Acta Bot. Neerl. 35: 457—467.

Rozema, J., Otte, M. L., Broekman, R. A. & Wezenbeek, J. M., 1986b. The uptake and translocation of heavy metals by salt marsh plants from contaminated salt marsh sediment: possibilities for bioindication. Proc. Int. Conf. Environmental Contamination, Amsterdam, CEP Consultants, Edinburgh: 123—125.

Rozema, J., Otte, M. L., van Schie, C. & Ernst, W. H. O., 1988. Foliar uptake of heavy metals by estuarine plants in response to contaminated sea water flooding. Proc. Int. Conf. Environmental Contamination, Venice: 73—75.

Rozema, J. & Roosenstein, J., 1985. Effects of zinc, copper and cadmium on the mineral nutrition and ion secretion of salt secreting halophytes. Vegetatio 62: 554—556.

Salomons, W. & Förstner, U., 1984. Metals in the Hydrocycle. Springer Verlag, Berlin.

Schierup, H.-H. & Larsen, V. J., 1981. Macrophyte cycling of zinc, copper, lead and cadmium in the littoral zone of a polluted and non-polluted lake I. Availability, uptake and translocation of heavy metals in *Phragmites australis* (Cav.) Trin. Aquat. Bot. 11: 197—210.

Taylor, G. J. & Crowder, A. A., 1983. Uptake and accumulation of copper, nickel and iron by *Typha latifolia* grown in solution culture. Can. J. Bot. 61: 1825—1830.

Tinker, P. B. & Gildon, A., 1983. Mycorrhizal fungi and ion uptake. In: D. A. Robb & Pierpoint, W. S. (eds.). Metals and micro nutrients: uptake and utilization by plants. Academic Press, London, New York: 21—32.

Trolldenier, G., 1988. Visualization of oxidizing power of rice roots and possible participation of bacteria in iron deposition. Z. Pflanzenernähr. Bodenk. 151: 117—121.

Van Duin, W. E., Rozema, J. & Ernst, W. H. O., 1990. Seasonal and spatial variation in the occurrence of vesicular arbuscular (VA) mycorrhiza in salt marsh plants. Agric. Ecosyst. & Environ. 29: 107—110.

Author's Address
M. L. Otte
Vrije Universiteit Amsterdam
Department of Ecology and Ecotoxicology
De Boelelaan 1087
1081 HV Amsterdam
The Netherlands

133

Phase of succession from formerly fertilized grassland to species-rich hay-meadow; recovery after disturbance. (Photograph: J. van Andel.)

CHAPTER 13

Population ecology of plant species in disturbed forest and grassland habitats

J. VAN ANDEL, J. VAN BAALEN & N. A. M. G. ROZIJN

Abstract. Disturbed habitats can be defined only with reference to undisturbed systems. Therefore, we compare species in undisturbed and disturbed forest and grassland habitats. In the present chapter, patch dynamics and environmental heterogeneity are considered characteristic of undisturbed steady state systems. For successional communities the direction of succession and the fate of populations of characteristic species are useful criteria to distinguish between undisturbed and disturbed development of a community. Population ecological processes of co-occurring species are useful determinants of disturbance in both steady state and successional communities.

The understorey plant community in undisturbed steady state forests shows variability in population processes among the co-occurring species. A change to either a uniform climax or to a clear-cut area induces disturbance of the community. A uniform climax levels off the patch dynamics of the existing species and clear-cutting results in species of quite other plant communities becoming established. The latter species may, again, vary in population ecology, because all of them are adapted to disturbed forest sites.

We reject using the term "equilibrium" for steady state grasslands and "non-equilibrium" for successional grasslands, because both types of grassland can be disturbed or undisturbed. Steady state grasslands are comparable to forests in that patch dynamics are considered part of the "normal functioning" of the system and that uniform succession, either progressive or regressive indicates disturbance. Successional grasslands may be regarded as disturbed if they show no sign of developing patches. Knowledge of population dynamics and population genetics of characteristic species is indispensable to judge whether succession proceeds undisturbed.

1. Introduction

Disturbance has been defined as "a change in conditions which interferes with the normal functioning of a biological system" (Van Andel & Van den Bergh 1987). In the present contribution we treat the population ecology of plant species *within* disturbed habitats in temperate woodlands and grasslands. In order to interpret population processes in a proper context, we should consider both undisturbed and disturbed communities. This implies that the scale of disturbance is at the level of the community, for example (i) forest clear-cutting, inducing a secondary succession of species other than understorey herbs of woods; (ii) stabilization of blowing sand in coastal dune grasslands, inducing succession from communities of mainly annuals to others dominated by perennial grasses; (iii) fertilization of successional grassland communities, inducing dominance of a few competitive species. In the present chapter, patch dynamics are considered signs of "normal functioning" of non-successional communities, while in successional communities species richness and survival of sparse species are important parameters to judge between normal functioning and disturbance of a community. This approach implies that a species is regarded as "adapted to disturbance" only if it is capable of persisting after disturbance of the community in which it exists. Generally, disturbed and undisturbed communities are composed of different plant associations, with species or ecotypes having specific ecological optima.

J. Rozema and J. A. C. Verkleij (eds.), Ecological Responses to Environmental Stresses, 136—148.

2. Woodlands

According to Whitmore (1982) forests throughout the world are fundamentally similar in their patterns in space and time, because the same processes of succession and maintenance operate. He stated that no forest is stable in the sense of being unchanging; all are in a continuous state of flux. Forests are considered to be in a "shifting mosaic steady state" (Bormann & Likens 1979), consisting of patches of gap species, pioneers and climax species (cf. also Bongers & Popma 1988). The aggregate of successional processes in a forest landscape produces an equilibrium between events that return advanced successional states to earlier states and the process of succession itself (Loucks *et al.* 1985).

Runkle (1985), defining disturbance as a force that kills at least one canopy tree, came to the somewhat surprising conclusion that different mesic forests probably do not show very great differences in their average rates of disturbance. Collins & Pickett (1987) showed that neither single-tree gaps ($33-37$ m^2) nor multi-tree gaps ($51-151$ m^2) in a hardwood forest had pronounced effects on the distribution of soil moisture or of soil air temperature. The D : H ratio (gap diameter : height of the trees) might be a useful criterion to distinguish clearings (D : H $> 2 : 1$) and gaps (Table I, Collins *et al.* 1985).

If such small-scale "disturbances" are a normal phenomenon, it is questionable whether these processes should be considered disturbances of the community (cf. Van Andel & Van den Bergh 1987). The understorey of woodlands and forests normally consists of "inflexible sun plants", "inflexible shade species" and "light-flexible herbs" (Collins *et al.* 1985). Plant species colonizing woodland clearings (Van Andel & Ernst 1985) belong to quite another type of plant associations than the ground flora of woodlands. From this point of view woodland clearings should be regarded as an indication of real disturbance of the woodland system, albeit that recovery on the long term may be possible, whereas gap species are to be considered a normal part of undisturbed forests. Species in clearings should have mechanisms to tide over a forest cycle until the next disturbance, either by dispersal in time (persistent seed bank) or by dispersal in space (other neighbouring clearings).

First we will survey understorey species, thereafter species in clearings, with emphasis on population ecological processes and life cycles.

2.1. Undisturbed woodlands

Many characteristic woodland species take advantage of gaps or ephemeral openings in the forest, either occurring naturally or through traditional coppicing. After closure of the canopy they persist as adults, in contrast to species of woodland clearings (disturbed forests) like, e.g., *Cirsium palustre* in ash coppice (Pons & During 1987) and *Digitalis purpurea* in clearings (Van Baalen 1982b), which persist as seed bank under a closed canopy, light being the principle stimulus for germination. Barkham (1980) showed that the relative proportion of adults and subadults in populations of *Narcissus pseudonarcissus* changed according to canopy conditions. Though the half-lives of adults varied from 18 to 12 years in shaded and open sites respectively, the probability of an adult *Narcissus* flowering was greater in open sites than under hardwood canopy. More seeds were produced in the open, and there was a greater chance that vegetative offspring would reach adult size in their first year. The proportion of vegetative offspring was three times greater in open sites. Results from a simulation model (Barkham & Hance 1982) suggested that genet density rises with an increased rate of clonal growth, at least until very high densities are reached. An increase in the proportion of vegetative offspring furnishes established clones with a higher probability of survival, while occasionally new genets are added to the population as a result of seed

Table I. Ratio of gap diameter to intact canopy height (D : H) in forests (from Collins *et al.*, 1985).

Gap size (D : H)	Ambient light (%)
0 : 1	10
0.5 : 1	20—45
1 : 1	45—70
2 : 1	65—90

production and germination. Barkham & Hance (1982) concluded that an oscillating environment such as that imposed by traditional coppicing may allow a population to maintain a relatively high genetic variability and to persist in sites where under a constant (shade) regime it would not. They suggested that the local abundance of other woodland species such as *Anemone nemorosa, Allium ursinum, Galeobdolon luteum* and *Mercurialis perennis* could also be the result of an advantage gained from a cyclically varying coppice environment (cf. also Mitchell & Woodward 1988). A study of *Allium ursinum* (Ernst 1979) indicated that increased light in the more open forest may correlate with increased growth of the *Allium* bulbs. Plants under a relatively open Stellario-Carpinetum canopy (65% of full radiation) had bulbs five and a half times heavier than those of plants on a south-facing Asperulo-Fagetum (40% of full radiation), and these individuals had bulbs twice as heavy as those of *Allium* on a north-facing slope (6% of full daylight). In *Fragaria vesca*, both sexual and asexual reproduction were confined to medium and high light intensity regimes, and the proportion of biomass allocated to flowering was greatest in high light intensity (Chabot 1978). Tamm (1972) reported the results of a long-term demographic study on *Primula veris*. He showed that populations of this species remained stable in an annually mown "dry meadow" inside a forest (half-life 50 years), whereas in an increasingly shaded "ash grove", the population declined rapidly (half-life 2.9 years) and a sharp decline in flowering resulted from canopy closure.

In contrast to the aforementioned understorey species, others perform better in a shaded environment. Whigham (1974) transplanted populations of *Uvularia petiolata*, a liliaceous "pseudo-annual" (Kawano 1985), and found that 22% of the emerged plants in the hardwood control forest flowered, as opposed to 13.9% in the pine wood and meadow sites, and significantly more plants reproduced vegetatively under the hardwood canopy. Seeds were produced at the hardwood and pine sites, but none of the transplants in the open field sites produced seeds, due to lack fo staphylinid beetle pollinators. For *Scilla non-scripta*, a woodland species absent from open sites, Blackman & Rutter (1950) reported that shading to 0.5–0.6 and 0.2 daylight (over a 4-year period) led to heavier bulbs (1.64 and 1.80 final/initial dry weight, versus 0.84 for control unshaded plants).

Herbaceous species in temperate woodlands have diverse phytogeographical origins; the coexistence of such different phytogeographic elements is possibly a consequence of the reshuffling of vegetation by repeated late-glacial and/or post-glacial migrations (Kawano 1985). Whether woodland species are found in woodlands or in the open, "there is no doubt that the longevity and expression of the life cycle, the development of unique production systems, reproductive behaviour and reproductive efficiency of woodland herbs have all differentiated in intimate relation with the rather regular yearly, cyclical change in environmental factors in temperate woodlands" (Kawano 1985). Apart from different groups of evergreen perennials, deciduous perennials are represented by typical spring plants, semi-shade plants and typical shade plants. Seasonal fluctuations play a part both in the understorey and in gaps.

A few studies on woodland herbs suggest that soil moisture and air humidity are of greater ecological significance than insolation or shade *per se*. Packham & Willis (1976) showed that shade-species like *Oxalis acetosella* and *Galeobdolon luteum* have a marked plasticity in their production of sun and shade leaves, but are more strictly confined to woodlands in regions of low rainfall in particular. This may also be related to temperature effects (cf. Schulze 1972). Givnish (1988) showed that the effective light-compensation point is profoundly influenced by the cost of night leaf respiration and the construction of leaf, support and root tissue, at least in trees. Moreover, shade tolerance is influenced by soil pH, probably related to nitrate vs. ammonium supply (Peace & Grubb 1982). In general, selection for efficient utilization of the resource in lowest *relative* supply, rather than single-resource limitation, has been a strong driving force behind the physiological adaptation of species to their environment (cf. Jonasson & Widerberg 1988).

2.2. Disturbed woodlands (clearings)

Disturbance of woodlands does not only damage trees, but also the understorey and the soil system. With regard to the phase of colonization (response to forest disturbance) we will first mention aspects of bacterial and soil invertebrate response, because these have received little attention so far. Thereafter we will consider population processes of plant species during secondary succession of clear-cut areas.

Soil organisms

Niemalä & Sundman (1977) showed that quite a large part of the structure of the bacterial populations was unaffected by clear-cutting of a spruce stand in southern Finland (the soil type is a moraine with a podzol profile and an uppermost layer of 4–5 cm of raw humus; the pH-H_2O of the different strata varied between 3.5 and 5.0). It seemed that the bacterial populations of the humus and of the mineral layers were not very different in the undisturbed sampling site (control). Because of clear-cutting the populations of the two layers appeared to drift apart, with the difference culminating somewhere near 7 years after clear-cutting. The populations in the mineral layer seemed to be more profoundly disturbed than those of the humus layer. Evidently, the cessation of tree root activity was a more important influence on the soil bacterial flora than the addition of litter to the surface. In the mineral soil layer in particular, clear-cutting had caused a significant relative increase in caseolytic and lipolytic, rhamnose-negative organisms which also had a low acid tolerance. Independent of this change, a short-lived increase of acid producers (from sucrose), also capable of $CaHPO_4$ dissolution, took place in the populations of the mineral layers and, to a lesser extent, in the humus layer. These organisms were lipolytic, but rhamnose-positive. After 13 years, signs of return to the original state became evident. The authors stated that, if bacterial populations contribute to successful establishment of seedlings, then at least 10 years of recovery after clear-cutting are necessary for suc-

cessful reforestation. Such an effect has, however, not been established.

A similar trend of disturbance and recovery prevailed in the invertebrate community studied by Huhta (1976), in spruce forests at two latitudes in Finland. During the first few years after clear-cutting there was a strong increase in the total biomass and community respiration of the soil animals, but from the 4–6th year onwards a return to the original state took place. By the 9–13th year after felling, the total biomass had returned to the original level. In southern Finland, the Lumbricids and Enchytraeids were mainly responsible for the temporary increase in biomass. In the north, where the Lumbricids are practically absent, it was attributable to the Enchytraeids alone. One species was strongly dominant in each of these families. The author suggested that the biomass and performance of soil animals are largely regulated by the amount of available nutrients, which is increased after clear-cutting but becomes gradually exhausted as the production of new leaf litter is negligible. It is not the actual amount of organic matter that is the principal factor, but the rate of biological processes taking place in it (Huhta 1976).

Secondary succession

The majority of woodland species is not capable of surviving after large-scale tree felling, and plant species characteristic of open areas start to colonize, either from anemochorous seeds (e.g., *Senecio sylvaticus, Chamaenerion angustifolium*) or from a persistent seed bank (e.g., *Digitalis purpurea*), or from plants which had been able to survive vegetatively under the forest canopy (e.g., *Deschampsia flexuosa, Senecio fuchsii*). These species profit from mineral nutrients in the litter layer in particular (Ernst & Nelissen 1979), but internal allocation of nutrients is species-specific (Van Andel & Jager 1981).

Population processes of plant species in woodland clearings have been reviewed earlier (Van Andel & Ernst 1985). In the present paper *Digitalis purpurea* may serve as an example of response to forest disturbance, i.c. clear-cutting. Van Baalen (1982a) studied 68 woodland clearings in the southern part of the Netherlands and adjacent parts of Belgium and Germany. *Digitalis purpurea* was one of the major components of the colonizing vegetation, frequently accompanied by *Teucrium scorodonia, Scrophularia nodosa, Rubus idaeus, Luzula luzuloides, Senecio fuchsii, Pteridium aquilinum*, and remnants or saplings of *Sambucus racemosa, Betula pendula*, and *Quercus robur*. After tree felling, low densities of *Digitalis purpurea* seedlings emerged from a buried seed bank, full daylight triggering germination (Table II, Van Baalen 1982b). Recruitment of only a small fraction of the buried seed bank (*c.* 900 seeds · m^{-2}) resulted in a mean rosette density of one or two plants per m^2, which came into flower in the next and following years. As a result, such a population may expand explosively. While early established individuals can produce secondary rosettes and flower in subsequent years, a high density of rosettes leads to a diminished rate of both vegetative growth and seed production (Van Baalen & Prins 1983). In the third year after clear-cutting, the early established large plants filled the "flowering gap" which would occur in the case of strict bienniality. A sharp decline in inflorescence density of *Digitalis purpurea* populations could be observed from the fourth or fifth year after clear-cutting onwards. Within these few years the size of the buried seed bank increased at least fifty to hundred-fold. This may be considered the major response to forest disturbance. *Digitalis purpurea* behaves as a paucennial ruderal species.

How do plant species, established in the early phase of succession in woodland clearings, respond to changing conditions of nutrient availability and light intensity during later successional stages? Growth and seed production in *Digitalis purpurea* were found to be strongly reduced and germination was inhibited in early/mid- successional vegetation. A remarkable shift occurred from fast and repeated flowering of compounded rosettes in the colonization phase to delayed flowering of monocarpic rosettes in later phases (Van Baalen & Prins 1983). This species produces seeds in which dormancy is induced by burial; these seeds may remain viable for some decades and require light for germination (Van Baalen 1982b).

In another member of the Scrophulariaceae,

Scrophularia nodosa, seed viability had decreased to 23% after one year of burial. How do adult plants of this species respond to differences in light and nutrient supply?

In a radiation gradient (Fig. 1, Van Baalen 1982a) a marked trade-off between sexual reproduction and vegetative propagation was shown: more flowering and seed production at higher radiation intensity (40 and 100% of full radiation), and exclusively rhizome production at the lowest radiation level (8% of full daylight). The effect of soil fertility was dependent on the level of radiation (interaction) as far as rhizome production is concerned; at low light intensity increasing soil fertility did not stimulate rhizome production, whereas at medium and high radiation levels it did. Capsule and seed production increased, independent of the radiation level, by the addition of nutrients. Under shaded conditions the concentration of many nutrients increased strongly in most plant parts. Apparently, *Scrophularia nodosa* is much more adapted to survive under an increasingly closed canopy than

Table IIa. Germination in the dark at 20° C (optimal soil moisture) of freshly harvested *Digitalis purpurea* seeds originating from different populations (from Van Baalen 1982b).

Population	Filter paper		Soil
	X̄	s	X̄
Vaals (clearing)	4.3	2.5	0.0
Louveigne (wood)	3.0	2.2	0.0
Epen (clearing)	67.5	7.8	0.0
Ter Graat (wood)	20.0	1.4	0.0
Winterberg (clearing)	0.0		0.0

Table IIb. Germination percentages after 30 days (20° C) on moist filter paper and moist soil at different light intensities and light qualities. (n = 4, 100 seeds per replication) (population Vaals). (from Van Baalen 1982b).

	Filter paper		Soil	
	X̄	s	X̄	s
20% daylight	99	1.4	97	2.1
5% daylight	98	1.8	59	1.8
5% leaf-filtered daylight	84	4.4	0	

Digitalis purpurea. Still more tolerant of shading during forest rehabilitation seems to be *Senecio fuchsii* (Compositae), which occurs in forest clearings and woodlands. Seeds of this species were no longer viable after one year of burial (Van Baalen et al. 1984a). The flowering response of *Senecio fuchsii* plants was more sensitive to soil fertility than to radiation intensity. Only 3–7% of the plants produced an inflorescence at the low fertility treatments under three regimes of radiation (8, 40 and 100% of full daylight). Under fertile conditions, the percentage of flowering plants amounted to 37, 67 and 90% at radiation levels of 8, 40 and 100% respectively. The total dry matter production of *Senecio fuchsii* (including rhizome production) also showed a higher sensitivity to soil fertility than to radiation intensity. This species is relatively tolerant of shade, and is also adapted to take advantage of increasing nutrient availability (and light) in cleared forest areas.

3. Grasslands

Pickett (1980) distinguished "equilibrium" and "non-equilibrium" plant communities, the former being established by uninterrupted successions, the latter being "disturbed" by factors preventing establishment of a competitive equilibrium. He stated that "disturbance occurs frequently enough in many systems to destroy or disadvantage the competitive dominants of late successional communities and so allows the coexistence of species with many degrees of competitive ability". Fox (1981) tested and confirmed Connell's (1978) "intermediate disturbance hypothesis", which states that intermediate levels of disturbance, either in space or in time, increase species diversity, due to microsuccession of species in disturbed patches. Until present, this hypothesis represents a central problem in community ecology (cf. Armstrong 1988). Even in low-productive grasslands, damage of potential dominants plays an important part in the maintenance of species diversity (Grime et al. 1987). The latter authors provided evidence that, in a microcosm in which plant communities were allowed to develop from seeds on a nutrient-poor

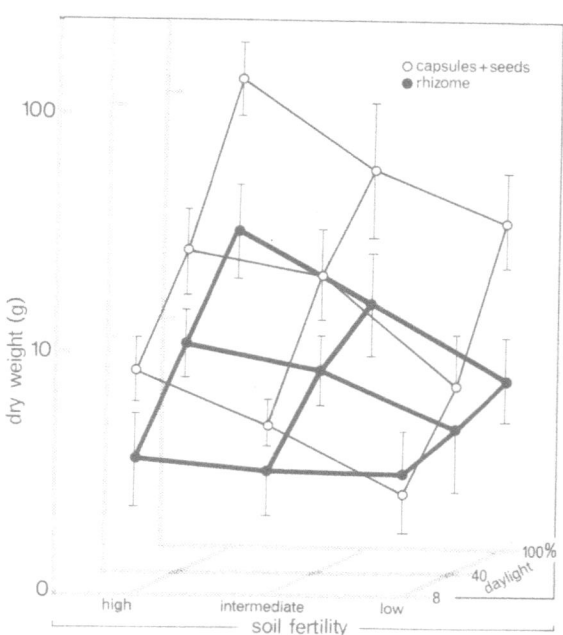

Fig. 1. Dry weight (log scale) of rhizomes (●) and of capsules + seeds (○) of *Scrophularia nodosa* grown in a 3 × 3 factorial design of radiation intensity and soil fertility (N = 10 per treatment). (from Van Baalen 1982a)

calcareous sand, artificial grazing was an important factor preventing competitive exclusion. Moreover, transfer of resources from canopy dominants (source) to subordinate species (sinks) through VA-mycorrhizal connections (cf. also Whittingham & Read 1982) advanced the coexistence of species.

The theme "disturbance in grasslands" has been thoroughly treated in an earlier volume (Van Andel, Bakker & Snaydon, eds 1987). We rejected the view of considering *a priori* "non-equilibrium grasslands" being disturbed. For example, as far as grazing by herbivores is an integral part of long-term processes in grasslands, these communities are not disturbed (cf. Drent & Prins 1987) as long as they are not overgrazed (Fresco et al. 1987). Disturbance of stable wet meadows may result from a sudden lowering of the groundwater table, thus inducing mineralization processes and the establishment of ruderal species (Grootjans et al. 1985). On the other hand, vegetation succession is the normal case in reclaimed polder areas (Beeftink 1987) or in the case of nature management to restore degraded fertilized grassland to species-rich meadows (Bakker 1987). Therefore, we will consider species abundance and population processes in both successional and steady state grasslands, each of which can be disturbed or undisturbed.

3.1. Steady state grasslands

Reference systems

Steady state natural and semi-natural grasslands may, on the analogy of forest systems, be regarded as fine-textured mosaics of pioneer and "climax" species, as was shown for prairies (Loucks et al. 1985). While in forests the trees are the matrix-forming species of the community, in grasslands the perennial herbs and grasses form the matrix, in the interstices of which short-lived plants come and go (cf. Grubb 1985). The latter author showed a remarkable year-to-year constancy in relative abundance of matrix-forming perennials in short-turf chalk grasslands. Such a rather strict hierarchy could have been maintained through interference competition (cf. also Sterk 1975). In general, the sparse plants had their leaves low in the canopy. Their persistence was not due to shade-tolerance, but to activities of grazing animals. With regard to the short-lived plants in these chalk grasslands, Grubb (1985) concluded that there were marked fluctuations in abundance with time, reflecting differences in the regeneration niche in respect to sensitivity of turf height, drought, grazing by molluscs, and the extent of the seed bank in the soil (cf. also Schenkeveld & Verkaar 1984). Small-scale disturbances should be recognized when considering the population ecology of particular species (see Grubb 1988). In the present paper they are regarded as a normal phenomenon in undisturbed grassland communities.

In low-productive coastal dune grasslands, i.e., Tortulo-Phleetum communities under the influence of blowing sand, species diversity may result from a lack of matrix-forming species, the "interstitial species" being the dominants. The majority of the latter species has an annual life cycle, and VA-mycorrhiza is usually absent (Ernst et al. 1984). Life history traits differ both among and within the annual species (Rozijn 1984). Species-specific mechanisms of germination (Table III, Rozijn &

Van Andel 1985) result in a common "drought avoidance syndrome" in the seedling phase. In the adult phase, flowering of *Aira caryophyllea* is accelerated by dry circumstances, thus avoiding drought, whereas *Aira praecox* shows drought tolerance in the flowering phase (Rozijn & Van der Werf 1986). In *Erophila verna* and *Cerastium semidecandrum*, low soil moisture content leads to a lower reproductive effort. Flowering time of *Phleum arenarium* mainly depends on soil fertility (Ernst 1983b); nutrient surplus increased the period of vegetative growth. Increased soil fertility led to a lower production of seeds in *Erophila verna*, in contrast to *Cerastium semidecandrum*. All these aspects of niche-differentiation indicate that coexistence of species in open dune grasslands depends on environmental heterogeneity. Loss of environmental fluctuations, due to, e.g., stabilization of blown sand, results in vegetation succession and dominance of perennial species. In this type of habitat, vegetation succession should be considered a sign of disturbance, environmental "unpredictability" being the normal condition to which the existing plants are adapted (Ernst 1983a). Apart from differences among species, within-species variability is an important part of adaptation to environmental fluctuations. In many winter annuals, polymorphism can be maintained through selfing, e.g., in *Phleum arenarium* (Ernst 1981) and *Erophila verna* (Van Andel et al. 1986). In the perennial outbreeding *Polygala vulgaris*, occurring on continually shifting sand around the edges of dune slacks, polymorphism among populations is also maintained due to genetic drift, resulting from founder effects and recolonization of denuded areas (Lack & Kay 1987; cf. also Wade & McCauley 1988).

Disturbance

Watt's (1947) classic studies stand out as a model for understanding how species diversity can be maintained within habitats through the process of patch dynamics. If species are arranged in patches, they may coexist much longer then when they occur in homogeneous mixtures (cf. Van Andel & Nelissen 1981; Van Andel & Dueck 1982; Van Baalen et *al.* 1984b). Within a single patch, selection may be variable in both space and time, as was shown for patches of the umbellifers *Pastinaca sativa* and *Lomatium grayi* (Thompson 1985). This means that constancy in the hierarchy of species abundance in a community may coincide with a change in the genetic composition of a population (patch), which eventually might indicate disturbance of the population, or even the community.

Schaal & Levin (1976) proposed that if the habitat deteriorated, heterosis in *Liatris cylindrica* would maintain genetic variation in the face of population bottlenecks. Similarly, Grootjans *et al.* (1987) suggested that hybrids, though they may be indicators of "hybridized habitats" and of species decline, could help to expand the genetic composition of endangered plant populations, allowing them to expand and occupy newly created habitats. Meanwhile, it is questionable which type of genetic variation is related to fitness in an ecological context. Kik (1987) showed, by applying reciprocal transplant experiments, that a huge amount of genetic variability within and among populations of *Agrostis stolonifera* (allozyme variation, ploidy levels, plasticity in relative growth rate, morphology, reproductive allocation) was overruled by environmental effects on survival and performance. Similarly, Biere (1990) found that neither germination and growth differences between seedlings from experimental crossings among genotypes of *Lychnis flos-cuculi*, nor genotypic variation in plasticity for growth, were sufficient to allow the species to become established and flower under suboptimal field conditions. Therefore, a decrease in genetic variability of a declining, isolated population (which may eventually become extinct), might be the consequence of a decline due to ecological disturbance of a habitat, rather than being the cause of extinction. This may be illustrated by a recent study on *Phyteuma nigrum*, an outbreeding perennial becoming extinct in many grassland habitats in catchment areas in the Netherlands. In a particular case the ecological causes of the decline of a small population were fairly well known (habitat deterioration). We grew progeny groups from seeds, collected from individual parent plants from this small population, which was close to extinction, and from another relatively large population as well (Van Andel *et al.* 1988). The mean overall performance of progeny from the large population was higher than that from the small one (Table IV). Though the mean performance of the latter group was lower, the variability among the progeny families was higher, due to a low within-family variability (Fig. 2). Corresponding to these results, the genetic variability of the progeny from the small population was lower, compared to the large population's progeny, at least as measured for the allozymes of LAP, GPI and GOT (Table V, Ter Steege 1988).

The important question to be asked is not whether the low genetic variability of the small population might be the cause of decline, but whether it would still be sufficient for re-establishment and

Table III. Life history characteristics during the early stages of the life-cycle of seven annual species from the open dune habitat (+ = present; − absent; * = not measured). (From Rozijn 1984).

	Persistent seedbank	Innate dormancy	Necessary germination conditions	
			Warm period	id. followed by cold temperature
Aira praecox	+	+	−	+
Cerastium semidecandrum	+	+	+	−
Erophila verna	+	+	−	+
Myosotis ramosissima	+	−	−	−
Saxifraga tridactylites	*	+	−	+
Senecio vulgaris	+	−	−	−
Veronica arvensis	+	−	−	−

expansion, provided that the conditions could be re-directed towards the optimum for this species.

3.2. Successional grasslands

In successional grassland communities, large fluctuations of abundance of species with time are the rule rather than the exception. This has been shown in many registrations of permanent plots, e.g, in developing saltmarshes (Beeftink 1987; Olff *et al.* 1988), abandoned fields (Pickett 1982), and managed semi-natural grasslands (Van den Bergh 1979; Bakker 1987). In such cases, a change in abundance of species should not *a priori* be considered disturbance of the community. It depends on the criteria for development, chosen before the start of management. In former agricultural valley grasslands, managed to become restored to species-rich communities, Bakker (1987) detected during a period of 15 years an increase in the ratio of abundance between species indicating nutrient-poor conditions, relative to those indicating nutrient-rich conditions. In general, after cessation of fertilizer applications in these grasslands, a successional series of communities can be expected, characterized by respectively *Lolium perenne*, *Holcus lanatus*, *Festuca rubra* and, in the long run, *Juncus acutiflorus*. In the early phase of succession, i.e., in the *Holcus lanatus* community, fluctuations in cover percentages of species are more frequently related to wheather conditions than those in the *Juncus acutiflorus* community (Fresco *et al.* 1989). This indicates that the early successional community is more sensitive to changes in conditions, following the "disturbance" of the *Lolium perenne* community, than later successional stages. However, peak

density of the interstitial, annual species *Rhinanthus angustifolius*, was always reached 6–12 years after mowing for hay without fertilizer application (Ter Borg 1985), thus indicating a normal successional phenomenon. The regeneration niche of *Rhinanthus angustifolius* is an important aspect of the response of this species to a change in the structure of the canopy. In the early phase of management, large plants produce relatively large seeds which germinate earlier, compared to the smaller seeds produced in later phases of succession, due to plasticity of individual plants (De Hullu 1985; Ernst *et al.* 1987). Expanding populations of this species were also monitored in a species-rich wet fen meadow, as a result of drainage activities (De Hullu & Grootjans 1987). In this case, the *Rhinanthus* population indicated disturbance of the grassland community, which should normally have remained stable.

Results from the Park Grass Experiment at Rothamsted (U.K.) suggest that hay-meadow grasslands reach "equilibrium" after a certain period of management, due to both exogenous and endogenous factors (Silvertown 1980). He suggested that populations were regulated by processes operating within individual years. Dramatic increases in the populations of individual species following years of exceptional drought, were rapidly stabilized by a subsequent decrease of the species concerned. This example illustrates that, as soon as successional grasslands have reached a certain steady state as a result of a long-term management regime, fluctuations in abundance are part of the normal functioning of the community; in these

Table IV. Mean total dry weight (mg ± 1 SE) of progeny from two field populations of *Phyteuma nigrum*, after 10 weeks growth in the glasshouse in differently sized pots (with 400 and 60 g soil, respectively) (from Van Andel *et al.* 1988).

Treatment	Population	
	Meander (large)	Diepveen (small)
Large pots	1167.5 ± 62.8	876.4 ± 47.2
Small pots	434.5 ± 20.2	342.5 ± 19.7

Table V. Allozymes detected by means of electrophoretic analyses of progeny from two populations of *Phyteuma nigrum*: c. 90 seedlings from 19 parent plants in a large population, and c. 30 seedlings from 9 parent plants in a small population. S: slow, I: intermediate, F: fast (from Ter Steege 1988).

Enzyme	Population	
	Meander (large)	Diepveen (small)
L A P	S, I, F	F
G P I	S, I, F	I
G O T	S, I, F	I, F

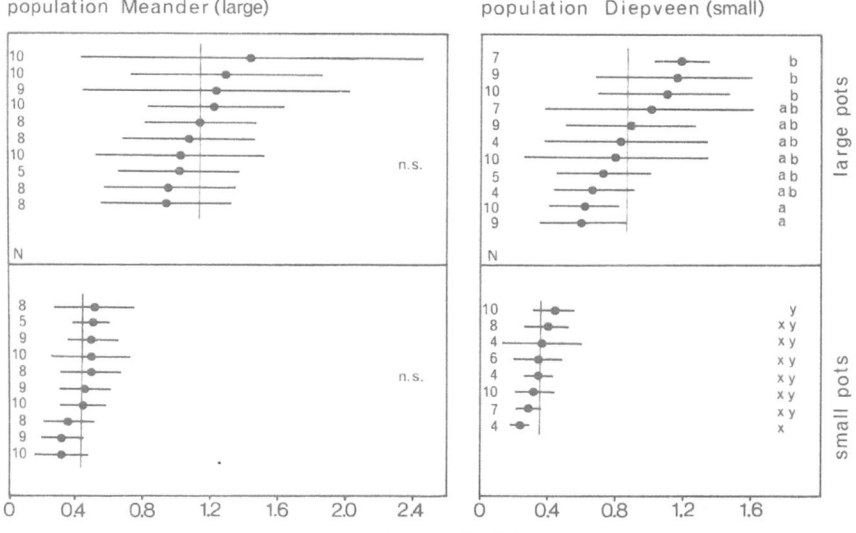

population Meander (large)   population Diepveen (small)

Fig. 2. Mean production (± 1 SD) of progeny groups, grown from seeds of individual parents in two differently sized field populations of *Phyteuma nigrum*. Plants were grown individually in pots of two different sizes. Mean value per treatment is given by a vertical bar. Significant differences between progeny groups are indicated (a *vs* b, x *vs* y). (from Van Andel *et al.* 1988)

types of grasslands, succession should be regarded as indicating community disturbance (see section 3.1).

4. Concluding remarks

Whitmore (1982) stated: "Where conservation values prevail, mature phase forest and a range of patches are all needed, for this maximizes ecosystem and hence niche diversity and species richness of plants as well as animals." This ecologically important conclusion does not only hold for forest communities, but — on the analogy — also for grassland communities maintained at "climax" conditions. Disturbance results from levelling off the differentiation in patch dynamics, either through regressive damage to the existing vegetation or through induction of progressive succession of all of the patches. The fate of a species in a community can help evaluate whether a community is "normally functioning" or "disturbed", but this should be considered in the context of knowledge of the plant associations concerned.

Many understorey species of woodlands, in par-

ticular those profiting by full day-light, can also be found in meadows. Part of these may originate from primary stands on the margins of woods. European grasslands expanded as man destroyed the forest by cutting and fire. At first pastures developed gradually by grazing and browsing of cattle in woodlands. Scholz (1975) stated that, for a proper understanding of grassland evolution, it is not sufficient to consider cultural effects alone, but also the course of events since the pleistocene glaciation, e.g. immigration and hybridization-differentiation cycles. Independent of the evolutionary history of meadow and pasture plants within the region of the summergreen deciduous forest, grasslands should not be regarded as disturbed forests at the present. If a plant species is capable of occurring in both woodlands and meadows, sometimes in different geographic regions, it can be "normally functioning" or "disturbed" in each of these community types. For example, the biotopes of *Orchis mascula* in Sweden vary from open, dry meadows to shady groves (Nilsson 1983), while in the mediterranean region this species is found in woodlands (E. Dijk, pers. comm.). *Primula veris* persisted for a long time in an undisturbed wooded meadow, whereas it declined (became disturbed) after canopy closure (Tamm 1972). These two observations can also be interpreted as an indication that this species is capable of persisting in undisturbed forests containing patches of ephemeral gaps. Something similar may hold for *Senecio fuchsii* and *Phyteuma nigrum*.

Lovett Doust (1987) investigated clones of *Ranunculus repens* in adjacent woodland and grassland sites, and showed that plants from each population maintained locally specialized patterns of dry matter allocation, though this was modified by particular combinations of light intensity and nutrient supply. The grassland plants seemed to consolidate low supplies of resources in the parent rosette, whereas the woodland plants consolidated them in daughter ramets. This exaxmple suggests that differences in vegetation density between grassland and woodland communities may be of crucial importance with regard to sensitivity to disturbance. If competition is stronger in grasslands than in the understorey of woodlands, grasslands may be more sensitive of small changes in conditions, e.g. in light intensity inside the vegetation.

References

Armstrong, R. A., 1988. The effects of disturbance patch size on species coexistence. J. theor. Biol. 133: 169—184.

Bakker, J. P., 1987. Restoration of species-rich grassland after a period of fertilizer application. In: J. Van Andel, J. P. Bakker & R. W. Snaydon (eds.). Disturbance in Grasslands, pp. 185—200. Junk Publ., Dordrecht.

Barkham, J. P., 1980. Population dynamics of the wild daffodil (*Narcissus pseudonarcissus*). I. Clonal growth, seed reproduction, mortality and the effects of density. J. Ecol. 68: 607—633.

Barkham, J. P. & Hance, C. E., 1982. Population dynamics of the wild daffodil (*Narcissus pseudonarcissus*). III. Implications of a computer model of 1000 years of population change. J. Ecol. 70: 323—344.

Beeftink, W. G., 1987. Vegetation responses to changes in tidal inundation of salt marshes. In: J. Van Andel, J. P. Bakker & R. W. Snaydon (eds.). Disturbance in Grasslands, pp. 97—117. Junk Publ., Dordrecht.

Biere, A., 1990. Phenotypic variation in *Lychnis flos-cuculi*. Thesis Univ. Groningen.

Blackman, G. E. & Rutter, A. J., 1950. Physiological and ecological studies in the analysis of plant environment. V. An assessment of factors controlling the distribution of the bluebell (*Scilla non-scripta*) in different communities. Ann. Bot. (London) 14: 407—420.

Bongers, F. J. J. M. & Popma, J. A. M., 1988. Trees and gaps in a Mexican tropical rain forest. Thesis Univ. Utrecht.

Bormann, F. H. & Likens, G. E., 1979. Pattern and Process in a Forested Ecosystem. Springer Verlag, Berlin.

Chabot, B. F., 1978. Environmental influences on photosynthesis and growth in *Fragaria vesca*. New Phytol. 80: 87—98.

Collins, B. S., Dunne, K. P. & Pickett, S. T. A., 1985. Responses of forest herbs to canopy gaps. In: S. T. A. Pickett & White (eds.). The Ecology of Natural Disturbance and Patch dynamics, pp. 217—234. Acad. Press, London.

Collins, B. S. & Pickett, S. T. A., 1987. Influence of canopy opening on the environment and herb layer in a northern hardwoods forest. Vegetatio 70: 3—10.

Connell, J. H., 1978. Diversity in tropical rain forests and coral reefs. Science 199: 1302—1310.

De Hullu, P. C., 1985. Population dynamics of *Rhinanthus angustifolius* in a succession series. Thesis Univ. Groningen.

De Hullu, P. C. & Grootjans, A. P., 1987. Population responses of *Rhinanthus angustifolius* to disturbance of grassland communities. In: J. Van Andel, J. P. Bakker & R. W. Snaydon (eds.). Disturbance in Grasslands, pp. 265—271. Junk Publ., Dordrecht.

Drent, R. H. & Prins, H. H. T., 1987. The herbivore as prisoner of its food supply. In: J. Van Andel, J. P. Bakker & R. W. Snaydon (eds.). Disturbance in Grasslands, pp. 131—147. Junk Publ., Dordrecht.

Ernst, W. H. O., 1979. Population biology of *Allium ursinum* in Northern Germany. J. Ecol. 67: 347—362.

Ernst, W. H. O., 1981. Ecological implications of fruit variability in *Phleum arenarium*, an annual dune grass. Flora 171: 387—398.

Ernst, W. H. O., 1983a. Anpassungsstrategien einjähriger Dünenpflanzen. Verh. Ges. Ökol. 10: 485—495. Göttingen.

Ernst, W. H. O., 1983b. Element nutrition of two contrasted dune annuals. J. Ecol. 71: 197—209.

Ernst, W. H. O. & Nelissen, H. J. M., 1979. Growth and mineral nutrition of plant species from clearings on different horizons of an iron-humus podzol profile. Oecologia 41: 175—182.

Ernst, W. H. O., Van Duin, W. E. & Oolbekking, G. T., 1984. Vesicular-arbuscular mycorrhiza in dune vegetation. Acta Bot. Neerl. 33: 151—160.

Ernst, W. H. O., Nelissen, H. J. M. & De Hullu, P. C., 1987. Size hierarchy and mineral status of *Rhinanthus angustifolius* populations under different grassland management regimes. Vegetatio 70: 93—103.

Fox, J. F., 1981. Intermediate levels of soil disturbance maximize alpine plant diversity. Nature 293: 564—565.

Fresco, L. F. M., Van Laarhoven, H. P. M., Loonen, M. J. J. E. & Moesker, T., 1987. Ecological modeling of short-term plant community dynamics under grazing with and without disturbance. In: J. Van Andel, J. P. Bakker & R. W. Snaydon (eds.). Disturbance in Grasslands, pp. 149—165. Junk Publ., Dordrecht.

Fresco, L. F. M., Strijkstra, R. & Bakker, J. P., 1989. Fluctuations in coverage of grassland species in relation to wheather conditions. Acta Bot. Neerl. 38: 355–356.

Givnish, T. J., 1988. Adaptation to sun and shade: A whole-plant perspective. Austr. J. Plant Physiol. 15: 63—92.

Grime, J. P., Mackey, J. M. L., Hillier, S. H. & Read, D. J., 1987. Floristic diversity in a model system using experimental microcosms. Nature 328: 420—422.

Grootjans, A. P., Schipper, P. J. & Van der Windt, H. J., 1985. Influence of drainage on N-mineralization and vegetation response in wet meadows. I. Calthion palustris stands. Acta Oecologia/Oecol. Plant. 6: 403—417.

Grootjans, A. P., Allersma, G. J. R. & Kik, C., 1987. Hybridization of the habitat in disturbed hay meadows. In: J. Van Andel, J. P. Bakker & R. W. Snaydon (eds.). Disturbance in Grasslands, pp. 67—77. Junk Publ., Dordrecht.

Grubb, P. J., 1985. Problems posed by sparse and patchily distributed species in species-rich plant communities. In: J. Diamond & T. J. Case (eds.). Community Ecology, pp. 207—225. Harper & Row, New York.

Grubb, P. J., 1988. The uncoupling of disturbance and recruitment, two kinds of seed bank,and persistence of plant populations at the regional and local scales. Ann. Zool. Fennici 25: 23—36.

Huhta, V., 1976. Effect of clear-cutting on numbers, biomass and community respiration of soil invertebrates. Ann. Zool. Fennici 13: 63—80.

Jonasson, S. & Widerberg, B., 1988. The resource balance of Milium effusum with emphasis on environmental resource supply. Oecologia 76: 11—19.

Kawano, S., 1985. Life history characteristics of temperate woodland plants in Japan. In: J. White (ed.). The Population Structure of Vegetation, pp. 515—549. Junk Publ., Dordrecht.

Kik, C., 1987. On the ecological genetics of the clonal perennial Agrostis stolonifera. Thesis Univ. Groningen.

Lack, A. J. & Kay, Q. O. N., 1987. Genetic structure, gene flow and reproduction ecology in sand-dune populations of Polygala vulgaris. J. Ecol. 75: 259—276.

Loucks, O. L., Plumb-Mentjes, M. L. & Rogers, D., 1985. Gap processes and large-scale disturbances in sand prairies. In: S. T. A. Pickett & P. S. White (eds. 1985). The Ecology of Natural Disturbance and Patch Dynamics, pp. 71—83. Acad. Press, London.

Lovett Doust, L., 1987. Population dynamics and local specialization in a clonal perennial (Ranunculus repens). III. Responses to light and nutrient supply. J. Ecol. 75: 555—568.

Mitchell, P. L. & Woodward, F. J., 1988. Responses of three woodland herbs to reduced photosynthetically active radiation and low red to far-red ratio in shade. J. Ecol. 76: 807—825.

Niemalä, S. & Sundman, V., 1977. Effects of clear-cutting on the composition of bacterial populations of northern spruce forest soil. Can. J. Microbiol. 23: 131—138.

Nilsson, L. A., 1983. Antheology of Orchis mascula (Orchidaceae). Nord. J. Bot. 3: 157—179.

Olff, H., Bakker, J. P. & Fresco., L. F. M., 1988. The effect of fluctuations in tidal inundation frequency on a salt-marsh vegetation. Vegetatio 78: 13—20.

Packham, J. R. & Willis, A. J., 1976. Aspects of the ecological amplitude of two woodland herbs, Oxalis acetosella L. and Galeobdolon luteum Huds., J. Ecol. 64: 485—510.

Peace, W. J. H & Grubb, P. J., 1982. Interaction of light and mineral nutrient supply in the growth of Impatiens parviflora. New Phytol. 90: 127—150.

Pickett, S. T. A., 1980. Non-equilibrium coexistence of plants. Bull. Torrey Bot. Club 107: 238—248.

Pickett, S. T. A., 1982. Population patterns through twenty years of old field succession. Vegetatio 49: 45—59.

Pons, Th. L. & During, H. J., 1987. Biennial behaviour of Cirsium palustre in ash coppice. Holarctic Ecol. 10: 40—44.

Rozijn, N. A. M. G., 1984. Adaptive strategies of some dune annuals. Thesis Free University, Amsterdam.

Rozijn, N. A. M. G. & Van Andel, J., 1985. Analysis of the germination syndrome of dune annuals. Flora 177: 175—185.

Rozijn, N. A. M. G. & Van der Werf, D. C., 1986. Effect of drought during different stages in the life-cycle on the growth and biomass allocation of two Aira species. J. Ecol. 74: 507—523.

Runkle, J. R., 1985. Disturbance regimes in temperate forests. In: S. T. A. Pickett & P. S. White (eds.). The Ecology of Natural Disturbance and Patch Dynamics, pp. 17—33. Acad. Press, London.

Schaal, B. A. & Levin, D. A., 1976. The demographic genetics of Liatris cylindrica Michx. (Compositae). Amer. Nat. 110: 191—206.

Schenkeveld, A. J. M. & Verkaar, H. J. P. A., 1984. On the ecology of short-lived forbs in chalk grasslands. Thesis Univ. Utrecht.

Scholz, H., 1975. Grassland evolution in Europe. Taxon 24: 81—90.

Schulze, E.-D., 1972. Die Wirkung von Licht und Temperatur auf den CO_2-Gaswechsel verschiedener Lebensformen aus der Krautschichte eines montanen Buchenwaldes. Oecologia 9: 235—258.

Silvertown, J., 1980. The dynamics of a grassland ecosystem: Botanical equilibrium in the Park Grass Experiment. J. Appl. Ecol. 17: 491—504.

Sterk, A. A., 1975. Demographic studies of Anthyllis vulneraria L. in the Netherlands. Acta Bot. Neerl. 24: 315—337.

Tamm, C. O., 1972. Survival and flowering of perennial herbs. III. The behaviour of Primula veris on permanent plots. Oikos 23: 159—166.

Ter Borg, S. J., 1985. Population biology and habitat relations of some hemiparasitic Scrophulariaceae. In: J. White (ed.). The Population Structure of Vegetation, pp. 463—487. Junk Publ., Dordrecht.

Ter Steege, M. W., 1988. Genetische variatie in populaties van Phyteuma nigrum. Internal report Depts. of Plant Ecology and Pop. Genetics. Univ. Groningen.

Thompson, J. N., 1985. Within-patch dynamics of life histories, populations, and interactions: Selection over time in small spaces. In: S. T. A. Pickett & P. S. White (eds.). The Ecology of Natural Disturbance and Patch Dynamics, pp. 253—264. Acad. Press, London.

Van Andel, J. & Van den Bergh, J. P., 1987. Disturbance of grasslands. Outline of the theme. In: J. Van Andel, J. P. Bakker & R. W. Snaydon (eds.). Disturbance in Grasslands, pp. 3—13. Junk Publ., Dordrecht.

Van Andel, J. & Dueck, T., 1982. The importance of the physical pattern of plant species in replacement series. Oikos 39: 59—62.

Van Andel, J. & Ernst, W. H. O., 1985. Ecophysiological adaptation, plastic responses, and genetic variation of annuals, biennials and perennials in woodland clearings. In: J. Haeck & J. W. Woldendorp (eds.). Structure and Functioning of Plant Populations II, pp. 27—49. North-Holland Publ. Comp., Amsterdam.

Van Andel, J. & Jager, J. C., 1981. Analysis of growth and nutrition of six plant species of woodland clearings. J. Ecol. 69: 871—882.

Van Andel, J. & Nelissen, H. J. M., 1981. An experimental approach to the study of species interference in a patchy vegetation. Vegetatio 45: 155—163.

Van Andel, J., Rozijn, N. A. M. G., Ernst, W. H. O. & Nelissen, H. J. M., 1986. Variability in growth and reproduction in F_1-families of an *Erophila verna* population. Oecologia 69: 79—85.

Van Andel, J., Wesselingh, R. A. & Van Donk, H. J., 1988. The performance of progeny groups from two populations of *Phyteuma nigrum*, with particular reference to the chance of survival or extinction. Acta Bot. Neerl. 37: 165—169.

Van Andel, J., Bakker, J. P. & Snaydon, R. W. (eds.), 1987. Disturbance in Grasslands. Causes, effects and processes. 316 pp. Junk, Publ., Dordrecht.

Van Baalen, J., 1982a. Population biology of plants in woodland clearings. Thesis Free University, Amsterdam.

Van Baalen, J., 1982b. Germination ecology and seed population dynamics of *Digitalis purpurea*. Oecologia 53: 61—67.

Van Baalen, J. & Prins, E. G. M. N., 1983. Growth and reproduction of *Digitalis purpurea* in different stages of succession. Oecologia 58: 84—91.

Van Baalen, J., Nelissen, H. J. M., Ernst, W. H. O., Wattel, J. & Vooys, R., 1984a. Reproductive processes in *Senecio fuchsii* (partly in comparison with *Eupatorium cannabinum*) as affected by temperature, irradiance and soil fertility. Flora 175: 81—90.

Van Baalen, J., Kuiters, A. Th. & Van der Woude, C. S. C., 1984b. Interference of *Scrophularia nodosa* and *Digitalis purpurea* in mixed seedling cultures, as affected by the specific emergence date. Acta Oecologica/Oecol. Plant. 5 (19): 279—290.

Van den Bergh, J. P., 1979. Changes in the composition of mixed populations of grassland species. In: M. J. A. Werger (ed.). The Study of Vegetation, pp. 57—80. Junk Publ., Den Haag (Dordrecht).

Wade, M. J. & McCauley, D. E., 1988. Extinction and recolonization: Their effects on the genetic differentiation of local populations. Evolution 42: 995—1005.

Watt, A. S., 1947. Pattern and process in the plant community. J. Ecol. 35: 1—22.

Whitmore, T. C., 1982. On pattern and process in forests. In: E. J. Newman (ed.). The Plant Community as a Working Mechanism, pp. 45—59. British Ecological Society.

Whittingham, J. & Read, D. J., 1982. Vesicular-arbuscular mycorrhiza in natural vegetation systems. III. Nutrient transfer between plants with mycorrhizal interconnections. New Phytol. 90: 277—284.

Whigham, D., 1974. An ecological life history study of *Uvularia petiolata* L. Amer. Midl. Nat. 91: 343—359.

Authors' addresses
J. van Andel
Laboratorium voor Plantenoecologie
Rijksuniversiteit Groningen,
Postbus 14
9750 AA Haren
The Netherlands

J. van Baalen
Ministerie van Landbouw en Visserij
Directie NMF
Postbus 20401
2500 EK 's-Gravenhage
The Netherlands

N. A. M. G. Rozijn
Vrije Universiteit Amsterdam
Postbus 7161
1007 MC Amsterdam
The Netherlands

Chara globularis, occurring in Botshol, containing oögonia and antheridia (Photograph. A. P. van Beem).

CHAPTER 14

Ecological responses of macro- and microphytic algae to water pollution

J. SIMONS & P. J. R. DE VRIES

General Introduction

In The Netherlands there are many small fresh-water habitats where non-planktonic algae represent an important and often dominating biotic community. In recent time extensive blanketing algal mats and closed layers of duckweeds have become very common due to increased contamination by agricultural wastewater and manuring of pastures. Prolific algal growth may outcompete the aquatic macrophytes by reducing irradiance (Phillips *et al.* 1978) and by directly hindering photosynthesis by increasing pH and reducing the CO_2 content of the water (Simpson & Eaton 1986).

In terms of biomass the most important compounds of the filamentous algal community in shallow stagnant waters are species of *Cladophora* (mostly *C. glomerata*), *Enteromorpha intestinalis*, *Hydrodictyon reticulatum*, *Oedogonium*, *Mougeotia*, *Spirogyra*, *Zygnema*, *Tribonema*, *Vaucheria*. *Spirogyra* and *Oedogonium* are represented by the largest number of species. Some members of the Charales, of which *Chara vulgaris* and *C. globularis* are the most common ones, also occur, especially at sites which are not strongly contaminated.

In the period from 1980–1987 we investigated several aspects of biology and distribution of the zygnematacean green algae, especially *Spirogyra* and accompanying filamentous algae (Simons 1987; Simons *et al.* 1982, 1984, 1990). Algal mats of *Spirogyra* grow rapidly in spring and early summer. There is a peak in conjugation and zygospore pro-duction from late May to early June. Alkalinity is an important factor for the distribution of species. Only few species occur in low alkaline and slightly acid (pH 5.5–7) waters. In this type of habitat *Spirogyra* species are replaced by representatives of *Mougeotia* and *Zygnema*. Most *Spirogyra* species are rather sensitive to eutrophication.

Besides macroscopic filamentous algae, we also studied different aspects of the smaller benthic algae belonging to the periphyton community. An important member of this community is the heterotrichous chaetophoralean green alga *Stigeoclonium*. We clarified the confusing taxonomy by reducing the number of species to three (Simons *et al.* 1986), and in one strain we observed sexuality indicating a diplontic instead of the assumed haplontic type of life cycle (Simons & Van Beem 1987). In earlier studies some nice examples of ecotypic differentiation within a species were discovered (Francke & Rhebergen 1982; Francke & Den Oude 1983). *Stigeoclonium* occurs over a wide range of habitats. Therefore this alga was investigated for its use as bioassay organism by De Vries (1986). An application of this approach is presented in this chapter.

In The Netherlands, eutrophication of shallow peat lakes in the central western part of the country has impoverished the aquatic communities which were once rich and very diversified with among others many species of Characeae. In this chapter, an example is described of the nature reserve area de Botshol, where in recent years the characean

J. Rozema and J. A. C. Verkleij (eds.), Ecological Responses to Environmental Stresses, 150–169.

community has nearly disappeared. Restoration measures are planned and the process will be followed by monitoring research.

References

De Vries, P. J. R., 1986. Bioassays on water quality using the attached filamentous alga *Stigeoclonium* Kütz. Thesis Free University, Amsterdam.

Francke, J. A. & Rhebergen, L. J., 1982. Euryhaline ecotypes in some species of *Stigeoclonium* Kütz. Br. phycol. J. 17: 135—145.

Francke, J. A. & Den Oude, P. J., 1983. Growth of *Stigeoclonium* and *Oedogonium* species in artificial ammonium-N and phosphate-P gradients. Aquat. Bot. 15: 375—380.

Phillips, G. L., Eminson, D. F. & Moss, B., 1978. A mechanism to account for macrophyte decline in progressively eutrophicated freshwaters. Aquat. Bot. 4: 103—126.

Simons, J., Van Beem, A. P. & De Vries, P. J. R., 1982. Structure and chemical composition of the spore wall in *Spirogyra* (Zygnemataceae, Chlorophyceae). Acta Bot. Neerl. 31: 359—370.

Simons, J., De Vries, P. J. R. & Van Beem, A. P., 1984. Induction of conjugation and spore formation in species of *Spirogyra* (Chlorophyceae, Zygnematales). Acta Bot. Neerl. 33: 323—334.

Simons, J., Van Beem, A. P., & De Vries, P. J. R., 1986. Morphology of the prostrate thallus of *Stigeoclonium* (Chlorophyceae, Chaetophorales) and its taxonomic implications. Phycologia 25: 210—220.

Simons, J. & Van Beem, A. P., 1987. Observations on asexual and sexual reproduction in *Stigeoclonium helveticum* Vischer (Chlorophyta) with implications for the life history. Phycologia 26: 356—362.

Simons, J., 1987. *Spirogyra* species and accompanying algae from dune waters in The Netherlands. Acta Bot. Neerl. 36: 13—31.

Simons, J. & Van Beem, A. P., 1990. *Spirogyra* species and accompanying algae from pools and ditches in The Netherlands. Aquat. Bot. 37: 247–269.

Simpson, P. S. & Eaton, J. W., 1986. Comparative studies of the photosynthesis of the submerged macrophyte *Elodea canadensis* and the filamentous algae *Chladophora glomerata* and *Spirogyra* sp. Aquat. Bot. 24: 1—12.

Assessment of eutrophication using the periphytic alga Stigeoclonium as bioassay organism

P. J. R. DE VRIES

Abstract. Bioassays using *Stigeoclonium tenue* Kütz., an epiphytic filamentous green alga, were conducted monthly on water from 22 sites in ditches situated in the central western part of The Netherlands, including 6 sites near discharges of waste water from farms and sewage treatment plants as a pollution source. Considerable concentrations of N and P were measured at sites near the discharge, usually resulting in high algal yields. However, in some tests growth inhibition occurred, presumably by toxic substances. At distant sites with lower yields, a primary and frequently a secondary limiting nutrient could be detected. In most cases algal yields were enhanced by the addition of nitrogen during July—September, and by phosphorus addition during the winter. The yields measured in winter were mostly higher than those from summer. Phosphorus limitation was observed when ratio's of total N to total P were higher than 12. The information from bioassays was related to that derived from the periphytic algal species composition at 19 out of the 22 sites. It appeared that the ranking sequence of sites based on bioassay yields was significantly correlated with the ranking sequence of sites based on the periphytic algal species composition. The lowest yields occurred at sites which were characterized as habitats with relatively good water quality, based on the algal species assemblage.

1. Introduction

Eutrophication defined as the overenrichment of surface water with nutrients, has become a serious water management concern. The resulting increase in fertility causes symptoms such as plankton blooms, heavy growth of periphyton and of certain aquatic plants. These effects often influence adversally the vital uses of water, such as supply of drinking water, fishering, recreation, irrigation and aesthetic qualities.

For the objective assessment of the significance of eutrophication in water bodies, it is essential that adequate studies are undertaken in order to plan the measures necessary to improve water quality to the standards required. For a good planning, background data for measuring and monitoring eutrophication are needed. One of the recommended methods to achieve ecological background information is the algal bioassay. In the broadest sense,

a bioassay is a determination of the biological effects of some substance or environmental condition, and includes the use of organisms to detect or to measure the concentration of substances or to indicate the nature of physical conditions in the environment (Weber 1973).

The algal assay may be used for investigation of water to determine its nutrient status and sensitivity to change, for evaluation of materials and products to determine their potential effects including toxicity on algal growth in receiving waters, or for assessment of effects of changes in wastewater treatment processes on receiving waters.

Most algal assays are very similar in theory. After elimination of the indigenous organisms, algae are inoculated into water samples with or without enrichments. Incubation is for days under standard conditions. Growth, or lack of it, is used as an indication of available nutrients or toxic substances.

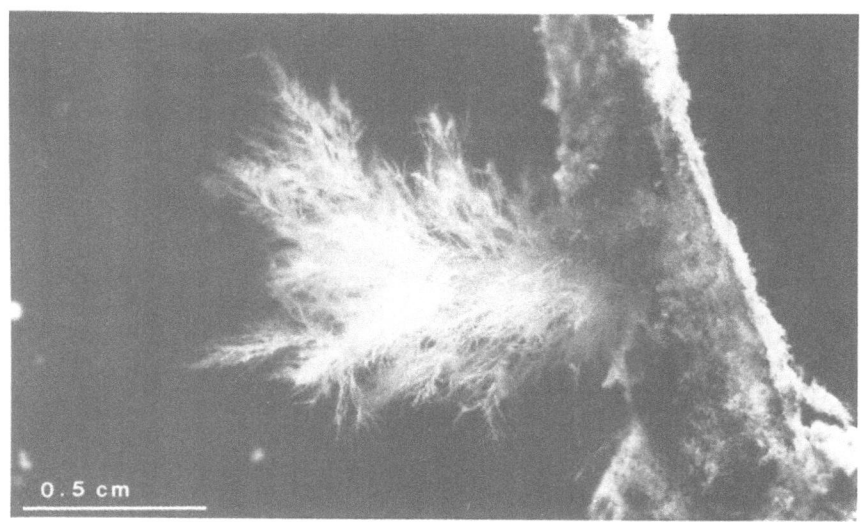

Fig. 1. Macroscopical visible plant of *Stigeoclonium helveticum* on a reed stem.

Bioassays, using planktonic algae, are mainly applied in studies on lakes and rivers. Small water bodies with an extensive littoral zone, like ditches, ponds and canals, have not often been studied. In bioassays with samples of small waterbodies the use of a dominant member of the local community as a test organism seems more advisable. Therefore a bioassay procedure has been developed using the attached filamentous alga *Stigeoclonium* Kütz (Chlorophyceae) as test organism (De Vries 1986). This periphytic alga is a typical compound of small waterbodies. In the present study the assay procedure is applied to examine its suitability for monitoring aspects of eutrophication in small water bodies, in this case ditches situated in cattle pasture areas.

2. The testalga

The genus *Stigeoclonium*, established by Kützing (1843), includes all attached, branched, uniseriate, filamentous green algae of which the cells of the main axis and the branches are similar in size. In the field the alga may be visible as small pencil-like tufts on macrophytes or other objects (Fig. 1). The thallus is heterotrichous and consists of a prostrate and an erect system of filaments. The terminal cells of the erect filaments may produce multicellular, hyaline hairs. Both prostrate and erect cells may produce rhizoids. Reproduction is predominantly a-sexual by quadriflagellate zoospores. Recently (Simons *et al.* 1986) a taxonomic revision based on features of the prostrate system was presented resulting in the recognition of only three species, viz. *Stigeoclonium helveticum*, *S. tenue* and *S. farctum*.

Stigeoclonium tenue, a common alga in the periphyton throughout the year and tolerant over a wide range of environmental factors (Francke 1982), is used in this study.

3. The bioassay procedure

Stigeoclonium can be isolated from the field by means of artificial substrates. In the laboratory unialgal strains are cultured from zoospores which can easily be induced. Stock cultures are maintained in a synthetic medium at $16°C$, $50–80\,\mu E$ $m^{-2}\,s^{-1}$ (PAR), a $12:12$ light-dark regime and pH 7–7.5. Prior to experiments the algal material is placed in medium without nitrogen and phosphorus to avoid carry over of luxury uptaken nutrients to the assay. Samples of testwater are autoclaved to

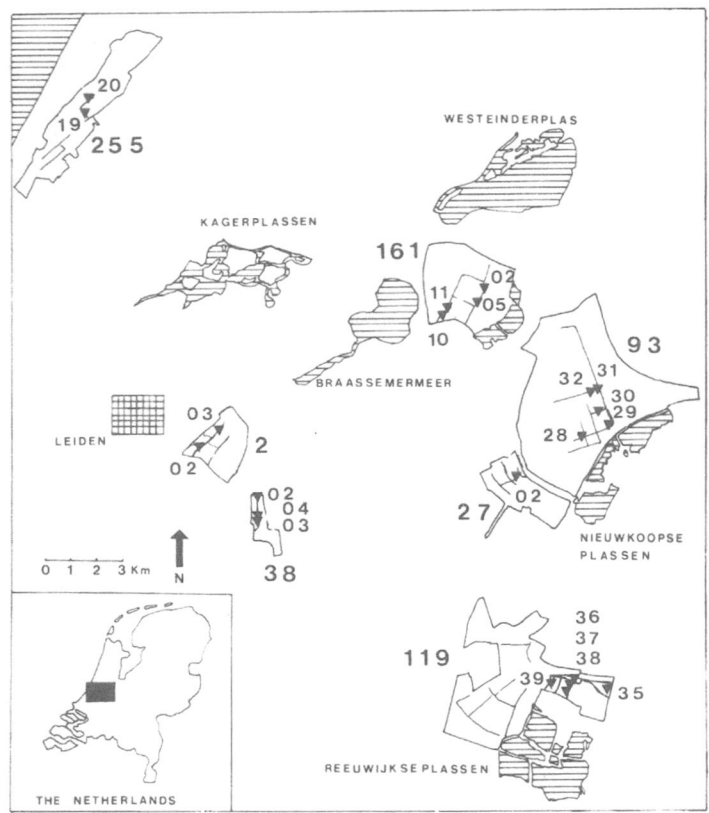

Fig. 2. Map of sampling area with location of sites. Site numbers are observation sites of the waterboard "Hoogheemraadschap Rijnland".

inactivate indigenous algae and bacteria. Assays are carried out using culture tubes. The yield is determined when the stationary growth phase is reached, after 3 weeks of incubation. A rapid and accurate quantification of the biomass is achieved by sonification of the algae in the culture tube for 30 sec, and measuring the optical density at 798 nm. For further details, see De Vries 1986.

4. Study sites and procedure

The sample sites were located in the central western part of the Netherlands (Fig. 2) and were selected by the waterboard agency "Hoogheemraadschap Rijnland" with respect to differences in water quality. When sites were located in ditches with

discharge of wastewater from farms or sewage treatment plants, one sample was taken near the point of discharge and another at some distance from the pollution source (Table I). Testwater was collected monthly in 1984 and preserved by deep freezing. Data on physical and chemical parameters of the testwater were kindly provided by the waterboard "Hoogheemraadschap Rijnland".

5. Results

5.1. Correlation between P- en N-concentrations and bioassay growth yields

Judging from the results as shown in Fig. 3, it appears that the correlation between algal growth yields and P- and N-concentration in the ditch waters is not consistent. At several sites the trend in P- and N-concentrations considered over a whole year is more or less reflected in the growth yields (e.g., the sites 119.35, 119.37 and 119.39; 38.04; 161.10; 93.29). At some sites there is a stronger correlation with the N-concentrations than with the P-concentrations (e.g., the sites 225.19, 161.05, 161.02, 93.29).

5.2. Seasonality in growth yields

Most sites show a decrease in the concentration of nitrogen and/or phosphorus during the spring and summer months and an increase during autumn and winter (e.g., the sites 2.04, 38.04, 27.02, 161.02, 93.30). Mostly, such sites have a lowered growth potential for *Stigeoclonium* during late spring and summer (e.g., sites 119.35, 119.38, 2.04, 255.19, 38.03, 161.10, 161.11). The average monthly yield and standard deviation of all relevant samples are presented in Fig. 4. The yields declined steadily from January to July, followed by an increase from August till December.

5.3. Limiting nutrients and their seasonality

Growth limitation by nitrogen, resulting in positive growth responses only after nitrogen addition, occurred in 66 of the 264 assays, e.g., at site 161.05 in January, March, and May–October. A positive growth response after the addition of phosphorus, indicating P-limitation, was obtained in only 7 cases out of the 264 assays (e.g., sites 93.30 and 38.04 in January, sites 119.38, 93.31, and 38.04 in February, site 30.02 in March and site 93.29 in October).

In most of the assays the algal yields increased when nitrogen was added to the testwater, but the production was less than the maximum yields as obtained with phosphorus and nitrogen added together. This indicates that in waters primarily limited by nitrogen, phosphorus became the next limiting nutrient when sufficient nitrogen was provided. These results occurred in 87 of the 264 tests (e.g., site 161.11 and 93.32).

In 20% of the tests phosphorus was the primarily limiting nutrient and nitrogen the next one (e.g., site 93.28 in April–June and October–December, site 93.29 in February–May, November and December, site 93.30 in February–June and October–December).

Limitations, in fig. 2 indicated as (N/P), for instance at site 119.37 in January, November and December, imply that no significant increase in yield was recorded after single addition of each nutrient, while the combination of both nitrogen and phosphorus increased the yield significantly. Therefore in these cases the primarily limiting nutrient could not be determined with the method used.

In 27 assays, the yield did not increase even after addition of both nutrients. This suggests that some

Table I. Location of sites and their expected waterquality. The site numbers are observation locations of the waterboard "Hoogheemraadschap Rijnland". Water quality is based on data of macrophytes (Smit pers. comm.)

Polder	Water quality			Gradient	
	good	intermediate	poor	near point discharge	distant site
Wassenaarse polder		161.10	161.11	161.05	161.02
Polder Achthoven				2.04	2.03
Noordzijder polder				255.19	255.20
Polder Reeuwijk	119.35	119.38	119.39	119.37	119.36
Polder Nieuwkoop	93.28	93.30	93.29	93.31	93.32
Gemenewegse polder	38.03	38.04	38.02		
Polder Aarlanderveen				27.02	

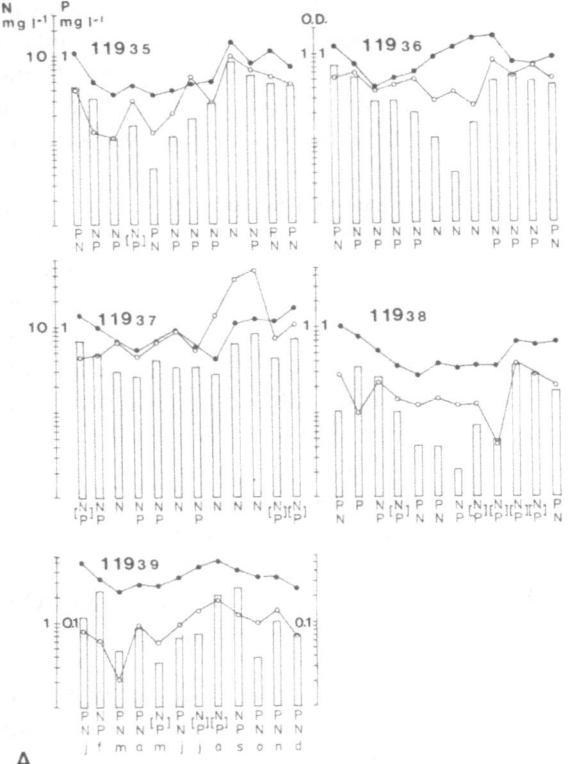

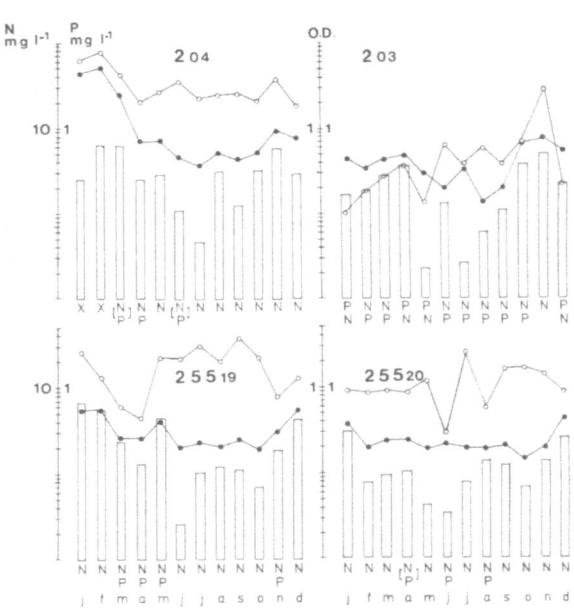

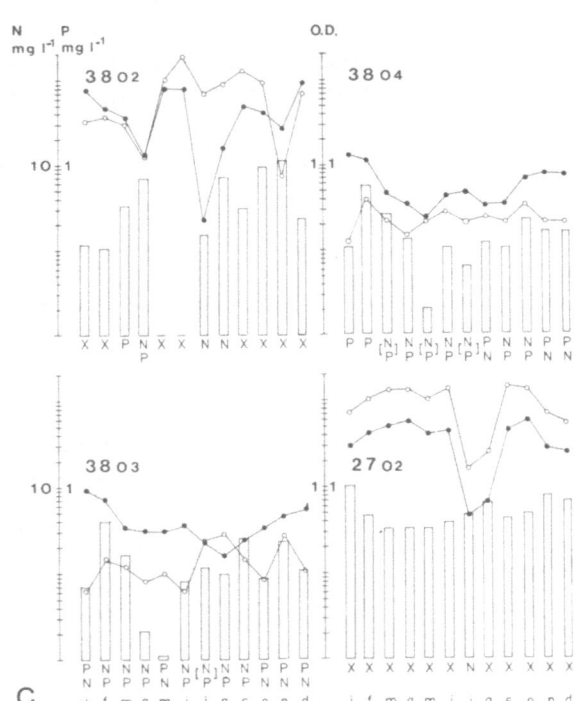

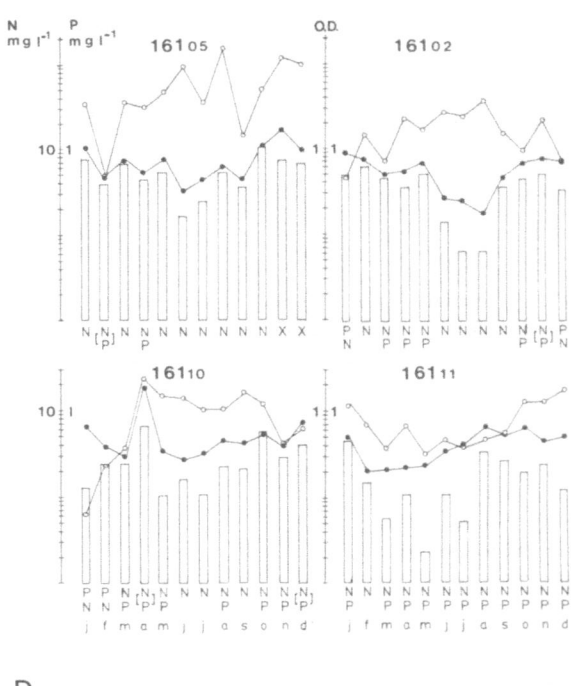

A

B

C

D

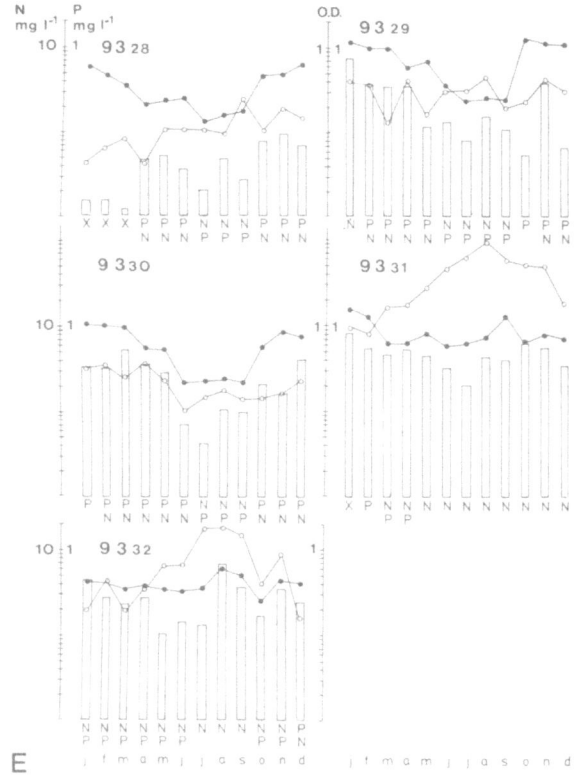

←

Fig. 3. Concentration of total N (mg l⁻¹) ●—●, total P (mg l⁻¹) ○—○ of the watersamples and yields (o.d. 798 nm) of *Stigeoclonium tenue* after 3 weeks of incubation. The letters indicate primary (first row), and if detectable secondary limiting nutrient (second row). N = nitrogen, P = phosphorus, N and P in brackets indicate no primarily N or P limitation, X = an unknown factor determines the yield.

other nutrient is limiting or that the yield is inhibited by an unknown factor (X) (e.g., sites 38.02 and 27.02).

During the course of the year a shift from one limiting nutrient to another could occur. Phosphorus limitation occurred frequently in autumn and winter, while nitrogen limitation became more important during late spring and summer. For each month, the percentages of assays with a primary nitrogen or phosphorus limitation are presented in Fig. 5. Assays with an unknown or primary phosphorus limitation (N/P) were not taken into account. Up to 95% of the assays from July to September were N-limited, whereas in January and December 63 and 71% respectively, were P-limited.

5.4. *Average yields in relation to trophic status*

The average yields, obtained in the assays may provide information on the productivity (trophic status) of the sites. The yields for all sampling sites are presented in Fig. 6.

The results show that sites near a point discharge (e.g., sites 161.05, 2.04, 255.19, 119.37, 93.31, 27.02) have higher mean yields than the more distant sites. However, no clear correlation was found between the expected water quality and the mean yield in the assays, at least using the water quality data based on macrophytes (see Table I).

5.5. *Evaluation of bioassays in relation to periphytic species assemblage*

In order to compare the bioassay results with the species composition of the periphytic algal community, microscope slides as artificial substrate were exposed at each site for a period of 2–4 weeks from February till July or November (Ten Cate 1985, internal report). The presence and abundance of the algal species were recorded to describe sites with good, intermediate, and poor water quality, according to the data on the macrophytes as provided by the water agency "Hoogheemraadschap Rijnland" (Table I).

It was obvious that a group of green algae, e.g., *Chaetosphaeridium globosum, Dicranochaete reniformis, Palmodictyon* spp. and chrysophytes as *Chrysopyxis* spp., and *Dinobryon* spp. were restricted to sites having a relative good water quality, whereas sites of apparent bad quality, mostly situated near a point discharge, were characterized by algae that occur at nearly all sites as the greens *Ectogeron elodae, Pseudendoclonium prostratum, Sphaerobotrys fluviatilis, Stigeoclonium tenue,* and chryso-xanthophytes as *Chrysophaera botryoides, Ophiocytium arbuscula.*

Based on the algal composition, a ranking of all sites was made from number 1 (best water quality) to number 19 (worst quality). This ranking was

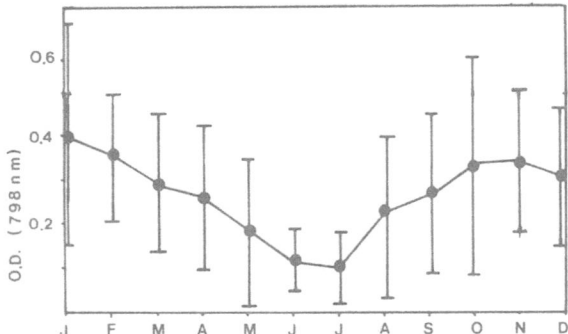

Fig. 4. Mean yield and standard deviation obtained in the assays during the year. Assays with X as limiting or inhibiting factor were omitted.

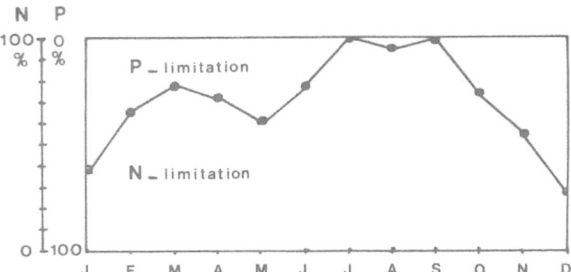

Fig. 5. Percentages of assays with primary N or P limitation during the year. Assays with X or [N/P] as limiting factors were omitted.

compared with a ranking sequence based on the average yields in the bioassays (Fig. 7).

Spearman's rank correlation coefficient (Sokal & Rohlf 1969) was used to test both series. A highly significant correlation ($r = 0.87$, $P < 0.01$) was found.

6. Discussion

Although there was not a very sharp correlation between bioassay growth yield and nutrient content of the testwater, general trends were reflected very well. The relatively strong correlation which was found at several sites between growth yield and N-content can be a consequence of nitrogen being the primary or only limiting nutrient at such sites. This was indeed the case at most of these sites.

Seasonal changes of algal yields have been reported previously (Davis & Dacosta 1980; De Vries *et al.* 1983). The highest yields occurred during winter whereas the lowest yields were recorded during summer. The present results are consistent with these observations. In the ditches, the nutrients are locked up in blanketing algal mats and macrophytes which show maximum growth in spring and summer. Algal yields in the assays were therefore inversely related to the biomass of the primary production in this aquatic environment.

With regard to limiting nutrients, one should be cautious to draw conclusions from a bioassay approach. A shift in limiting nutrients during the seasons is not often reported. Storch & Dietrich (1979) found limitation by nitrogen and phosphorus in summer, but not in midwinter. A shift from phosphorus limitation in winter to nitrogen limitation in summer has been observed in bioassays of water from canals in The Netherlands (De Vries *et al.* 1983). The often observed decrease of bio-available nutrients, especially of nitrogen, during summer may cause a decrease in the N/P ratio to below the critical level of 12 for *Stigeoclonium* (De Vries 1986) where N becomes the primary limiting nutrient.

The trend of relatively high average yields at sites near discharges and lower yields at more distant sites, is as to be expected. However, at each site there was a large variation over the seasons. Consequently, there is a risk that wrong conclusions are drawn when bioassay tests are based on few data, at least in small water bodies as ditches with a large variation in many factors.

The strong correlation between the ranking of sites based on average bioassay yields and a ranking based on periphytic algal composition, supports the hypothesis that the species composition of the periphytic algal community reflects the trophic state of the water (Weitzel 1979).

An advantage of the bioassay method is that it measures the biological responses to the pool of nutrients and substances which are actually available for algal growth. Also the presence of toxic substances can be detected with a bioassay. Based on the above mentioned statements, it appears

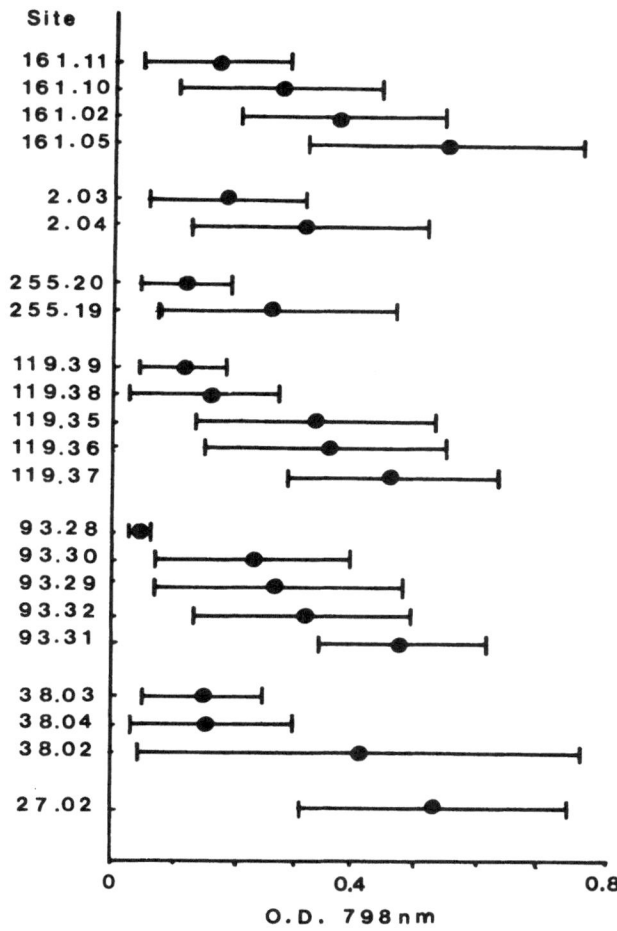

Fig. 6. Average yields and standard deviations obtained in test-waters of the sites during the year. Mean yields at 38.02 and 27.02 are underestimated due to the frequent inhibition of the yields.

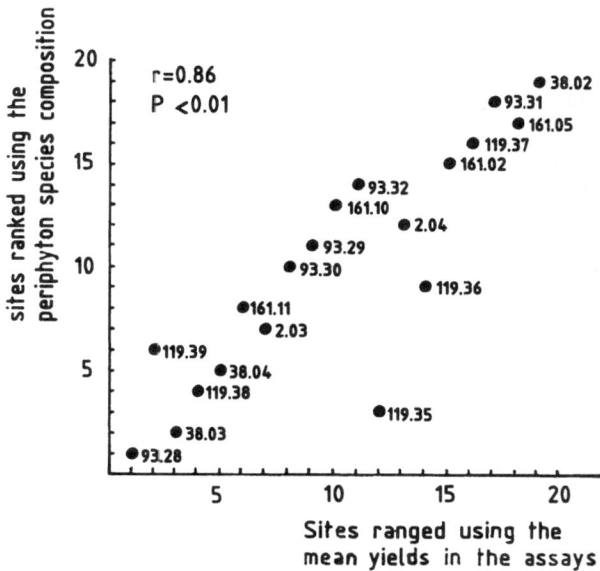

Fig. 7. Rank order of the mean yields plotted against the rank order of the periphytic species composition.

References

Davis, J. & Da Costa, J., 1980. The use of algal assays and chlorophyll concentrations to determine fertility of water in small impoundments in West Virginia. Hydrobiologia 71: 19—34.

De Vries, P. J. R., Torenbeek, M. & Hillebrand, H., 1983. Bioassays with *Stigeoclonium* Kütz (Chlorophyceae) to identify nitrogen and phosphorus limitations. Aquat. Bot. 17: 95—106.

De Vries, P. J. R., 1986. Bioassays on water quality using the attached filamentous alga *Stigeoclonium* Kütz. Thesis Free University, Amsterdam.

Francke, J. A., 1982. Morphological plasticity and ecological range in three *Stigeoclonium* species (Chlorophyceae, Chaetophorales). Br. Phycol. J. 17: 117—133.

Kützing, F. T., 1843. Phycologia Generalis, Leipzig.

Simons, J., Van Beem, A. P. & De Vries, P. J. R., 1986. Morphology of the prostrate thallus of *Stigeoclonium* (Chlorophyceae, Chaetophorales) and its taxonomic implications. Phycologia 25: 210—220.

Sokal, R. R. & Rohlf, F. J., 1969. Biometry — the principles and practice of statistics in Biological Research. Freeman, San Francisco.

Storch, T. A. & Dietrich, G. A., 1979. Seasonal cycling of algal nutrient limitation in Chautangua Lake, New York. J. Phycol. 15: 399—405.

Weber, C. I., 1973. Recent developments in the measurements

plausible that the yield of bioassays indicating the trophic level, is a useful addition to the range of methods already used to measure water quality. The results also demonstrate that *Stigeoclonium*, an alga which is representative for the periphytic community, is successful as a bioassay organism. Thus it is a useful extra means for assessing biological water quality of shallow waters where the planktonic algal community is of minor importance.

159

of the response of plankton and periphyton to changes in their environments. In: G. E. Glass (ed.) Bioassays techniques and environmental chemistry. Ann. Arbor Science Publishers Inc. Michigan pp. 119—138.

Weitzel, R. L., 1979. Methods and Measurements of Periphyton Communities: a review. ASTM, Philadelphia.

Author's Address
P. J. R. De Vries
Unie van Waterschappen
Joh. van Oldebarneveldlaan 5
P.O. Box 80200
2508 GE The Hague
The Netherlands

Decline of the Characeae community in the shallow peat lake Botshol

J. SIMONS

Abstract. A review is given about the reduction in diversity of the different aquatic communities in the nature reserve area Botshol, a shallow peat lake in the province of Utrecht, The Netherlands. The decline in species richness was most pronounced in the characean community with a reduction from 6 species, among which *Chara hispida* and *Nitellopsis obtusa* with abundant occurrence in 1969, till only one species (*Chara globularis*) in 1988. The oligohaline macrophyte *Najas marina* also strongly declined. Some species as the coenocytic filamentous alga *Vaucheria dichotoma* and in particular the moss *Fontinalis antipyretica* have reached dominance at many localities. The reduction in species richness was accompanied by only minor changes in measured parameters. One of the reasons for this phenomenon may be the buffering capacity of the system of reed marshes and small ponds near the inlet of water from the strongly contaminated rivulet Oude Waver. Measures were taken to reduce the phosphorus load from $0.6\,\mathrm{g\,P\,m^{-2}y^{-1}}$ to $0.1\,\mathrm{g\,P\,m^{-2}y^{-1}}$. The prognoses for recovery ae optimistic on account of the fact that the system is relatively small and as yet in a beginning stage of eutrophication.

1. Introduction

Chara species generally occur in stagnant waters with high transparency, calcium- and chloride-rich hard water, and low phosphate load. Some species (e.g., *Chara canescens*) grow in brackish water. Several species take up HCO_3-ions as source of inorganic carbon for photosynthetic CO_2-fixation, which are abundant in hard waters. As a result, calcium carbonate is deposited on the thallus surface. Most species grow fixed with rhizoids on silt, mud, peat, or sand with a little silt covering. Recently it was shown (Andrews 1987) that the rhizoids have also a nutrient uptake ability. Sexual spores with decay-resistant walls are formed in spring and summer. The oospores in the sediments are able to remain dormant, but viable, for some years (Olsen 1944; Krause 1981). Water birds are important agents of dispersal (Imahori 1954). The geographic distribution is world-wide, but most species occur in temperate areas. It is well known and documented that most species of Characeae are more or less sensitive to eutrophication (Forsberg 1965; Hutchinson 1975).

In the Netherlands 21 species of *Characeae* have been recorded, 9 of which belonging to *Chara*, 8 to *Nitella*, 3 to *Tolypella* and 1 to *Nitellopsis* (Maier 1972). Several of these species have become rare or almost disappeared by increasing pollution in recent years. Examples are the Loosdrecht Lakes (Best *et al.* 1984) where in 1942 a rich characean vegetation existed which was in 1980 restricted to 2 species (*Chara vulgaris* and *C. globularis*) at only one locality. Another documented example is the shallow peat lake Naardermeer where extensive growth occurred of *Nitellopsis obtusa* until 1984 (Spruijt 1986, internal report). Other species recorded at that time are *Chara aspera*, *C. connivens*, *C. contraria*, *C. globularis*, *C. vulgaris var. longibracteata*, *Tolypella glomerata*, and *Nitella mucro-*

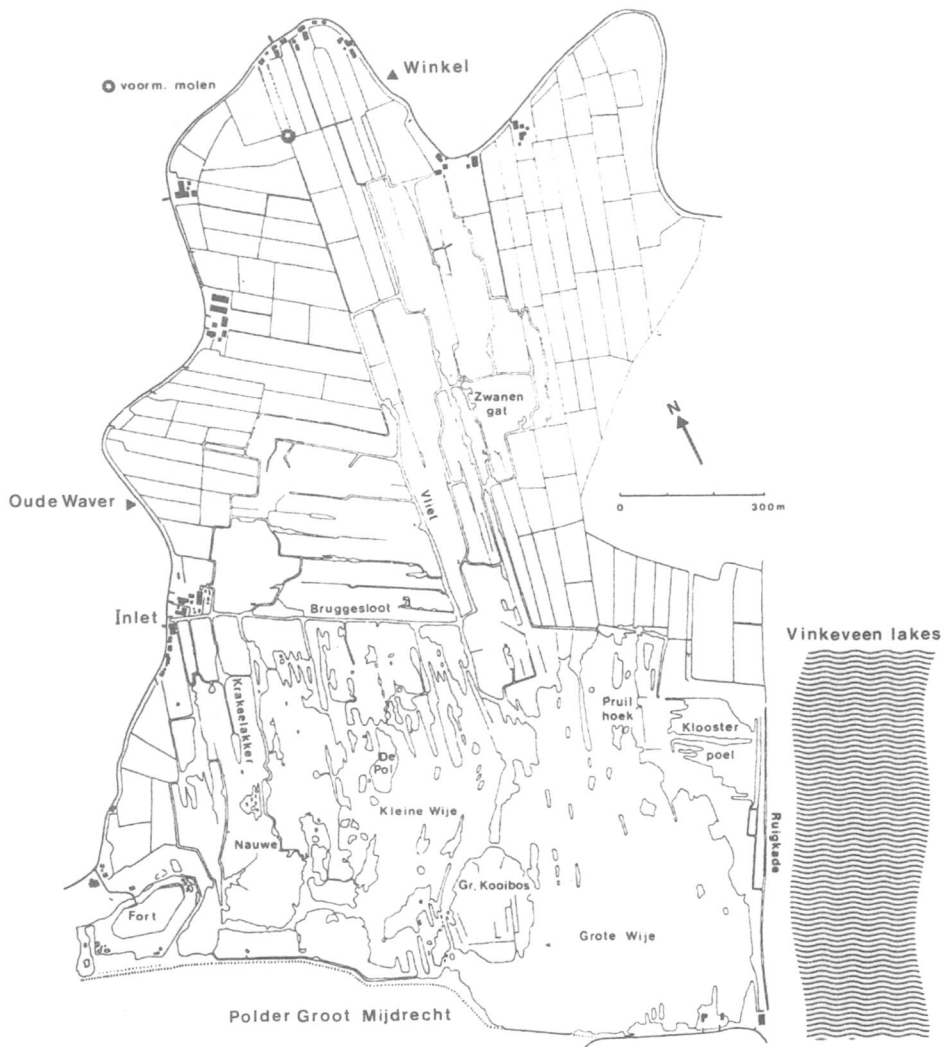

Fig. 1. Map of the Botshol area.

nata. Since 1980 a steady decline of *Chara* species was observed resulting in only 5 species (*Chara aspera, C. contraria, C. globularis, Nitella mucronata, Nitellopsis obtusa*) with scattered and restricted occurrence in 1986. In the year 1985 measures were taken to improve water quality. The most important measure was reduction of phosphor load of the inlet water.

In this chapter a detailed account is given of the recent history of the small shallow peat lake Botshol near Abcoude in the province of Utrecht.

2. Description of the polder Botshol

2.1. Hydrology and water characteristics

Botshol (52°15′ N, 4°26′E) is an area with shallow (depth 1.50–2.50 m) peat lakes, ditches, and pools surrounded by marshes and agricultural grassland. Digging of peat which started in the 18th century resulted in two relatively large lakes: Grote Wije and Kleine Wije (Fig. 1). The water level in the polder Botshol is at 2.45 m below NAP (dutch ord-

162

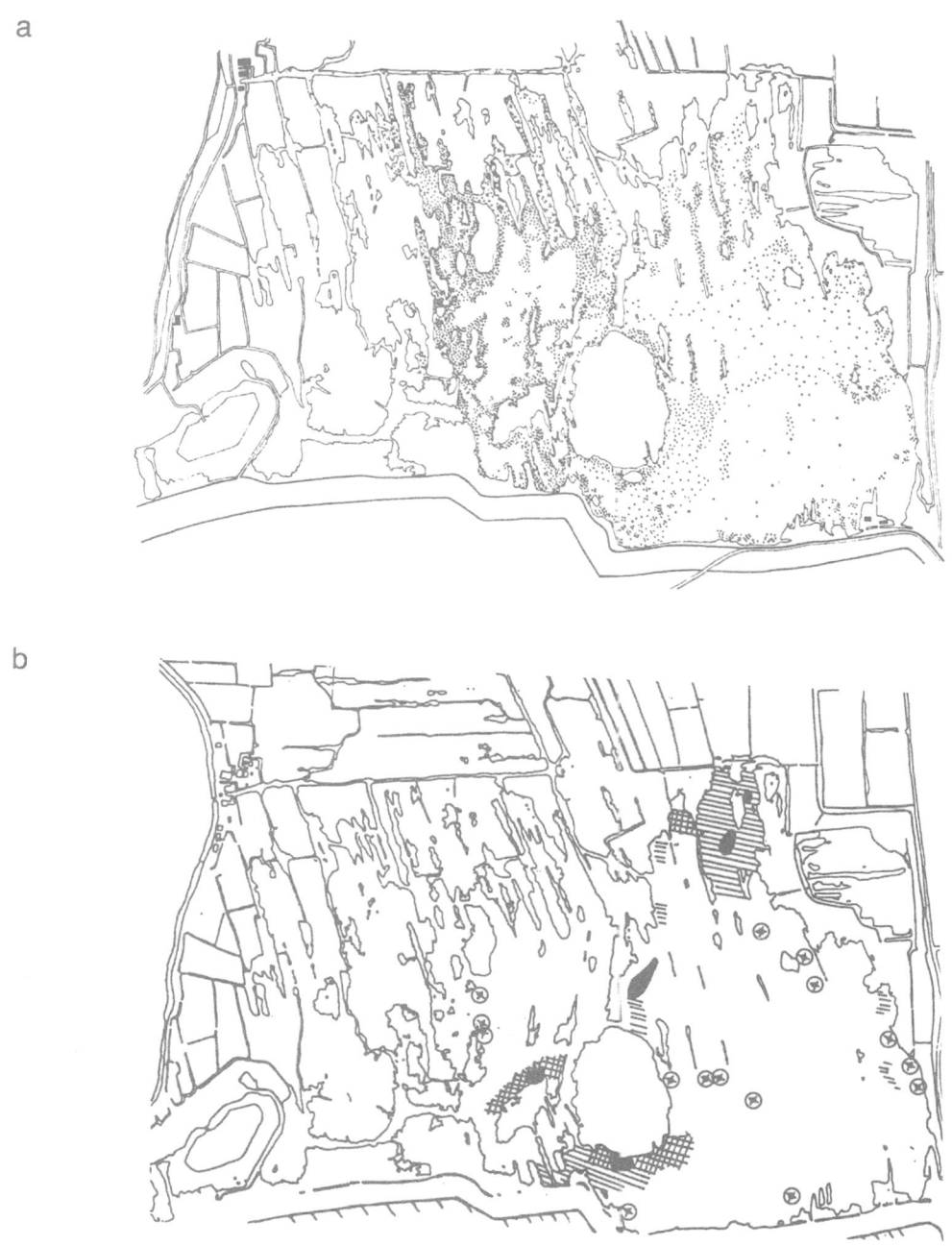

a

b

Fig. 2a. Distribution of *Najas marina* in 1969; *2b.* Distribution of *Najar marina* in 1987.

nance level). The former rivers Oude Waver and Winkel which adjoin to Botshol at the western and northern sides are at 0.40 m below NAP. In the deep polder Groot Mijdrecht, at the southern side, the water level is 6.50 m below NAP. The Vinke-veen lakes, at the eastern side, are at 2.15 m below NAP. Inflow of water depends on rainfall, inlet from the Oude Waver, and seepage from the Oude Waver, Winkel and Vinkeveen. Much water leaves the area by a downward seepage flow towards the

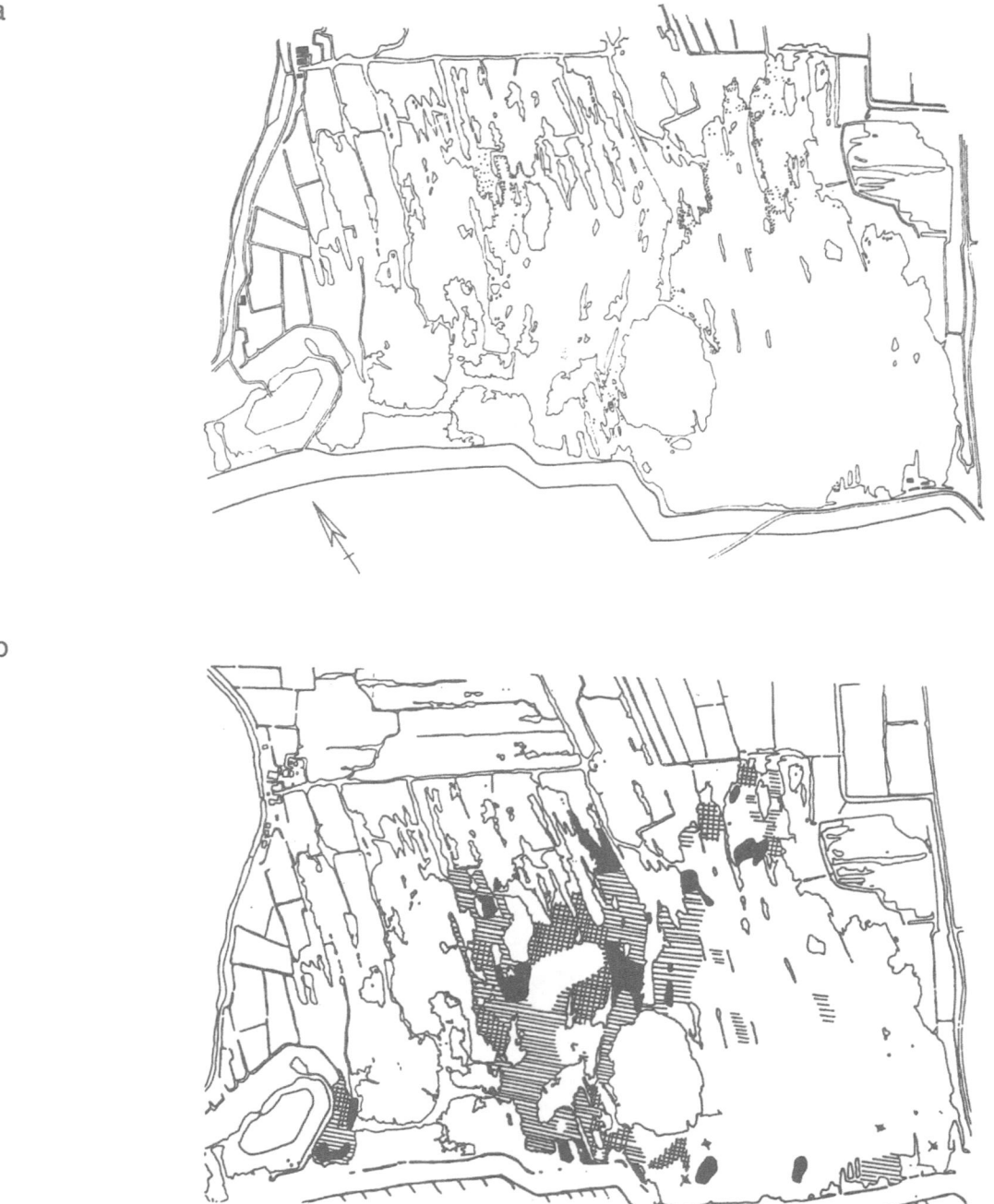

Fig. 3a. Distribution of *Fontinalis antipyretica* in 1969; *3b*. Distribution of *Fontinalis antipyretica* in 1987.

deeper level of the polder Groot Mijdrecht. The regulated inlet of water in summer from the strongly eutrophicated and brackish river Oude Waver causes a gradient of chloride in the lake area from NW to SE.

The lake water has a relatively high chloride level

a

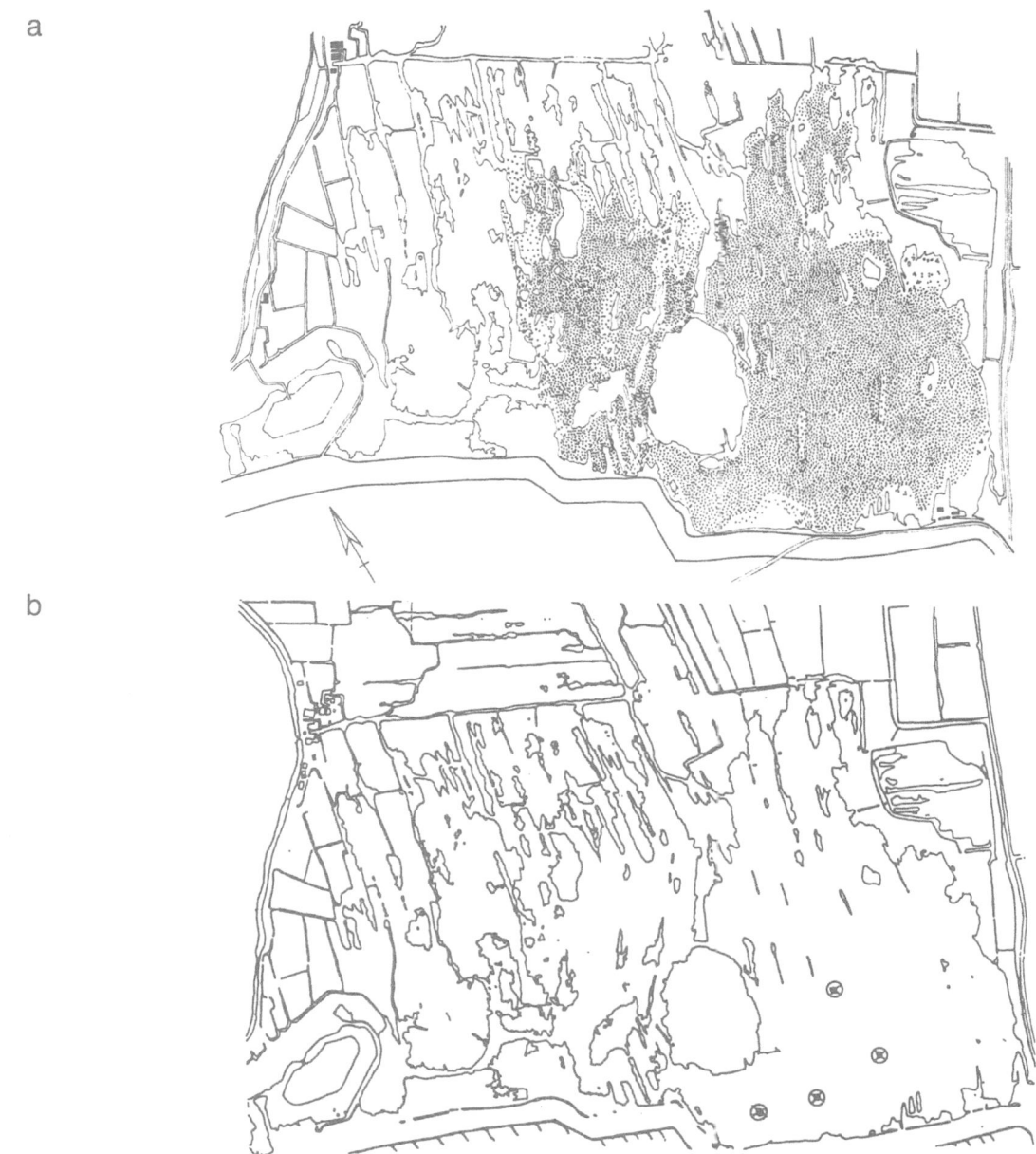

b

Fig. 4a. Distribution of *Nitellopsis obtusa* in 1967; *4b*. Distribution of *Nitellopsis obtusa* in 1987.

(382–447 mg l⁻¹); summer transparency (secchi disk): 1.00–2.10 m; Ca: 89–111 mg l⁻¹; Kjeldahl N (mean summer value): 1.60 mg l⁻¹; total P (mean summer-value): 0.05 mg l⁻¹; chlorophyll-a: 3–18 mg m⁻³. These data refer to estimations in the year 1988 (Rip & Everards 1988, report PU*). The sediments consist of peaty sand or clay with varying amounts of silt.

* Province of Utrecht, Department of water quality management.

2.2. The phytoplankton

Historical data about the plankton are scarce. Mur (1976) described the phytoplankton and stated that it consisted of many chrysophytes and dinoflagellates (e.g., *Ceratium hirundinella* and *Peridinium borgei*). Bijwaard & Timmerman (1982, internal report) characterized the phytoplankton as a chlorococca-

165

Fig. 5. Distribution of *Chara hispida* in 1967.

lean/diatom type with presence of bloom forming bluegreens as *Microcystis aeruginosa*. Bos & Meekel (1987, internal report) found a comparable type, but with more bluegreens, especially in summer.

2.3. Aquatic macrophytes, epiphytic alge and Characeae

According to Westhoff (1949) the Botshol lakes Grote and Kleine Wije were formerly dominated by *Chara hispida* while also the rare *Najas marina* was abundant, especially in Grote Wije. As accompanying Characeae are mentioned by Westhoff (1949): *Chara aspera, C. globularis, C. vulgaris* and the rare *Nitellopsis obtusa* (abundant in Grote Wije).

The epiphytic algae were studied by Hillebrand (1977). Remarkable green algae at that time are *Chaetophora incrassata* and *Batrachospermum moniliforme*, which are described as an algal association by Den Hartog (1959). The xanthophyte *Vaucheria dichotoma* is mentioned for Grote Wije. *Cladophora glomerata* occurs at nearly all sites and during all seasons on reed stems and other objects.

In a study of Vroman (1976), detailed distribution maps are presented of the macrophytes and Characeae as present from 1967 to 1969. *Najas marina* (Fig. 2) is very common, especially in Kleine Wije, and *Fontinalis antipyretica* (Fig. 3) occurs at many sites along both lakes. *Nitellopsis obtusa* (Fig. 4) is very abundant in both lakes. Other characean species are *Chara aspera, C. contraria, C. globularis*, and *C. hispida*. Except *Chara hispida* (Fig. 5) which is common in Kleine Wije, these species have a scattered distribution.

About 12 years later, the situation was again described (Wijngaard & Zloch 1981, internal report). It appeared that the mass occurrence of Characeae was no longer present. *Nitellopsis obtusa* had disappeared from Kleine Wije and *Najas marina* was strongly declined. *Fontinalis antipyretica* had reached dominance at many sites and also *Vaucheria dichotoma* was very abundant.

The most recent data were gathered in 1985 (Spruijt, internal report), and 1987, 1988 (Daalder & Ohm, internal reports). From these studies the conclusion can be drawn that the most recent situation has not much changed since 1980. *Chara globularis* (Fig. 6) is the only characean species reaching considerable abundance at some sites in Grote Wije, while *Nitellopsis obtusa* (Fig. 4) can scarcely be found. *Fontinalis antipyretica* (Fig. 3) is very abundant, especially in Kleine Wije. Also

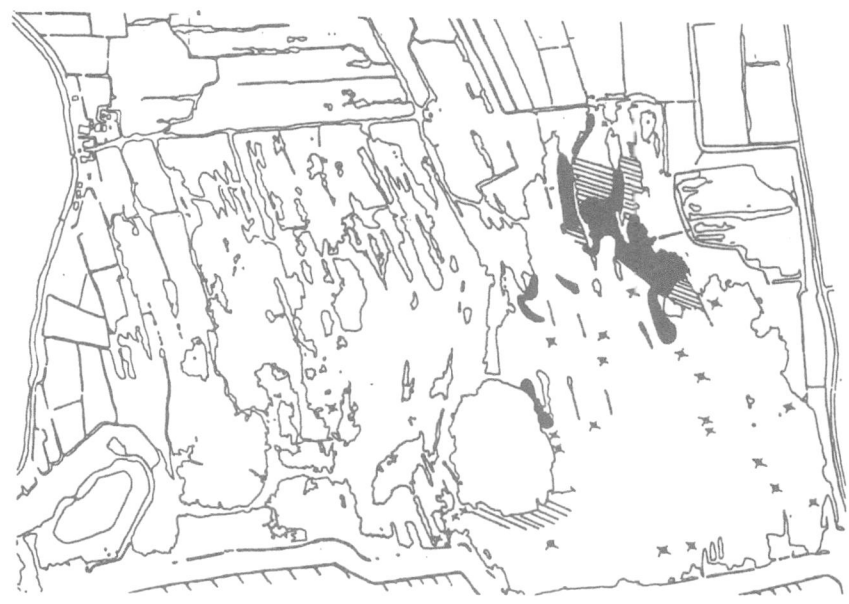

Fig. 6. Distribution of *Chara globularis* in 1987.

Vaucheria dichotoma (Fig. 7) forms large masses at shallow quiet places with much silt in the sediment. *Najas marina* (Fig. 2) has maintained itself at only four sites in considerable quantity.

3. Discussion

The strong decline of Characeae which took place in the period 1970–1980 was scarcely accompanied by significant changes in measured parameters. Only a slight increase of total-P was measured in that period. Transparency decreased periodically, especially in summer. Unfortunately there are no sufficient measures of chlorphyll-a.

Decline of the Characeae, especially of *Nitellopsis obtusa* was followed by a strong increase of the bottom dwelling *Fontinalis antipyretica* and *Vaucheria dichotoma*, both species being known as resistant against eutrophication (Melzer 1976). *Fontinalis antipyretica* is known as a species which uses predominantly CO_2 for photosynthesis (Maberly & Spence 1983). Therefore the luxurious growth of *Fontinalis* on the sediment of Kleine Wije and parts of Grote Wije seems strange in this hardwater lake. According to J. G. M. Roelofs (personal communication), the reason could be that the water layer directly above the sediment contains much CO_2 which presumably originates from a high rate of microbial decomposition of organic substances in a very reductive sediment.

As expected, and known from literature (John *et al.* 1982; Krause 1981) the most sensitive Characeae disappeared first, e.g., *Chara hispida, C. aspera* and *Nitellopsis obtusa*. The remaining *Chara globularis* is known to be rather euryoec and resistant to eutrophication (Krause 1981). The disappearance of Characeae is presumably not directly caused by increase of phosphorus (Blindow 1988), but by decrease of water transparency related to increased plankton growth, which is presumably directly related to a higher phosphorus level. For Botshol there are no indications that bottom dwelling fishes as bream which may increase their numbers in eutrophicating systems (Andersson *et al.* 1978) have caused the decrease of Characeae.

The rare macrophyte *Najas marina* could maintain itself presumably by the lack of competition with other submerse plants and by the slightly brackish water which is optimal for this species (Agami *et al.*

167

Fig. 7. Distribution of *Vaucheria dichotoma* in 1987.

1984). This brackish character may be one of the reasons for the former abundance of Characeae and the absence of many other macrophytes.

Regarding the apparent eutrophication phenomena it is difficult to appoint one main factor as responsible. The available data support the assumption that a shift from the inlet location in the northern part of the area to the present strongly polluted Oude Waver location in 1967, combined with increrased manure discharge from the northern agricultural grasslands, are the main causes of the observed deleterious effects.

4. Restoration measures and future development

In recent years, policy makers and nature management institutions developed strategies to reduce the phosphore load in inland surface waters by integrated measures (Jagtman 1988). Botshol was appointed as one of the test cases of eutrophication control in wetland areas. The measures planned for Botshol (Rip *et al.* 1989) imply dephosphorilation of the inlet Oude Waver water and a hydrological

isolation of the lake area from the neighbouring agricultural grasslands. After realisation of these measures (June 1989), the external phosphorus load will be reduced from $0.6\,g\;P\;m^{-2}y^{-1}$ to $0.1\,g\;m^{-2}y^{-1}$.

The prognoses for recovery are optimistic for several reasons: the small lake area has maintained a relatively good water quality with low nutrient concentrations and a moderate plankton density (max 18 mg chlorophyll-a m^{-3}). In the lake sediments low concentrations of P (c. $0.32\,mg\,P\,g^{-1}$ dry weight) are measured (Rip & Everards 1988, report PU). Available data on limiting nutrients (Bos & Meekel 1987, internal report; Rip & Everards 1988, report PWU) indicate that P is the first limiting factor, at least in spring.

The main reason for this relatively good situation despite the long period of increased contamination may be the buffering capacity of the area near the inlet where a complicated system exists of small ditches and reed land and much silt in the sediment. Moreover the iron content of the inlet water is considerable (about 17 mg Fe l^{-1}, according to Van Liere & Hillebrand 1976) causing also high iron contents in the sediment near the inlet point (72–93 mg Fe kg^{-1}), according to Rip & Everards 1988, report PU), which promotes the phosphate binding capacity of the sediment.

The reference situation is well documented and therefore changes in parameters and species composition can and will be followed in detail by research at several institutes. One of the most interesting questions is whether and when the Characeae community will return, and if so in what species composition. During the summers of 1989 and 1990 there was already an increase in coverage of Characeae and number of species (*Chara hispida* and especially *C. connivens*), while *Fontinalis antipyretica* decreased and *Vaucheria dichotoma* nearly disappeared. Detailed documentation of this process will increase our knowledge on the dynamics and flexibility of aquatic vegetations which is useful for water management purposes.

References

Agami, M., Eshel, A. & Waisel, Y., 1984. *Najas marina* in Israel: is it a halophyte or a glycophyte? Physiol. Plant. 61: 634–636.

Andersson, G., Berggren, H., Cronberg, G. & Gelin, C., 1978. Effect of plantivorous and benthivorous fish on organisms and water chemistry in eutrophic lakes. Hydrobiologia 59: 9–15.

Andrews, J., 1987. Phosphate uptake by the component parts of *Chara hispida*. Br. phycol. J. 22: 49–53.

Best, E. P. H., De Vries, D. & Reins, A., 1984. The macrophytes in the Loosdrecht Lakes: A story of their decline in the course of eutrophication. Verh. Internat. Verein. Limnol. 22: 868–875.

Blindow, I., 1988. Phosphorus toxicity in *Chara*. (Short communication) Aquat. Bot. 32: 393–395.

Forsberg, C., 1965. Environmental conditions of Swedish charophytes. Symb. bot. Upsal., 18: 1–67.

Hartog, C. den, 1959. The *Batrachospermeto-Chaetophoretum*, a remarkable algal association in The Netherlands. Acta Bot. Neerl. 8: 247–256.

Hillebrand, H., 1977. Periodicity and distribution of the multicellular green algae in two lake areas in The Netherlands. Thesis Free University, Amsterdam.

Hutchinson, G. E., 1975. A treatise on limnology. Vol. III. Limnological Botany. New York.

Imahori, K., 1954. Ecology, phytogeography and taxonomy of the Japanese Charophyta. Kamarawa.

Jagtman, E., 1988. Eutrofiëringsbeleid in ontwikkeling: 1987 omslagjaar? H$_2$O 21: 468–469.

John, D. M., Champ, W. S. T. & Moore, J. A., 1982. The changing status of Characeae in four marl lakes in the Irish Midlands. J. Life Sci. R. Dubl. Soc. 4: 47–71.

Krause, W., 1981. Characeen als Bioindikatoren für den Gewässerzustand. Limnologica, Berlin 13: 399–418.

Maberly, S. C. & Spence, D. H. N., 1983. Photosynthetic inorganic carbon use by freshwater plants. J. Ecol. 71: 705–724.

Maier, E. X., 1972. De kranswieren (Charophyta) van Nederland. Wetensch. Meded. K.N.N.V. 93, Hoogwoud.

Melzer, A., 1976. Makrophytische Wasserpflanzen als Indikatoren des Gewässerzustandes oberbayerischer Seen. Diss. Bot. 34, Cramer Vadux.

Mur, L. R., 1976. Enige gegevens omtrent het plankton van de Botshol. In: De Noordelijke Vechtplassen, Vlaardingen pp. 347–359.

Olsen, S., 1944 Danish Charophyta. Chorological, ecological and biological investigations. K. dansk. Vidensk. Selsk. Biol. Skr. 3: 1–240.

Rip, W. J., Pluim, J. W. & Simons, J., 1989. Integrale eutrofieringsbestrijding in de natuurgebied Botshol. In: L. van Liere, R. M. M. Roijackers, P. J. T. Verstraelen (eds.). Integraal waterbeheer in de Goois/Utrechts Stuwwallen- en Plassengebied, CHO-TNO, 's-Gravenhage pp. 229–245.

Van Liere, E. & Hillebrand, H., 1976. Het water in de Botshol. In: De Noordelijke Vechtplassen. Vlaardingen pp. 263–277.

Vroman, M., 1976. De verspreiding van waterplanten in de Botshol. In: De Noordelijke Vechtplassen. Vlaardingen pp. 317–331.

Westhoff, V., 1949. Landschap, Flora en Vegetatie van de Botshol. Baambrugge.

Author's Address
J. Simons
Department of Ecology and Ecotoxicology
Vrije Universiteit Amsterdam
De Boelelaan 1087
1081 HV Amsterdam
The Netherlands

Lake ecosystem with abundant vegetation of reed (*Phragmites australis*). (Photograph M. van der Werff.)

CHAPTER 15

Common reed

M. VAN DER WERFF

Abstract. In this review physiological and ecological aspects of the common reed, *Phragmites austrlis* are elucidated. As a true polymorphic species two basic ecotypic variations are distinguished: optimal and restricted populations. This ecotypic differentiation is reflected in the physiological requirements. As a species common reed can occur in a wide range of environmental conditions, determined by a number of physical and chemical factors like oxygen, water, temperature, pH, macro- and micronutrients, and inorganic and organic contaminants.

1. Introduction

The common reed, *Phragmites australis* (Cav.) Trin. is a helophytic grass having a world-wide distribution. It generally inhabits wetlands, where under the proper conditions it may flourish and become the dominant plant species. Its survival and reproduction strategies also make it, in most temperate zones, one of the first colonizers in newly reclaimed wetlands, like dredged material disposal sites or lakes drained for agricultural uses (new-born polders). Despite its growth habit, apparently aggressive, reeds do not normally invade established wetlands and displace other plant species. Reeds dominate a wetland only after a disturbance such as drainage or pollution. The common reed is able to withstand extreme environmental conditions, including the presence of toxic contaminants.

In Europe, but also in Japan, reeds have always been regarded as a valuable resource. Roof thatching is probably the best known application. The high quality reed that was needed, could only be produced by specific management. Harvest and management also created favorable conditions to

specific forms of wildlife, like meadow birds. These economic and natural aspects probably created a basis to look for wider applications tuned to present day needs. Such needs were delineated in dewatering and ripening of soil, water purification, raw material for industry and as an energy source.

2. Ecology

2.1. Plant communities

Phragmites australis grows in marshes, swamps, wet waste areas, along bayous, streams, lakes, ponds, and ditches. In Europe it belongs to those plant communities being syntaxonomically classified as Phragmitetea due to the high frequency of *Phragmites australis* in phytosociological reviews (Tuxen 1982). *Phragmites australis* occurs on almost every soil type as long as it is moist enough but appears to do best in firm mineral clays in proximity to the water table. Bibby & Lunn (1982) describe 5 moderately distinct groups of habitats, in which reed can form extensive mon-dominant stands:
- artificial habitats (clay pits, silt settling lagoons,

J. Rozema and J. A. C. Verkleij (eds.), Ecological Responses to Environmental Stresses, 172—182.

industrial waste lands, dredged material disposal sites, new-borne polders)
- lakesides
- estuaries (upper reaches of estuaries, tidal sections of rivers, with a periodic inundation with brackish water)
- coastal zones (reed encroachment on occlusions of tidal creeks, rivers and flood plains with a shallow flooding)
- inland wetlands

2.2. Ecotypic variation

The size of the aerial stems and rhizomes, stem coloration, inflorescence size, and number and weight of viable seeds are very variable, because its phenotype is influenced by the environment, *Phragmites australis* is a true polymorphic species. For this reason previously proposed taxonomic variations such as *gigantissima, stolonifera*, and *flavescens* are apparently ecotypes (Haslam 1972). Haslam (1970b) distinguished two basic ecotypic variations:

1. optimal populations
2. restricted populations.

The optimal populations represent the typical form of *P. australis*. These populations occur in more favorable environmental conditions (Haslam 1970b, 1970c, 1972).

The restricted stands occur because one or more factors are creating environmental stress. The plants are usually short and depauperate. The restricted stands occur on higher grounds where the rhizomes are not submerged in water. Rhizomes are short because of lack of water and nutrients (Haslam 1972). Upland stands are exposed to frost that may kill buds (Van der Toorn 1971). *Phragmites* in a harsh environment tends to have limited carbohydrate reserves (Buttery & Lambert 1965), which account in part for the less viable rhizomes. Shading produces a somewhat distinctive ecotype with flaccid leaves and an increased blade weight, while the characteristic ribbing is absent. Excess salinity is also known to produce depauperate plants. The smallest *Phragmites* recorded by Has-

lam (1970a) had nearly prostrate shoots about 5 to 15 cm long. These plants were growing in hard, rather salty, sandy soil. Waisel & Rechaw (1971) distinguished halophytic and glycophytic populations. Seeds from the halophytic population germinated in salt water of up to 2.3% Cl. Germination of seeds from a glycophytic population was severely inhibited at this concentration.

Van der Toorn (1972) distinguished two different ecotypes based on transplant studies:

1. a peat ecotype consisting of reed with short shoots and a high shoot density occuring in peat marshes.
2. a riverine ecotype consisting of reed with long shoots and limited shoot density, occuring mainly in fresh water tidal areas.

The riverine ecotype compared with the peat ecotype has (1) a higher above-ground production, (2) a greater tolerance for tidal submersion, (3) less tolerance for ground frosts in spring and (4) a greater sensitivity to infestation by the moth *Archanara geminipunctata*.

2.3. Propagation

Reeds flower in summer and wind dispersable seeds are set in autumn. Seeds disperse at the beginning of winter and germinate in early spring. The amount of seeds per panicle can be very large. Numbers of seed (Szczepanski 1978), viability and germination can be very variable. Gustafsson & Simak (1963) used X-ray photography to determine viability of seed; they found only 5% of all florets photographed to be viable. Seed germination ranged from very poor to 100% (Haslam 1972).

Seed germination and proper seedling development require specific conditions. The soil must be wet to saturated. Soils even may be submerged with approximately 1 cm of water (Haslam 1971b, 1972). Szczepanski (1978) still found germination with a water cover of almost 1 m, although it was retarded to 45 days. After germination started severe frosts can be very harmful. Possibly as a natural protection, Szczepanski (1978) found that it was not possible to induce germination before mid February.

Optimal germination requires a high temperature (25–35° C) (Van der Toorn 1972) and intensive light. For this reason, populations develop explosively in open, wet areas, like newly drained lakes and dredged material disposal sites. Rate of germination is temperature dependent; at low temperature germination time is prolonged (Szczepanski 1978). Application of varied temperatures (20° C during 8 h light and 10° C during 16 h dark) gave a maximal germination (Van der Toorn 1972). Germination also requires a high nitrogen and phosphate content of the soil. Water is essential for a fast phosphate transport to the seedling, since at water levels more than 3 cm below a seedling, phosphorus deficiency occurs.

Seedlings may also occur in areas recently disturbed by man or in upland areas where plants have been killed by frost. But these are usually not successful, as competitors appear later in the year and shade them out (Haslam 1971b, 1972).

Chapman (1960) obtained optimum germination in 1% NaCl. From other experiments it was shown that seedlings, however, require less than 0.5% salt water for adequate survival (Haslam 1971b, Harradine 1982). Differences in optimum and sensitivity can be explained by genetic variation.

Establishing stands from seeds is not advised because of the low germination rates (Haslam 1972). However, in The Netherlands *Phragmites* has been sown highly successful, on extensive areas (Van der Toorn 1982). Commercial generative propagation is also discussed by Veber (1978).

Vegetative reproduction normally takes place in two ways, firstly from buds in the rhizomes and secondly regenerative from meristemic tissue in shoot nodes. The last way is common in places where reed is grazed or damaged (i.e. by the lepidopterian parasite *Archanara*). Additional shoots may also emerge from broken stems floating in water. In principle, each node can develop into a new shoot (Szczepanski 1978).

Vegetative extension is strongly promoted by stolons, the creeping shoot at the periphery of the reed plant. These runners even grow on rather bare ground. As an example of the importance of vegetative propagation, dense and vast reed beds were reported to be formed within three years, by means of stolons. They developed from low densities of seedlings after sowing from airplanes (Van der Toorn 1972).

It is important for young plants to develop 10–12 shoots with subsequent rhizomal development to survive frosts. The young shoots may be killed, but if frost-tolerant horizontal rhizome development has occured, new shoots will be produced.

Reed beds may be very old (ca. 1000 years). Such populations are maintained through vegetative propagation (Haslam 1973).

Commercial propagation can be accomplished by seed or portions of rhizomes (with or without roots) before spring growth is initiated. Young green shoots are to be planted in May or June. Rhizomes bearing dead aerial stems and transplanted in spring provide the best results (Haslam 1972). Veber (1978) gives an extended survey on the artificial propagation and cultivation of *Phragmites*. Different types of vegetative propagation are discussed:

1. dividing reed stolons
2. layering reed shoots
3. stem cuttings, where the nodal meristems remain intact
4. rhizome cuttings

3. Physiological requirements

3.1. Substrate

Stands of *P. australis* are found in soil textures varying from gravel to silt and clay (Van der Toorn 1972, Boorman & Fuller 1981). It grows better in a fine texture soil (Haslam 1973). The soil organic content may range from 1–97% (Misra 1938, Haslam 1972). It often occurs in peat soils (Van der Toorn 1972, Boorman & Fuller 1981). However, in some environments contact with the soil is unnecessary, as the reed forms thick mats (0.5–1.2 m) floating on the surface water (Szczepanski 1978). *Phragmites australis* grows at a soil or water surface pH ranging from 3.6 to 8.6, but the more robust stands appear within a pH range of 5.5 to 7.5 and in mesotrophic or eutrophic environments (Haslam 1972).

3.2. Waterlevel

Phragmites is usually found in saturated soils with a low redox potential. Under these conditions the oxygen supply to the roots is low. In order to overcome a long-term anaerobic condition *Phragmites* has an aerenchymatous system which can transport atmospheric oxygen to the roots. In temperate climates the occurence of the plant was found to be restricted between 2 m above, and about 1 m below, ground water levels (Haslam 1970c). In warmer climates, *P. australis* may occur at greater depths (In Uganda as deep as 4 m; Haslam 1972). In Britain, the maximum depth of approximately 1.5 m is associated with mineral-rich water. Because only few competitive plant species have found to penetrate to this depth, an insufficient oxygen influx (Yamasaki 1984) and nutrient status seem to control this lower boundary (Haslam 1970c, 1971a).

At the upper boundary, *P. australis* usually occurs as a restricted population. Competition with other plant species, infestations and nutrient status, are controling factors. Rhizomes can reach almost 2 m below the soil surface with roots penetrating even deeper, allowing the plant to reach deep into the soil profile (Haslam 1970c). Water level effects bud width and consequently shoot height. In unflooded soil the buds are narrow and tend to occur close to the soil surface (between slightly above and 13 cm deep), while those in flooded areas are wider and occur deeper in the soil (Haslam 1970c).

Under flooded conditions *P. australis* also tends to have the highest production and the largest development of new rhizomes (Yamasaki 1981). Rezk and Edany (1979) reported optimum growth at a water level of 0.1 m below soil surface. Shoot density was not considered.

3.3. Temperature

Haslam (1972) stated that a hot summer can increase reed height with up to 0.5 m. This increase is caused by internodal growth. In general however, short-term hot and cold spells have little effect on the overall growth-rate (Haslam 1969). Seedling establishment, bud emergence, and the timing of the growth cycle are the areas of possible temperature related ecotypic variation (Haslam 1975).

Optimum growth for young shoots is at approximately 25° C (Haslam 1971b, 1973). Frosts or cold will increase shoot density, crop weight, and pre-emergence period (Haslam 1969, 1972) and will stimulate bud development, but in turn will decrease stem height and diameter. Severe frosts of − 14° C or below may kill up to 100% of the shoots. Buds cannot survive more than three consecutive days of frozen soil (Haslam 1972). A high water level tends to protect new shoots from frost in cooler climates (Haslam 1970c). Haslam (1972) found that recovery from frost damage may be very slow. This last author also noted an increase in height of reeds with decreasing latitude and altitude as well as with warm summers.

For these reasons *Phragmites australis* is found in temperate climates. It is not found in artic regions and at high altitudes with arctic conditions.

3.4. Photosynthesis and evapotranspiration

Oxygen content of rhizomes, stems, and leaves was studied by Krasovskii & Chashckukhin (1974). The oxygen content of the rhizomes in flooded ground, increased from 4–9% in spring to 15–19% in midsummer. Oxygen content of stems varied from 19–22% and of leaves from 12–23%. It was presumed that oxygen resulting from photosynthesis was transferred to the rhizomes.

The transpiration rate for *P. australis* is considered high, varying with the ecotype of the plant and temperature. Transpiration occurs between May and September in Great Britain with peak periods in July and August (Haslam 1970c). In Manitoba, Canada, Phillips (1976) found the peak period of transpiration also occurred in July.

Khashes & Bobro (1971) found daily transpiration characterized by a single peak at 12:00–13:00 hours. They suggested that the morning transpiration value is correlated with air temperature and deficit of saturation while the afternoon drop in transpiration is conditioned by the physiological state of the leaves.

On the average transpiration accounts for 75–90% of evapotranspiration. Several Central European plant physiologists reported gravimetric measurements of evapotranspiration in *Phragmites*. In summer the daily evapotranspiration for whole shoots ranges from 5.5 mm to 11.4 mm (Kvet 1973, Rychnovska & Smid 1973, Smid 1975) and is for single leaves about 13 mm (Tuschl 1970).

Evapotranspiration of vigorous *P. australis* stands was considered to remove a large amount of subsurface water and effect the hydrological condition of the sand deserts of the Northern Aral region of the U.S.S.R. (Vostkova 1975). Because of its ability to extract large amounts of water from the soil, *Phragmites* is often introduced in polders after initial reclamation in the Netherlands (Van der Toorn 1982) and in Japan (Kamio 1982). Both active transpiration and deep rooting of *P. australis* plants are of importance to reduce soil water to extensive depth. In this way, mud flats and dredged material disposal sites can be stabilized.

Additional factors regulate the evapotranspiration rate. Krolikowska (1971) examined *Phragmites* at terrestrial and aquatic sites around Midolajski Lake, Poland, and found evapotranspiration increases in proportion to increases in fresh weight, air temperature, and solar radiation. As expected a rise in relative humidity decreased transpiration, as did inflorescence formation. The intensity of transpiration rose rapidly in the morning and decreased in the afternoon. The younger leaves of the upper portion of the aerial shoots transpired more than the lower (older) leaves. Pazourek (1973) found a stomatal gradient did exist from basal to apical leaves.

3.5. Nutrient uptake

As mentioned above, *Phragmites australis* can grow in an exceptionally wide range of soil and water chemistry. It can be limited by different nutrients, depending on the nutrient status of the area concerned. In general *Phragmites* flourishes in mesotrophic and eutrophic habitats. It is also found more on coarse-grained than on fine-grained substrates.

Haslam (1973) learned that nutrient uptake is mainly from the upper 0.5 m of the soil, by the branched horizontal roots of the upper rhizomes. But, the hydropotes of the submerged aboveground part of the aerial shoot may also be of importance in nutrient uptake, even many times higher than the uptake by the underground roots (Luther 1983). In *Phragmites* stands a large number of analyses were performed on nutrients in soil and water and on nutrients in the plant tissue.

Nitrogen is absorbed by plant roots in the form of nitrate or ammonium ions. In Lake Gnadensee, Banoub (1975) followed differences in water chemistry outside and within a *Phragmites* stand. In the growing season the pH, the concentrations of total inorganic N, of NO_2-N and especially of NO_3-N decrease from the border towards the center of a reed bed, whereas the NH_4-N concentration increases. *Phragmites* shows a strong internal retranslocation of nitrogen. The N concentration in plant tissue is highest during the growing season. In leaves and stem the percentage N decreases from 2.61 in June to 0.64 in January (Schlott & Malicky 1984) and from 2.77 in May to 1.00 in October (Dykyjova 1979). The N content of the shoots only decreases after August. With the decrease of N in shoots, the N content of the rhizomes increases. In the following first half of the growing season, the appearance of new shoot biomass coincides again with a decrease of N in the rhizomes (Van der Linden 1980). However the concentration of N in the rhizomes also decreases with age (Ulehlova *et al.* 1973). Available N is essential for reed production, since Allen & Pearsall (1963) found that shoot production is related with leaf nitrogen. Another important macronutrient is phosphorus. The concentration of P in water can fluctuate heavily during season (Bayly & O'Neill 1972). Like for ammonium, Banoub (1975) also found higher P concentrations in the surface water inside the reed stand than outside. The concentrations of P in tissues of *Phragmites* vary with season. Bayly & O'Neill (1972), Dykyjova (1979), Best *et al.* (1981) and Schlott & Malicky (1984), found the highest P levels (0.3–0.8% dry weight) in the shoots in spring, decreasing to the lowest (0.15–0.20% dry weight) levels during autumn and winter.

Potassium is the third macronutrient. At the peak of the growing season, plants contain maximum concentrations of 1% K dry weight, which level decreases to 0.4% K dry weight in autumn (Ulehlova *et al.* 1973, Dykyjova 1979).

As *Phragmites* occurs in fresh water habitats as well as on the upper edges of salt marshes, it will be exposed to a large variation in the Na concentration of soil and surface water. In a peat water regime, Daniels (1975) found a good correlation between the quantity of sodium in the water and that in the aerial portion of the plant. Such a correlation was not observed with other ions in the water. It may be assumed that sodium uptake can not be regulated in contrast to other ions.

There are only a few data on the Cl concentrations in the habitat of *Phragmites*. In a fresh water habitat Banoub (1975) reported the same range in Cl concentrations (3.1–9.3 mg l^{-1}) in water within as well as outside a reed stand. Kaul (1984) reported a range in the water of 9–12 mg Cl l^{-1}. However, since this species also occurs in salt marshes, its tolerance to Cl is of interest. The below values are based on concentrations in soil. Ranwell *et al.* (1964) reported that *Phragmites* can tolerate up to 1.2% chlorinity in the 10–12 cm soil layer. Juvenile plants seem to be more sensitive, having a critical limit at 0.9%, whereas adult plants show no inhibition at this concentration (Bakker & Biewinga 1957). Haslam (1971b) discovered that plants of 10 cm high transplanted to a 2% chloride value remained small. At a 1% concentration, the transplants varied from small to the size of the controls. At a 0.5% value, plants achieved a size comparable to controls. Presumably the controls were grown in fresh water. Van der Toorn (1972) reported a range of 0.395–1.230% Cl (dry weight) in the shoots of *Phragmites australis* as the range over 29 sites in August.

Because pH can vary greatly in a reed stand, so a wide range of Ca concentrations can be expected. Ranges from 3–270 mg l^{-1} in the surface water (Allen & Pearsall 1963) have been reported and from 500–40,000 mg g^{-1} dry weight soil (Ho 1981, Kaul 1984). However, unlike the other nutrients, Bayly & O'Neill (1972) did not find a clear trend in the Ca concentration in water during the growing season. Banoub (1975) reported a slightly lower Ca concentration (aver. 49.3 mg l^{-1}) outside the reed stand, compared with within (aver. 53.3 mg l^{-1}). Ca concentrations in soil, the total as well as in the extract, can vary enormously. Also in soils no trends were found during the season (Bayly & O'Neill 1972). In the plant tissue, Dykyjova (1979) measured a increase of 0.54% to 1.60% in Ca (dry weight) in the shoot from June to August. Bayly & O'Neill (1972) found an increase in Ca from 0.14% in May to a peak in July of 0.29% and thereafter a decrease to 0.18% in August.

Although Bayly & O'Neill (1972) found a slight increase in the Mg concentration in the water from May to July. This is probably not related to the activity of reeds, since Banoub (1975) found the same Mg levels within and outside of the reed stand. The maximum concentration of solved Mg in July corresponds with a minimum in the extractable Mg concentration in the soil by the end of this month (Bayly & O'Neill 1972). These authors reported a decrease in the Mg concentration in shoots from 0.86–0.97% to 0.46–0.73% (dry weight) from May to August. From May to October, Dykyjova (1979) found at lower concentrations, a decrease of comparable magnitude, from 0.14 to 0.08%. In contradiction of comparable magnitude, from 0.14 to 0.08%. In contradiction with the above results, Wallentinus (1973) reported that the Mg content in the leaves in highest at the termination of the growing season.

Little is known about iron and manganese related to the habitat of *Phragmites*. Lanning & Eleuterius (1985) reported for 5 sites the concentration of Fe and Mn in water. The range for Fe is 0.15–0.42 mg l^{-1} and for Mn 0.05–0.09 mg l^{-1}. No data are available for the Fe and Mn concentrations in soils in a *Phragmites* stand. Iron concentrations were found to be greater in leaves than in stems but were substantially lower than those in roots (Kovacs *et al.* 1978, Ho 1981). However possibly a large part of the iron, reported in roots, is deposited on the roots as a plaque of ferric hydroxide (Crowder & MacFie, 1986). These plaques occur especially in July and August, when photosynthesis is high and roots loose oxygen. Auclair (1979) reported that the Fe and Mn contents of plant tissues were highly correlated with soil organic matter. Also a high correlation was found between iron in tissues and K, Na and Ca in the soil. The concentrations of Mn, in the shoot of reeds were lowest at the beginning of growth and continued to rise until seed set, whereas the underground portions showed the opposite trend (Baudo & Varini 1976). Ho (1981) reported similar observations for Fe and other nutrients in the stem and leaves of reeds but found no consistent seasonal pattern for roots and rhizomes.

The essential micro-nutrients zinc, copper, molybdenum, and also silicon are discussed with regard to their concentrations in the water in reed stands and to their concentrations in *Phragmites* tissues. Although it is not sure that silicon is considered to be an essential nutrient, it is beyond dispute, that Si in the form of silicic acid contributes greatly to the firmness of the cell wall of monocotyledons, like grasses (Baumeister & Ernst 1978). In this way Si contributes to the physical stability of stems and leaves.

In an unpolluted lake, Schierup & Larsen (1981) found three times higher concentrations of Zn and Cu in the interstitial water of the upper 10 cm of soil in a reed stand than in the surface water. Below 10 cm, the concentrations decreased. The Zn concentration in reed (about $20\,\mu g\ g^{-1}$ dry weight in stems and leaves) was reported to be negatively correlated with Na and K in the sediment (Auclair 1979). No significant correlations were found between Cu (0.4–$30\,\mu g\ g^{-1}$ dry weight in stems and leaves) and edaphic factors (Auclair 1979). Kufel & Kufel (1980) also found no relations between concentrations of Cu and Mo (about $0.2\,\mu g\ g^{-1}$ dry weight in stems and leaves) in plant tissues and those in substrates.

Silicon appears to be a major element in reeds. The Si concentration represents 65% of total ash in the leaves, 19% of total ash in the stem and 83% of total ash in the panicle (Lanning & Eleuterius 1985).

Seasonal influences in metal content may be important. Zn in leaves and stems and Cu in stems was reported to be at the maximum in May and declined thereafter (Kufel 1978; Larsen & Schierup 1981). But Cu in leaves was relatively constant throughout the season (Larsen & Schierup 1981). On the contrary Baudo & Varini (1976) reported that the concentrations of Cu in the shoot of reeds were lowest at the beginning of growth and conformed to rise until seed set whereas the underground portions showed the opposite trend.

There is no indication that situations of deficiency in micronutrients arise, not even in a peat soil. This is possibly due to the very extended root system with the numerous highly ramified adventivous roots, which arise from the vertical rhizomes. Especially in a peat soil these roots form a dense mass and have an important function in nutrient uptake (Van der Toorn 1972).

4. Contaminant uptake

Phragmites australis can withstand environmental extremes (Kufel & Kufel 1980) including the presence of toxic contaminants and, thus, may be the first plant to invade and vegetate successfully areas which cannot support the growth of other plant species (Ricciutti 1983). The invasion of disturbed areas by reed beds represents an important stage in hydroseral succession (Ingram *et al.* 1980, Bibby & Lund 1982) which entraps sediment and organic debris and prevents erosion, thus forming the initial process of natural restoration of a disturbed wetland to productivity. The establishment of reed beds, thus, may represent a significant step in the reclamation of wetland creation and confined upland disposal sites containing contaminated dredged material. Very little, however, is known about the uptake of contaminants by reeds, the physiological effects of contaminants on the growth and development of reed beds. Although the potential for contaminant entry into the food chain via reed beds can also not be assessed, grazing is less than 0.5% of total production (estimated from Imhof 1973). Moreover, biomobility of contaminants may be greatly reduced by covering with layers of litter and complexation of contaminants by humus (Van der Werff 1981).

4.1. Heavy metals

Most of the literature on heavy metal accumulation by *Phragmites australis* is concerned with trace metals essential for plant nutrition, especially copper, iron, manganese and zinc, of which only Cu and Zn are commonly cited as being toxic in higher concentrations.

Concentrations of Cd, Cr, Cu, Pb, and Zn generally were higher in the roots and/or rhizomes (maximum values were 1.2, 16, 190, 14 and 240 μg g^{-1} dry weight respectively) than in the above-ground plant parts (maximum values were 0.03, 3.6, 26, 3.5 and 48 μg g^{-1} dry weight respectively) (Chiaudani 1969, Baudo and Varini 1976, Kovacs *et al.* 1978, Kufel 1978, Larsen 1983). The essential trace metals proved to be more mobile between plant organs in contrast to Cd and Pb, which apparently are translocated very poorly. Surprisingly, Baudo & Varini (1976) reported highest mean concentrations of Cr (6.23 μg g^{-1}) and Cu (1466 μg g^{-1}) in the inflorescence and the highest Mn (93.2 μg high Zn content (150 μg g^{-1}) in the upper 20–40 cm of the stem, which contains the inflorescence.

The Cu content of leaves apparently is well regulated by the plant at environmental levels below the acute toxicity threshold, as Cu levels were similar in the leaves of plants growing on sediments of different Cu content (Chiaudani 1969). Kufel & Kufel (1980) found that Pb in tissues tended to increase with increased Pb in the substrate. Larsen & Schierup (1981) concluded that "the amount of heavy metals present in the rhizosphere of sediments may considered to be the amount potentially available to plants". Data were unavailable for the overall distributions of Hg, Mo, and Ni within *Phragmites* tissues.

High concentrations of metals in plant tissues reflected both the relative plant availability of the metals in sediment (Schierup & Larsen 1981). Plants growing in more contaminated areas generally had elevated levels of heavy metals in the tissues in comparison to plants growing in less contaminated areas, as would be expected (Baudo & Varini 1976, Chiaudani 1969, Larsen & Schierup 1981).

Substrate characteristics, however, may significantly influence the uptake of trace metals by reeds. Schierup & Larsen (1981) suggested that the higher pH, lower redox values and higher organic contents (thus high CEC) of Lake Sorteso (Denmark) rendered the heavy metals less available than those in Lake Hampen; thus, edaphic factors formed greater plant uptake of heavy metals from lake Hampen sediments, even tough metal concentrations in the sediments were lower than at Lake Sorteso.

Seasonal variation may be related to physiological processes and to contaminant input (atmospheric deposition). Kufel (1978) reported that the highest concentrations of Pb and Cu in aboveground portions of reed were found during early summer and declined thereafter; Co declined only slightly throughout the season. Larsen & Schierup (1981) and Larsen (1983) found that the Pb content of the leaves increased not only during but also after the growing season, presumably as the result of atmospheric deposition. Another example of atmospheric impact is given by Odum & Driffmeyer (1978). They found higher levels of Pb on *Phragmites* leaves at shorter distances from the highway. These levels were increased with a factor 6 in the standing dead leaves and with a factor 24 in the leaf litter. Since grazing is mostly limited to the young sprouts in spring, a proper assessment of concentrations in this period is needed in order to estimate passage of contaminants in the food chain. Unfortunately detailed and reliable information is not available.

As compared with another helophytic plant, *Typha latifolia*, maximum levels of Zn and Cu in the underground organs of *Phragmites* are lower, however, between species the levels in leaves and shoots are comparable. Sr seems to be better translocated to the shoots by *Typha* than by *Phragmites* (Kovacs 1982, Taylor & Crowder 1983a, 1983b). Whether this also holds for other elements was not reported.

4.2. Organic contaminants

Literature pertaining to the uptake by and effects of organic contaminants from *P. australis* is almost non-existent. Sarkka *et al.* (1978) reported the uptake of chlorinated hydrocarbons by reeds and other plants in Lake Paijanne, Finland. The ranges of concentrations reported for reeds were 0–267 μg kg^{-1} (PCB), 0–7 μg kg^{-1} (total DDT), and 0–4 μg kg^{-1} (aldrin). The organochlorine content of reeds varied widely with sampling stations (4) and dates (3) with no consistent pattern for plant uptake.

Physiological effects of pesticides on *Phragmites* have been reported by Merezhko and Shokod'ko (1978). The accumulation of DDT by reed was dependent upon the concentration of DDT in the environment and DDT was absorbed to the level causing inhibition of metabolic activities. The levels of pesticide accumulation in the plants were unclear in both of the latter publications, however. Shokod'ko *et al.* (1978) reported that DDT and BHC (lindane) decreased $^{14}CO_2$ assimilation in *Phragmites* and suggested that these pesticides had a specific effect in photosynthesis as well as a "total toxic effect". *Phragmites* was more resistant to DDT than to BHC. The pressure of DDT and BHC promoted ^{14}C "outflowing into roots" in direct proportion to the pesticide concentrations. The "intensification of assimilates into roots in the presence of DDT and BHC" was indicated as a "protective reaction of *Phragmitis communis* Trin. to the action of the toxic agent".

In general it is assumed that organic contaminants adsorb to the roots, but are not translocated to above-ground plant parts. As for lead (Larsen 1983), atmospheric deposition is the most likely source for contamination of leaves and stems with organic contaminants.

References

Allen, S. E. & Pearsall, W. H., 1963. Leaf analysis and shoot production in *Phragmites*. Oikos 14: 176—189.

Auclair, A. N. D., 1979. Factors affecting tissue nutrient concentrations in a scirpus-equisetum wetland. Ecology 60: 337—348.

Bakker, D. & Biewinga, D. T., 1957. Reeds in the Noordoostpolder. Van Zee tot Land, Nr 21, Directie van de Wieringermeer (Noordoostpolderwerken), Zwolle 55 pp.

Banoub, M. W., 1975. The effect of reeds on the waterchemistry of Gnadensee (Bodensee). Arch. Hydrobiol. 75: 500—521.

Baudo, R. & Varini, P. G., 1976. Copper, manganese and chromium concentrations of five macrophytes from the delta of river Toce (Northern Italy). Mem. Ist. Ital. Idrobiol 33: 305—324.

Baumeister, W. & Ernst, W., 1978. Mineralstoffe und Pflanzenwachstum. Gustav Fischer Verlag. Stuttgart, New York. 416 pp.

Bayly, I. L. & O'Neill, T. A., 1972. Seasonal ionic fluctuations in a *Phragmites communis* community. Can. J. Bot. 50: 2103—2109.

Best, E. P. H., Zippen, M. & Dassen, J. H. A., 1981. Growth and production of *Phragmites australis* in Lake Vechten. Hydrobiol. Bull. 15: 165—173.

Bibby, C. J. & Lund, J., 1982. Conservation of reed beds and their avifauna in England and Wales, UK. Biol. Cons. 23: 167—186.

Boorman, L. A. & Fuller, R. M., 1981. The changing status of a reedswamp in the Norfolk Broads. J. Appl. Ecol. 18: 241—269.

Buttery, B. R., & Lambert, J. M., 1965. Competition between *Glyceria maxima* and *Phragmites communis* in the region of Surlingham Broad. I. The competition mechanism. J. Ecol. 53: 163—181.

Chapman, V. J., 1960. Salt Marshes and Salt Deserts of the World. L. Hill, London 392 pp.

Chiaudani, G., 1969. Normal contents and accumulations of copper in *Phragmites communis* as a response to those in the sediments of 6 Italian lakes. Mem Ist Ital Idrobiol 25: 81—95.

Crowder, A. A. & MacFie, S. M., 1986. Seasonal deposition of ferric hydroxide plaque on roots of wetland plants. Can. J. Bot. 64: 2120—2124.

Daniels, R. E., 1975. Observations on the performance of *Narthecium ossifragum* (L.) Huds. and *Phragmites communis* Trin. J. Ecol. 63: 965—977.

Durska, B., 1970. Changes in the Reed (*Phragmites communis* Trin.) condition caused by diseases of fungal and animal origin. Pol. Arch. Hydrobiol. 17: 373—396.

Dykyjova, D., 1979. Selective uptake of mineral ions and their concentration factors in aquatic higher plants. Folia Geobot. Phytotax 14: 267—325.

Gustafsson, A. & Simak, M., 1963. X-ray photography and seed sterility in *Phragmites communis* Trin. Hereditas 49: 442—450.

Harradine, A. R., 1982. Effect of salinity on germination and growth of *Pennisetum macrourum* in Southern Tasmania. J. Appl. Ecol. 19: 273—282.

Haslam, S. M., 1969. The development and emergence of buds in *Phragmites communis* Trin. Ann. Bot. N.S. 33: 289—301.

Haslam, S. M., 1970a. The development of the annual population in *Phragmites communis* Trin. Ann. Bot. N.S. 34: 571—591.

Haslam, S. M., 1970b. The performance of *Phragmites communis* Trin. Ann. Bot. N.S. 34: 147—158.

Haslam, S. M., 1970c. The performance of *Phragmites communis* Trin. in relation to water-supply. Ann. Bot. N.S. 34: 867—877.

Haslam, S. M., 1971a. Community regulation in *Phragmites communis* Trin. II. Mixed stands. J. Ecol. 59: 75—88.

Haslam, S. M., 1971b. The development and establishment of young plants of *Phragmites communis* Trin. Ann. Bot. N.S. 35: 1059—1072.

Haslam, S. M., 1972. *Phragmites communis* Trin. Biol. Flora Brit. Isles. J. Ecol. 60: 585—610.

Haslam, S. M., 1973. Some aspects of the life history and autecology of *Phragmites communis* Trin. A Revies. Pol. Arch. Hydrobiol. 20: 79—100.

Haslam, S. M., 1975. The performance of *Phragmites communis* Trin. in relation to temperature. Ann. Bot. 39: 881—889.

Ho, Y. B., 1981. Mineral composition of *Phragmites australis* in Scottish lochs as related to eutrophication. I: Seasonal changes in organs. Hydrobiologia 85: 227—237.

Imhof, G., 1973. Aspects of energy flow by different food chains in a reed bed. A review. Pol. Arch. Hydrobiol. 20: 65—168.

Ingram, H. A. P., Barclay, A. M., Coupar, A. M., Glover, J. G., Lynch, B. M. & Sprent, J. I., 1980. *Phragmites* performance in reed beds in the Tay estuary. Proceed. of the Roy. Soc. of Edinburgh. 78B: 89—107.

Kamio, A., 1982. Studies on the drying of marshy and heavy clay soil ground by means of vegetation on the process of polder land drainage and structural changes of *Phragmites communis* community in the Hachirogata central polder Akita prefecture, Japan. JPN J. Ecol. 32: 357—364.

Kaul, S., 1984. Biomass and mineral composition of aquatic macrophytes in the Hygam wetland, Kashmir with reference to substrate nutrients. Acta hydrochim. et hydrobiol. 12: 81—91.

Khashes, T. M. & Bobro, V. I., 1971. Diurnal and seasonal transpiration dynamics in *Phragmites communis* Trin. UKR Bot. Zh. 28: 521—524.

Kovacs, M., 1982. Chemical composition of the lesser reedmace (*Typha angustifolia* L.) in lake Balaton. Acta Bot. Acad. Sci. Hung. 28: 297—307.

Kovacs, M., Precsenyi, I. & Podani, J., 1978. Anhaufung von elementen in balatoner Schilfrohr (*Phragmites communis*). Acta Bot. Acad. Sci. Hung. 24: 99—111.

Krasovskii, L. I. & Chashckukhin, V. A., 1974. Oxygen regime of the rootstocks of common reed. Fiziol. Rast. 21: 315—319.

Krolikowska, J., 1971. Transpiration of reed (*Phragmites communis* Trin.). Pol. Arch. Hydrobiol. 18: 347—358.

Kufel, I., 1978. Seasonal changes of Pb, Cu, Mo and Co in above-ground parts of *Phragmites australis* Trin. ex Steudel and *Typha angustifolia* L. Bull. Acad. Pol. Sci. 26: 765—770.

Kufel, I. & Kufel, L., 1980. Chemical composition of reed (*Phragmites australis* Trin. ex Steudel) in relation to the substratum. Bull. Acad. Pol. Sci. Ser. Sci. Biol. 28: 563—568.

Kvet, J., 1973. Transpiration of South Moravian *Phragmites communis*. In: Littorial of the Nesyt Fishpond, edited by

Kvet, J. pp 143—146. Studie CSAU 15/1973. Academia, Praha.

Lanning, F. C. & Eleuterius, L. N., 1985. Silica and ash in tissues of some plants growing in the coastal area of Mississippi USA. Ann. of Bot. 56: 157—172.

Larsen, V. J., 1983. The significance of atmospheric deposition of heavy metals in 4 Danish Lakes. Sci. Total Environ. 30: 111—128.

Larsen, V. J. & Schierup, H., 1981. Macrophyte cycling of zinc, copper, lead and cadmium in the littoral zone of a polluted and non-polluted lake. II Seasonal changes in heavy metal content of aboveground biomass and decomposing leaves of *Phragmites australis*. Aquat. Bot. 11: 211—230.

Luther, H., 1983. On life forms and aboveground and underground biomass of aquatic macrophytes. Acta Bot. Fenn. 123: 1—23.

Merezhko, A. 1. & Shokod'ko, T. I., 1978. Characteristics of DDT uptake by higher aquatic plants. Gidrobiol Zh 14: 84—91.

Misra, R. D., 1938. Edaphic factors in the distribution of aquatic plants in the English Lakes. J. Ecol. 26: 411—451.

Odum, W. E. & Driffmeyer, J. E., 1978. Sorption of pollutants by plant detritus: a review. Environ. Health Perspectives. 27: 133—137.

Pazourek, J., 1973. The density of stomata in leaves of different ecotypes of *Phragmites communis*. Folia Geobot. Phytotax. 8: 15—21.

Phillips, S. F., 1976. The relationship between evapotranspiration by *Phragmites communis* Trin. and water table fluctuations in the Delta Marsh, Manitoba. Diss. Abstr. Int'l. B. The Sciences and Engr. Vol. 37 (6).

Ranwell, D. S., Bird, E. F. C., Hubbard, J. C. E. & Stebbens, R. E., 1964. Tidal submergence and chlorinity at Poole Harbour. J. Ecol. 52: 627—642.

Rezk, M. R. & Edany, T. Y., 1979. Comparitive responses of two reed sp. to water table levels. Egypt. J. Bot. 22: 157—172.

Ricciutti, E. R., 1983. The all too common, common reed. Aubudon 85: 65—66.

Rychnovska, M & Smid, P., 1973. Preliminary evaluation of transpiration in two *Phragmites* stands. Ecosystem study on wetland bioma in Czechoslovakia. Czechosl. IBP/PT-PP Report No. 3 (Hejny, S. ed.) pp. 111—119, Trebon.

Sarkka, J., Hattula, M. L., Sjanatuinen, J. & Paasivirta, J., 1978. Chlorinated hydrocarbons and mercury in aquatic vascular plants of lake Paijanne, Finland. Bull Environm. Contam. Toxicol 20: 361—368.

Schierup, H. & Larsen, V. J., 1981. Macrophyte cycling of zinc, copper, lead and cadmium in the littoral zone of a polluted and non-polluted lake. I Availability, uptake and translocation of heavy metals in *Phragmites australis* (Cav.) Trin. Aquat. Bot. 11: 197—210.

Schlott, G. & Malicky, G., 1984. Biomass and Phosphorus content of the macrophytes of the northeast bay of the Lunzer Untersee Austria in relation to nutrient-rich inflows and to the sediment. Arch. Hydrobiol. 101: 265—277.

Shokod'ko, T. I., Merezhko, A. I. & Lyashenko, A. H., 1978.

Effect of DDT and BHC on assimilation and outflow of carbon-14 in *Phragmites communis*. Gidrobiol Zh. 14: 105—109.

Smid, P., 1975. Evaporation from a reedswamp. J. Ecol. 60: 299—309.

Szczepanski, A. J., 1978. Ecology of macrophytes in wetlands. Pol. Ecol. Studies 4: 45—94.

Taylor, G. J. & Crowder, A. A., 1983a. Uptake and accumulation of heavy metals by *Typha latifolia* in wetlands of the Sudbury, Ontario region. Can. J. Bot. 61: 63—73.

Taylor, G. J. & Crowder, A. A., 1983b. Uptake and accumulation of copper, nickel and iron by *Typha latifolia* grown in solution culture. Can J. Bot. 61: 1825—1830.

Tuschl, P., 1970. Die Transpiration von *Phragmites communis* Trin. in Geschlossenen bestand des Neusiedler Sees. Wiss. Arb. Burgenld. 44: 126—186. Tuxen, R. (ed), 1982. Phragmitetea. Bibliographia Phytosociologica Syntaxonomica. Vol. 36. Cramer, J. Vaduz.

Ulehlova, B., Husak, S. & Dvorak, J., 1973. Mineral cycles in reed stands of Nesyt Fishpond in southern Moravia. Pol. Arch. Hydrobiol. 20: 121—129.

Van der Linden, M. J. H. A., 1980. Nitrogen economy of reed vegetation in the Zuidelijk Flevoland Polder. I. Distribution of nitrogen among shoots and rhizomes during the growing season and loss of nitrogen due to fire management. Acta Oecol., Oecol. plant. 1: 219—230.

Van der Toorn, J., 1971. Investigations on the ecological differentiation of *Phragmites communis* Trin. in the Netherlands. Hydrobiologia 12: 97—106.

Van der Toorn, J., 1972. Variability of *Phragmites australis* (Cav.) Trin. ex. Steudel in relation to the environment. Van zee tot land 48: 122 pp.

Van der Toorn, J., 1982. Invloed van beschadigingen op de groei van riet en vegetatie ontwikkeling in the IJsselmeerpolders. Vakblad voor biologen 20: 394—397.

Van der Werff, M., 1981. Ecotoxicity of heavy metals in aquatic and terrestrial higher plants. Thesis Vrije Universiteit, Amsterdam, 1981.

Veber, K., 1978. Propagation, Cultivation and Exploitation of Common Reed in Czechoslovakia. In: Pond littoral ecosystems, edited by Dykyjova, D. & Kvet, J. Springer Verlag, Berlin, Heidelberg, New York. pp. 416—423.

Vostokova, E. A., 1975. Ecologic and hydrogeologic role of sheet grass communities in the sand deserts of the northern Aral Region. Ekologiya 6: 83—89.

Waisel, Y. & Rechaw, Y., 1971. Ecotypic differentiation in *Phragmites communis* Trin. Hydrobiologia 12: 259—266.

Wallentinus, H. G., *et al.*, 1973. The reed *Phragmites communis* Trin., in the Brunnsviken. Sven. Bot. Tidskr. 67: 81—96.

Yamasaki, S., 1981. Effects of water level on the development of rhizomes of three hygrophytes. Jap. J. Ecol. 31: 353—359.

Yamasaki, S., 1984. Role of plant aeration in zonation of *Zizania latifolia* and *Phragmites australis*. Aquat. Bot 18: 287—297.

Authors' Address
M. Van der Werff
Dept. Ecology and Ecotoxicology
Vrije Universiteit Amsterdam
De Boelelaan 1087
1081 HV Amsterdam
The Netherlands

Oosterkwelder salt marsh on the island of Schiermonnikoog. Salt marsh vegetation may be disturbed by pollution of seawater with oil. (Photograph: P. C. Leendertse.)

CHAPTER 16

The impact of oil pollution on salt marsh vegetation

M. C. Th. SCHOLTEN & P. C. LEENDERTSE

Abstract. Salt marshes run a high risk of pollution in the event of an oil spill in coastal regions. Oil polluted salt marshes can be perturbed for a long time. These perturbations are not only determined by the amount of oil and retention rates, but also by shifts in competitive interactions and consequent shifts in vegetation structure. Experimental studies have shown that species that respond in an opportunistic way (potential for recovery) become dominant and suppress the recovery of the other species. Vigorous growth of *Puccinellia*, following oil spills in the field, can be explained by changed competitive interactions which favour it. *Festuca* and *Plantago* respond differently to oil pollution when growing with different assemblages of species.

1. Introduction

Tidal flats and salt marshes run a high risk of pollution in the event of an oil spill in coastal regions. At low tide an oil slick can strand in the intertidal zones. A strong and long lasting environmental impact is common in many oil spill cases in such shallow, coastal sedimentation areas (Stebbings 1970; Burns & Teal 1978; Hampson & Moul 1978; Géhu & Géhu-Franck 1981; Levasseur *et al.* 1981).

The most obvious damage to the salt marsh vegetation is due to deposition and adsorption of oil on the leaves and buds of the plants, and the associated reduction in gas exchange, transpiration and light perception (Scholten *et al.* 1987b; Huiskes & Rozema 1988). This, together with physiological strains caused by oil compounds taken up through stomata or roots, results in the death of green tissues (Baker, 1971a). However, the net photosynthetic rate per unit of green leaves is not reduced (Scholten *et al.* 1987b).

Heavy crude oils (e.g., Arabian, Nigerian or Louisiana crude) mainly cause *smothering* effects, whereas light crude (e.g., Forties: a North Sea

crude) and refined oils (e.g., no. 2 fuel oil: diesel) are mainly responsible for toxic effects due to penetration of aromatics into plant tissues (Baker *et al.* 1984; Alexander & Web 1985). Aromatics are also responsible for a reduction of the seed production when flowers are polluted, and an inhibition of the embryological development of endosperms when seeds are affected.

Next to its direct impact on plant growth and survival, oil can have indirect effects via changing environmental conditions. It is known that oil polluted sediments generally become anaerobic (Stebbings 1970). A reduced bioturbation of the sediments due to benthic faunal mortality may enhance this and in any event the degradation of oil compounds is extremely low in anaerobic sediments (Mago *et al.* 1978).

Some authors suggest a better nitrogen supply in slightly oil polluted sediments, due to an increase of the nitrogen fixation rate (Thomson & Webb 1984). However, in general the consequences of these oil mediated changes in environmental conditions on plant growth and vegetation dynamics are poorly understood.

J. Rozema and J. A. C. Verkleij (eds.), Ecological Responses to Environmental Stresses, 184—190.

2. The perturbation of salt marsh vegetation after oil pollution

The duration of the perturbation of salt marsh vegetation after an oil spill depends on the opportunity for recovery for salt marsh plants, determined by (1) the amount and depuration rates of oil from the sediments, (2) the life history and growth form of individual halophytic species and (3) the changed competitive situation between the halophytes due to alterations of the vegetation structure.

Experimental oil spills in the field indicate a complete recovery of a Spartina marsh within 2 years (Baker *et al.* 1984, Alexander & Webb 1985). However, longer recovery periods are reported for actual spills (Hampson & Moul 1978; Levasseur *et al.* 1981). Burns & Teal (1979) found no recovery of *Spartina* and *Salicornia* on sediments containing more than 1000 ppm of oil.

They found a gradual degradation of no. 2 fuel oil from 2700 ppm to 1900 ppm within six years. Polynuclear aromatic compounds form a substantial part of residual fraction of the oil in sediments, implying a long lasting toxic exposure of the salt marsh community. A chronic exposure to oil results in a drastic, lasting perturbation of the salt marsh vegetation structure (Baker 1971c; Hershner & Lake 1980).

3. The effects of oil pollution on competitive relationships

In experimental studies with coastal plankton and zoobenthos communities it was shown that the mortality or the reduced reproduction of bivalves and crustaceans results in a complete shift in the ecological interactions. Food organisms (e.g., algae, snails) benefit from the reduced consumption, whereas opportunistic competitors (e.g., protozoans, worms) benefit from this excess of food sources (Scholten & Kuiper 1987). Such shifts in structure can retard or even prevent a complete recovery of the community. Analogous shifts can occur within a salt marsh vegetation, where oppor-

tunistic plants can benefit from the light and nutrients that is left by death or stunted growth of strong competitors.

Some salt marsh plant species exhibit an opportunistic strategy with a high potential for recruitment either by seed dispersal or by lateral vegetative clonal growth, whereas other species have a more conservative strategy forming sustainable perennial structures by which they defend occupied niches without rapid expansion (Scholten in prep.). An opportunistic response to the mortality or growth reduction of salt marsh plants after an oil spill can favour the plants with a high potential regrowth, and thus can hinder the regrowth of the more conservative species for a long time.

4. The impact of oil pollution on life strategies

Salt marsh plants with a fast, opportunistic vegetative expansion (e.g., *Puccinellia* and *Agrostis*) can easily recover and benefit from an oil pollution at any time (Golombek & Neugebohrn 1984). Other perennial species can only recover when vegetative regrowth is possible, e.g., during the initial phases of the growing season. Rosette plants with rhizomes or taproots (e.g., *Aster*, *Limonium*, *Triglochin* or *Plantago*) show a rapid regrowth from the below ground organs during spring. An oil spill in late spring — early summer causes severe growth reduction in that year. During the winter period there is no damage at all as a result of the fact that the main hibernating organs are below ground. A similar response is seen for salt marsh grasses, such as *Spartina*, which regrow from rhizomes (Baker 1971b; Alexander & Webb 1985).

Annual species (e.g., *Salicornia, Suaeda* and *Spergularia*) can reinvade polluted sites by seed transport from unpolluted marshes. The only longer term effects that are recorded for these species is a reduced seedling density in the year after the spill, when plants are inundated during the flowering period (Baker 1971b).

Species with limited opportunities for vegetative regrowth or seedling recruitment, especially those with relatively low growth rates (e.g., *Festuca* and

Juncus) or those with above ground buds (e.g., *Halimione* and *Artemisia*) are the most sensitive to oiling.

The ranking of salt marsh species in order of their sensitivity to oil pollution, given by Baker (1979) on the basis of an experiment with repeated oiling of marsh vegetation, can be explained by the above mentioned interspecific differences in growth strategies.

5. The expected effect of oil pollution: experiments on "the low salt marsh" vegetation

It is difficult to understand the long term effects of an oil spill on coastal communities from field observations alone (Scholten & Kuiper 1987). Lack of comparable, unpolluted references and the predominance of ecological processes hampers an unambiguous interpretation of field data. Experimental ecosystem studies can bridge the gap between the observation of phenomena in the field and data from toxicological studies.

Some results obtained with experimental salt marsh vegetations can illustrate the described response of interacting salt marsh plants to oil pollution.

Twelve containers with a surface area of $0.25 \, \text{m}^2$ and a depth of 35 cm were filled with a natural salt marsh soil collected from the Wadden sea area in the Netherlands. In May, 1986 four salt marsh plant species (*Spartina anglica*, *Puccinellia maritima*, *Triglochin maritima* and *Halimione portulacoides*) were transplanted in natural densities in 8 pies. During the summer these twelve systems developed into replicate experimental (semi-natural)

low salt marsh vegetation. In October 1986, early March 1987, late April 1987 and mid June 1987, 250 ml of a light crude (Forties) oil was added on the soil surface of two systems on each occasion. The remaining two systems were left undisturbed and served as a control. In September 1987, the above ground biomass was harvested.

It was shown that the oil pollution reduced the total biomass production, especially when added during the winter-springtime period (Table I). *Halimione* proved to be the most sensitive species. *Spartina* suffered from oil pollution in the winter-springtime, when no large shoots were present. *Puccinellia* and *Triglochin* on the other hand clearly benefitted from the growth reduction of the other species. *Puccinellia* was seen to rapidly occupy the space left by *Halimione*. After the oil pollution in summertime the biomass of *Puccinellia* even exceeded that in the unpolluted references. After an oil spill in wintertime its biomass was reduced, but the recovery was good so that it eventually increased its dominance. *Triglochin* benefitted especially when *Halimione* was destroyed in the wintertime.

In 1988 a comparable experiment was performed with a more silty sediment, pretreated with 2600 ppm of Forties oil, representing 60 ml per container and thus 25% of the amount in 1986 and on this occasion homogeneously mixed to a depth of 35 cm. Two unpolluted systems served as a control. In May 1988 seven low salt marsh plant species were transplanted in seven pies. The same species as in 1986 were used together with *Aster tripolium*, *Limonium vulgare* and *Plantago maritima*. In September 1988 the above ground biomass was harvested.

Table I. Biomass (gram dry weight per m^2) of salt marsh plant species in september 1987 in a bioassay with application of 1000 ml Forties per m^2 at different times.

Application time	Total	Halimione	Spartina	Puccinellia	Triglochin
no application	360.0	182.0	1.2	165.8	11.0
June 1987	225.2	18.0	0.8	195.0	11.4
April 1987	191.2	52.4	0.0	124.2	14.6
March 1987	172.4	28.6	0.0	132.0	11.8
October 1987	169.6	10.4	0.0	137.0	22.2

In this experiment the oil polluted sediments did not reduce the total biomass. However, major shifts in species composition occurred (Table II). Again *Halimione* was the most sensitive species. *Plantago, Limonium* and *Aster* also suffered from the oil. Again *Puccinellia* and *Triglochin* benefitted from the contamination. *Spartina* also benefitted in this case, due to a better growth on the silty soil. The diversity of the vegetation was reduced in the oil polluted systems.

An additional greenhouse study gave more insight into the changed interactions between salt marsh species after an oil spill (Scholten *et al.* 1987b). When species were tested for oil sensitivity in a monoculture test, *Puccinellia* seemed to be more sensitive compared to *Halimione*. However, in a mixed culture a less strong growth reduction of *Puccinellia* was observed (Fig. 1). A consistently negative correlation of the biomass of the two mentioned species in a wide variety of experiments in which they grow together, provides evidence for a strong competitive interaction. It was observed that under unpolluted conditions *Halimione* has a competitive advantage over *Puccinellia*. This advantage progressively declines with increasing oil content of the sediment, because of a decreasing impact of *Halimione* on *Puccinellia* as a result of the reduced growth. It turned out that *Puccinellia* accumulated the nitrogen that was left by *Halimione* and thus benefits from the oil pollution (Fig. 2). Under unpolluted conditions the growth of *Puccinellia* was limited by nitrogen, as a consequence

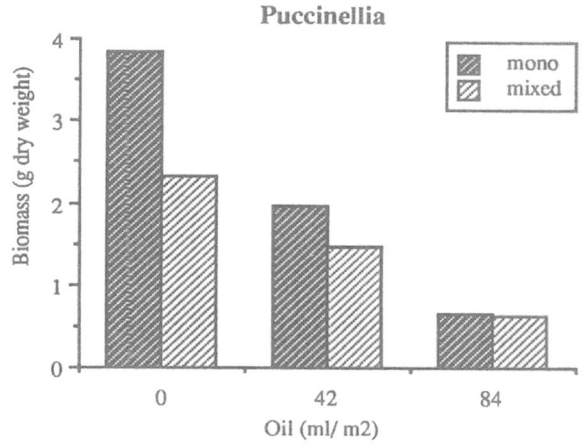

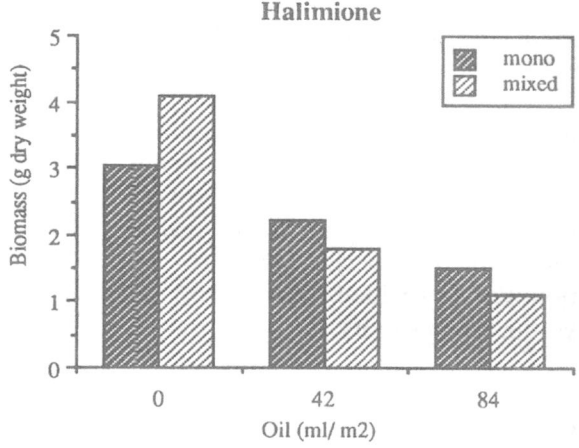

Fig. 1. Biomass production (g dry weight per 7 plant units) of *Puccinellia mariüma* and *Halimione portulacoides* in mono and mixed cultures after application of 0, 42 and 84 ml oil per m².

Table II. Biomass (gram dry weight per m²) of salt marsh plant species in September 1988 in a bioassay with application of 2600 ppm Forties oil in May 1988.

Species	Oil polluted	Reference
Puccinellia	47.2	26.0
Spartina	67.2	33.2
Triglochin	4.4	2.0
Aster	38.4	48.0
Limonium	3.6	7.6
Plantago	0.0	8.8
Halimione	0.8	20.0
Total	161.6	146.0
Diversity (Simpson)	0.68	0.75

of the efficient nitrogen accumulation of *Halimione* (Jensen 1985). The release of that competition-induced nitrogen limitation, due to the growth reduction of *Halimione*, compensates for the toxic reduction of the potential growth rate of *Puccinellia* by oil, as was seen in monoculture experiments. The extra growth of *Puccinellia* ultimately results in increasing growth reduction of *Halimione*, probably as a result of shading.

An excessive nitrogen supply repeals the nitrogen limitation of *Puccinellia*, and therefore the benefit gained from oil (Fig. 2) exposure.

The same greenhouse study also indicates that

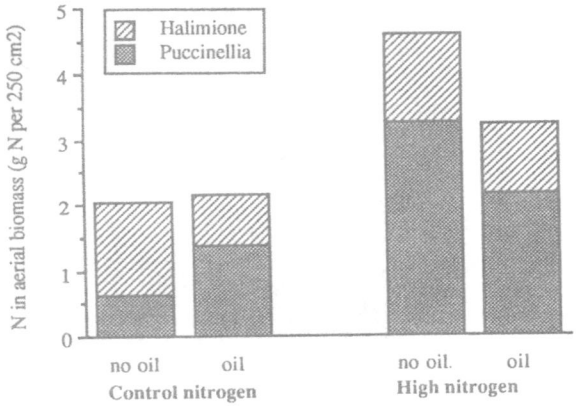

Fig. 2. Accumulation of nitrogen in aerial biomass (gram per 250 cm²) by *Puccinellia maritima* and *Halimione portulacoides* grown together at two nitrogen levels and treated with 0 and 84 ml oil per m².

the benefit of *Triglochin* on oil polluted sediment is similarly caused by reduced competition for nitrogen. *Spartina* is also known to be a poor competitor for nitrogen and light (Scholten *et al.* 1987a), and therefore this species may also benefit from oil pollution, provided that the oil does not cover young green shoots.

6. Experimental studies on the "high salt marsh" vegetation

In an experimental salt marsh vegetation study that was carried out parallel to the 1986/87 one described above, the effects of the application of 250 ml Forties oil on 0.25 m² of an high salt marsh vegetation, consisting of *Elymus pycnanthus, Fes-*

tuca rubra, Artemisia maritima and *Plantago maritima* were observed. The total biomass was only reduced after an oil spill in March (Table III). *Artemisia* showed short term biomass reduction after an oil spill in summer, but a good recovery from an oil spill during the winter or spring. *Festuca* turned out to be the most sensitive species. The biomass production of *Elymus* was reduced by an exposure in winter or early spring, when oil affected the young shoots. Then, *Plantago* could benefit from the reduced shading by *Elymus*. An oil spill in April-June did not affect *Elymus*, which benefitted from the reduced growth of *Festuca* and *Artemisia*, probably as a result of reduced competition for nutrients.

In additional greenhouse experiments in which *Festuca* was grown with *Juncus gerardii*, the former species seemed to benefit from a reduced competition with the latter in the event of an oil spill. In these experiments *Festuca* was also able to benefit from short term growth inhibition of *Artemisia*.

7. Conclusions

The experimental studies show that the long term perturbation of a salt marsh vegetation after an oil pollution is not only determined by the amount of oil and retention rates, but also by shifts in competitive interactions and consequent shifts in vegetation structure. Mortality or growth reduction of dominant plant species give other species the opportunity to flourish in the event of reduced competition.

When the life strategy enables a species to respond in an opportunistic way (potential for recov-

Table III. Biomass (gram dry weight per m²) of salt marsh plant species in September 1987 in a bioassay with application of 1000 ml Forties per m² at different times.

Application time	Total	Festuca	Artemisia	Plantago	Elymus
no application	313.0	72.4	41.6	8.8	190.4
June 1987	282.4	40.2	16.2	10.0	216.0
April 1987	321.0	44.8	32.8	8.4	235.0
March 1987	221.0	41.4	41.0	16.0	122.6
October 1987	265.6	40.0	36.6	16.6	172.4

ery) it can become dominant and suppress the recovery of the other species. Reported vigorous growth of *Puccinellia* following oil spills in the field (Baker 1971d, Stebbings 1970), may in fact be caused by changed competitive interactions favouring *Puccinellia*. Such changes can have long lasting or even irreversible effects on the structure and functioning of the once polluted or chronic polluted salt marsh vegetation.

The examples of *Puccinellia*, *Festuca* and *Plantago* make clear that a species can be sensitive to oil in some circumstances (*Puccinellia*: in monoculture; *Festuca*: growing with *Elymus* and *Artemisia*; *Plantago*: growing with low marsh plants), whereas it can benefit from oil when growing with other plants (*Puccinellia*: in a mixed vegetation; *Festuca*: growing with *Juncus*; also having a short term benefit when growing with *Artemesia*; *Plantago*: growing with high marsh plants). The sensitivity or tolerance to oil of individual species in the field cannot be predicted from monospecies tests alone. Their potential to profit from decreased competitive pressure under various conditions have to have be taken into account.

An oil polluted salt marshes can be perturbated for a long time, even after a small oil spill. Their protection from becoming polluted is essential. Booming a slick and pumping can help to defend the salt marshes from oil slicks. Once the oil is stranded, removal by cutting (provided that the cuttings will be removed adequately from the marsh) or flushing is recommended (Kiesling *et al.* 1988), as long as the bottom structure is not destroyed. The use of dispersants or burning will only aggravate the damage to the salt marsh vegetation.

Acknowledgements

Thanks are due to Martin Stroetenga (VU) and Gerard Hoornsman (TNO) for carrying out the experimental salt marsh vegetation studies, Henk Schat (VU) and Tim Bowmer (TNO) for correcting the text and Marie-Cécile Hendriks (TNO) for typing the manuscript.

References

Alexander, S. K. & Webb, Jr. J. W., 1985. Seasonal response of Spartina alterniflora to oil. In: Proceedings of the 1985 Oil Spill Conference (prevention, behaviour, control, clean up). EPA/API/USCG, Los Angeles pp. 355—357.

Baker, J. M., 1971a. The effect of oils on plant physiology. In: E. B. Cowell (eds.). Ecological effects of oil pollution on littoral communities. Appl. Sci. Publications, Barking, U.K. pp. 88—98.

Baker, J. M., 1971b. Seasonal effects. In: Ibid. pp. 44—51.

Baker, J. M., 1971c. Successive spillages. In: Ibid. pp. 21—32.

Baker, J. M., 1971d. Growth stimulation following oil pollution. In: Ibid. pp. 72—77.

Baker, J. M., 1979. Responses of salt marsh vegetation to oil spills and refinery effluents. In: R. L. Jefferies & A. J. Davy (eds.). Ecological processes in coastal environments. Blackweall Scientific Publ. pp. 529—543.

Baker, J. M., Crothers, J. H., Little, D. I., Oldham, J. H. & Wilson, C. M., 1984. Comparison of the fate and ecological effects of dispersed and nondispersed oil in a variety of intertidal habitats. In: T. E. Allen (ed.): Oil spill chemical dispersants: research, experience and recommendations, STP 840. American Society for Testing and Materials, Philadelphia. pp. 239—279.

Burns, K. A. & Teal, J. M., 1979. The West Falmouth oil spill: hydrocarbons in the salt marsh ecosystem. Estuarine Coastal Mar. Sci. 8: 349—360.

Géhu, J. M. & Géhu-Franck, J., 1981. Evolution des préssalés Nord-Armoricains sous l'impact de la Marée Noire. In: G. Gonan et al. (eds.): Amoco Cadiz, conséquences d'une pollution accidentele par les hydrocarbures. Centre National pour l'exploitation des Oceans, Paris. pp. 443—453.

Golombek, P. & Neugebohrn, L., 1984. Experimentelle Untersuchungen zur Wirkung van Rohol und Rohol Tensid-Gemischen im Okosystem Wattenmeer. XV Das Puccinellietum maritmae (Warning). Christiansen der Saltweise Senckenb. Marit. 16: 245—266.

Hampson, G. R. & Moul, E. T., 1978. No. 2 fuel oil spill in Boume, Massachusetts: immediate assessment of the effects on marine invertebrates and a 3 year study of growth and recovery of a salt marsh. J. Fish. Res. Board. Can. 35: 731—774.

Hershner, C. & Lake, L., 1980. Effects of chronic oil pollution on a salt marsh community. Mar. Biol. 56: 163—173.

Huiskes, A. H. L. & Rozema, J., 1988. The impact of anthropogenic activities on the coastal wetlands of the North Sea. In: W. Salomons, B. L. Bayne, E. K. Duursma & U. Forstner (eds.). Pollution of the North Sea: an assessment. Springer, Berlin pp. 455—473.

Jensen, A., 1985. On the ecophysiology of Haliomione portulacoides. Vegetatio 61: 231—240.

Kiesling, R. W., Alexander, S. K. & Webb, J. W., 1988. Evaluation of alternative oil spill clean-up techniques in a Spartina alterniflora salt marsh. Environ. Pollut. 55: 221—238.

Levasseur, J., Durand, M. A. & Jory, M. L., 1981. Aspects

biomorphologiques et floristique de la reconstiturion d'un couvert végétal phanérogamique doublement alteré par les hydrocarbures et les opérations subséquentes de nettoiement (cas particulier des marais maritimes de l'ile grand, cotes-du-nord). In: G. Gonan *et al.* (eds.) Amoco Cadiz, conséquences d'une polution accidentele par les hydrocarbures. Centre National pour l'exploitation des Oceans, Paris. pp. 455—473.

Mago, D. W., Page, D. S., Cooley, J., Sorenson, E., Bradley, F., Gilfillan, E. S. & Hanson, S. A., 1978. Weathering characteristics of petroleum hydrocarbons deposited in fine day marine sediments Searsport Maine. J. Fish. Res. Board Ca. 35: 552—562.

Scholten, M. C. Th., in prep. Competitive interactions amongst salt marsh plant species. Thesis in prep.

Scholten, M. & Kuiper, J., 1987. Assessment of oil pollution in the North Sea. In: P. J. Newman & A. R. Agg (eds.). Environmental protection of the North Sea. Heinemann, Oxford. pp. 446—455.

Scholten, M., Blaauw, P. A., Stroetenga, M. & Rozema, J., 1987a. The impact of competetive interactions on the growth and distribution of plant species in salt marshes. In: A. H. L. Huiskes, C. W. P. M. Blom & J. Rozema (eds.): Vegetation between land and sea. Junk Dordrecht. pp. 270—281.

Scholten, M., Leendertse, P. & Blaauw, P. A.,1987b. The effects of oil pollution on interacting salt marsh species. In: J. Kuiper & W. J. Van den Brink (eds.): Fate and effects of oil in marine ecosystems. Martinus Nijhoff, Dordrecht. pp. 225—228.

Stebbing, R. E., 1970. Recovery of salt marsh in Brittany sixteen months after heavy pollution by oil. Environm. Pollut. 1: 163—167.

Thomson, A. D. & Webb, K. L., 1984. The effect of chronic oil pollution on salt-marsh nitrogen fixation (acetylene reduction). Estuaries 7: 2—11.

Authors' Addresses
M. C. Th. Scholten
MT-TNO
Department of Biology
Laboratory for Applied Marine Research
P.O. Box 57
1740 AB Den Helder
The Netherlands

P. C. Leendertse
Vrije Universiteit Amsterdam
De Boelelaan 1087
1081 HV Amsterdam
The Netherlands

CHAPTER 17

Effects of air pollution on plants and vegetations

A. C. POSTHUMUS

Abstract. After a short introduction on the history of effects of air pollution on plants, the present day problem of "acidic rain" as a "pars pro toto" for air pollution in general is indicated. Effects of different air pollutants are described at several levels of organization, from the molecular to the ecosystem level. Also the mechanisms of effects of air pollution on plants are explicated, as far as possible, and some hypotheses are proposed, including the actions of oxidizing pollutants on phytohormones. Quantitative exposure-effect relationships, expressed as no (-adverse)-effect lines are shown to be very important for the development of criteria for air quality guidelines or standars. The problem of combination effects of different air pollutants is approached by using more-dimensional no (-adverse)-effect surfaces. Possible ways of prevention or diminution of effects of air pollution on plants are looked for and strategies are developed for the achievement of air quality objectives in the future.

1. Introduction

Since long time air pollution has influenced plants and vegetations adversely, because air pollutants from natural and man-made sources have been distributed over parts of the biosphere for millennia. Volcanic eruptions, natural forest fires and lightning for example have influenced the vegetations of the earth already during the prehistorical period. In the early days of the Egyptian Pharaoh's and of the Roman Emperors the iron, lead and other metal smelters produced a dirty smoke, not only affecting and disturbing human individuals and societies (Peters 1968), but also the plant life in the nearby surroundings. During the Middle Ages other extra sources of air pollution were active, for example the primitive, industrial activities in the cities, war fare and city fires. But just in the period of the first industrial revolution from ca 1850 on, antropogenic air pollution became a real threat for all living creatures, especially the more sensitive

plants in the surroundings of the local sources (Stöckhardt 1850).

Ultimately, after the 2nd world war air pollution caused by the explosive, industrial development, coinciding with a rapid increase of the human population and the motor traffic, has reached an extra high level, mostly on the local or regional scale. Abatement measures by high chimney applications, up to ± 400 m high, have diminished the local problems of high pollution levels, but have caused — in the meantime — widely spread, low pollution levels producing other, more hidden, chronic effects on remote forests and lakes (Overrein *et al.* 1980). "Acidic rain" has become the most well known phenomenon of pollution in the world, although this name may be an oversimplifying "pars pro toto" for air pollution in general. We always have to do with the very complicated influences of the total mixture of several fractions of the atmospheric deposition, as listed in Table I (Posthumus 1988). Acidic rain s.s. is only a small fraction of it.

J. Rozema and J. A. C. Verkleij (eds.), Ecological Responses to Environmental Stresses, 191—198.

2. Effects of different air pollutants at several levels of organization

As most of the research on effects of air pollutants has been performed with gases, attention will be focused on the well-known gaseous, phytotoxic air pollutants, listed in Table II. These components are different in phytotoxicity (decreasing from the top to the bottom of the table), but may all be involved in clear-cut, visible injury of the plants, sensitive to them. The symptoms of the effects, especially on the leaves and flowers, are morphological disturbances like curvatures and curlings, but also chlorotic and necrotic injuries of the leaves in different forms, and leaf, flower or fruit abortion, dependent on the affecting component and the plant species and variety (Jacobson & Hill 1970). Sometimes these symptoms may be rather specific for a special component, and are used for biomonitoring of the effects of air pollutants, for example of HF and O_3 (Posthumus 1982a).

The basis for all these effects may be found in changes of the plants at the molecular, cellular or higher level. Changes in the outer surface structures (cuticular wax layers for example), in the inner, cellular and other membranes, or in special coding, regulating, or catalysing substances in the plants are the first biochemical lesions, leading to more

Table I. Different fractions of the atmospheric deposition.

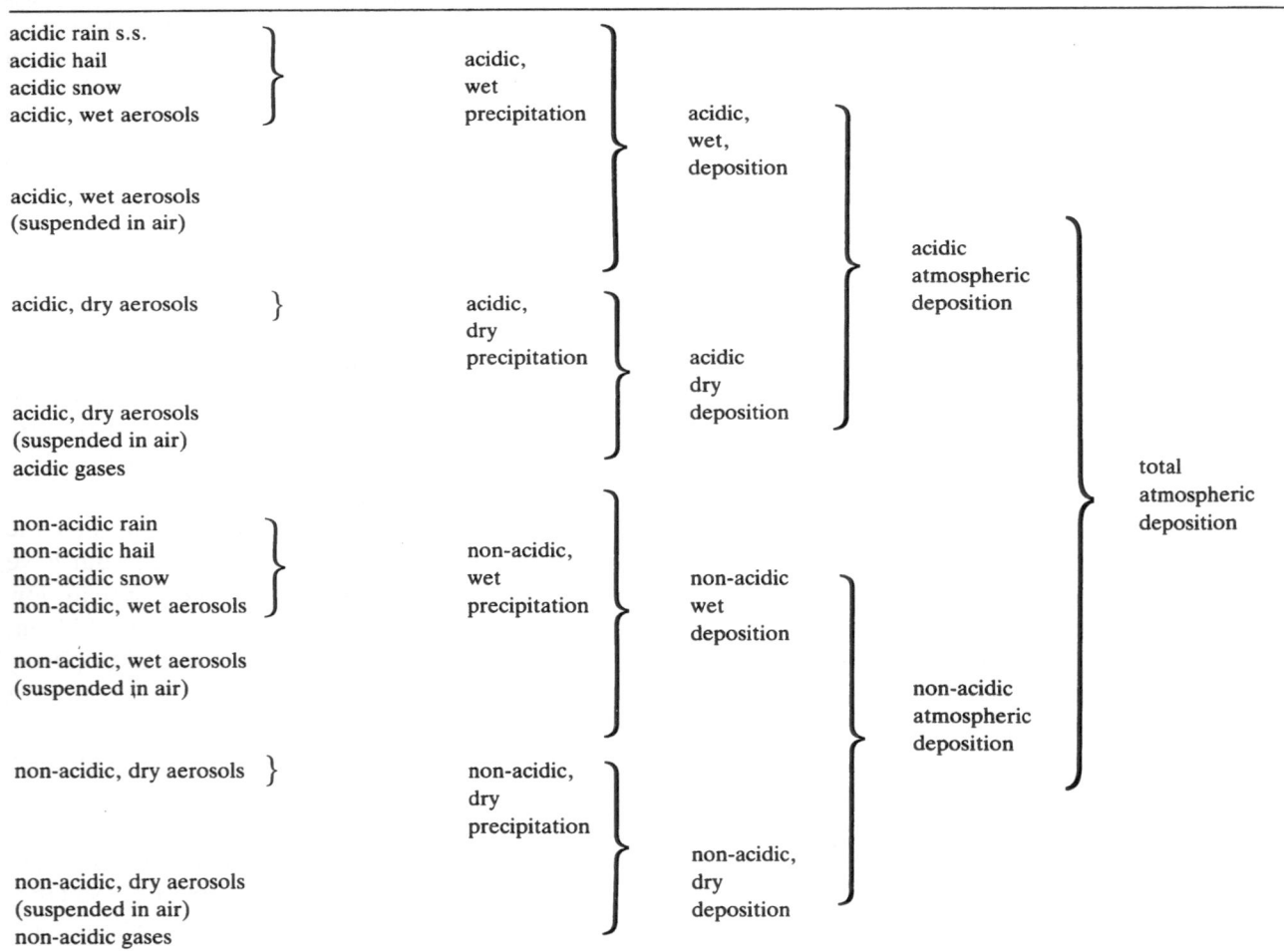

complicated secondary effects (Koziol & Whatley 1984). As a result, also effects on processes of photosynthesis, respiration, growth, development and reproduction of plants have been caused by air pollution in practice (Unsworth & Ormrod 1982). Specially negative effects on these plant processes may influence the plant performances at the higher levels of organization, for example in populations, vegetations and ecosystems. Changes in competitive relationships between varieties and species may be the result of the influences of air pollutants, ultimately leading to the disappearance of the most sensitive ones. At extreme high pollution levels the total vegetation and ecosystem may be extinguished.

3. Mechanisms of effects

Although effects of air pollutants on plants may be clear at all levels of organization (from molecule, via organel, cell, tissue, organ, plant, populations, and vegetation to ecosystem), it is evident that the origin of all effects is laying at the molecular level, where the reactions with air pollutants take place.

Mostly the oxidative properties of the air polluting components or the transformation products of these are responsible for the primary action at the targets. Peroxidation of the fatty acids in membranes, disturbance of disulphide bonds in proteins, oxidation of sulphhydrile groups in amino acids etcetera are involved in the breakdown of the cell integrity and the inhibition of enzyme activities (Mudd 1982). Changes in structures and metabolic activities are the results, and ultimately cells may lose their turgor, pigments and physiological functions. It may be stated that in the mechanisms of effect development in plants phytohormones can play an important role, although their involvement has not yet been proven undoubtedly. Many processes, which have been shown to be influenced by air pollution, are regulated by phytohormones in general. For example the translocation of photosynthates from the leaves to the roots and other storage organs is inhibited by O_3 and SO_2, and it is well known that the sink function is regulated by hormones. The same holds for the observed changes in the apical dominance phenomenon in tree branches. Possible changes in hormonal balances under the influence of air pollutants are postulated now to be the driving forces. Air pollutants like O_3 and other oxidants are well equiped to destruct the chemical components acting as phytohormones, for example indole acetic acid and cytokinins. Disturbances of the hormonal balance may result in several subtle, but also drastic changes in plants.

A special case of effects of air pollution on plants is the mutagenic effect of SO_2 in high concentrations. In this case SO_2 reacts with pyrimidines of DNA, causing disturbances of the basic structure of the genetic material. This may be compared with the action of UV-B irradiation on DNA, causing different forms of cancer in human beings.

Table II. List of phytotoxic gases with an indication of just effective concentrations for chronic and acute effects in ppb.

Component	formula	chronic effects	acute effects
hydrogen fluoride	HF	0.1–0.5	1–5
peroxyacetyle nitrate			10–20
ozone	O_3		25–40
sulphur dioxide	SO_2	5–10	70
ethene	C_2H_4	10–50	50
chlorine	Cl_2		100
hydrochloric acid	HCl	30	100
nitrogen dioxide	NO_2	100	600
ammonia	NH_3	150	1000

4. Exposure-effect relationships

As a central theme in air pollution effect studies on plants, the relationships between exposures of the plants and the resulting effects are important for several reasons. Specially the quantitative relationship between the amount of pollutant that is influencing the plant and the resulting effect intensity is very important to know, for example in connection with criteria for air quality standards or guidelines. On the basis of these criteria, produced by the phytotoxicological research, the national and su-

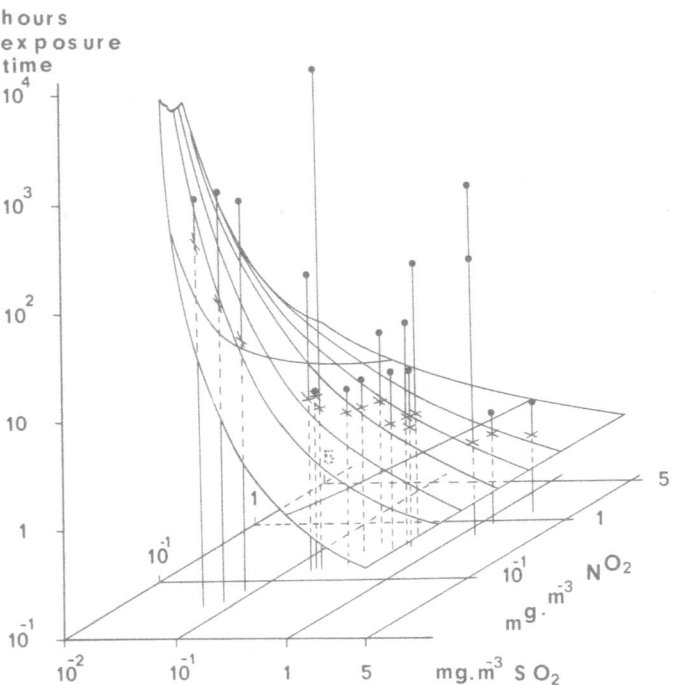

Fig. 1. Diagram showing the 3-dimensional concentrations/time model for defined effects of SO_2 + NO_2 on plants. Effective combinations of SO_2 concentration, NO_2 concentration and exposure duration are only to be found above the no(-adverse)-effect surface.

pranational, legislative commissions may make their political decisions about this matter.

The quantitative determination of the exposure of a plant to air pollution, however, is a difficult task. The most important parameter to know would be the real dose of the component X in the plant, expressed as the amount of X per plant or per unit of plant mass, causing the measured effect after some time. A good way to approach this dose is the measuring of the flux of the component X into the plant. But as flux density measurement (expressed as $\mu g\ m^{-2}s^{-1}$) is very complicated, most investigators only use the product of the ambient concentration of the component and the exposure time as a measure of apparent dose (expressed as $\mu g\ m^{-3}$ h). In this respect the concentration has been proven to be more important for the effects than the exposure time (Guderian 1977).

In practice, the development of criteria for air quality standards is based on the construction of the no(-adverse)-effect lines for different compo-

nents, discriminating between effective and non-effective combinations of exposure concentrations and exposure time periods. As concentrations are used the mean concentrations over the total exposure time periods. The effects studied should be of practical importance and have to be defined very clearly. For different kinds of effects, and different effect intensities also, the no(-adverse)-effect lines will be differently located.

A great problem is that air pollution mostly contains more than one air polluting component, and that the combination effect of the components together is not always additive, but sometimes less than additive or even more than additive. This depends on the kind and the concentration of the components and on the exposure time. For these combinations of air pollutants special air quality standards should be derived on the basis of more-dimensional no(-adverse)-effect surfaces, as shown for example for SO_2 and NO_2 in Fig. 1 (Posthumus 1982b; Van der Eerden & Duym 1988).

Fig. 2. Example of an open-top chamber of the Research Institute for Plant Protection, Wageningen, as used for long-term fumigation experiments with plants.

When several components of air pollution are acting on the vegetation, simultaneously or sequentially, it is very difficult to relate this multiple exposure to the ultimately resulting effects after some time.

Another complication of the exposure-effect relationship for effects of air pollutants on plants is the drastic influence of several environmental conditions and of other abiotic and biotic stress factors on the effects. Meteorological conditions, sometimes causing really strong stresses in the plants, soil and nutrient conditions, biotic stress factors like pathogens and pests, all may influence the integrated effects of the multi-causal complex. It is clear that the problem of the novel forest decline is a very good example of this multiple-stress phenomenon (Moseholm *et al.* 1988).

To investigate exposure-effect relationships several experimental fumigation facilities have been used. From fully conditioned fumigation cabinets, via fumigation greenhouses and different types of open-top chambers to really open-air field fumigation systems have been developed at the Research Institute for Plant Protection in Wageningen. One type of open-top chamber is shown in Fig. 2.

5. Possible ways of prevention or diminution of effects

It will be clear that the best way to prevent or diminish effects of air pollution on plants is to prevent or diminish the production and/or emission of air pollution as far as possible. But to stimulate this difficult, social-economic and technical process it is important to produce the data of the no(-adverse)-effect levels for several components, separate and in combinations, to show what low concentrations of pollutants may cause damage. On the basis of this knowledge, good air quality standards or guidelines may be set respectively proposed.

Other approaches to alleviate the pollution burden on the vegetation are theoretically possible, because special conditions and chemicals are known, which may counteract the processes of effect generating. For example dry climatic conditions, inducing water stress in plants, may reduce the uptake of air pollutants through the stomata of the leaves and thus minimize the effects. In extremely threatening situations of high pollution levels on a local scale, artifical induction of water stress in manipulated vegetation systems might be a method for the reduction of the effect intensities. Special chemicals, for example the anti-oxidans EDU (ethylene diurea) may be sprayed on plant leaves to counteract the negative influences of O_3 by disturbing the O_3 working mechanism. But the disadvantage of this method is that it is very impractical and economically not attractive, because the sprayings have to be repeated very frequently.

For agricultural, horticultural and forestry crop plants the possible way of breeding for resistance against air pollution has also been practised, sometimes even unintendedly. Under the selection pressure of ambient air pollution breeders have selected the best performing cultivars, including the resistence against prevailing air pollution, in respect to visible injury and reduction of growth and production. But it is generally agreed in our country that it should never become a common practice to breed for resistence against air pollution, because it is not good to adapt the cultivated vegetation to pollution artificially, in contrast with the natural vegetation. Furthermore, resistence against pollution will never be without costs of reduced productivity and/or loss of genetic variation within the species.

6. Strategies for the future

Because of the vast amount of information needed for the set-up of really good air quality standards, a strong selection of the right test plants to do a lot of concentrated research on, is very important. One needs a kind of a "guinea pig" plant to perform internationally coördinated, phytotoxicological research, to produce a scientifically sound basis for the air quality management in relation to vegetation. Especially the growing interest in trees and forests under the influence of environmental pollution and changing global climate, makes it desirable to select one or a few tree species to study more intensively. Douglas fir, summer oak and poplar may be a good selection for the Netherlands to concentrate the research on. Also special bioassays for observations in detail may be a promising approach.

In the meantime it will be necessary to pay more attention to the possible interactions of air pollution with other abiotic and biotic stress factors. Realistic combinations of air pollutants and other chemical, physical and biological stress factors should be studied on plants in natural conditions. For example combinations of air and soil pollution, and of photochemical air pollution and UV-B irradiation (expected to increase by stratospheric ozone depletion) are important for the future. One should try to integrate the effects of air pollution with the effects of other environmental pollutants into effects of the total pollution of all compartments of the environment. Also interactions between air pollutants and plant pahtogens and pests, as studied already for O_3 and weak pathogens by researchers at the Research Institute for Plant Protection at Wageningen (Leone & Tonneijck 1988), are very important for risk evaluation for crops in practice.

To investigate all previously mentioned pollution effects, the need for simulation of situations from practice in conditions as realistic as possible is clear. Till now open-top chambers are the most frequently used facilities to approach the realistic situations as far as possible. But it is clear that really open-air field manipulations are still better and will be needed to evaluate results gained in open-top chamber (Mooi & Van der Zalm 1986).

Also better co-ordinated research programmes in this field have to be planned and performed. Co-operating research institutes, specialized in different, basic disciplines, are needed to tackle the multidisciplinary problems of environmental pollution effects.

Even international co-operation in the EC and the ECE is foreseen to study the common problems of negative influences of transboundary air pollution on crops, forests, natural vegetations and ecosystems.

7. Conclusions

Air pollution in general has been shown to be of considerable importance for plants and vegetation in a negative sense. Although a lot of information has been gained already in the field of phytotoxicology of air pollutants, still quite a few problems have to be solved. Really good, quantitative exposure-effect relationships for several air pollutants in respect to many plant species and varieties are lacking. Knowledge on the multiple interactions of different air pollutants and on the combination effects of air pollution and other abiotic and biotic stress-factors on plants is largely insufficient. For these reasons it is difficult at the moment to propose well founded criteria for air quality standards, needed for the protection of the vegetation in all possible conditions.

A lot of research in well known, but realistic conditions has to be performed to determine the risks for plants and vegetations, imposed by the total atmospheric deposition in combination with other stress factors. This may be the aim of nationally and internationally co-operating, scientific institutions, bringing their own specialisms into the general field of environmental stress research on plants and vegetations.

In addition to the traditional, meteorological and other stress factors, also the expected global changes in the climate conditions on the basis of antropogenic emissions will have to be taken into account. The increasing concentration of atmospheric CO_2 and the expected increase of UV-B irradiation by stratospheric O_3 depletion will have their positive and negative influences in combined effects with several air pollutants. So, also these factors have to be included in more sophisticated combination effect studies with phytotoxic air pollutants.

It should be the aim of the phytotoxicological research to produce clear evidence for the risks to the vegetation by the expanding activities of mankind. Doing this it may be possible to set limits to the deterioration of that part of the global biosphere mankind is depending on.

References

Guderian, R., 1977. Air pollution. Phytotoxicity of acidic gases and its significance in air pollution control. Ecological studies 22. Springer-Verlag, Berlin.

Jacobson, J. S. & Hill, A. C., 1970. Recognition of air pollution injury to vegetation: a pictorial atlas. Air Pollution Control Association, Pittsburgh, Pennsylvania.

Koziol, M. J. & Whatley, F. R., 1984. Gaseous air pollutants and plant metabolism. Butterworths, London.

Leone, G. & Tonneijck, A. E. G., 1988. Interactions between ozone and weak pathogens in causing injury to horticultural crops. Interim report, R379, Research Institute for Plant Protection, Wageningen.

Mooi, J. & Van der Zalm, A. J. A., 1986. Research on the effects of higher than ambient concentrations of SO_2 and NO_2 on vegetation under semi-natural conditions. The developing and testing of a field fumigation system. Final report, R317. Research Institute for Plant Protection, Wageningen.

Moseholm, L., Andersen, B. & Johnsen, J., 1988. Acid deposition and novel forest decline in central and northern Europe. Nordic Council of Ministers, Copenhagen.

Mudd, J. B., 1982. Effects of oxidants on metabolic function. In: M. H. Unsworth & D. P. Ormrod (eds.), Effects of gaseous air pollution in agriculture and horticulture. Butterworths, London, pp. 189–203.

Overrein, L. N., Seip, H. M. & Tollan, A., 1980. Acid precipitation effects on forest and fish. Final report of the SNSF project 1972–1980, Agricult. Univ. of Norway, Oslo-Ås.

Peters, H., 1968. Op weg naar Nova Atlantis. Openbare les TH Twente, Enschede.

Posthumus, A. C., 1982a. Biological indicators of air pollution. In: M. H. Unsworth & D. P. Ormrod (eds.), Effects of gaseous air pollution in agriculture and horticulture. Butterworths, London, pp. 27–42.

Posthumus, A. C., 1982b. Ecological effects associated with NO_x, especially on plants and vegetation. In: T. Schneider & L. Grant (eds.), Air pollution by nitrogen oxides. Studies in Environmental Science 21. Elsevier Scientific Publishing Company, Amsterdam, pp. 45–60.

Posthumus, A. C., 1988. Na zure regen komt O_3-zonneschijn. Oratie VU, Amsterdam. VU Boekhandel/Uitgeverij, Amsterdam.

Stöckhardt, J. A., 1850. Über die Einwirkung des Rauches von

Silberhütten auf die benachbarte Vegetation. Polyt. Centr. Bl. 16: 257–278.

Unsworth, M. H. & Ormrod, D. P., 1982. Effects of gaseous air pollution in agriculture and horticulture. Butterworths, London.

Van der Eerden, L. J. & Duym, N., 1988. An evaluation method for combined effects of SO_2 and NO_2 on vegetation. Environ. Pollut. 53: 468–470.

Authors' Address
A.C. Posthumus
Dept. of Ecology and Ecotoxicology
Vrije Universiteit Amsterdam
De Boelelaan 1087
1081 HV Amsterdam
The Netherlands

Open top chambers used to fumigate plants and trees in a semi open environment. (Photograph: A. Koedam)

CHAPTER 18

Air pollution and reproductive processes in natural plant species

Th. A. DUECK

Abstract. At low ambient levels of air pollution, invisible injury to plants such as reductions in vegetative and generative reproduction may lead to reduced fitness of the population. Although current ambient air pollution concentrations are variable in time, they can influence the natural species richness and vegetation composition by reducing growth and seed production. The data available on this subject indicate that air pollutants can indeed inhibit flower production, pollen germination and development, reduce seed filling and increase seed abortion, thus reducing the genetic variation within the population. In the presence of air pollution, seeds often fail to germinate and when germinated, seedling mortality can increase. The nature of seedling spacing and frequency of reseeding grass fields increases plant sensitivity to air pollution.

The available data on air pollutant effects on reproduction in relation to the additional influence of other abiotic or biotic stresses is discussed.

1. Introduction

In the past much air pollution research has been centered around agricultural and horticultural species. Indeed, the economic importance of the effects of ambient air pollution was recently pointed out by Van der Eerden *et al.* (1988), who estimated a 5% (Hfl. 6.4×10^8 or $\$3.2 \times 10^8$) annual loss in BNP in The Netherlands due to ozone, sulphur dioxide and fluoride, which is quite considerable. However, when Bleasdale (1973) demonstrated that substantial growth reductions in *Lolium perenne* due to ambient concentrations of SO_2 in the U.K. occurred, interest began to shift to air pollutant influences on grasses and other natural plant species. At the same time, studies carried out exposing plants to chronic levels of air pollution instead of acute, extremely short term fumigations, showed the lack of a correlation between the two resulting effects (Ayazloo & Bell 1981), and stimulated this line of research. It is at these low ambient levels of air pollution that invisible injury may occur, leading to reduced fitness of a species or population and thus to changes in the growth habitat, species richness and composition of the vegetation. The influence that air pollutants have on plant populations can be considerably influenced by factors other than the pollutant itself, i.e., abiotic factors such as nutrition, light intensity, humidity and other climatological conditions (Mansfield *et al.* 1986) or by interactions with biotic factors such as pathogenic fungi, insects (McNeill *et al.* 1986) or with the genetically determined tolerance of the plant species (Ernst *et al.* 1985; Dueck 1986; Scholz *et al.* 1989).

Most of the research on plants in the past has been focussed on loss of photosynthetically active area due to morphological or physiological injury to leaves and eventually growth reductions. However, the exposure of natural vegetation to air pollutants affects plant populations differently at various life stages. The phases of growth too, are differentially sensitive to air pollutants or pollutant combinations, whether to exposures to chronic levels or

J. Rozema and J. A. C. Verkleij (eds.), Ecological Responses to Environmental Stresses, 200—207.
© *1991 Kluwer Academic Publishers.*

to peak concentrations of aerial contaminants. Thus, the development and maintenance of a particular plant species may well depend on the survival or productivity of the particular life stage most susceptible to these pollutants. Compared with other, usually persistent pollutants, air pollutants may appear to exert less environmental pressure, especially at chronic concentrations. However, they are clearly strong enough to affect plant processes and to interest scientists in them.

In the following, I will attempt to briefly review some aspects of research on natural plant species, stressing the importance of reproduction for these species and the role air pollutants play in determining it.

2. Germination and seedling growth

The obvious problems in creating natural conditions in which the effect of air pollutants on germination can be studied has left this life stage largely unexamined. The choice of an environmental regime controlling temperature, humidity, soil and light conditions would fail to simulate natural conditions. Fluctuations of these environmental variables, even under laboratory conditions are likely to affect the germination of plants. Such fluctuations occurred in an experiment where irregularities in the soil moisture content and probably a lower pH influenced the germination of seeds more than exposure to SO_2. Still, differences between populations of *Silene cucubalus* with respect to the germination behaviour were found following exposure to $200 \mu g \, m^{-3} SO_2$ (Dueck 1986). Although the viability of seeds of all four populations of *S. cucubalus* used in this experiment was high, two populations with small seeds germinated poorly, as smaller seeds of a particular species generally do, reducing the reproductive potential of the population. Also, seeds of the population found to be tolerant to SO_2 (Ernst *et al.* 1985), germinated earlier than seeds of the same size from a sensitive population during exposure to SO_2. Because SO_2 may be absorbed by seeds moistened on the soil surface and can acidify the soil surface when absorbed, the earlier germination of SO_2-tolerant seeds suggests a need for

sulphur for germination or more resilience when confronted with a lowered pH.

The influence of air pollutants on germination and seedling survival in *Calluna vulgaris* during an 8 month fumigation period in open-top chambers shown in Table I and Fig. 1 gives an indication of the extent to which air pollutants affect these early life stage processes (Dueck 1990). The effect of ammonia on *C. vulgaris* was similar to that of ambient air by reducing survival to 20% below that realized in charcoal-filtered air. SO_2 however, and the combination of SO_2 and NH_3 reduced survival even more, to respectively 50% and 83% below that in charcoal-filtered air. In addition, a very strong effect on the number of seedlings which germinated during the fumigation was found. The number of newly germinated seedlings was significantly lower in the polluted air treatments than the number in charcoal-filtered air.

The effect of air pollution on seedlings appears to be much stronger than on adult plants. There could be several reasons for this: (i) seedlings are less able to detoxify air pollutants; (ii) they are compelled to absorb and take up relatively more pollutant than adult plants; (iii) genetic variation in a seedling population is higher, consequently there are more sensitive individuals. Both Bell *et al.* (1979) working with *Lolium perenne* and Crittenden & Read (1979), who studied the effects of sulphur dioxide on *Lolium multiflorum* and *Dactylis glomerata,* mentioned that SO_2 affected the relative growth rate (RGR) of these seedlings more than the RGR of older plants and more than the other parameters studied. Jones & Mansfield (1982), who noted that the response of reduced growth in *Phleum pratense* seedlings occurred earlier than in adult plants, stated that the lower RGR was due to reduced root growth, which they found to be more inhibited by SO_2 than shoot growth. This effect on seedling growth in grass species may well be the result of increased pollutant uptake due to increases in specific leaf area and due to the lower boundary layer resistance of spaced plants. Although a grass field does not resemble a vegetation of spaced plants, newly sown grass fields likely approach that situation, in contrast to the grass swards which will follow (Bell 1982). In a natural

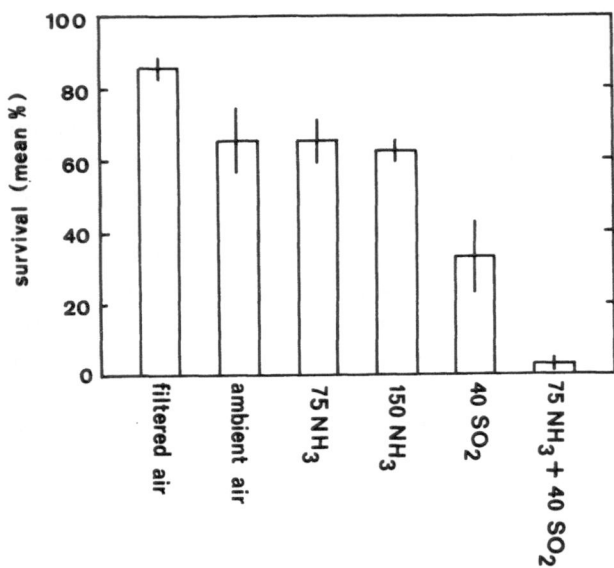

Fig. 1. Mean proportions of *Calluna vulgaris* seedlings that survived exposure to charcoal-filtered air, to ambient air, to ammonia and to sulphur dioxide. Pollutant concentrations in ppb, vertical vars denote S.E., n = 10. (Dueck, 1990).

mixed vegetation, grasses especially are inclined to form swards resulting in a vegetation structure with a much higher boundary resistance layer, diminishing pollutant uptake. Species in a closed vegetation may also be exposed to severe intraspecific and interspecific competition, as well as to the selecting force of air pollutants, which will be spoken of later in this chapter. In closed vegetation or in grass swards, ecotypes adapted to air pollution may be able to compensate for inhibited growth of sensitive individuals. In addition, selection will naturally decrease the amount of variation in an adult population, retaining only the most tolerant individuals. Thus, the seedling population with a large gene pool will be relatively more sensitive to air pollutants. The frequency of reseeding or plowing grass fields is therefore likely to affect productive grasses in that the most frequently reseeded fields will be the most affected by air pollution (Bell 1982; Dueck 1986).

3. Adult plant growth

Because of their economic importance, most of the

studies done on mature plants have been confined to agricultural species. Until the 1970s, relatively few fumigation experiments had been carried out with natural plant species, especially experiments using low or ambient concentrations of gaseous pollutants in long term experiments. The first published work in the U.K. on pasture grasses exposed to ambient concentrations of SO_2 insufficient to realize visible injury was from Bleasdale (1973), and in the U.S.A. from Heitschmidt *et al.* (1978), who reported on effects of prolonged exposure of natural prairie grasslands to low levels of SO_2.

The following remarks will be confined to low level air pollution effects on vegetative growth which are likely to influence generative reproduction at a later stage.

At low pollutant levels, usually SO_2 and NO_2 effect shoot stimulation and reduced root growth (Walmsley 1980; Whitmore & Mansfield 1983; Mansfield *et al.* 1986), resulting in increased shoot : root ratios. Low levels of nitrogen oxides and ammonia stimulate leaf growth, while reducing flowering and root growth, but in combination with SO_2 reduce growth more than SO_2 alone (Whitmore & Mansfield 1982; Dueck unpubl.). Inhibition of the photosynthetic efficiency by ozone (Tingey & Taylor 1982; Okano *et al.* 1984), sulphur dioxide (cf. Black 1982) and even ammonia (Dueck unpubl.) may be compensated for by increased leaf area (Walmsley 1980) due to a greater portion of

Table I. The mean number of *Calluna vulgaris* seedlings (new) per pot that germinated during continuous exposure to SO_2 and/or NH_3 for 8 months, as well as the effect of fumigation in open-top chambers on the mean number of seedlings (old) already germinated before the fumigation commenced. n = 10. (Dueck, 1990).

	New seedlings	old seedlings	
		t = 0	t = 8 months
filtered air	39	37	31
ambient air	12	21	16
75 ppb NH_3	5	21	14
150 ppb NH_3	11	39	26
40 ppb SO_2	6	25	10
75 ppb NH_3 + 40 ppb SO_2	< 1	31	1

assimilates being partitioned to leaves at the cost of roots and stems as sinks (Okano *et al.* 1984). This can indeed minimize immediate negative effects of air pollution to shoot production and may even stimulate total photosynthesis. It may, however, be a costly investment on the long term, rendering plants with a smaller root system relative to the shoot prone to drought and decreasing their potential for regrowth following grazing or reproductive capacity later in the growing season. Even though some species may not be directly threatened by air pollution, the survival of a particular population may well be. The following example shows that the competitive capacity of rare species growing on oligotrophic soils can be influenced by air pollutants, thus endangering the species. *Viola canina* was allowed to compete with *Agrostis capillaris* while being exposed to 40 ppb SO_2 and 75 ppb NH_3. Shoots of *V. canina* grew better, though not always significantly better when exposed to intraspecific competition and both air pollutants. Root growth however, was unaffected by pollution in the monocultures, but was strongly reduced in mixed cultures, due not only to interspecific competition, but to SO_2 and to a lesser degree, to NH_3 (Dueck unpubl.). Freer-Smith (1984) found that the degree of responses of several deciduous tree species to low levels of SO_2 and NO_2 altered after overwintering, with the injury done by sulphur dioxide becoming more marked and the beneficial effects of nitrogen dioxide being lost.

A number of plant populations have been able to adapt to air polluted conditions (Wilson & Bell 1985) and thus form differentiated populations able to grow and reproduce in a polluted atmosphere. Horsman *et al.* (1979) found that populations of *Lolium perenne* growing in urban sites where higher concentrations of sulphur dioxide prevail, had evolved populations better able to grow while fumigated with SO_2 than rural populations (Fig. 2). This can occur only if the air pollutant stress is maintained to such a degree that differences between ecotypes are brought to light that influence the fitness of the population, resulting in selection (Roose *et al.* 1982).

A factor probably of more importance to plants in an open vegetation than to species in a closed vegetation influencing the maintenance of the pop-

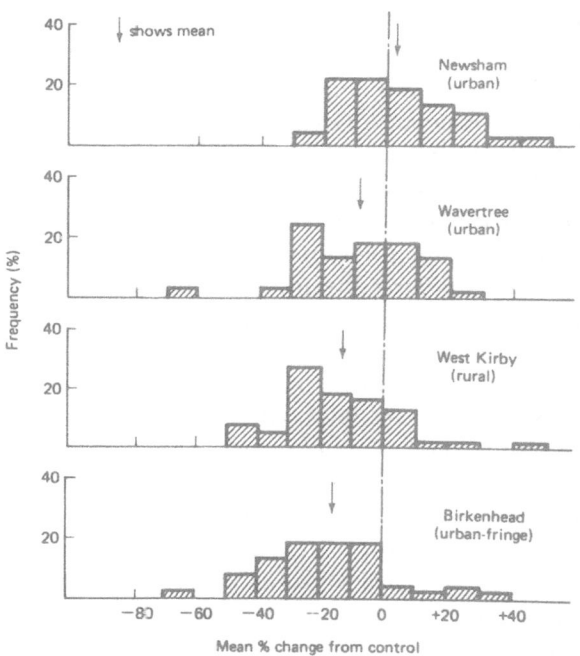

Fig. 2. The response of four ryegrass populations from Merseyside to SO_2 measured by growth in $650\,\mu g$ SO_2 m^{-3} as % of growth in control ($35\,\mu g$ m^{-3}): there has been evolution of SO_2 tolerance in the city populations. (Horsman *et al.* 1979).

ulation and thus its fitness is the sexual reproductive capacity. The development of tolerance in grass swards would appear to differ from that of spaced plants in an open vegetation, especially as swards are less dependent on sexual reproductive capacity for maintenance of the population.

4. Seed production

Processes mostly directly influencing seed quality and quantity are flower production, the process of fertilization and the maturation of seed on the plant. The most often studied pollutants ozone, sulphur dioxide and nitrogen dioxide, alone or in combination, have been shown able to influence seed production. Air pollutants generally seem to lead to foliar injury to which yield loss is often correlated (Dubay 1981; Benoit *et al.* 1983). The vegetative growth and seed production of several understorey species fumigated with ozone was re-

203

duced (Harward & Treshow 1975), and in another study, ozone alone or combined with SO₂ also reduced growth but completely hampered flowering in *Silene cucubalus* (Ernst *et al.* 1985).

With respect to fertilization and seed set, nearly all relevant literature is confined to the effect of sulphur dioxide. While not taken into account in studies of SO₂ tolerance in *Lolium perenne* and *Agrostis capillaris*, attention has been paid to seed production in species such as *Geranium caroliniaunum* (Taylor & Murdy 1975), *Lepidium virginicum* (Murdy 1979) and *Silene cucubalus* (Ernst *et al.* 1985; Dueck *et al.* 1987). Although pollen germination and pollen tube elongation are known to be inhibited by air pollution, pollen germination is the more sensitive stage of fertilization (Dubay 1981), because it is more directly exposed to pollutant effects (Wolters & Martens 1987). A reduction in the number of viable pollen, seeds per cone and number of filled seeds in *Pinus strobus* and *Pinus resinosa* exposed to air pollution was found by Houston & Dochinger (1977). However, Murdy (1979), who found that fruit sterility in *Lepidium virginicum* increased with SO₂ exposure, postulated that the acidifying effect of SO₂ on the stigma inhibited pollen tube growth (Table II). Indeed, a factor such as moisture on the stigma of coniferous species reduced the sulphur dioxide concentration necessary to affect these stages almost by a factor 10 (Karnosky & Stairs 1974). Cox (1988) also found wet acidic deposition to inhibit pollen germination *in vivo* and increase fruit abortion in *Populus tremuloides*.

After fertilization has taken place, whether or not the fertilization capacity has become impaired by air pollution, the development of the generative reproduction potential commences. Dubay & Murdy (1983) found that seed set in *Geranium carolinianum* exposed to SO₂ was significantly correlated to the number of germinated pollen grains per stigma, but although SO₂ reduced pollen germination, seed set may well remain unaffected due to an overabundance of pollen grains.

Thereafter, seed abortion or insufficient development resulting in smaller seeds with less reserve can also impair seed production potential. Sulphur dioxide alone, which stimulated growth of a *Silene cucubalus* population tolerant to SO₂, did not influence the number of seeds produced, but the individual seed weight was reduced by more than 20% (Ernst *et al.* 1985). The effects of air pollutant combinations, often additive or more-than-additive as found for growth parameters in *Poa pratensis* (Whitmore & Mansfield 1983), may become confounded when a second stress factor, i.e., soil pollution becomes involved. When four ecotypes of *Silene cucubalus* were exposed to heavy metals and a combination of low levels of SO₂ and NO₂ combined with daily fluctuating concentrations of ozone up to 120 μg m⁻³ (Table III), the mean number of flowers per plant was significantly reduced (Dueck *et al.* 1987). However, when exposed to soil as well as air pollution, the effect on the mean number and weight of seeds was influenced in such a manner that the large differences in the potential seed production appeared to be related to the interaction of each ecotype with air and/or soil pollutants. (Table IV). When grown on nutrient-rich soil, the effect of air pollution on the flower production of the same ecotypes was lost (Tonneijck pers. comm.). It is conceivable that tolerant offspring may be relatively more fertile after exposure to air pollution than without it.

Table II. Percentage of sterility of *Lepidium virginicum* plants from inside and outside the Copper Basin in control and SO₂ treatments. (Murdy 1979).

Treatment	Inside Copper Basin			Outside Copper Basin			P between populations
	No.	$\bar{X}$	SD	No.	$\bar{X}$	SD	
Control	28	6.3%	0.101	35	8.1%	0.092	NS
SO₂ (2,130 μg/m³ per 9 h)	29	13.8%	0.122	33	24.5%	0.174	< 0.01
P within populations		< 0.02			< 0.001		...

Table III. The mean number of flowers + S.E. produced per plant in four populations of *Silene cucubalus* grown with zinc or copper in the soil, alone or combined with sulphur dioxide, ozone and nitrogen dioxide (SON). (Dueck *et al.* 1987).

population	air pollutants	flowers per plant (n)		
		metal pollutant		
		control	zinc	copper
sensitive	control	53 ± 3 (3)	56 ± 3 (3)	39 ± 4 (3)
	SON	38 ± 3 (3)	28 ± 11 (3)	43 ± 5 (3)
Zn-tolerant	control	20 ± 4 (3)	47 ± 6 (3)	25 ± 3 (3)
	SON	17 ± 5 (3)	32 ± 6 (3)	20 ± 4 (3)
Cu-tolerant	control	24 ± 2 (3)	43 ± 15 (3)	62 ± 8 (3)
	SON	24 ± 4 (3)	29 ± 6 (3)	36 ± 3 (3)
SO_2-tolerant	control	40 ± 16 (3)	65 ± 8 (3)	48 ± 2 (3)
	SON	33 ± 15 (3)	77 ± 8 (3)	34 ± 5 (3)

5. Conclusions

Which conclusions are we able to make with the knowledge we have about air pollutant effects on reproductive processes in natural plant species? The data presently available are largely qualitative, but indicate that reproductive processes can be influenced by air pollution. There are clearly too few data available to quantify effects resulting in exposure-response relationships for reproduction processes in natural plant species.

There are however, some general conclusions that I believe can be made. When comparing the importance of ambient air pollution with other stress factors as selective agent, as long as the selection pressure is of a chronic nature and fluctuates

normally, air pollution is not an especially strong force. Yet, it has been able to change the genetic nature of plant populations (Verkleij *et al.* 1989).

Until more quantitative data concerning the influence of air pollutants on reproductive processes becomes available, data on yield reduced by air pollution in agricultural species may have to be utilized to indicate the effective level of pollutants on the reproduction of natural plant populations in the field. However, it is no more than an indication, and the effects of air pollutants on reproduction in natural plant species definitely requires more attention.

From available data we can conclude that at sites where pollution levels are high enough to eradicate the less adaptable species, evolution of tolerant

Table IV. The potential seed biomass production (mg) per plant in four populations of *Silene cucubalus* when exposed to sulphur dioxide, ozone and nitrogen dioxide (SON) and zinc or copper in the soil. (nd = not determined). (Dueck *et al.* 1987).

population	air pollutants	soil pollutants		
		control	zinc	copper
sensitive	control	1898	3145	1849
	SON	2918	1723	3322
Zn-tolerant	control	586	1343	899
	SON	565	1389	601
Cu-tolerant	control	620	866	2501
	SON	1063	744	1300
SO_2-tolerant	control	1360	3426	1848
	SON	nd	3189	1386

plant populations may be a rapid process, probably less than 5 years. This is similar to that which occurs with persistent stress factors with a high selection pressure, i.e., soil polluted with heavy metals. A very possible result is the pollution-induced alteration of an entire plant community. Among other responses, sexual reproduction in plants has been shown to be changed by air pollutant stress. Such a population has a smaller gene pool, providing it with less capacity to adapt to changes in its environment.

Irregular peak concentrations (duration in hours) or episodes (duration in days) of air pollutants such as those of ozone that can occur during the summer months might affect the flowering process and consequently, seed production, in species having a short flower production period, especially endangering them.

Due to the low levels of air pollution in rural areas at greater distances from point sources, one might expect that the selection pressure exerted on plant populations is low. The more sensitive individuals in such a population will be longer retained than in populations exposed to acute pollutant stress. Yet, plant populations do become tolerant to chronic air pollution. Following alleviation of air pollutant selection pressure, a plant population that has become tolerant to pollution may recover, or become a more sensitive population with a higher degree of variation. The degree of selection pressure will determine the size of the gene pool, making recovery in severely selected populations a slower process than adaptation. Yet recovery can occur within a matter of years, depending on the rate of generative reproduction and differences in the relative fitness of genotypes. Tolerant genotypes are more competitive in the presence of polluted air, but are less fit than sensitive genotypes in clean air.

The current interest in the influence of soil pollution as well as pathogens on natural plant populations and in reproduction in particular, has increased considerably. The few data on the influence of environmental factors such as humidity, soil fertility and pathogens on air pollutant effects, both following exposure to and alleviation from air pollutant stress, have only complicated results.

Therefore, future research should focus more on interactions with such biotic and abiotic stress factors to enable scientists to gain insight into the pollution problems facing us now.

Acknowledgements

The author wishes to thank L. J. van der Eerden and A. E. G. Tonneijck for critically reading the manuscript.

References

Ayazloo, M. & Bell, J. N. B., 1981. Studies on the tolerance to sulphur dioxide of grass populations in polluted areas. I. Identification of tolerant populations. New Phytol. 88: 203—222.

Bell, J. N. B., 1982. Sulphur dioxide and the growth of grasses. In: M.H. Unsworth & D.P. Ormrod (eds.), Effects of Gaseous Air Pollution in Agriculture and Horticulture. pp. 225—246. Butterworths, London.

Bell, J. N. B., Rutter, A. J. & Relton, J. 1979. Studies on the effects of low levels of sulphur dioxide on the growth of Lolium perenne L. New Phytol. 83: 627—643.

Benoit, L. F., Skelly, J. M., Moore, L. D. & Dochinger, L. S., 1983. The influence of ozone on Pinus strobus L. pollen germination. Can. J. For. Res. 13: 184—187.

Black, V. J., 1982. Effects of sulphur dioxide on physiological processes in plants. In: M.H. Unsworth & D.P. Ormrod (eds.), Effects of Gaseous Air Pollution in Agriculture and Horticulture. pp. 67—91. Butterworths, London.

Bleasdale, J. K. A., 1973. Effects of coal-smoke pollution gases on the growth of ryegrass (Lolium perenne L.). Environ. Pollut. 5: 275—285.

Cox, R. M., 1988. Sensitivity of forest plant reproduction to long-range transported air pollutants: the effects of wet deposited acidity and copper on reproduction of Populus tremuloides. New Phytol. 110: 33—38.

Crittenden, P.D. & Read, D. J., 1979. The effects of air pollution on plant growth with special reference to sulphur dioxide. III. Growth studies with Lolium multiflorum Lam. and Dactylis glomerata L. New Phytol. 83: 645—651.

Dubay, D. T., 1981. Interspecific differences in the effect of sulfur dioxide on angiosperm sexual reproduction. Ph.D. Thesis, Department of Botany, Emory University, Atlanta, Georgia.

Dubay, D. T. & Murdy, W. H., 1983. Direct adverse effects of SO_2 on seed set in Geranium carolinianum L.: A consequence of reduced pollen germination on the stigma. Bot. Gaz. 144: 376—381.

Dueck, Th. A., 1986. Impact of heavy metals and air pollutants on plants. Ph.D. Thesis, Department of Ecology & Ecotoxicology, Free University, Amsterdam.

Dueck, Th. A., 1990. Effect of ammonia and sulphur dioxide on the survival and growth of *Calluna vulgaris* (L.) Hull seedlings. Funct. Ecol. 4: 109–116.

Dueck, Th. A., Wolting, H. G., Moet, D. R. & Pasman, F. J. M., 1987. Growth and reproduction of *Silene cucubalis* Wib. intermittently exposed to low levels of air pollutants, zinc and copper. New Phytol. 105: 633—645.

Ernst, W. H. O., Tonneijck, A. E. G. & Pasman, F. J. M., 1985. Ecotypic response of *Silene cucubalus* to air pollutants (SO_2, O_3). J. Plant Physiol. 118: 439—450.

Freer-Smith, P. H., 1984. The response of six broadleaved trees during long-term exposure to SO_2 and NO_2. New Phytol. 97: 49—61.

Harward, M. & Treshow, M., 1975. Impact of ozone on the growth and reproduction of understorey plants in the aspen zone of western U.S.A. Environ. Cons. 2: 17—23.

Heitschmidt, R. K., Lauenroth, W. K. & Dodd, J. L., 1978. Effects of controlled levels of sulfur dioxide on western wheatgrass in a southeastern Montana grassland. J. Appl. Ecol. 14: 859—868.

Horsman, D. C., Roberts, T. M. & Bradshaw, A. D., 1979. Studies on the effect of sulphur dioxide in perennial ryegrass (*Lolium perenne* L.). J. Exp. Bot. 30: 495—501.

Houston, D. B. & Dochinger, L. S., 1977. Effects of ambient air pollution on cone, seed and pollen characteristics in eastern white and red pines. Environ. Pollut. 12: 1—5.

Jones, T. & Mansfield, T. A., 1982. The effect of SO_2 on growth and development of seedlings of *Phleum pratense* under different light and temperature environments. Environ. Pollut. A 27: 57—71.

Karnosky, D. F. & Stairs, G. R., 1974. The effects of SO_2 on in vitro forest tree pollen germination and tube elongation. J. Environ. Qual. 3: 406—409.

Mansfield, T. A., Davies, W. J. & Whitmore, M. E., 1986. Interactions between the responses of plants to pollution and other environmental factors such as drought, light and temperature. In: How are the effects of air pollutants on agricultural crops influenced by the interaction with other limiting factors? pp. 2—15. COST Workshop 1986, Denmark.

McNeill, S., Bell, J. N. B., Aminu-Kano, M. & Mansfield, P., 1986. SO_2, plant, insect and pathogen interactions. In: How are the effects of air pollutants on agricultural crops influenced by the interaction with other limiting factors? pp. 108—115. COST Workshop 1986, Denmark.

Murdy, W. H., 1979. Effect of SO_2 on sexual reproduction in *Lepidium virginicum* L. originating from regions with different SO_2 concentrations. Bot. Gaz. 140: 299—303.

Okano, K., Ito, O., Takeba, B., Shimizu, A. & Totsuka, T., 1984. Alteration of ^{13}C-assimilate partitioning in plants of *Phaseolus vulgaris* exposed to ozone. New Phytol. 97: 155—163.

Roose, M. L., Bradshaw, A. D. & Roberts, T. M., 1982. Evolution of resistance to gaseous air pollutants. In: M.H. Unsworth & D.P. Ormrod (eds.), Effects of Gaseous Air Pollution in Agriculture and Horticulture. pp. 379—409. Butterworths, London.

Scholz, F., Gregorius, H.-R. & Rudin, D., 1989. Genetic Effects of Air Pollutants in Forest Tree Populations. Springer-Verlag. Berlin, Heidelberg.

Taylor Jr., G. E. & Murdy, W. H., 1975. Population differentiation of an annual plant species *Geranium carolinianum*, in response to sulphur dioxide. Bot. Gaz. 136: 212—215.

Tingey, D. T. & Taylor Jr., G. E., 1982. Variation in response to ozone: A conceptual model of physiological events. In: M.H. Unsworth & D.P. Ormrod (eds.), Effects of Gaseous Air Pollution in Agriculture and Horticulture. pp. 113—138. Butterworths, London.

Van der Eerden, L. J., Tonneijck, A. E. G. & Wijnands, J. H. M., 1988. Crop loss due to air pollution in The Netherlands. Environ. Pollut. 53: 365—376.

Verkleij, J. A. C., Bast-Cramer, W. B. & Koevoets, P., 1989. Genetic studies in populations of *Silene cucubalus* occurring on various polluted and unpolluted areas. In: F. Scholz, H.-R. Gregorius & D. Rudin (eds.) Genetic Effects of Air Pollutants in Forest Tree Populations. pp. 107—114. Springer-Verlag. Berlin, Heidelberg.

Walmsley, L., Ashmore, M. R. & Bell, J. N. B., 1980. Adaptation of radish *Raphanus sativus* L. in response to continuous exposure to ozone. Environ. Pollut. A 23: 165—177.

Whitmore, M. E. & Mansfield, T. A., 1983. Effects of long-term exposures to SO_2 and NO_2 on *Poa pratensis* and other grasses. Environ. Pollut. A 31: 217—235.

Wilson, G. B. & Bell, J. N. B., 1985. Studies on the tolerance to SO_2 of grass populations in polluted areas. III. Investigations on the rate of development of tolerance. New Phytol. 100: 63—77.

Wolters, J. H. B. & Martens, M. J. M., 1987. Effects of air pollutants on pollen. Bot. Rev. 53: 372—414.

Author's Address
Th. A. Dueck
Department of Ecology & Soil Ecology
Research Institute for Plant Protection
P.O. Box 9060
6700 GW Wageningen
The Netherlands

Many environmental problems are due to anthropogenic activities that increase the availability of potentially toxic chemicals. The picture shows a large lead smelter that has emitted enormous quantities of metals; concentrations of cadmium in the soil surrounding this smelter (up to 200 mg/kg) have affected the soil community and have caused some soil invertebrates to become genetically adapted.

CHAPTER 19

Developments and present status of terrestrial ecotoxicology

E. N. G. JOOSSE & N. M. VAN STRAALEN

Abstract. Ecotoxicology is a new science, combining expertise from ecology, toxicology and environmental chemistry. Although the roots of these disciplines are very different, there is a growing tendency for an integrated approach; this is stimulated by the need for ecological risk assessment of chemicals in the environment. Research on metal toxicity in soil invertebrates provides an illustration of this development. On the basis of estimated frequency distributions of sensitivity of soil organisms, soil quality criteria have been derived for cadmium, lead, lindane and atrazine. Contaminant levels in soil exceeding these levels may select for increased resistance in soil animal populations. Physiological and genetic research has shown that metal tolerance can be achieved by decreased assimilation (Zn tolerance in *Porcellio scaber*) and by increased metal excretion (Cd tolerance in *Orchesella cincta*). Metal concentrations differ greatly between the various species of the soil arthropod community. As to metal accumulation, soil animals can be classified in two or three groups which do not coincide with feeding habits, but follow the gross taxonomic classification. Development of ecotoxicological theory for species variation in accumulation potential and sensitivity will contribute to a strengthening of the basis for risk assessment of chemicals in the environment.

1. Historical perspective

Changes in ecosystems as a consequence of human activities have generated a rapidly growing concern about the environment. In the scientific world this development has given rise to a new discipline: ecotoxicology. Ecotoxicology is concerned with the effects of influences of potentially toxic chemicals on individual organisms and especially with the consequences of these effects on the functioning of the fundamental units in ecology: populations and ecosystems. Although the term "ecotoxicology" was introduced already in 1969 by Truhaut (Truhaut 1977), the awareness of the "eco" aspects became only apparent around the 80-ties. These aspects are extensively discussed by Moriarty (1983) with emphasis on population processes and by Sheehan et al. (1984) with emphasis on ecosystem evaluation. Ernst & Joosse (1983) made a survey of the symptoms resulting from various human

activities in the natural environment and discussed the corresponding toxic mechanisms in populations of plants and animals.

The roots and the scientific knowledge of ecotoxicology are found both in toxicology and in ecology. Toxicology, which is concerned with the influences and the interactions of chemicals with biological systems (Pascoe 1983), is originally embedded in the medical and veterinary sciences; by using the techniques derived from these sciences, the toxic action of chemicals is studied in detail (Koeman 1989). The approach is essentially directed to individual organisms primarily serving the welfare of humans and domestic animals.

The environmental problems at first had an agricultural origin. The use of pesticides and herbicides is a post war problem (Mellanby 1972). Later the industrial pollution effects (e.g., Koeman 1971) broadened the scope of toxicology with environmental toxicology and focused on free-living ani-

J. Rozema and J. A. C. Verkleij (eds.), Ecological Responses to Environmental Stresses, 210—218.

mals. Direct relationships between environment and individual health of animals concerning the effects of contaminated herbage and other food have already been described as early as 1907, when Haywood for the first time discussed the effects of wastes from copper smelters in the US on vegetation and cattle and described that the cattle were killed by arsenic given off in the fumes and deposited on the surrounding pastures where cattle were browsing, or living on hay cut from it (Haywood 1907). It was however, only 1970 when Show published alarming data about lead concentrations in samples of roadside grass resulting from lead alkyl additives in petrol (Chow 1970). Sewage sludge (Dijkshoorn & Lampe 1975) and copper high tension lines (Kraal & Ernst 1976) were found further to contaminate herbage used as animal food. A recent example is the common seal infected by the canine distemper virus (Osterhaus & Vedder 1988) which appears to be related to a suppressed immune system and disturbed vitamin A circulation caused by environmental contaminants (Brouwer et al. 1989).

Attention for health of invertebrates appeared still later. Williamson & Evans (1972) were the first to describe high concentrations of lead in roadside invertebrates. Coughtrey (1975) drew attention to the effects of smelters and mining on invertebrate animals and found very high concentrations of cadmium, lead and zinc in snails.

This type of studies on individual basis, followed a single species approach and reflected the aims of ecotoxicology, which are to find and understand early symptoms (danger signs) as a basis for ecological effects assessment. Early changes on ecosystem level are difficult to detect among natural variations and are considered to be relatively insensitive phenomena. Individual organisms are more susceptible and are easily confronted with toxic influences. Toxicological experience can be very helpful in finding early changes. An example is the standard bioassay to demonstrate lead intoxication in humans (Burch & Siegel 1971; Berlin & Schaller 1974). The essense is that δ-amino-levulinic dehydratase (δ-ALAD), an enzyme that functions in the synthesis of haemoglobine and the haem-protein cytochrome P450, is inhibited by Pb. Activity of δ-ALAD has been demonstrated in all vertebrates, in several worm species and in *Planaria*. Effects of lead on this enzyme have been demonstrated in invertebrates such as *Lumbricus terrestris* (Ireland & Fischer 1978). It is a very sensitive test, which can be applied with success in free-living birds (Hutton 1980; Grue et al. 1986) and fishes (Jackim 1973).

Although data about toxic effects in individual organisms contribute to early signalization and evaluation of environmental hazards, the perception of possible effects in successive generations remains obscure. Adapted (tolerant) genotypes present in the gene pool of populations can be selected for, leading to genetic changes, which have to be considered as a basic ecotoxicological parameter. Among plant species many examples exist about development of resistance against surplus of heavy metals, which appears to be species and metal-specific (Ernst 1984, 1985). Among terrestrial animals evidence for population differentiation is scarce and only known from the snail *Helix aspersa* (Beeby & Richmond 1987), the woodlouse *Porcellio scaber* (Joosse et al. 1981; Van Capelleveen 1987) and the springtail *Orchesella cincta* (Van Straalen et al. 1987; Posthuma 1990). Generally, these adaptations appear to have consequences for growth and reproduction, both in plant species (Ernst 1982) and animal species (Van Capelleveen 1987; Joosse & Verhoef 1987; Posthuma 1990), possibly as a consequence of energy investments in tolerance mechanisms. In section 3 of this chapter the present state of knowledge concerning resistance to metals in terrestrial invertebrates will be presented.

The second pillar of ecotoxicology is ecology. This discipline has a different origin and input. The "eco"-part of ecotoxicology can boast on a respectable amount of basic plant ecological research. The history of this research learns that two different methods have been developed, primarily serving exploitation of metal deposits. The geobotanical prospecting method uses plant species and associations of species that indicate the presence of abnormally high metal concentrations (Ernst 1974a). Ecophysiological research on uptake and accumulation mechanisms in these specific plant species (i.e., Ernst 1969, 1974b, 1985; Lolkema *et al.* 1984, concerning copper) has contributed to the revegetation developments of devastated metalliferous mine spoils (Smith and Bradshaw 1970; Ernst 1981). It has also been of basic importance for the insight in the so called biogeochemical prospecting method, which uses the chemical analysis of plant samples. This method assumes that high concentrations of metals in plants will correlate with high concentrations in soil.

Animal ecology followed this development only slowly. Terrestrial animals or their tissues have seldom been used as indicators (Martin & Coughtrey 1982). The destructive nature of the techniques, damage of the environment and mainly the complex relationship between concentrations in animal tissues and environment (Janssen 1988; Janssen *et al.* 1990) caused by different foraging areas and physiology, restrict the use of animals. Thus unlike in animals, a vast amount of plant-ecological knowledge has laid the basis for a critical use of plants as accumulator species indicating environmental pollution (Ernst 1972, 1984; Ernst *et al.* 1983). The animal studies on uptake and accumulation, however, have laid a basis for an important ecotoxicological parameter to establish the risk of toxic substances for natural ecosystems: biomagnification. The results have given insight in the role of different animal species in the mobility of toxic elements and have learned that the accepted ideas about food chain effects are of minor value (Van Straalen 1988). Van Straalen & Van Wensem (1986) demonstrated for a terrestrial food chain that the physiological equipment of animals is decisive for the accumulation of pollutants. Some species accumulate more than others, so that critical pathways can be distinguished. Some aspects on accumulation of metals by soil animals will be discussed in the fourth section of this chapter.

The most recent developments in ecotoxicology emphasize the ultimate objective to establish risks in various constituents of ecosystems. Starting with sensitive responses of individual organisms to pollutants, the extrapolation of these data to higher organisation levels has to be performed. This requires the development of ecotoxicological theory and research methodology (Van Straalen 1988).

The essential question is whether individual harm, and what type of harm, has to be considered as serious for the survival of the population; this forces us to relate the consequences of individual injuries to population processes. Population parameters offer a more realistic indication, since within a population, mechanisms may exist to compensate or to magnify toxic effects on individuals (Van Straalen *et al.* 1989). Premature death and reduced reproduction success may be reflected in lower abundances, but in some organisms mortality is more effective, in others reproduction. Diminishing activities of organisms possibly will be reflected in changing dynamic ecosystems. Van Capelleveen (1987) demonstrated for instance that a decreased consumption rate of woodlice causes a reduced decomposition of litter.

This type of ecotoxicological effect studies can importantly contribute to estimate the risks for an undesirable impact on the environment (Van Straalen & Denneman 1989). Risk management is now an important aspect of environmental legislation. Besides existing limits to protect human health (ADI, MAC), ecotoxicologists have to provide data to propose quality standards to protect the natural environment.

The historical perspective shows that, although toxicology and ecology have a very different origin, there is a growing tendency to integrate aspects from both branches into a new discipline. Research in our department on metal toxicity in soil communities provides an illustration of this development and is summarized below.

2. Toxicity of metals to soil animals

Metals emitted into the air by human activities accumulate in the soil because of their affinity to clay minerals and humus. Especially in undisturbed soil profiles, where there is an upper organic layer of partly decomposed leaf litter, concentrations of several metals have reached levels far above their background values, even in areas remote from emission sources. Concern has grown about the possible threat to the ecological function of the soil, emanating from the irreversible process of accumu-

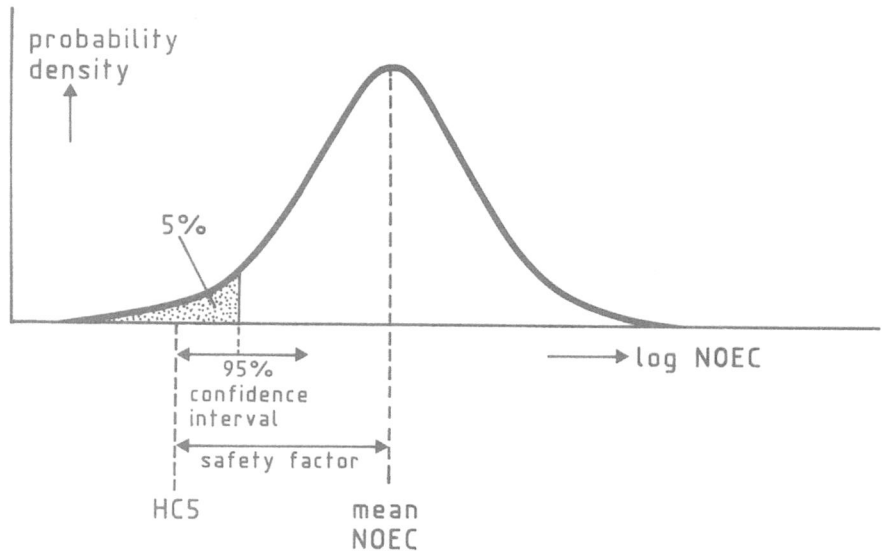

Fig. 1. Theoretical distribution of NOEC-values of species in a large community, showing the estimation of HC5 as the lower 95% confidence limit of the 5-percentile of the frequency distribution of sensitivities.

lation. Since environmental protection policy in The Netherlands is directed to protecting soil functions, there is a question of how to indicate levels of metal contamination which would not endanger the long-term functioning of soil life.

A new approach to assessing the risk of chemicals in the environment has been developed in recent years and can be applied to soil pollution problems. Instead of hazard assessment, which is essentially qualitative and leads to a ranking of chemicals, attention has now been given to risk assessment, which is a quantitative estimation method. The ecological risk of a chemical is conceived as the probability that, as a consequence of the chemical's presence, a certain undesirable event will happen. Once the event has been specified in an ecologically relevant way, the aim of ecotoxicological risk assessment is to estimate the probability of the event occurring, or to calculate a concentration level such that the probability is reduced below a certain threshold.

In the context of soil pollution, several events can be specified which must be avoided if the soil is to be protected. For example, Domsch *et al.* (1983) have, on the basis of natural fluctuations of soil microbial activity, postulated that concentrations

of pesticides should not increase beyond a level causing an inhibition of microbial activity from which there is no recovery within a certain period. Since the parameters studied in microbial studies are often less sensitive to contamination than the activities of single animal species, a different approach has been developed for soil animals (Van Straalen & Denneman 1989).

Protection of animal life in soil is achieved if there is only a small chance that a species from the soil community is exposed to a level greater than its no effect concentration (NOEC). Assuming a statistical distribution of sensitivities, the probability that an NOEC is exceeded can be calculated as a function of the concentration in soil. The parameters of the distribution are estimated from toxicity data on a certain number of test species. Since the battery of species is usually only a small sample from the community to be protected, the margins of safety depend on the number of species tested.

This model has been applied to available data on toxicity of metals and pesticides for soil animals (Denneman *et al.* 1989). The calculations were aimed at calculating the HC5 (hazardous concentration for 5% of the species), which is a concentration level such that there is a 5% probability that a

Table I. Soil quality criteria (HC5) for cadmium, lead, lindane and atrazine. The data have been adjusted to a so-called standard soil (25% lutum, 10% humus). For details see Denneman *et al.* (1989).

substance	number of species tested	HC5 (mg/kg)
cadmium	7	0.16
lead	5	141
lindane	4	0.00085
atrazine	4	15.1

species selected at random from a large community is exposed above its NOEC. The model is visualized in Fig. 1. Table I summarizes the results for cadmium, lead, lindane and atrazin. The HC5-value for Cd is estimated as $0.16\,\mu g/g$, which is somewhat below the value of $0.8\,\mu g/g$, recently proposed by the Dutch Ministry as a soil quality reference value. Consequently, more than 5% of the soil community is estimated to be unprotected if the soil concentration would equal $0.8\,\mu g/g$.

Although calculations of this type have a great appeal to policy-makers, they should be handled with great care. One of the main problems is the unknown distribution of sensitivities over the various functional groups in the soil community. Why are some groups more sensitive than others? Can soil animals increase their resistance by adaptation? These questions have been analysed recently to some extent and will be discussed below.

3. Resistance to metals in soil animal populations

It may be expected that, when concentrations of metals in the soil reach levels that are toxic to soil organisms, selection will favour the more resistant types. If there is genetic variation in resistance, populations in contaminated areas will come to differ from their conspecifics in unpolluted areas. This differentiation of populations has been well documented for soil microflora and higher plants, but has been described only recently for soil inhabiting animals (Beeby & Richmond 1987; Van Capelleveen 1987; Van Straalen *et al.* 1987). For the aquatic environment, a recent review (Klerks & Weis 1987) has demonstrated that many (but not all) species can develop metal resistance; however, in the majority of cases it has not been proven that development of resistance was achieved by genetic changes. Compared to the rapid evolution of pesticide resistance in many arthropods, development of metal resistance in animals can be considered as a slow process.

Development of metal resistance possibly requires several interrelated metabolic changes. Along with resistance, various other aspects are modified, such as growth and reproduction. The work of Van Capelleveen (1987) on *Porcellio scaber* provides a nice illustration of this point. A population inhabiting an area contaminated by zinc smelter emissions was compared to a reference population and several differences were noted (see Table II). Some of these differences are evidence of an adaptive response in the smelter population (e.g., optimal growth at increased Zn-exposure), while other differences point to a decreased ecological performance (e.g., smaller body size).

The physiological basis for metal resistance in *Porcellio* requires further investigation. Results obtained recently in our group (Donker unpublished) demonstrate that Zn-assimilation efficiency is

Table II. Summary of differences between two populations of *Porcellio scaber*. Based on Van Capelleveen (1987).

	control population	population close to Zn smelter
Zn-concentration in food resulting in optimal growth efficiency ($\mu mol/g$)	6	15
Zn-concentration in food resulting in lowest (optimal) assimilation efficiency ($\mu mol/g$)	6	30
Brood development time (days) under unstressed conditions	38—42	30—35
Median female weight in field population (mg)	20.9	10.5
Average brood size at female weight = 50 mg in the field, in spring	47	64
Average brood size at female weight = 50 mg in the fields, in summer	22	37

lower in metal resistant isopods, especially when exposed to uncontaminated food. This may explain why resistant isopods cannot extract enough zinc from normal food and therefore grow better when given dietary Zn in intermediate concentrations.

In the collembolan *Orchesella cincta* resistance to cadmium has been demonstrated in populations originating from metal-contaminated sites (Posthuma 1990). Since this concerned the laboratory-reared F_1-generation, there is evidence for genetic differences between populations. Up to now this seems to be the only well documented case for genetic adaptation in soil animals under heavy metal stress.

Metal resistance in *Orchesella cincta* is achieved by increased metal excretion. The excretion efficiency of individual Collembola can be measured by analysis of the excretion product, the degenerated exfoliated gut epithelium, shed after each moult. Comparing four populations in which both metal resistance (expressed as a tolerance index) and excretion efficiency has been measured, it appears that there is a clear correlation between these two measures (Fig. 2). The position of the Collembola population from Budel is remarkable. Although the soil at this place is heavily contaminated by recent zinc factory emissions, it has a low tolerance to Cd, as well as a low excretion efficiency. Short-term selection may not have been effective enough to increase tolerance.

Although the average tolerance is increased in populations subjected to long-term selection (Fig. 2: Plombières, Stolberg), the frequency distributions of sensitivities show that less efficient individuals are not eliminated from adapted populations (Van Straalen *et al*. 1986). Metal resistance in animals seems to be of degree, rather than of kind. The extreme selection which has produced heavy metal vegetation types has not been described for soil animals. This apparent lack of a specific heavy metal fauna may relate to the different ways in which soil animals are exposed, in comparison to plants (the intestinal wall forming a natural barrier), and to the flexibility provided by behavioural responses. In terms of the theory developed by Ernst (1983), a soil animal population subjected to high levels of metal contamination, is in a continuous state of stress; although selection increases

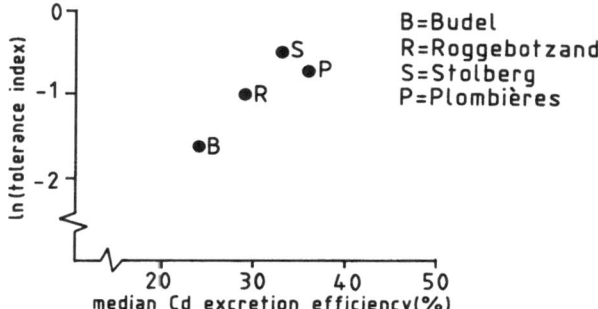

Fig. 2. Resistance of four *Orchesella cincta* populations to cadmium, expressed as the mean tolerance index (logarithmically transformed) and as the median Cd-excretion efficiency. The tolerance index measures the extent of growth reduction under Cd stress. The excretion efficiency expresses excretion of Cd by means of intestinal exfoliation as a percentage of assimilation. Data were taken from Van Straalen *et al*. (1987) and Posthuma (1990).

the average resistance, the population does not achieve a new ecological optimum sensu Ernst.

4. Accumulation of metals by soil animals

The residues of chemicals in animals are determined by the balance between uptake, degradation and excretion. These mechanisms show significant variation between groups. For example, the high levels of persistent organochlorines in cormorants and shags may be related to the low degradation capacity of the microsomal enzyme system in the liver of these and other fish-eating birds (Walker 1980).

In the context of soil animals, little comparative data are available. However, recently developed micromethods for analysis of metals have revealed significant variation between species in the soil community. From data given in Janssen (1988) and Van Straalen & Van Wensem (1986) two main groups of soil arthropods can be constructed, one demonstrating high levels of cadmium and one demonstrating low levels (Table III and Fig. 3). In general, arthropods with an arachnid morphology (spiders, mites, harvestmen) belong to the high-level group, while insects and related groups (springtails, carabids) belong to the low-level group. Isopods invariably contain the highest

215

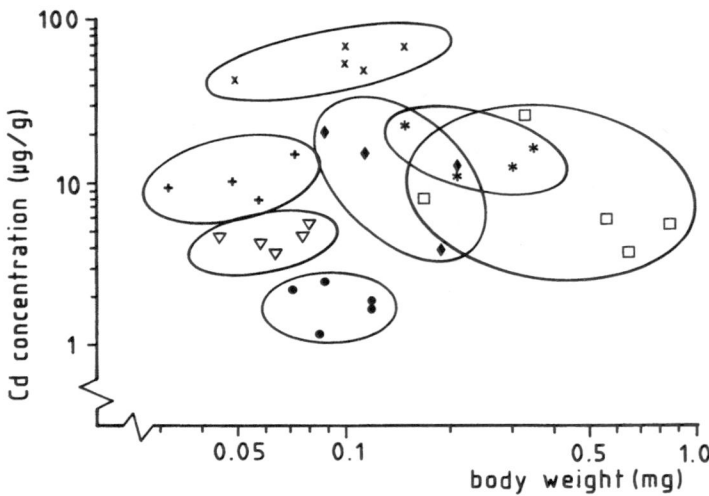

Fig. 3. Cadmium levels of several microarthropods in a contaminated pine forest soil at Stolberg-Binsfeldhammer, FRG. Cadmium concentrations of individual animals are plotted as a function of body-weight. The average soil concentration was 189 μg/g. ● = Entomobryidae, $\triangledown$ = Sminthuridae, + = Onychiuridae, × = Diplura, ◆ = Gamasidae, ★ = Pseudoscorpiones, □ = Linyphiidae.

amounts of metal among all soil arthropods. The group of oribatid mites seems to be rather diverse; it contains typical high-level species (e.g., *Platynothrus peltifer*), as well as species with low amounts (*Chamobates cuspidatus, Trichoribates trimaculatus*). The reason for this diversity is not known. The precise route of exposure of different species may be important, especially in relation to the animal's position in the soil profile. Both within the Collembola and within the centipedes, deeper living species seem to contain higher amounts of cadmium than species living in the superficial soil layer.

Residues of metals in soil animals are often strongly correlated; that is, if a species contains a high amount of one metal, it also contains relatively high amounts of other metals. Different divalent cations are probably taken up by the same mechanisms. Thus, if an animal has a high need for trace elements such as Fe, Mn, Zn and Cu, then it might also more easily accumulate xenobiotic metals such as Cd and Pb. There are some examples to support this hypothesis.

Isopods contain very high concentrations of copper. This elements is concentrated in the "small cells" of the hepatopancreas, where numerous intracellular granules are located. It has been shown

Table III. Division of soil arthropod fauna on the basis of cadmium residues at contaminated sites. Based on Janssen (1988), Van Capelleveen (1987), Van Straalen & Van Wensem (1986) and Fig. 3.

High cadmium concentrations	Low cadmium concentrations
Isopods	Oribatid mites from family Notaspididae
Harvestmen	Lithobiid centipedes
Linyphiid spiders	Entomobryid Collembola
Pseudoscorpions	Carabid beetles
Gamasid mites	Staphylinid beetles
Oribatid mites from family Camisiidae	Crickets
Geophilid centipedes	Grasshoppers
Diplurans	
Onychiurid Collembola	

recently that the granules are part of the lysosomal system and that the deposition of copper is facilitated by acid and alkaline phosphatase activities (Prosi & Dallinger 1988; Dallinger & Prosi 1988). The large copper store in the hepatopancreas probably relates to the presence of copper in hemocyanine, the major haemolymph protein in crustaceans, which functions in oxygen transport. Because of their high need for copper, isopods might also be prone to accumulating other metals.

Another example comes from the oribatid mites. The species *Platynothrus peltifer* appears to contain high amounts of manganese, compared to collembola such as *Orchesella cincta* (Van Straalen *et al.* 1987). On a molar basis, it even contains more Mn than Fe, which is a rather unusual situation in the animal kingdom. A functional explanation for this cannot, as yet, be given. The high need for trace elements could nevertheless explain the accumulation of Cd by this species.

The soil animal community comprises hundreds of species. To explain the accumulation of metals and to identify critical pathways of transfer, one is in need of a general theory. The above stated demonstrates that such a theory should be based on the physiological equipment of species, i.e., on their mineral nutrition, storage capacity and mechanisms of excretion.

References

Beeby, A. & Richmond, L., 1987. Adaptation by an urban population of the snail *Helix aspersa* to a diet contaminated with lead. Environ. Pollut. (A) 46: 73—82.

Berlin, A. B. & Schaller, K. H., 1974. European standardized method for the determination of δ-aminolevulinic acid dehydratase activity in blood. Z. Klin. Chem. Klin. Biochem. 12: 389—390.

Brouwer, A., Reijnders, P. J. H. & Koeman, J. H., 1989. Polychlorinated biphenyl (PCB) – contaminated fish induces vitamin A and thyroid hormone deficiency in the common seal (*Phoca vitulina*). Aquat. Toxicol. 15: 99—106.

Burch, H. B. & Siegel, A. L., 1971. Improved method for measurement of delta-aminolevulinic acid dehydratase-activity of human erythrocytes. Clin. Chem. 17: 1038—1041.

Chow, T. J., 1970. Lead accumulation in roadside soil and grass. Nature 225: 295—296.

Coughtrey, P. J., 1975. Cadmium in terrestrial ecosystems: a case study at Avonmouth, Bristol (U.K.) Ph.D. Thesis, University of Bristol.

Dallinger, R. & Prosi, F., 1988. Heavy metals in the terrestrial isopod *Porcellio scaber* Latreille. II. Subcellular fractionation of metal-accumulating lysosomes from hepatopancreas. Cell Biol. Toxicol. 4: 97—109.

Denneman, C. A. J., Traas, T. P., Van Straalen, N. M. & Joosse, E. N. G., 1989. Ecotoxicologische advieswaarden voor stofgehalten in de bodem. Milieu 4: 8—14.

Domsch, K. H., Jagnow, G. & Anderson, T.-H., 1983. An ecological concept for the assessment of side-effects of agrochemicals on soil microorganisms. Res. Rev. 86: 65—105.

Dijkshoorn, W. & Lampe, J. E. M., 1975. Availability for ryegrass of cadmium and zinc from dressings of sewage sludge. Neth. J. Agric. Sci. 23: 338—344.

Ernst, W. H. O., 1969. Zur Physiologie der Schwermetallpflanzen. Subzelluläre Speicherungsorte des Zinks. Ber. Dtsch. Bot. Ges. 82: 161—164.

Ernst, W. H. O., 1972. Zink- und Cadmium-Imissionen auf Böden und Pflanzen in der Umgebung einer Zinkhütte. Ber. Dtsch. Bot. Ges. 85: 295—300.

Ernst, W. H. O., 1974a. Schwermetallvegetation der Erde. Gustav Fischer Verlag, Stuttgart.

Ernst, W. H. O., 1974b. Mechanismen der Schwermetallresistenz. Verh. Ges. Ökol. Erlangen: 189—197.

Ernst, W. H. O., 1981. Problem bei den Begrünung und Aufforstung von Schwermetallhalden. Ber. Int. Symp. Int. Ver. Vegetationsk.: 237—248.

Ernst, W. H. O., 1982. Schwermetallpflanzen. In: H. Kinzel (ed.) Pflanzenökologie und Mineralstoffwechsel, Ulmen, Stuttgart, pp. 472—499.

Ernst, W. H. O., 1983. Ökologische Anpassungsstrategien an Bodenfaktoren. Ber. Dtsch. Bot. Ges. 96: 49—71.

Ernst, W. H. O., 1984. Indicatoren van een overmaat aan zware metalen in terrestrische ecosystemen. In: E.P.H. Best & J. Haeck (eds.) Ecologische indicatoren. Pudoc, Wageningen: 109—120.

Ernst, W. H. O., 1985. Schwermetallimmissionen-Ökophysiologische und populations-genetische Aspekte. Düsseldorfer Geobot. Kolloq. 2: 43—57.

Ernst, W. H. O., Dueck, Th. A. & Lolkema, P. C., 1985. Genetische effecten van emissies van zware metalen op planten. Lucht en Omgeving: 69—72.

Ernst, W. H. O. & Joosse, E. N. G., 1983. Umweltbelastung durch Mineralstoffe. VEB Gustav Fischer Verlag, Jena.

Ernst, W. H. O., Verkleij, J. A. C. & Vooijs, R., 1983. Bioindication of a surplus of heavy metals in terrestrial ecosystems. Environ. Monit. Assessment. 3: 297—305.

Grue, G. E. D., Hoffman, B. J., Nelson Beyer, W. & Franson, L. P., 1986. Lead concentrations and reproductive success in European starlings *Sturnus vulgaris*, nesting within highway roadside verges. Environ. Pollut. 42: 157—182.

Haywood, J. K., 1907. Injury to vegetation and animal life by smelter fumes. J. Am. Chem. Soc. 29: 998—1009.

Hutton, M., 1980. Metal combination of feral pigeons *Columba livia* from the London area: part 2 — Biological effects of lead exposure. Environ. Pollut. 22: 281—293.

Ireland, M. P. & Fisher, E., 1978. Effect of Pb^{2+} on Fe^{3+} tissue concentrations and delta-aminolaevulinic acid dehydratase activity in *Lumbricus terrestris*. Acta Biol. Acad. Sci. Hung. 29: 395—400.

Jackim, E., 1973. Influence of lead and other metals on fish δ-aminolevulinate dehydratase activity. J. Fish. Res. Board Can. 30: 560—562.

Janssen, M. P. M., 1988. Species dependent cadmium accumulation by forest litter arthropods. In: Proc. Int. Conf. Environmental Contamination, Venice 1988. CEP Consultants, Edinburgh, pp. 436—438.

Janssen, M. P. M., Joosse, E. N. G. & Van Straalen, N. M., 1990. Seasonal variation in the cadmium concentration of litter arthropods from a cadmium contaminated site. Pedobiologia (in press).

Joosse, E. N. G., Wulffraat, K. J. & Glas, H. P., 1981. Tolerance and acclimation to zinc of the isopod *Porcellio scaber* Latr. In: Proc. Int. Conf. Heavy metals in the environment, Amsterdam. CEP Consultants, Edinburgh, pp. 425—428.

Joosse, E. N. G. & Verhoef, H. A., 1987. Developments in ecophysiological research on soil invertebrates. Adv. Ecol. Res. 16: 175—248.

Klerks, P. L. & Weis, J. S., 1987. Genetic adaptation to heavy metals in aquatic organisms: a review. Environ. Pollut. 45: 173—205.

Koeman, J. H., 1971. Het voorkomen en de toxicologische betekenis van enkele chloorwaterstoffen aan de Nederlandse kust in de periodes van 1965—1970. Ph.D. Thesis, Rijksuniversiteit Utrecht.

Koeman, J. H., 1989. Ecotoxicology: Present status. In: H. Løkke, H. Tyle & F. Bro-Rasmussen (eds.) Proc. 1st. European Conf. Ecotoxicology. Technical University Denmark, Lyngby, pp. 5—20.

Kraal, H. & Ernst, W. H. O., 1976. Influences of copper high tension lines on plants and soils. Environ. Pollut. 11: 131—135.

Lolkema, P. C., Donker, M. H., Schouten, A. J. & Ernst, W. H. O., 1984. The possible role of metallothioneins in copper tolerance of *Silene cucubalus*. Planta 162: 174—179.

Martin, M. H. & Coughtrey, P. J., 1982. Biological monitoring of heavy metal pollution. Applied Science Publishers London.

Mellanby, K., 1972. The biology of pollution. Studies in Biology no. 38. Edward Arnold, London.

Moriarty, F., 1983. Ecotoxicology. The study of pollutants in ecosystems. Academic Press, London.

Osterhaus, A. D. M. E. & Vedder, E. J., 1988. Identification of virus causing recent seal deaths. Nature 335: 20.

Pascoe, D., 1983. Toxicology. Studies in Biology no. 149. Edward Arnold, London.

Posthuma, L., 1990. Genetic differentiation between population of *Orchesella cincta* (Collembola) from heavy metal contaminated sites. J. Appl. Ecol. 27: 609—622.

Prosi, F. & Dallinger, R., 1988. Heavy metals in the terrestrial isopod *Porcellio scaber* Latreille. I. Histochemical and ultrastructural characterization of metal-containing lysosomes. Cell Biol. Toxicol. 4: 81—96.

Sheehan, P. J., Miller, D. R., Butler, G. C. & Bourdeau, Ph., 1984. Effects of pollutants at the ecosystem level. Scope 22, John Wiley & Sons, Chicester.

Smith, R. A. H. & Bradshaw, A. D., 1970. The reclamation of toxic metalliferous wastes. Nature 227: 376—377.

Truhaut, V. R., 1969. Ecotoxicology: objectives, principles and perspectives. Ecotox. Environ. Saf. 1: 151—173.

Van Capelleveen, H. E., 1987. Ecotoxicity of heavy metals for terrestrial isopods. Ph.D. Thesis, Vrije Universiteit Amsterdam.

Van Straalen, N. M., 1988. Ecotoxicologische theorievorming over opname, effecten en doorgifte van stoffen in dierpopulaties. Milieu 3: 40—45.

Van Straalen, N. M., Burghouts, T. B. A., Doornhof, M. J., Groot, G. M., Janssen, M. P. M., Joosse, E. N. G., Van Meerendonk, J. H., Theeuwen, J. P. J. J., Verhoef, H. A. & Zoomer, H. R., 1987. Efficiency of lead and cadmium excretion in populations of *Orchesella cincta* (Collembola) from various contaminated forest soils. J. Appl. Ecol. 24: 953—968.

Van Straalen, N. M., De Goede, R. G. M. & Schobben, J. J. M., 1989. Population consequences of cadmium toxicity in soil microarthropods Ecotox. Environ. Saf. 17: 190—204.

Van Straalen, N. M. & Denneman, C. A. J., 1989. Ecotoxicological evaluation of soil quality criteria. Ecotox. Environ. Saf. 18: 241—251.

Van Straalen, N. M., Geurs, M. & Van der Linden, J. M., 1987. Abundance, pH-preference and mineral content of Oribatida and Collembola in relation to vitality of pine-forests in The Netherlands. In: R. Perry, R.M. Harrison, J.N.B. Bell & J.N. Lester (eds.) Acid Rain: Scientific and Technological Advances. Selper Ltd, London, pp. 674—679.

Van Straalen, N. M., Groot, G. M. & Zoomer, H. R., 1986. Adaptation of Collembola to heavy metal soil contamination. In: Proc. Int. Conf. Environmental Contamination, Amsterdam 1986. CEP Consultants, Edinburgh, pp. 16—20.

Van Straalen, N. M. & Van Wensem, J., 1986. Heavy metal content of forest litter arthropods as related to body-size and trophic level. Environ. Pollut. (A) 42: 209—221.

Walker, C. H., 1980. Species variations in some hepatic microsomal enzymes that metabolize xenobiotics. Progr. Drug. Met. 5: 113—164.

Williamson, P. & Evans, P. R., 1972. Lead levels in roadside invertebrates and small mammals. Bull. Environ. Contam. Toxicol. 8: 280—288.

Authors' Address
Department of Ecology and Ecotoxicology
Vrije Universiteit Amsterdam
De Boelelaan 1087
1081 HV Amsterdam

218

Open top chamber for outdoor carbondioxide enrichment studies. (Photograph: Foto en Tekenkamer, Faculty of Biology, Vrije Universiteit, Amsterdam.)

CHAPTER 20

Global change, the impact of the greenhouse effect (atmospheric CO$_2$ enrichment) and the increased UV-B radiation on terrestrial plants

J. ROZEMA, G. M. LENSSEN, W. J. ARP & J. W. M. VAN DE STAAIJ

Abstract. Atmospheric enrichment of CO$_2$ will favour growth of C$_3$ plant species and as a result the competitive balance between C$_3$ and C$_4$ plant species may markedly change. The greenhouse effect consists, however, of both an increase of atmospheric CO$_2$ and global warming, with an expected increase of the global temperature of 1.5–4.5° C with a doubling of the atmospheric concentration of carbon dioxide. Such a rise of temperature will prove advantageous to C$_4$ plants. It is also indicated that below a mean air temperature of 18.5° C no positive growth response to CO$_2$ enrichment will occur.

Increased UV-B radiation will negatively affect the growth of many plant species, monocots possibly being less sensitive than dicot plants. Both the causes of physiological damage by increased UV-B and adaptations to increased UV-B are incompletely understood. There is special need for assessment of UV-B effects on plants in long term field studies. The combined effect of CO$_2$ enrichment, global warming, UV-B increase, and soil and air pollution (ozone, SO$_2$, acid rain etc.) on terrestic and aquatic ecosystems is unknown.

The combined effects of climatic change factors and the soil and air pollution factors need to be studied in the near future.

1. Global change

Currently, global change indicates a number of events caused by human activities that lead to changes in the global environment, more in particular climatic changes are concerned. There is a continuous increase of the content of carbon dioxide in the atmosphere, and a world wide depletion of the stratospheric ozone layer. At the same time, air pollution and acid rain affect the quality of forests and other natural and agro-ecosystems. Soil pollution may threaten even human health in some cases and represents a major economic factor when for example costs for cleaning, treatment or removal are considered. Consideration of the effects of the rise of atmospheric CO$_2$ may not only include the direct effects on plants, and the changed solar radiation balance, but also a variety of secondary consequences such as a shift in the competitive balance between C$_3$ and C$_4$ plant species, changes

in the qualities of crops as a result of an increased C/N ratio, changes in the occurrence and damage by plagues such as pathogens and herbivorous insects consuming leaves and other parts of crops. The rise of the atmospheric temperature at the same time as a result of the greenhouse effect, affects growth of plants and animals and will change the geographical distribution of plant and animal species with unknown effects on the functioning of ecosystems as a whole. Some reports predict shifts of climate and vegetation zones towards the poles. There is much dispute on the climatic change that will take place as a result of the increased global surface temperatures. Precipitation patterns will change the Greenland icecap and may melt partially, the volume of the ocean water will expand and the sea level will rise. Erosion of sand dunes, damage to dikes and a new major financial investment to coastal defense in The Netherlands seems to be unavoidable. The drought

J. Rozema and J. A. C. Verkleij (eds.), Ecological Responses to Environmental Stresses, 220–231.

over North America in the summer of 1988 has seriously reduced the yield of crops. In Europe the mild winter of 1988/1989 and the long hot summer of 1989 provide further evidence for the gradual arrival of the greenhouse effect. Of course, the summer drought in America can be explained by other meteorological phenomenas than by the rise of atmospheric carbon dioxide and the mild winter in Western Europe of 1988/1989 is within the statistical range of the prevailing atlantic temperate sea climate. But both these two events have significantly contributed to the increased public concern for the cause and consequences of the climatic and global change.

2. Solar energy, global atmosphere, ozone and UV-B

Life on earth is dependent on the solar energy, heating the global atmosphere and enabling green plants to fix carbon dioxide from the air or from the water. From the solar energy spectrum, ultraviolet radiation forms a small part, but it has an enormous impact on the evolution and development of plant and animal life on earth.

It is believed that the early global atmosphere was reducing, devoid of oxygen and ozone. Plant and animal life developed in the oceans and seas, where UV radiation may have helped to form important biological macromolecules like aminoacids as precursors for proteins. The filtering of UV-B in the surface layers of marine aquatic ecosystems protected plant and animal life in deeper layers from the destructive influence of UV-radiation. With the development of early photosynthesis in the oceans, the oxygen content of the atmosphere increased. However, the atmospheric oxygen content during the Precambian was on the order of 10^{-3} of the present level of roughly 20%. Oxygen in the atmosphere effectively absorbs solar UV-C and as a result ozone is being produced in the upper layer of the atmosphere. This stratospheric ozone layer absorbs solar UV-C completely and a significant part of the solar UV-B radiation (Kaplan 1978).

3. Increased UV-B radiation due to stratospheric ozone depletion

3.1. Occurrence and characteristics of UV-B radiation

The spectrum of solar radiation covers a wide range of wavelengths from long waves including the infrared part of the spectrum, the photosynthetic active radiation band (PAR, 700–400 nm) and the shorter wavelengths including ultraviolet radiation.

The ultraviolet part of the spectrum is generally subdivided into three bands. UV-C (100–280 nm), UV-B (280–320 nm) and UV-A (320–400 nm) (Caldwell 1981). The shortest band (UV-C) is used for sterilization purposes (killing of bacteria) in kitchens, hospitals and microbiological laboratories). UV-B radiation affects immune system in the human skin. As a result, humans are more sensitive to infectious diseases. UV-B radiation also plays a role in the formation of skin tumors and damage to the human eye. UV-B radiation is the cause of skin affections and diseases, including skin cancer. The longest wave band UV-A is not visible for the human eye, but functions in the vision of insects. The stratospheric ozone is the primary layer attenuating UV radiation. UV-radiation shorter than 280 nm including UV-C and part of the UV-B band is absorbed by the non-disturbed stratospheric ozone. Radiation with a wavelength greater than 320 nm is not absorbed by ozone. Therefore UV-A and part of the UV-B radiation reaches the surface of the earth (Green 1983).

The protective ozone layer is at a distance of about 15–40 km from the earth's surface and recently, measurements have revealed periodical fluctuations of ozone concentrations and subsequent variation of UV-B reaching the earth's surface. The thickness of the ozone layer varies with the geographical latitude. Minimum values of the thickness of the ozone column have been found for the tropical areas of South America and Africa. Also, a hole in the ozone layer has been discovered at the Antarctic since about 1975 (Farman *et al.* 1985). Depletion of ozone has been reported for the Arctic, that might result in an ozone "hole" over Northern-Europe (Pearce & Anderson 1989). The nature of these differences and variation (fluctuations?) is only partly understood because of the poor and young data set. Ozone is being produced in a series of more than 200 (photo)chemical reactions. Ozone is produced when solar radiation causes the splitting of oxygen molecules, whereby oxygen radicals develop. These oxygen radicals give rise to the formation of ozone. Halogen compounds and nitrogen oxides in low natural concentrations play a major role in these photochemical reactions. Changes of the above concentrations by anthropogenic activities may lead to drastic shifts in the equilibrium reactions. On the other hand the variable and massive production of bromine containing compounds by marine (brown) algae might affect the outcome of photochemical reactions in the troposphere and stratosphere. The ozone concentrations in the stratosphere represent the net outcome of reactions that lead to production or breakdown of ozone. There is natural variation of these processes with the seasonal and annual change of the weather in the different climate types on earth. Recently, evidence has accumulated to assume that man-made atmospheric contaminants such as chlorofluorocarbons (CFCs) are the cause of stratospheric ozone depletion. Chlorofluorocarbons are being used as spray propellants ($CFCl_3$ and CF_2Cl_2), refrigerants and foam-blowing agents. The highly inert CFCs move slowly to the upper stratosphere, undergo photodissociation by high energy solar radiation, whereby free chlorine atoms are released. This chlorine produced is believed to destroy ozone. The Montreal protocol of 1987 consists of an international agreement on the reduction of emission of CFCs. This measure is technically feasible because there are suitable and non-hazardous alternatives for the application of CFCs. However, the chemicals destroying ozone have a long residence time in the stratosphere during which they proceed to destroy ozone. There is a lag of 100 years, which means that those atmospheric pollutants now being released will lead to depletion of stratospheric ozone for the next 100 years (Teramura 1987). It may also be questioned how effective the Montreal protocol will be.

Emission of ozone at the earth's surface does not affect the stratospheric ozone layer. Atmospheric ozone is being destroyed rapidly by water vapour present in the atmosphere. Increased UV-B radiation at the level of the earth's surface may produce ozone, in addition to anthropogenically caused development of ozone. One percent depletion of ozone will increase UV-B radiation to such an extent that six percent more cases of skin cancer will occur (Slaper & van der Leun 1989).

Measurements of levels of ultraviolet radiation at different geographical locations (Caldwell *et al.* 1980) revealed that the total fraction of UV-B (280–320 nm) reaching the earth's surface in the tropics exceeds the sites near the polar circle only by a factor of 1.6, despite the fact that the ozone layer is usually much thicker in polar areas. However, the quality of the UV-B radiation differs for the two locations compared. In the tropics the fraction of shorter UV-B wavelengths is larger. As a result, biological effectiveness of the UV-B radiation reaching the tropics is 7 to 10 times larger than that reaching the polar regions (Caldwell *et al.* 1980). UV-radiation reaching the surface of the earth may be of three kinds: i. direct solar UV-B, ii. skylight UV-B (up to 40–70% of total UV-B), iii. reflection via the soil surface, for snow surfaces reflected radiation may contribute 60% of the total UV-B radiation level, for wet, dark soil it is only 10% (Caldwell 1981).

3.2. Effects of increasing UV-B radiation on terrestrial ecosystems

Outdoor studies on the effects of increased levels of UV-B radiation due to ozone depletion are scanty. In an experimental study Van de Staaij, Rozema & Stroetenga (1990) found a 28% growth reduction and depression of photosynthesis in the halophytic dicot *Aster tripolium* and only small decreases of biomass growth and photosynthesis in the C_4-grass species *Spartina anglica*. In a recent report (Hoffmann 1987) it is documented that of 10 major types of terrestrial plant ecosystems (biomes) only the agricultural, temperate grassland and forest, and the tundra ecosystem, have been studied slightly (with most research attention being paid to the agricultural ecosystem). Therefore, it must be concluded that at the moment no detailed predictions can be made on the effects of increasing UV-B on natural ecosystems. Barnes *et al.* (1988) performed a six year field study and report reduced plant height and leaf length in *Triticum* and *Avena*. The total biomass production was not affected. *Triticum* increased its competitive ability towards *Avena* under increased UV-B radiation. These effects were not associated with reduced rate of photosynthesis.

In addition, it is not only the issue of increasing UV-B due to ozone depletion that threatens life on earth, global change comprises the simultaneous rise of atmospheric carbon dioxide and atmospheric temperature in a system with seriously local water, air and soil pollution. It is therefore far from trivial to state that the problem of global environmental change is crucial and urgent both from a point of view of human health and of environmental quality.

3.3. Causes of growth reduction of plants as a response to increased UV-B radiation

Reduced leaf growth of *Rumex patientia* was reported to be due to an UV-B effect on cell division and not on cell expansion (Dickson & Caldwell 1978). Reduced plant growth and photosynthesis in response to UV-B relates to disturbance of the reaction centre of photosynthesis system II (PS2), in that no longer energy is being available for the transport of electrons from PS2 to PS1. Therefore no re-reduction of chlorophyll and production of ATP is possible. UV-B radiation damage may also consist of membrane injury, leading to changes permeability (leakage of K^+, Na^+, Cl^- and HCO_3^- for example) (Bornman *et al.* 1983). UV-B radiation effects on cell division may relate to UV-B damage to DNA chains causing single strand breaks (SSB) or the production of dimers of pyrimidine. UV-C and UV-B induce the production of pyrimidine dimers, while UV-A causes single strand breaks (Rothman & Sellow 1979). For higher plants, production of pyrimidine dimers has been demonstrated in response to increased UV-B radiation for *Lathyrus sativus*, *Hordeum vulgare*,

Nicotiana tabacum and *Ginkgo biloba* (Beggs *et al.* 1985).

3.4. Adaptations of plants to increased levels of UV-B radiation

Among the mechanisms that protect plants to UV-B damage are the following (Beggs *et al.* 1986):
a. Repair of damage caused by UV-B radiation. Pyrimidine dimers may be removed by the action of enzymes, so that the correct DNA sequence remains. Also the quenching and scavenging of free radicals produced as a result of oxygen singlets by photo-oxidation has been described as a way of reducing the injury due to UV-B. Postreplication repair implying the replication and combination of new DNA strains to replace damaged ones, has also been described.
b. Plants may be capable of minimizing the injuring effects of UV-B by delay of growth so that no UV effects on cell division are possible.
c. Plant adaptations that reduce the level of UV-B radiation, so that sensitive plant structures or processes can not be injured. Attenuation of UV-B radiation by cuticle and cell wall seems to be of minor importance since they do not absorb a significant fraction of the UV-B radiation (Caldwell *et al.* 1983). Most probably pigments (flavonoids and related phenolic compounds) in the vacuoles of epidermal cells, located in outer tissue layers may be capable of absorption of UV-B. These compounds have high absorption coefficients for UV-B and a high transmittance of photosynthetically active radiation (Teramura 1987).

Plants differ greatly in their response to natural and artificially enhanced levels of UV-B radiation. Some species appear to be sensitive to ambient levels of solar UV-B, while other species show no symptoms of damage at enhanced levels of UV-B (Teramura 1987). In addition, differential sensitivity to UV-B radiation exists within one species when cultivars or subspecies are among the plants that are relatively sensitive to UV-B radiation. Surveys of effects of UV-B radiation on (crop) plants have been produced by Teramura (1983) and lead to the conclusion that serious reductions of the yield of various crops are likely to occur with ozone depletion. It must be noted that the outcome of many experiments on the effects of increased UV-B on plant performance must be critically considered. Firstly the effect of UV-B on plants is dependent on other environmental factors. At low levels of photosynthetically active radiation (PAR), the effect of UV-B is much more pronounced than at high levels of PAR (Teramura 1980). Secondly, and perhaps more important, is the technical set up and facilities used for reaching enhanced levels of UV-B in greenhouse and field studies. When no adequate filter system is used in addition to the UV-tube lamps, UV-C radiation may occur and the damaging effect of UV-B radiation will be overestimated (Teramura 1987).

Recently Tevini *et al.* (1989) used an ozone filter system to reduce the solar ultraviolet radiation, but it can only be applied in small growth chambers in an environment with high solar input of ultraviolet radiation.

4. Rise of the atmospheric content of carbon dioxide

4.1. Carbon dioxide in the atmosphere

The content of carbon dioxide of the atmosphere is rising and expected to double near the end of the twenty-first century. Atmospheric levels of carbon dioxide have increased approximately 25% from 1800 to 1985, mainly due to human activities such as deforestation as a result of expansion of agriculture and more recently, resulting from burning of fossil fuels. When a doubling of the CO_2 level of 1800 occurs, the increase of the mean global temperature, will probably be larger than during any period in the last 100.000 years (Trabalka 1985). Measurements of Keeling of the atmospheric CO_2 at the Mauna Loa Observatory in Hawaii demonstrated a rise from 316 ppm (1958) to 345 ppm in 1985. Not all sources and sinks and exchanges of carbon between the compartments of the global carbon system, including the biosphere, the atmosphere and the oceans are quantified, neither can the climatic effects of increasing carbon dioxide be projected

and detected, and research in this direction should be stimulated. Here the direct effects of increasing carbon dioxide on terrestrial ecosystems are discussed.

4.2. Historical changes in the atmospheric content of CO_2 and the response of higher plants

Recently Woodward (1987) suggested that stomatal numbers could decrease as a response to atmospheric CO_2 enrichment. This idea was based on short term experiments with plants growing outdoors under conditions of elevated CO_2. By analysis of historical (1800–1900) leaves obtained from herbarium stored plants and comparison with recent plant material (1987) it was concluded that a relative decrease of 40% of the stomatal density had occurred with the recent increase of atmospheric carbon dioxide from 280 ppm CO_2 (preindustrial records) to 340 ppm CO_2 (1987). If this relationship between stomatal density and atmospheric CO_2 content could be proved than perhaps the stomatal density of plant leaves could be used as an indication of changes of the atmospheric content of CO_2. The hypothesis and analyses have been checked by the Austrian scientist Körner (1988). In this study also historical and recent leaf material of over 200 plant species was analysed. In addition, the leaf material of alpine plant species from low and high altitudes was compared. High altitudes in mountain areas can also be characterized by a lowered partial CO_2 pressure. The result of this careful study was, that no statistically significant difference could be found between leaf material from 1890 and 1985.

In the CO_2 enrichment studies at the Free University, stomatal densities of leaves of *Aster tripolium* were estimated (Fig. 4). Stomatal densities were measured of the upper and lower leaf side. Stomatal density increases with increased salinity. When also changes of the Leaf Area Ratio ($m^2 \cdot g^{-1}$ dry weight) are being considered in relation to salinity and CO_2 enrichment, it appears that the increase of stomatal numbers with increased salinity is mainly the result of a decreased Leaf Area Ratio. This can be attributed to reduced expansion growth of leaves of *Aster tripolium* with increased salinity (Rozema *et al.* 1987). From our study it can be concluded that CO_2 enrichment does not significantly reduce stomatal density of the salt marsh halophyte *Aster tripolium*. The use of stomatal density as an indicator for changes in the atmospheric content of CO_2 seems to be doubtful.

4.3. Effects of atmospheric carbon dioxide enrichment on salt marsh plants

In a cooperative research programme of the Free University, Amsterdam and the Smithsonian Environmental Research Center, Edgewater, Maryland, USA analyses are made of effects of CO_2 enrichment in laboratory, greenhouse and experimental field studies on salt marsh plants (Free University, Amsterdam). The research group of the Smithsonian Environmental Research Center (Dr B.G. Drake) conducts more long term field studies of the growth and physiology of American salt marsh plant species. In addition, in the Free University, Amsterdam studies of the combined effects of CO_2 enrichment, increase of UV-B radiation and SO_2 and O_3 are carried out in greenhouse and fumigation chambers.

4.4. Greenhouse studies on the effects of carbon dioxide enrichment

At the Department of Ecology and Ecotoxicology of the Free University effects of carbon dioxide enrichment on plants are being assessed using different kinds of enclosures. Plant leaves in the leaf chamber (cuvette) of the Walz-BINOS-IRGA and ADC system allow measurements of photosynthesis, respiration and transpiration rate at different concentrations of carbon dioxide (Rozema *et al.* 1987).

Using this facility CO_2-response curves have been obtained for three salt marsh grass species (Fig. 1). More technical details are presented in Rozema *et al.* (1990a). *Scirpus maritima* and *Puccinellia maritima* are both C_3 species with a relatively high CO_2 compensation point (50–100 ppm CO_2) and maximum values of the photosynthetic rate at 400–500 ppm CO_2. The high CO_2 compensation point of these C_3 grass species is generally related to

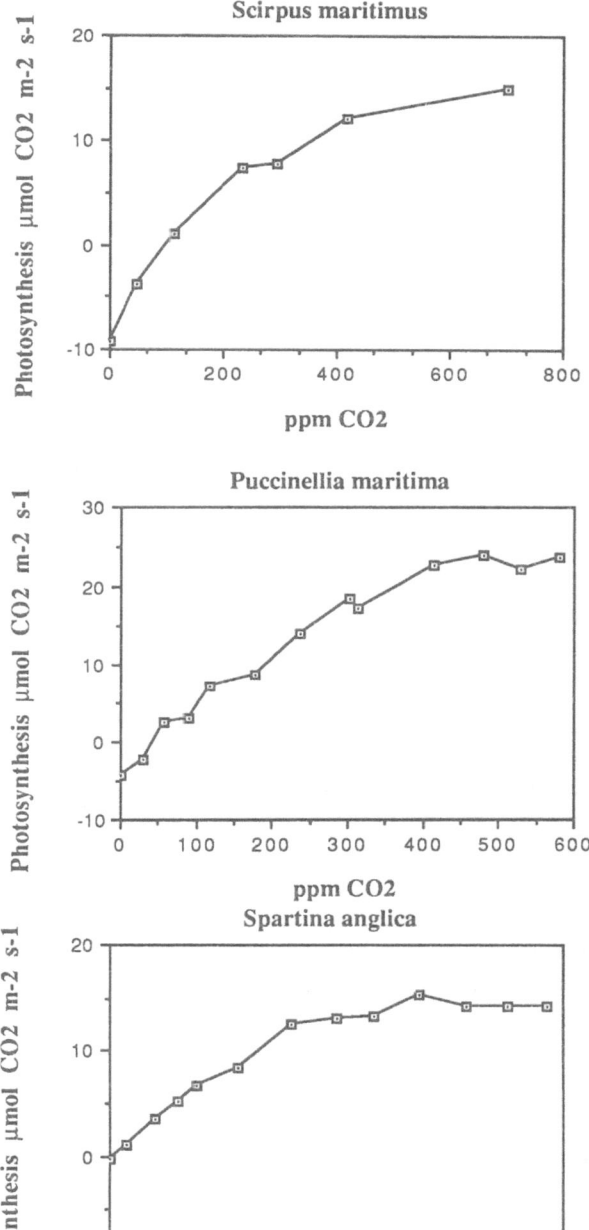

Fig. 1. Rate of net photosynthesis (μmol CO_2 m^{-2} s^{-1}) of two C_3 and one C_4 (*Spartina*) grass species of the salt marsh environment, with increasing concentrations of CO_2 (ppm) supplied and regulated using a Brooks Massflow controller and a BINOS (Heraeus-Hanau Western-Germany) IRGA System. Environmental conditions in the leaf chamber (cuvette): 20°C, 70% R.H., light intensity 1400 μE m^{-2} s^{-1} (PAR, measured with a Li-185 B quantumsensor). Flow rate of the air-stream through the gasexchange chamber was 1.5 l min^{-1}.

225

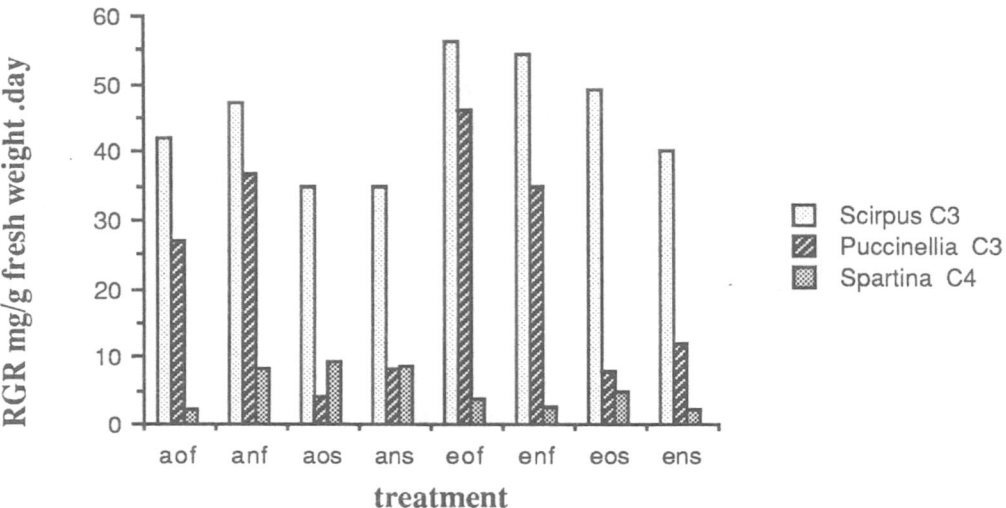

Fig. 2. Relative growth rate (mg g^{-1} day^{-1}, fresh weight basis) of C$_3$ and C$_4$ salt marsh grass species cultivated in nutrient solution with CO$_2$ enrichment, ambient and elevated, (A/E), aerated (O) or hypoxic (N) conditions, and with low (10 mM NaCl, F) and increased salinity (250 mM NaCl, S). Average values of four replicates.

the relatively high photorespiratory activity. This means high levels of the Ribulose biphosphate oxygenase. In the C$_4$-plant species *Spartina anglica* the balance between carboxylase and oxygenase reactions is more towards carboxylation. This is because in *Spartina anglica,* the phosphoenolpyruvate carboxylase enzyme helps to concentrate CO$_2$ at the site of carboxylation of the Rubisco enzyme. The low rate of photorespiration and the high affinity of PEP-case for CO$_2$ in the C$_4$ grass *Spartina anglica* allow the exploitation of much higher light intensities than C$_3$ plant species do. In the CO$_2$-response curves (Fig. 3) the C$_3$-plant species reach higher rates of photosynthesis because the C$_3$-Rubisco-enzyme may not be saturated with CO$_2$ at atmospheric CO$_2$ concentrations up to 800 ppm in the case of *Scirpus maritimus*. In the C$_4$-grass species the photosynthetic CO$_2$ fixation system appears to be saturated at atmospheric CO$_2$ concentrations of 250 ppm CO$_2$. The high maximum rate of photosynthesis of *Puccinellia maritima* may also be due to the underestimation of the leaf area of the enrolled leaves of this species.

In stainless steel, white coated growth chambers, illuminated with 400 Watt Philips HPIT lamps

(315 μE m^{-2} s^{-1}, 60% R.H., 21°C during the 12 h light period and 15°C in the dark period, salt marsh plants were grown in Hoagland's solution at ambient (340 ppm) and elevated (580 ppm) concentrations of CO$_2$ (Rozema *et al.* 1989).

The relative growth rate of the C$_3$ plant species was higher with elevated CO$_2$ in contrast to the growth response of the C$_4$ species (Fig. 2). The reduction of the growth rate of increased salinity was less with CO$_2$ enrichment in particular for *Scirpus maritimus*. Due to increased photosynthesis (*Scirpus*) or reduced transpiration (*Spartina*) the water use efficiency increased with CO$_2$ enrichment. For all the three species examined the total water potential of the plant, estimated using the Scholander's Pressure Bomb, tended to be less negative under saline, CO$_2$ enriched conditions (Rozema *et al.* 1989a and 1989b) (Fig. 3).

4.5. Field studies on the effects of carbon dioxide enrichment on salt marsh plants

The effects of short term treatments with elevated CO$_2$ on crops and other annual plants grown in controlled environments have been well docu-

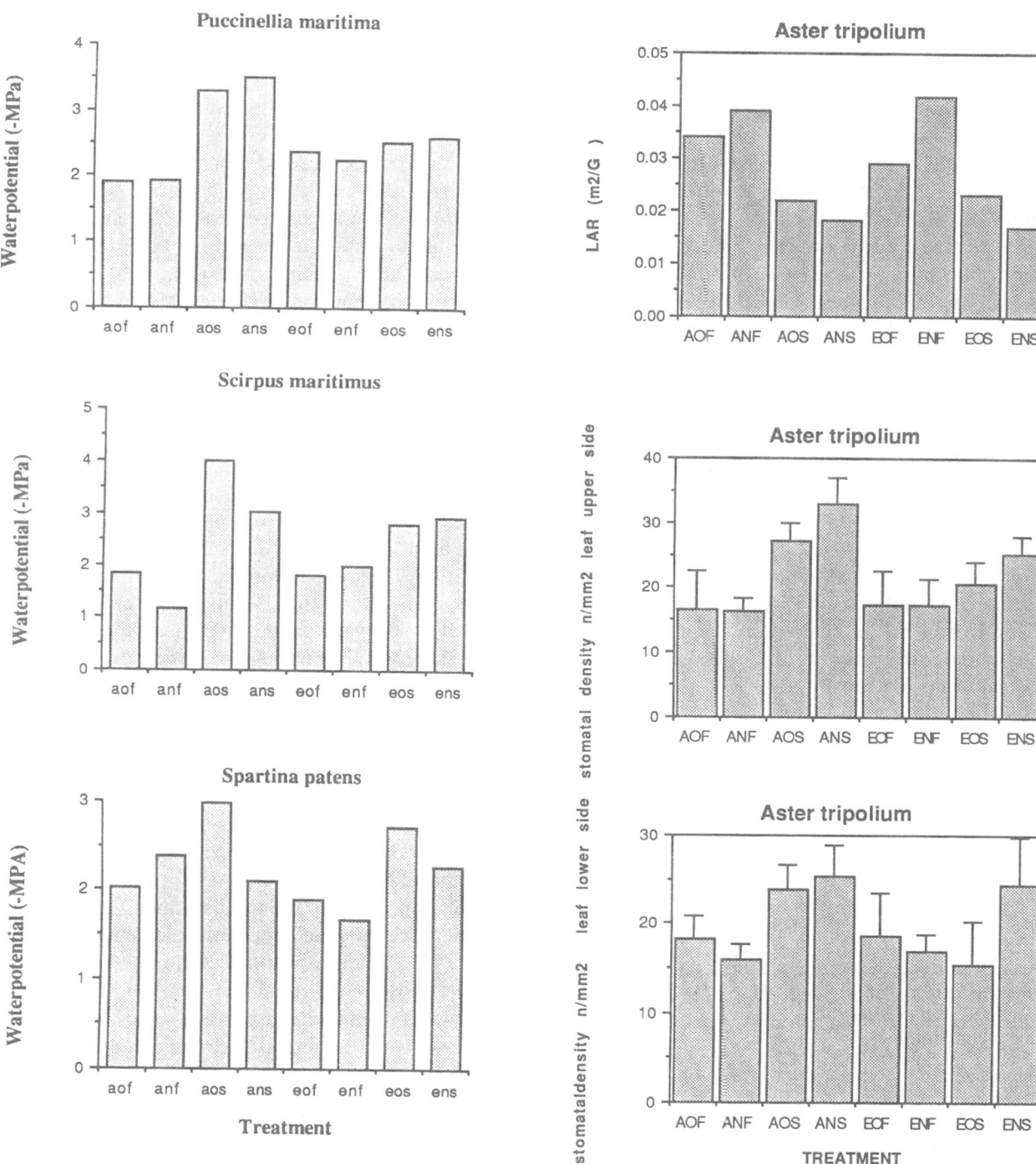

Fig. 3. Total waterpotential (MPa) of the shoot of three salt marsh grass species grown in growth chambers with ambient (A) (340 ppm CO_2) and elevated (E) (580 ppm CO_2), with aerated (O) or hypoxic (N) culture solution, with a low (10 mM NaCl, F) or increased (250 mM NaCl, S) salinity. Average values of three replications.

Fig. 4. Variation of stomatal density (n · mm^{-2}) of lower and upper leaf side of *Aster tripolium* plants with CO_2 enrichment (A/E), aeration (O/N) and salinity (F/S) (Average values of six replications with standard error of the mean. The Leaf Area Ratio was assessed of 5 plants.

mented. Very little, however, is known about the effects of long term treatment on natural ecosystems.

This section describes a long term study on the effects of elevated CO_2 on a brackish salt marsh vegetation. The study was carried out on an estuarine salt marsh located on the Rhode river, a sub-estuary of the Chesapeake Bay. A salt marsh was chosen because of the occurrence of monospecific stands of both a C_3 species (*Scirpus olneyi*) and a C_4 species (*Spartina patens*), in addition to a mixed vegetation with both the C_3 and C_4 species. Test atmospheres were created by placing 10 open top chambers on each of the three communities and raising the CO_2 concentration in half of the chambers. The day and night treatments continued during the entire growing season, from April through November (for more detailed technical information see Drake *et al.* (1989), Curtis *et al.* (1989a and 1989b).

By converting these chambers to temporarily closed top chambers, and by measuring the CO_2 concentration in the airflow entering and leaving the chamber, the amount of CO_2 taken up by the vegetation in the transpiration was measured. Total biomass, leaf area and number of shoots were

determined in a non-destructive way several times per season. The waterpotential of the plants and the salinity of the interstitial water were measured regularly throughout the season.

The main results after two growing seasons (1987 and 1988) are summarized as follows (see also Table I):

1. The carbon exchange of the C_3 community responded much more strongly to elevated CO_2 than the C_4 community (Fig. 1A). Due to a delayed senescence in the elevated chambers, there was a strong CO_2 effect on carbon exchange at the end of the season. Respiration was reduced in all communities. No acclimation of photosynthesis to elevated CO_2 was found after two seasons.

2. Elevated CO_2 reduced transpiration in all three communities, with the greatest reduction in the C_4 and mixed community (-30%) and the smallest in the C_3 community (-25%) (Fig. 5B). This reduction in transpiration corresponds with an increase in the water potential by about 0.5 MPa in both species. The water use efficiency, combining the effects on carbon uptake and water loss, showed a dramatic increase at elevated CO_2 (Fig. 5C).

   The increase in photosynthesis at elevated CO_2 resulted in an increase in biomass in the C_3 community. This was entirely due to an increase in the number of shoots because there was no effect on the weight per shoot. Root production doubled in the elevated CO_2 treatment. No effect on biomass was found in the C_4 community.

3. Senescence is delayed in the elevated chambers, in both the C_3 and the C_4 community.

This study shows that both carbon dioxide exchange and evapotranspiration can be measured accurately in the field at the plant community level. Elevated CO_2 greatly enhances photosynthesis in the C_3 community and to a smaller degree in the C_4 community. This is combined with a delay in senescence extending the growing season by a few weeks, and a reduction in respiration. Only in the C_3 community does the increase in the amount of carbon sequestered lead to an increase in the total biomass.

The reduction in water loss improves the water

Table I. A summary of the effects of elevated carbon dioxide, expressed as a percentage of the control (ambient) treatment, carbon exchange, water balance and biomass production of two salt marsh species in a Chesapeake Bay salt marsh.

	Scirpus olneyi (C_3)	*Spartina patens* (C_4)
Carbon exchange		
Daytime carbon uptake	+ 80%	+ 30%
Community respiration	− 40%	− 30%
Water balance		
Evapo-transpiration	− 25%	− 30%
Midday water potential	+ 0.5 MPa	+ 0.5 MPa
Water use efficiency	+ 155%	+ 80%
Biomass		
Shoot density	+ 20%	0%
Aboveground biomass	+ 20%	0%
Root biomass	+ 80%	0%
Delay in senescence	2 weeks	3 weeks

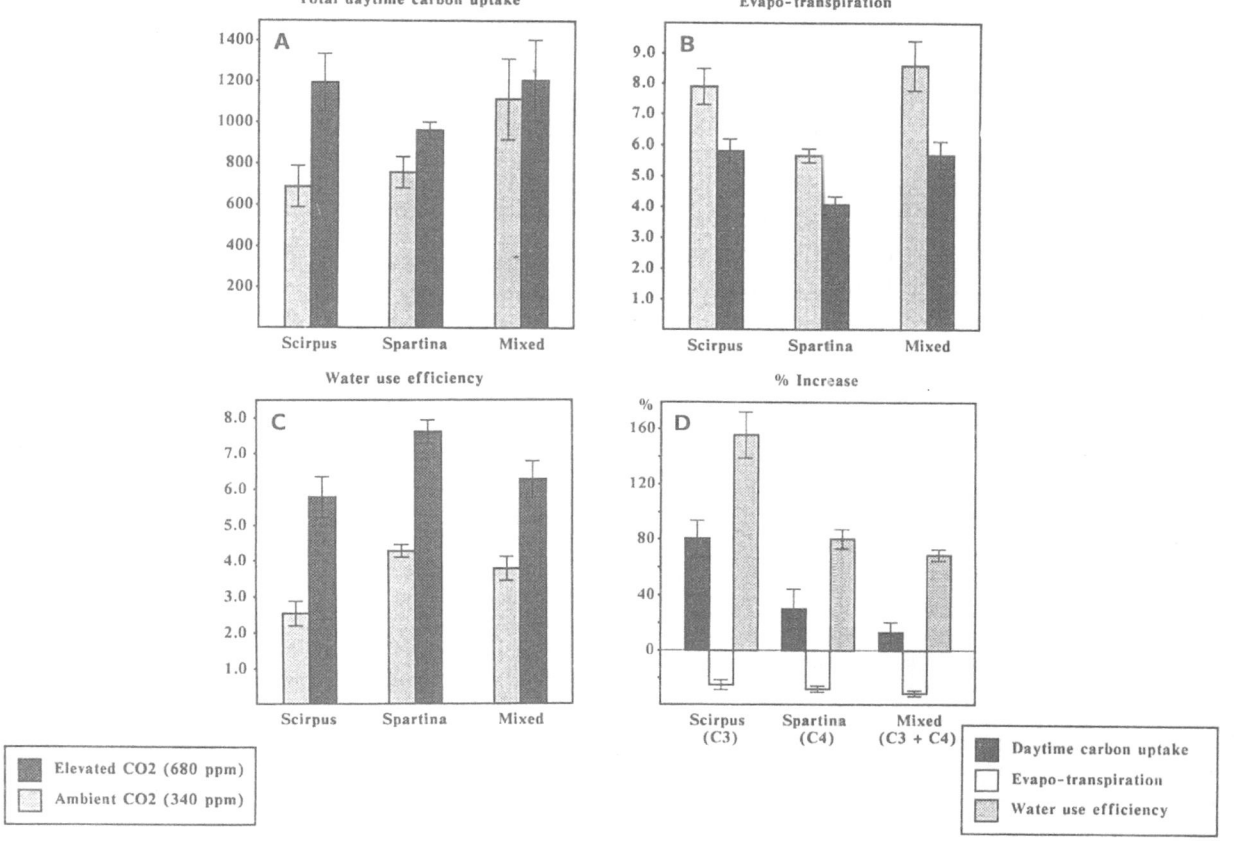

Fig. 5. Total daytime carbon uptake, evapotranspiration and water use efficiency of the *Scirpus* (C_3), *Spartina* (C_4) and the mixed community at ambient and elevated CO_2. Shown are the mean and standard error for the months June, July and August 1988. Total daytime carbon uptake is the total amount of carbon taken up during the light period, evapo-transpiration and water use efficiency were calculated for the time period 14:00–16:00 h.

balance and increases the water use efficiency in the C_3 and the C_4 community. This may help reduce salt and drought stress and may explain the reduction in respiration.

5. Combined effects of climatic and environmental changes on terrestrial ecosystems

It is likely that the atmospheric rise of carbon dioxide will favour the growth of C_3 plant species, and this could cause a shift in the competitive balance between C_3 and C_4 species. On the Northern Hemisphere, C_4 species prevail at northern latitudes with more temperature climate conditions. C_4 species

dominate in more southern latitudes. Based on the positive response to atmospheric carbon dioxide enrichment one could expect C_3 species to extend their geographical distribution more southward. At the same time, however, the greenhouse effect implies global warming, with an increase of the global temperature of 1.5–4.5° C when the atmospheric concentration of carbon dioxide is doubled during the next century (MacCracken & Luther 1985). Since C_4 plants have a higher temperature optimum for photosynthesis than C_3 plants (Ehleringer & Björkman 1977) growth of C_4 plants will increase with the predicted rise of the global temperature. On the other hand, Idso *et al.* (1987) have indicated that below a mean air temperature of

18.5° C C_3 plants will not respond positively to CO_2 enrichment.

Effects of increased UV-B radiation will be negative on many plant species, with a large range of variation between the species, monocot species possibly less sensitive to increased UV-B radiation than dicot plant species. When also other environmental factors such as soil and air pollution (ozone, SO_2, NO_x, etc.) are being considered, it will be difficult to predict the response of terrestric · ecosystems. In future research, analysis of the effects of a combination relevant climatic and environmental factors should be made.

References

Barnes, P. W., Jordan, P. W., Gold, W. G., Flint, S. D. & Caldwell, M. M., 1988. Competition, morphology and canopy structure in wheat (*Triticum aestivum* L.) and wild oat (*Avena fatua* L.) exposed to enhanced ultraviolet-B radiation. Funct. Ecol. 2: 319—330.

Beggs, C. J., Stolzer-Jehle, A. & Wellmann, E., 1985. Isoflavonoid formation as an indicator of UV-stress in bean (*Phaseolus vulgaris* L.) leaves. Plant Physiol. 79: 630—634.

Beggs, C. J., Schneider-Ziebert, U. & Wellmann, E., 1986. UV-B radiation and adaptative mechanisms in plants. In: R.C. Worrest & M.M. Caldwell (eds.) Stratospheric ozone reduction, solar ultraviolet radiation and plant life. Springer, Berlin. pp. 235—250.

Bornman, J. F., Evert, R. F. & Mierzwa, R. J., 1983. The effect of UV-B and UV-C radiation on sugar beet leaves. Protoplasma 117: 7-16.

Caldwell, M. M., 1981. Plant response to ultraviolet radiation. In: Encyclopedia of Plant Physiology, N. S. 12A Physiological Plant Ecology. Springer, Berlin, pp. 169—197.

Curtis, P. S., Drake, B. G., Leadley, P. W., Arp, W. J. & Whigham, D. F., 1989. Growth and senescence in plant communities exposed to elevated CO_2 concentrations on an estuarine marsh. Oecologia 78: 20—26.

Curtis, P. S., Drake, B. G. & Whigham, D. F., 1989. Nitrogen and carbon dynamics in C_3 and C_4 estuarine plants grown under elevated CO_2 *in situ*. Oecologia 78: 297—301.

Dickson, J. G. & Caldwell, M. M., 1978. Leaf development of *Rumex patientia* L. (Polygonaceae) exposed to UV radiation. Amer. J. Bot. 65: 857—863.

Drake, B. G., Leadley, P. W., Arp, W. J., Nassiry, D. & Curtis, P. S., 1989. An open top chamber for field studies of elevated atmospheric CO_2 concentration on salt marsh vegetation. Functional Ecology 3: 363—371.

Ehleringer, J. & Björkman, O., 1977. Quantum yields for CO_2 uptake in C_3 and C_4 plants: dependence on temperature, CO_2 and O_2 concentration. Plant Physiol. 59: 86—90.

Green, A. E. A., 1983. The penetration of ultra-violet radiation to the ground. Physiol. Plant. 58: 351—359.

Idso, S. B., Kimball, B. A., Anderson, M. G. & Mauney, J. R., 1987. Effects of atmospheric CO_2 enrichment on plant growth: the interactive role of air temperature. Agriculture, Ecosystems and Environment 20: 1—10.

Körner, Chr., 1988. Does global increase of CO_2 alter stomatal density? Flora 181: 253—257.

MacCracken, M. C. & Luther, F. M. (eds.), 1985. Detecting the climatic effects of increasing carbon dioxide. U. S. Dept. Energy, pp. 198.

Pearce, F. & Anderson, I., 1989. Is there an ozone hole over the north pole? New Scientist: 32—33.

Rothman, R. H. & Sellow, R. B., 1979. An action spectrum for killing and pyrimidine dimer formation in Chinese hamster V-79 cells. Photochem. & Photobiol. 29: 57—61.

Rozema, J., Arp, W., Van Diggelen, J., Kok, E. & Letschert, J., 1987. An ecophysiological comparison of the diurnal rhythm of the leaf elongation and changes of the leaf thickness of salt-resistant dicotyledonae and monocotyledonae. J. Exp. Bot. 188: 442—453.

Rozema, J., Dorel, F., Janissen, R., Lenssen, G., Broekman, R., Arp, W. & Drake, B. G., 1990a. Effect of elevated atmospheric CO_2 on growth, photosynthesis and water relations of salt marsh grass species. Aquatic Botany (in press).

Rozema, J., Lenssen, G. M. & Broekman, R. A., 1990b. Effects of atmospheric carbon dioxide enrichment on salt marsh plants. In: J.J. Beukema, W.J. Wolff & J.J.W.M. Brouns (eds.) Expected effects of climatic change on marine coastal ecosystems. Kluwer-Dordrecht, pp. 49—54.

Slaper, H. & Van der Leun, J. C., 1989. Schatting van het risico of huidtumoren bij blootstelling van de mens aan ultraviolette straling. Directie Stralenbescherming Ministerie van Volkshuisvesting, Ruimtelijke Ordening en Milieubeheer. pp. 51.

Staaij, J. van de, Rozema, J. & Stroetenga, M., 1990. Expected changes in Dutch coastal vegetation resulting from enhanced levels of solar UV-B due to stratospheric ozone depletion. In: J.J. Beukema, W.J. Wolff & J.J.W.M. Brouns (eds.) Expected effects of climatic change on marine coastal ecosystems. Kluwer, Dordrecht, pp. 211—217.

Strain, B. R., Curier, F. E., 1985. Direct effects of increasing carbon dioxide on vegetation. United States Department of Energy, DOE: 1—286.

Teramura, A. H., 1980. Effects of ultraviolet-B irradiance on soybean 1. Importance of photosynthetically active radiation in evaluating ultraviolet-B irradiance effects on soy bean and wheat growth. Physiol. Plant, 48: 333—339.

Teramura, A. H., 1987. Assessing the risks of trace gases that can modify the atmosphere. Vol. VIII. Technical support documentation ozone depletion and plants. US Environmental Protection Agency, pp. 1—77.

Tevini, M., Mark, U., Fieser, G. & Salle, M., 1989. Effects of enhanced solar UV-B radiation on growth and function of crop plants. Final report. Gesellschaft für Strahlen- und Umweltforschung, München.

Woodward, F.I., 1987. Stomatal numbers are sensitive to increases in CO$_2$ from pre-industrial levels. Nature 327: 617—618.

Worrest, R. C. & Caldwell, M. M. (eds.), 1986. Stratospheric Ozone Reduction, Solar Ultraviolet Radiation and Plant Life. NATO ASI Series G: Ecological Sciences 8, Springer Verlag, Berlin.

Authors' Addresses
Dr J. Rozema, G. M. Lenssen & J. W. M. van de Staaij
Department of Ecology and Ecotoxicology
Faculty of Biology
Vrije Universiteit
De Boelelaan 1087
1081 HV Amsterdam
The Netherlands

W. J. Arp
Smithsonian Environmental Research Center
Box 28
Edgewater MD21037
Maryland
USA

Snail eating leaves of *Trifolium repens*, a plant species used for the study of secondary plant products as protective agents. (Photograph: P. Kakes.)

CHAPTER 21

The genetics and ecology of variation in secondary plant substances

P. KAKES

Abstract. There is an enormous amount of variation in secondary plant metabolites. The variation, both quantitative and qualitative, can be exploited to study the genetics and the ecological functions of these substances. The present review discusses some selected groups of secondary metabolites, viz. those that have been studied both from the genetical and the ecological point of view. Variation in secondary metabolism may be caused by genetical differences or by the environment. The control and expression of two groups of secondary plant substances: cyanoglucosids and flavonoids are discussed as examples. The different functions of secondary metabolites in the ecology of the plant can be grouped as follows: Functions related to the primary metabolism, functions related to the abiotic environment and functions related to the biotic environment. One secondary metabolite can have different functions, belonging to different function groups.

One function of secondary metabolites that has received considerable attention in the past is protection against herbivores. There is certainly evidence that some secondary metabolites give protection against some herbivores. However, only a few combinations of plants and herbivores have been studied in detail. Less than a score of these studies have made use of variation in secondary plant substances with a known genetical base. Even if we can prove the defensive function of a particular secondary metabolite, this does not, in itself, constitute a proof of adaptedness. Only if and so far as the benefit of chemical defense exceeds the cost can we speak of adaptedness. A review of the literature on costs and benefits of chemical defense shows that both costs and benefits have been defined too narrowly in the past. Discussion of the ecological functions of cyanogenesis in white clover shows that both costs and benefits of cyanogenesis can be large, but that these effects are prominent in different stages of the life cycle of the plant. So in order to study the cost and benefit of a chemical defense system, the life history of the plant has to be taken into account.

1. Introduction

Secondary plant substances are compounds that have no direct role in the primary metabolism and occur erratically in the plant kingdom. Although the notion of "essential" and "non-essential" substances in plants is present in the works of the early plant physiologists Sachs and Pfeffer, the terms primary and secondary compounds were introduced by Albrecht Kossel (1891). In Kossel's view primary compounds are essential and consequently occur in every cell. Substances not occurring in every cell are called secondary. Kossel noted that it may be extremely difficult to decide to which category a substance belongs.

With thousands of secondary compounds described and new ones being found almost daily the problem of defining secondary plant substances is becoming more and more complex, especially since the function of newly described compounds (if any) is often not known. Nevertheless the concept of secondary plant substances has proved valuable and there is a general agreement about the major groups of secondary plant compounds. Since we now know that every cell of an organism has potentially the same biochemical capabilities, Kossel's restriction: not present in every cell, has been replaced by: not present in every plant species.

It should be stressed however, that both criteria given in the first sentence of the introduction

J. Rozema and J. A. C. Verkleij (eds.), Ecological Responses to Environmental Stresses, 234—249.
© *1991 Kluwer Academic Publishers.*

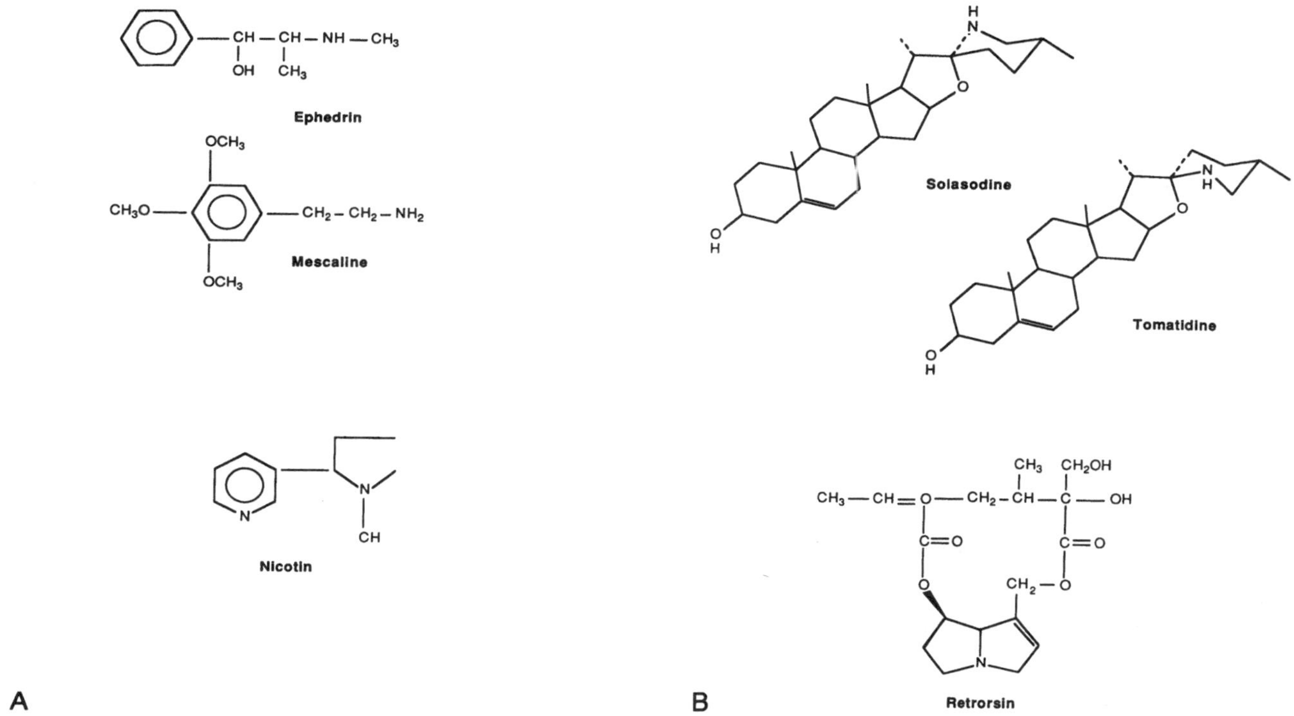

Fig. 1. A. Three types of alkaloids: Mescalin is a primary amine, ephedrin a secondary and nicotin a tertiary one. B. Solasodine and tomatidine as examples of isoprenoid alkaloids and retrorsin as an example of pyrrolizidine alkaloid. Note the internal ester bonds in retrorsin.

should be met. Substances with known primary function, but particular to certain plant groups, like inulin, replacing starch in some compositae are not considered secondary plant substances.

1.1. Types of secondary plant substances

The classification of the great number of secondary plant substances is difficult. A purely chemical classification is not useful, since historically recognized groups like the alkaloids, that belong functionally together, would be dispersed, and also not possible as the chemical structure of many secondary plant substances is currently insufficiently known. Consequently the classification used is not very homogeneous or consistent. This can be exemplified by comparing the alkaloids, a very diverse group, distinguished both for historical and functional reasons, and the cyanogenic glycosides, which are chemically homogeneous but may have different

functions in different species. Most, if not all secondary plant substances can be fitted into one of the following broadly defined groups:

Nitrogenous compounds:
Alkaloids
Non-protein aminoacids
Cyanogenic glucosides
Glucosinolates

Terpenoids:
Monoterpenes
Diterpenes
Sesquiterpene lactones
Saponins
Cardenolids

Phenolics:
Simple phenols
Flavonoids and isoflavonoids
Tannins

235

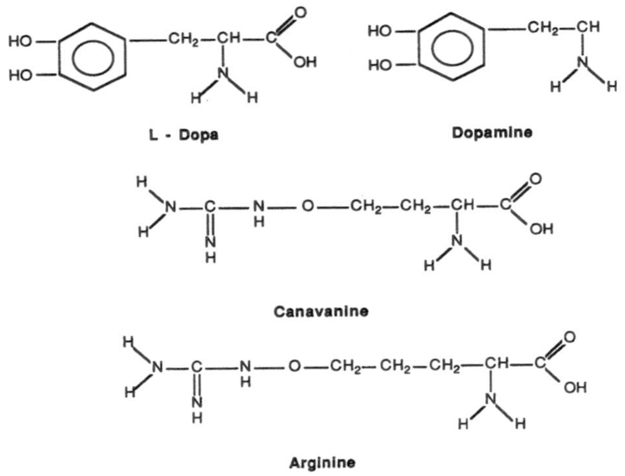

Fig. 2. Two non-protein amino-acids occurring in plants. L. dopa is converted in the brain to dopamine, canavanine is an analogon of arginine that is incorporated in proteins, leading to a partial loss of function of the protein.

1.2. General description of selected groups of secondary plant substances

Alkaloids

The alkaloids are probably the most diverse group of secondary plant compounds. Their name is derived from the fact that the first alkaloids studied behaved as weak bases in aqeous solution. The alkaloids are primary (mescalin), secondary (ephedrin) or tertiary (nicotin) amines (Fig. 1). The nitrogen atom(s) in alkaloids may be present in a pyrimidine ring, in a pyrrole ring, in one or more aliphatic chains or in conjugated heterocyclic ring structures. Many alkaloids have a strong physiological effect especially on animals, many are bitter to the human taste and are avoided as food by animals. Chemically related alkaloids often occur in unrelated taxonomic groups and vice versa, so the most logical classification is by structure and by biochemical pathway. Some examples are the tobacco alkaloids, characterized by a pyrimidine ring derived from nicotinic acid (Dawson *et al.* 1960), the pyrrolizidine alkaloids, characteristic for the genus *Senecio,* but also present in species of the Boraginaceae and the Leguminosae and the isoprenoid alkaloids, typical for wild species of *Solanum*

and *Lycopersicon.* In the cultivated species (potatoes, tomatoes) the level of alkaloids is low, especially in the parts used for human consumption, but recently selection programs, together with biotechnological experiments have been started to produce varieties of potatoes with high concentrations of isoprenoid alkaloids (solasodine, tomatine) as a source of chemicals for the production of steroid based drugs and herbicides. Alkaloids may be present in the plant in free form, as (internal) esters (the pyrrolizidine alkaloids) or as glycosides (the isoprenoid alkaloids).

Nonprotein aminoacids

A number of aminoacids with structures different from the twenty found in proteins and also different from intermediates in the primary metabolism, like ornithin, are found in the family of the Leguminosae (Fig. 2). Lathyrogens are toxic aminoacids present in the seeds of some *Lathyrus* species. When used as an important part of the human diet the seeds cause a characteristic neurological disorder: classical lathyrism. A number of unusual aminoacids have been isolated from *Lathyrus* species. All are toxic, but due to different mechanisms. In general however they seem to interfere with normal aminoacid metabolism, either by substrate inhibition or by endproduct inhibition.

Canavanine is an analogue of ornithin that is found in large amounts in the seeds of *Canavalia ensiformis* and *Dioclea megacarpa.* It is a potent growth inhibitor and probably acts by incorporation into proteins, producing enzymes with deviating spatial structure and reduced catalytic efficiency.

L,3,4-Dihydrophenylalanine (L-Dopa) is found in the seeds of various Leguminosae: in high concentration in several *Mucuna* species. The toxicity of L-Dopa is related to its ability to cross the blood-brain barrier. In the brain L-Dopa is converted to dopamine, a neurotransmitter. The former use of L-Dopa as drug in Parkinson's disease has been abandoned in view of the severe side-effects.

Cyanoglucosides

The cyanoglucoside form a rather small and homogeneous group. About 20 naturally occurring cya-

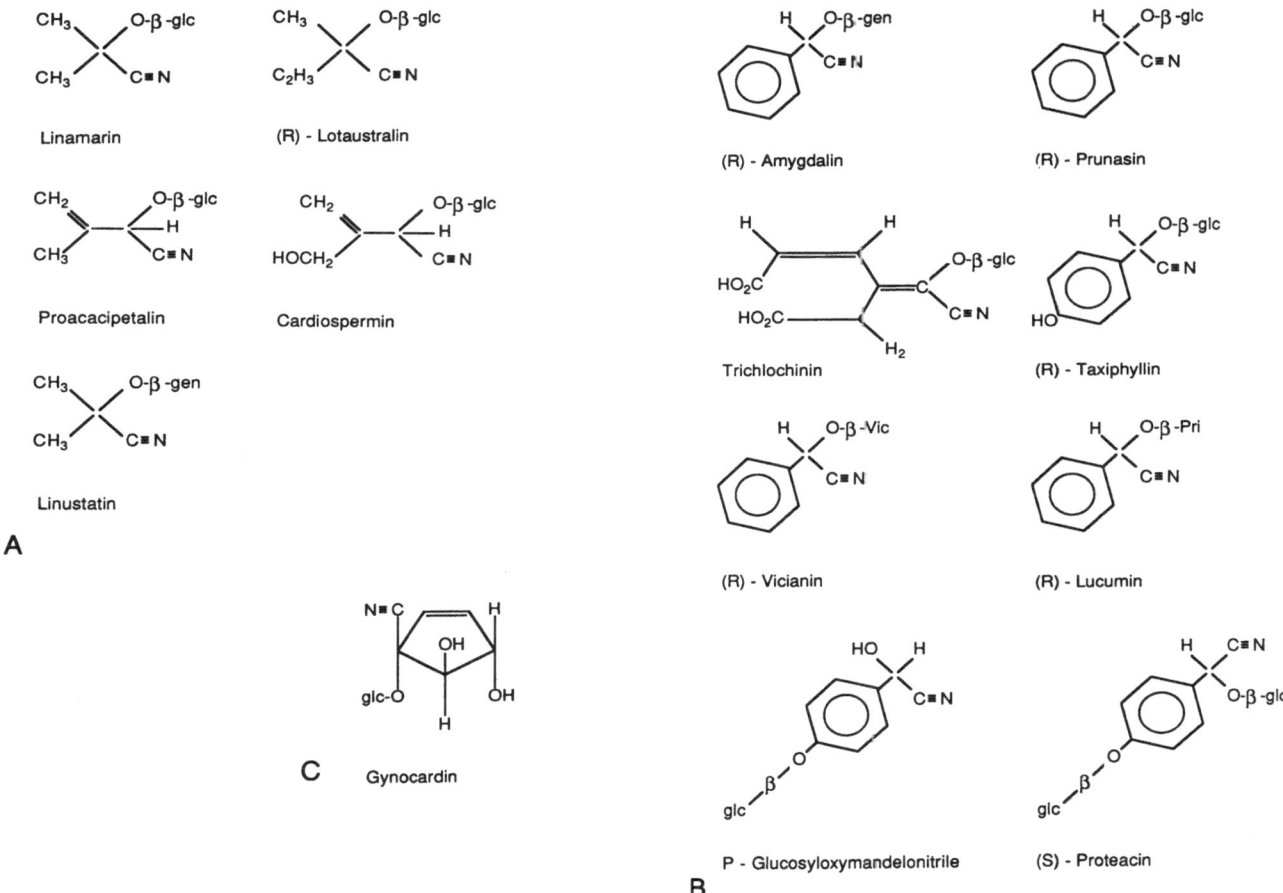

Fig. 3. A. Five cyanoglycosids derived from valine or isoleucine. Four are monoglycosids, linustatin is a diglycosid. B. Cyanoglucosids derived from aminoacids with cyclic structures. In trichlochinin the ring is opened during biosynthesis. C. Gynocardin as an example of a cyanoglucosid with cyclopentene structure.

noglucosides have been described. All are glucosides of α-hydroxinitriles, the variation found is a result of different rest groups R1 and R2 and of different (combinations of) sugars (see Fig. 3). The cyanoglucosids are more widespread than any other group of secondary plant products. In the plant kingdom they are found from the fungi to the angiosperms. Cyanoglucosids are also present in some bacteria and in Arthropoda (Millipedia, Lepidoptera). The cyanoglucosides are stable under physiological conditions, but readily hydrolized by specific hydrolases. The stability of the resulting a-hydroxinitriles is pH dependent. Bové & Conn (1961) have isolated from *Sorghum* leaves a α-hydroxinitrile lyase that breaks down the hydroxi-

nitrile to the corresponding aldehyd and HCN. The whole process, depicted in Fig. 4 for linamarin is known as cyanogenesis. In higher plant, HCN is only produced when the appropriate parts of the plants are damaged, either mechanically or by stress factors like frost. The reason is that the cyanoglucosides are present in the vacuole and the corresponding β-glucosidase(s) in the cell wall. So only after rupture of tonoplast and plasmalemma enzyme and substrate come into contact. The cyanoglucosides have been extensively studied since their discovery in 1830 by Robiquet & Boutron-Charlard (1830). Several reviews cover their biochemistry and ecological function (Conn 1980; Jones 1972; Nahrstedt 1985). Some aspects of cya-

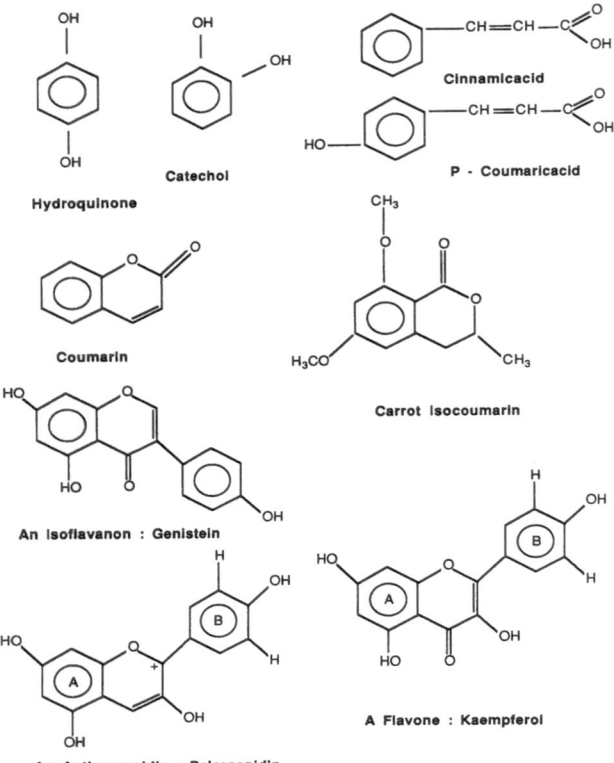

Fig. 4. Catabolism of linamarin in the presence of the β-glucosidase linamarase.

nogenesis will be dealt with in later sections of this paper.

Glucosinolates
Like the cyanoglucosides the glucosinolates are a uniform group (Fig. 5). There are today more than seventy glucosinolates described, occurring in eleven angiosperm families, five of which form the Order Capparales. This order contains the Brassicaceae, well known for their glucosinolate content. The glucosinolates are stable glycosides, that can be hydrolized by the enzyme myrosinase to yield glucose and a labile aglycone, that produces an isothiocyanate and sulfur (Fig. 6). The volatile isothiocyanates (mustard oils) have the distinctive taste and flavor associated with many of the species of Brassicaceae. Under different conditions a nitrile and elemental sulfur may be formed. Volatile isothiocyanates and nitriles are produced by macerated plant samples of *Brassica* and *Sinapis* spe-

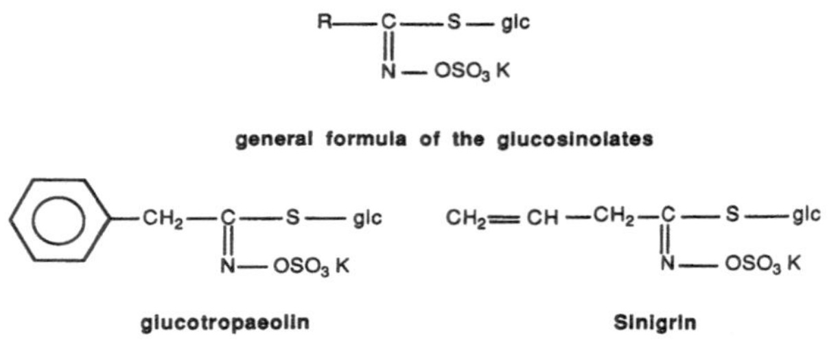

Fig. 5. General formula of the glucosinolates and the structures of a cyclic and an aliphatic glucosinolate.

cies. In the headspace of intact plants several isothiocyanates and one nitrile could be demonstrated by Tollstein & Bergström (1988) although in much lower concentration. Apparently some turnover of glucosinolates is taking place in the living plant.

In contrast to the situation in the cyanoglucosides all glucosinolates are hydrolyzed by one enzyme: myrosinase. Several isomyrosinases have been described in a number of *Brassica* species. These isozymes do not exhibit differences in substrate specifically however (Underhill 1980).

Several other catalytic processes of glucosinolates have been described, the enzymology of which is only partly understood. Notable end products are the nitriles, which are responsible for the unpleasant odor of Cabbage after prolonged cooking.

Rape seed is an important source of edible and industrial oils. After extraction of the oil a protein rich meal results that can be used for lifestock feeding. However, the hydrolysis products of glucosinolates, notably the thiocyanates are goitrogenic. Consequently much effort has been applied in producing cultivars with low glucosinolate content.

Phenolic compounds
As a group, phenolics are ubiquitous in the plant kingdom. It is a large and very diverse group and in the scope of this paper only a few examples can be given. The classification of the plant phenolics is by basic skeleton and we will arrange our examples also by that way: (see Fig. 7)

The simple phenols have one substituted aromatic ring. Phenol itself does not occur in living

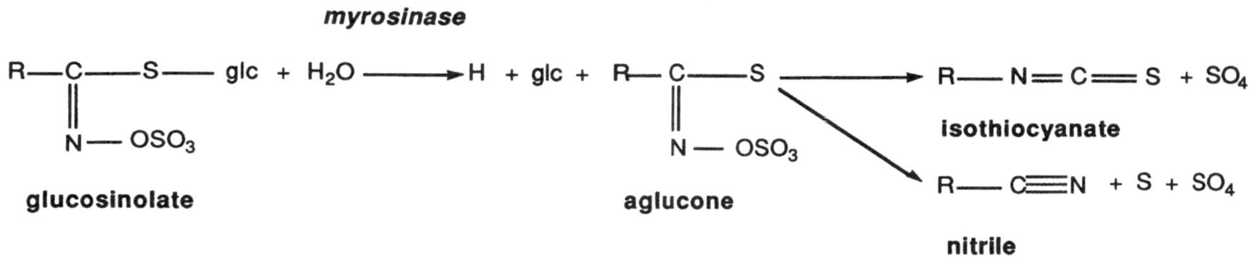

myrosinase

$$R\text{—}\overset{\displaystyle\|}{\underset{\displaystyle N\text{—}OSO_3}{C}}\text{—}S\text{—}glc + H_2O \longrightarrow H + glc + R\text{—}\overset{\displaystyle\|}{\underset{\displaystyle N\text{—}OSO_3}{C}}\text{—}S \longrightarrow R\text{—}N\!=\!C\!=\!S + SO_4$$

isothiocyanate

$$R\text{—}C\!\equiv\!N + S + SO_4$$

glucosinolate   **aglucone**   **nitrile**

Fig. 6. The action of myrosinase on a glucosinolate and two of the possible degradation products of the aglucone.

plants, but the dihydroxy-substituted isomeric substances catechol and hydroxyquinone are found in higher plants, the first one however, occurs only as a glucosid and is restricted to a few species.

A second important group is characterized by a C_6C_3 skeleton. The hydroxycinnamic acids, important intermediates for the more complex phenolic compounds, belong to this class together with the coumarins and isocoumarins. The third and probably best known group has a $C_6C_3C_6$ skeleton and contains the flavonoids and isoflavonoids. The last group is the polyphenolic one, which will be discussed in another part of this paper.

Phenolic compounds may be found in the plant in free form, but the majority is bound to one or more sugars by glycosidic linkage and/or esterified with one or more organic acids. The glycosids and glycosidic esters are generally soluble in water and are mainly found in the vacuole.

2. Control and expression of secondary metabolites

2.1. Genetical and environmental variation in the production of cyanoglucosides

A number of plant species have populations that are polymorphic for cyanogenesis, i.e. populations with cyanogenic and acyanogenic plants. *Trifolium repens* and *Lotus corniculatus* are examples particulary well studied, but other species (*Lotus alpinus*, Urbanska 1979, *Ranunculus montanus*, Dickenmann 1982, *Ranunculus repens*, Tjon Sie Fat 1979) have been reported. The actual number of polymorphic cyanogenic species should be much higher as in most species only a few plants from one location have been studied.

If cyanogenic and acyanogenic individuals of a species occur in one mendelian population one would expect a simple genetic difference between the types and this is indeed the case in *T. repens* and *L. corniculatus*. In both species the cyanotypes are the results of variation in two loci, one of them (Ac) regulating the presence/absence of the cyanogenic glucosides, the other (Li) the presence/absence of linamarase.

The level of cyanoglucosides in cyanogenic plants is in *T. repens* mainly controlled by Ac: AcAc plants contain approximately twice as much linamarin/lotaustralin compared to Acac plants (Hughes & Stirling 1982). Modifier genes have a minor effect in the families studied by Hughes *et al.* (1984) and by the present author (Kakes, unpublished results). Hughes *et al.* (1984) have found a third allele of Ac that causes a relatively low level of cyanoglucosids in both homozygous and heterozygous plants. Fig. 8 shows the biosynthetic pathway of linamarin and the possible steps that are influenced by the gene Ac.

In *Lotus corniculatus*, a tetraploid species that shows tetrasomic inheritance (Dawson 1946), the three heterozygotes and the homozygote for Ac form a series of increasing level of cyanoglucosids (Jones, personal communication).

Quantitative genetic variation in the level of cyanoglycosids has been found in cassave (*Manihot esculenta*) by De Bruijn (1971). In *Sorghum bicolor* and *Sorghum sudanense* there is also quantitative genetic variation (Nass 1972).

The effect of the environment on cyanogenesis has been studied by De Waal in *T. repens* (De Waal 1942). He found a periodicity in the HCN content on a dry matter base, with a maximum approximately 5 hr after sunrise and a lower maximum

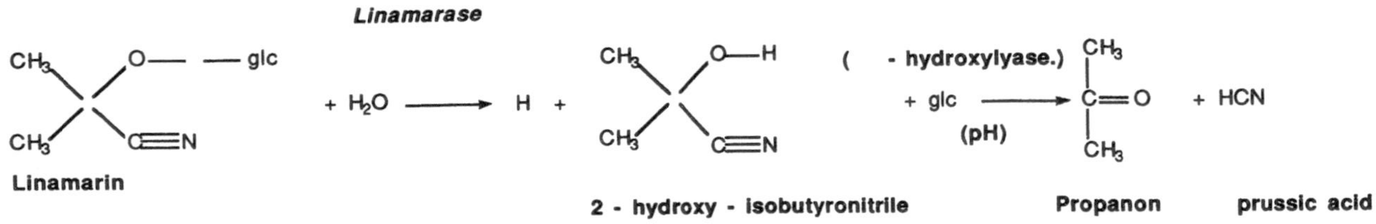

Fig. 7. Some examples of phenolic compounds. Cinnamic acid and P-coumaric acid are important intermediates in the biosynthesis of more complex phenols, like anthocyanidins and flavones.

during the night. The effect of temperature and developmental stage on the level of cyanoglucosids in *T. repens* was studied by Collinge & Hughes (1982). They found that cyanoglucosid production of germinating seeds starts at shoot emergence. In mature plants cyanoglucosid synthesis peaks in the "folded leaf" stage but goes on in the expanding leaf. Temperature is the most important environmental factor in cyanoglucosid production. The optimum temperature is 19°C. At 8°C and 27°C the cyanoglucosid content of the leaves is very low.

The present author has found that the expression of Li is temperature sensitive in some genotypes, but not in others. The temperature sensitive genotypes do not produce linamarase when the daily maximum temperature is over 25°C. Till (1987) has observed instability both for linamarase and for cyanoglucosid content in plants derived from natural populations and in controlled crossings between these plants. As the tests were done in the greenhouse in S. France during spring and summer, it is possible that lowered expression of Ac and Li was one of the causes of the intra-individual variability.

Ellis *et al.* (1977) have found that in *Lotus corniculatus* the cyanotype is unstable in some populations: low temperature reduces the expression of the genes Ac and Li.

2.2. *Genetical and environmental variation in flavonoid production*

There is a vast body of literature concerning the genetics and environmental variation of the anthocyanidins (Harborne *et al.* 1975). Genetic variation in anthocyanin production is particularly well stud-

ied in *Petunia hybrida.* (See Johnsson & Schram (1987) for a review.) The variation in flower colour in this species is caused mainly by variation in the anthocyanins and to a lesser extent by the type and concentration of flavonol co-pigments. The colour of the anthocyanins is dependent on substitution of the B-ring: hydroxylation at the 3′ and 5′ positions and on glycosylation of the C_3, and C_7-atom. Two series of colours can be distinguished: one starts with the red cyanidin 3′-glucoside and ends with the magenta paeonidin 3-(p-cumaroyl)-rutinosido-5-glucoside, the second starts with the grey delphinidin 3-glucosid and ends with the purple malvidin 3-(p-coumaroyl)-rutinosido-5-glucoside. The biochemical steps involved in the production of the different end products are: hydroxylation and methylation of the B-ring, glycosylation at C_3 and C_5, attachment of a second sugar molecular (leading to the formation of a rutinoside) and acylation of one or more of the sugars. The genes responsible for each of these reactions have been identified (Wiering & De Vlaming 1984) and part of the enzymes have been isolated. Generally the hydroxylation of the B-ring takes place very early in the biosynthesis, glycosylation, methylation and acylation later.

A number of genes have been found in *Petunia* that cause white (or pale yellow) flowers. Mutations in these genes block the biochemical pathway before anthocyanidin formation. An interesting gene is An 5, the gene for chalcone synthase, the enzyme that catalyzes the first step in the biosynthesis of the anthocyanidins and flavones. Some genotypes carrying the An 5 mutation are white flowering in winter (in the greenhouse) and lightly coloured in summer. In another genetic background, found in the cultivar Red Star, the flowers have alternating white and coloured sectors. Only the coloured sectors show chalcone synthase activity. The width of the coloured sectors and the intensity of the colour are dependent on temperature and light intensity. This is a very illustrative example of the way in which a major gene, the genetic background and environmental factors shape the developmental pattern of a flower. *Petunia hybrida* is a species hybrid and selection by breeders for aberrant flower colours and patterns has further increased the genetic variability. Nevertheless, flower colour variation can be found in natural populations of for instance *Silene dioica, Digitalis purpurea* and *Anagallis arvensis*. In natural populations there is generally one predominant form together with one or more much rarer deviant types. In garden populations with relaxed selection the deviant type can be much commoner (see Ernst 1987 for a discussion of selection for flower colour in *Digitalis purpurea*).

The effect of environmental factors on flavonoid production is well studied and we refer to Harborne (1980) for a review. Light, especially red and far red is a general requirement for the synthesis of phenolic compounds. Many factors that can produce stress conditions in plants stimulate the production of anthocyanins, especially in cells that normally have little or no anthocyanin.

3. The ecological significance of secondary plant substances

3.1. Introduction

The biochemistry of a plant is subject to selective pressure exercised by the biotic and abiotic environment just as any other genetically determined trait. Mutations, recombination events and possibly horizontal transfer of genes continually open and close new biochemical pathways. Whether a novel pathway will spread in the population is dependent on its effect on the array of genotypes in a particular environment, on the size of the population and on chance. Most of the innovations will rapidly go to extinction, a few will go to fixation and some will remain polymorphic in the popula-

tion for some time. Will the latter two categories contribute to the adaptedness (*sensu* Allard 1988) of their bearers? In large populations fixation of a new allele can only take place if that allele (or one very close to it) increases fitness. But once fixation is reached this condition does not apply any more. In other words, part of the biochemical phenotype may be the result of adaptation (Allard 1988) in the past. Polymorphic (*sensu* Ford 1964) populations give the best opportunity for the study of the ecological significance of a biochemical trait. Quantitative differences can also be used if the heritability of the variation is known. Polymorphic traits caused by allelic differences in one or a few genes pose a special problem: The effects found may be the result of pleiotropic effects of the gene under study or of genes linked to it (see Kakes 1989 for a discussion).

In studies of the effects on fitness of secondary plant compounds it is very important to take all stages of the life cycle into account. Furthermore, the environmental conditions during the experiments should be carefully monitored.

Whether secondary plant substances can contribute to the adaptedness of organisms is not a question anymore. Many examples of an established ecological function of secondary compounds belonging to different groups are to be found in the literature. {see Swain (1977) for a general review or Harborne (1976) and Jones (1972) for more specific review.} The ecological function of the majority of secondary plant substances has not been examined however, and many alleged functions rest on scanty evidence.

The possible functions of secondary plant substances may be roughly divided into four groups:
1. Functions related to the primary metabolism. Many secondary substances can be seen as by-passes of the primary metabolism. Primary metabolites with harmful effects in higher concentrations can be temporally stored in the form of less harmful derivates, to be reused or excreted.
2. Functions related to the abiotic environment. The accumulation of certain chemicals that accompanies adaptations to freezing, flooding or drought may be examples of this function, although the inclusion of polyhydric alcohols in

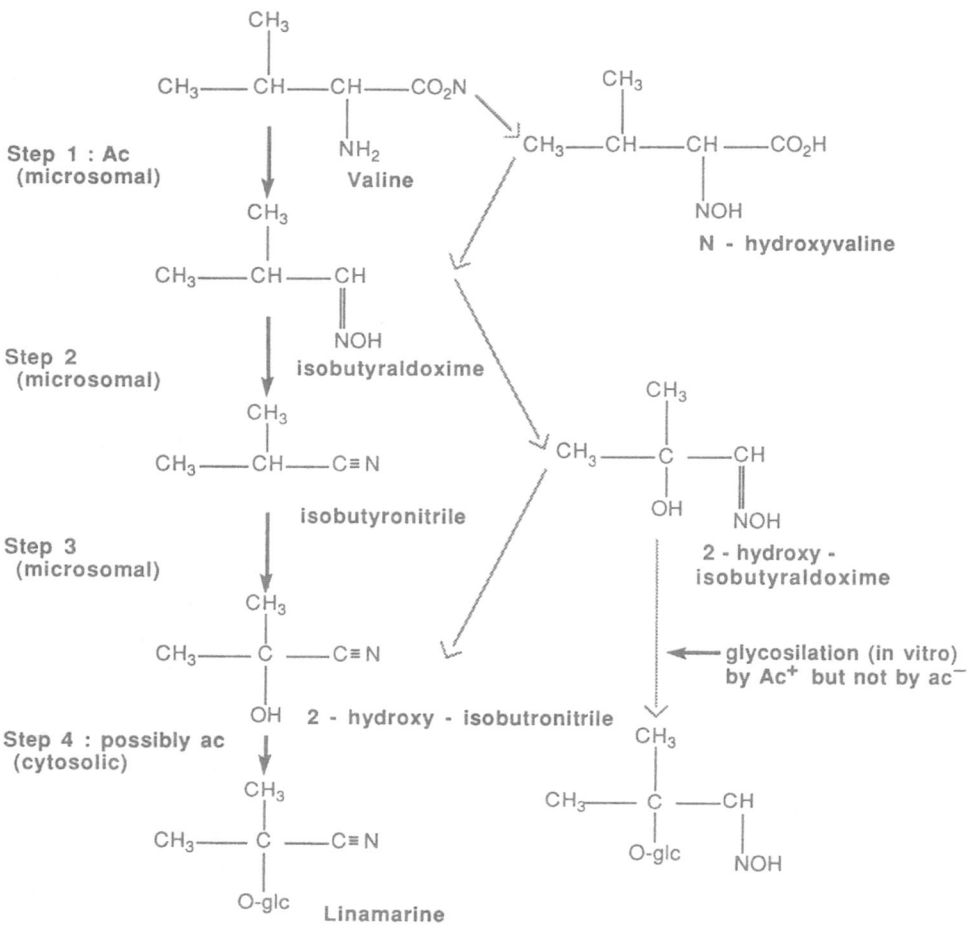

Fig. 8. The biosynthesis of linamarin, based on experiments on *Trifolium repens*. There is some evidence that the gene Ac affects both step 1 and step 4 of the pathway. If this is correct, it would be another argument for the complex nature of the Ac locus.

the secondary metabolites is somewhat doubtful as they occur in traces in the majority of higher plants. A number of secondary substances like anthocyanidins accumulate under general stress conditions, but it is not known if the response is adaptive.

3. Functions that regulate interactions with animals (pollinators, herbivores) or pathogens. The majority of established functions fall in this category and will be discussed later in this chapter.

4. Functions that regulate interactions with other higher plants. Allelopathy is the major if not only example. This topic will be discussed separately in the present volume (Kuiters).

It should be stressed that one secondary substance can have more than one function in an organism. It may even be possible that the presence of substance is slightly harmful in all stages of the life cycle except in one crucial stage.

The concept of different functions is also important for our understanding of the evolution of chemical defense. We will illustrate this concept by the biosynthesis of the cyanogenic glucosides linamarin and lotaustralin (Fig. 8). The precursors are valine and isoleucine, and all intermediates (aldoximes, nitriles, hydroxynitriles) are unstable and highly reactive compounds. The same intermediates are postulated in other pathways of nitrogenous substances. They are not found free in the

242

cell, presumably because their biosynthesis is channeled. The first mutation leading to cyanogenesis could be one that by blocking the conversion of a hydroxynitrile causes accumulation of this intermediate. Such a mutation would behave as a recessive lethal in a population, unless the organism would possess a glucosyltransferase active towards the hydroxynitrile. Such a glucosyltransferase (actually the last enzyme in the biochemical pathway) would produce a stable and harmless compound, accumulating in the vacuole (see Johnsson & Schram 1987, for a discussion of the relation between glucosilation and compartimentalization). The presence of cyanogenic glucosides alone confers some selective advantage (see later in this chapter). The novel glucosyltransferase would be the result of a mutation in a gene coding for one of the existing glycosyltransferases in the cell, causing the enzyme to recognise the hydroxynitrile as a substrate. The second mutation (in the glucosyltransferase gene) would change the effect of the first one (blocking the original pathway) from detrimental to slightly beneficial and thus would rapidly go to fixation in the population.

It should be stressed that there is no need to assume that the two mutations should occur in the same generation or in the same plant. Each one could be present in the population in a low frequency as a deleterious or even lethal recessive. Recombination would then produce the beneficial genotype referred to above.

3.2. Secondary plant products as protective agents

Introduction
The first one to study the protective function of plant chemicals in general was Ernst Stahl. In a long paper "Pflanzen und Schnecken" published in 1888, almost every aspect of the interactions of plants and herbivores is discussed, many of the interactions illustrated by simple but ingeneous experiments and astute field observations. After demonstrating that hungry slugs and snails often reject the leaves of particular plants, but eat them after extraction with water or alcohol, Stahl shows subsequently that the unpalatibility of the extracted leaves can be restored by spraying them with

crude extracts of the same plants. Stahl then continues:

"Die Pflanzenteile, welche aus den angegebenen Ursachen von den omnivoren Schnecken mehr oder weniger verschont bleiben, können wir als *chemisch geschützt,* die Substanzen, welche dies bewirken, als *chemische Schutzmittel* bezeichnen, ungeachtet der anderen Funktionen, welche diesen Substanzen außerdem noch im Haushalt der Pflanzen zukommen mögen."

Surprisingly, Stahls work had little follow up until 1960 when Jones started exploiting the polymorphism for cyanogenesis in *Trifolium repens* and *Lotus corniculatus* to study preferential eating. After 1960 much evidence has accumulated and several books and review papers have been published. We will discuss only a few examples to highlight the considerable progress that has been made as well as the pitfalls, unanswered questions and lack of quantitative understanding that challenge researchers in this area.

Chemically protection is not absolute
No matter how poisonous or unpalatable a plant may be for the majority of herbivores, it will be attacked, either by generalists eating low quantities in times of food shortage, or by specialists able to circumvent the barriers in one way or another. Generalist herbivorous species will often select those parts of the plants with the lowest concentration of protective compounds. Rabbits in winter eat the tuberous roots of *Senecio jacobea* that have a relatively low concentration of pyrrolizidine alkaloids (K. Vrielink, pers. comm.). On the other hand, specialists often use the general repellent substances as a cue to their host plants. A well known example form the glucosinolates in *Brassica* species that are an ovoposition cue for the butterfly *Pieris brassiceae*. Interestingly, *Cheiranthus × allionii* the Siberian wall flower that also contains glucosinolates, is avoided as host. Recently Rotschild *et al.* (1988) have shown that a cardenolide, strophantidin-glycoside acts as a contact deterrent in this case.

In cassave (*Manihot esculenta*) the generalist spider mite *Tetranychus urticae* selects the older leaves with relatively low content of cyanoglucosides, but a specialist: the cassave green mite (*Mononychellus tanajoa*) prefers the youngest leaves with high levels of cyanoglucosides (Bellotti & Van Schoonhoven 1978).

However, the palatibility of a plant is not only determined by secondary plant products. The food quality may be more important; if the quality is high, the amount consumed can be low and the absolute amount of toxic substances accordingly, permitting the animal to get rid of it. An example is given by Nayar & Frankel (1963). The byte response of the bean beetle *Epilachna varivestis,* a seed parasite on *Phaseolus lunatus* is dependent on the concentrations of linamarin and glucose in the food: linamarin inhibits the byte response, but glucose stimulates it.

Adaptations of herbivores

The adaptations present in specialist herbivores may be behavioural or biochemical. An example of a behavioural adaptation is the feeding of the winter moth (*Operophtera brumata*) on oak leaves: The caterpillars change from oak to other tree species when the concentration of condensed tannins raises in mid June. However, studies conducted over a couple of years using insectivorous birds show that the actual crop of caterpillars is mainly determined by food quality (Van Noordwijk pers. comm.).

The biochemical protection against cyanogenic glucosides can take place in several ways: The HCN evolved after hydrolysis may be bound to cysteine, producing asparagine. The enzyme responsible for the first step: β-cyanoalanine synthase has been found in a number of plants (Conn 1980) but not as yet in animals. The second possibility is a reaction of the CN^- ion with S to form the relatively harmless CNS^- ion. The sulfur donor may be anorganic sulfur (in salts like $NA_2S_2O_3$) or a sulfur containing amino acid. The enzyme involved, rhodanese (thiosulfate: cyanide sulphur-transferase), is present in the liver of warm blooded animals, but is also found in insects specialized on cyanogenic plants (Parsons & Rothschild, 1964).

Smit & Urbanska (1986) have demonstrated rhodanese activity in *Lotus corniculatus* and *L. alpinus*. Rhodanese activity in cyanogenic *L. alpinus* was much higher than in acyanogenic (Ac-lili and acacLi-) plants. If these data are correct, rhodanese activity in Lotus must be induced by the small amounts of HCN accidentally produced in the plant, as this is the only known effect of Ac and Li together.

The detoxification mechanism used by seed eating beetles specialized on L-canavanine containing seeds is particularly well studied. These beetles are able to convert L-canavanine to L-canaline and urea with arginase. A particularly active urease degrades urea to carbon dioxide and ammonia. It was thought (Rosenthal 1983) that the high activity of arginase with L. canavanine as a substrate is an adaptation peculiar to the canavanine eating seed predators, allowing them to degrade the L-canavanine quickly. In a recent paper Bleiler & Rosenthal (1988) showed that the ratio: maximum reaction velocity with arginine as substrate/maximum reaction velocity with L-canavanine as substrate is not different in canavanine feeding and non-canavanine feeding insects. There was a difference however, in urease activity. L-canavanine competes with arginine for the enzyme arginyl tRNA synthase. The canavanine containing proteins are structurally aberrant and show impaired function (Rosenthal 1986). The substitution error frequency (CEF) a measure for the incidence of aberrant proteins ranges from 1 : 75 to 1 : 500 in canavanine resistant insects. In an canavanine sensitive insect 1 : 6 was found. The authors conclude that the primary adaptation is the change in canavanine incorporation in proteins and that the degradation and re-use of canavanine is a general phenomenon not directly related to canavanin resistance.

Coevolution

The toxic chemicals present in the food plant may be sequestered in the herbivore, thereby making the herbivore itself unpalatable for its natural enemies. This often leads, by coevolution, to complex interactions between plants, herbivores and their predators, involving both Müllerian and Batesian mimicry. Surprisingly some species of lepidoptera that are cyanogenic and feed on cyanogenic plants are able to synthesize the cyanoglucosides themselves (Witthohn & Naumann 1987).

Alkaloids from the food plant taken up by the moth *Utetheisa ornatrix* may be used to protect the eggs of this species. Recently Dussourd *et al.* (1988) showed that a significant part of the alkaloids present in the eggs is actually contributed by the male. The alkaloids are concentrated in the male reproductive tract and transferred with the semen. The authors present preliminary evidence that the males advertize their alkaloid concentration by the amount of hydroxydanaidal (a courtship pheromone derived from alkaloids) they excrete. If the assumption is correct, investment in alkaloid in semen would raise the fitness of a male. On the other hand females would have the possibility of mating several times with "high alkaloid" males, which would have some advantage if the female herself has little alkaloid (owing to the marked variation in alkaloid content in the *Crotalaria* host plants).

The situation in *Utetheisa* has a remarkable parallel in the danaine butterflies. Species of this genus feed as caterpillars on Asclepiadaceae and use the cardenolides from these plants as a defense. The adult males, and to a lesser extent the females imbibe the fluid of damaged or senescent alkaloid bearing plant species. The male of *Danaus gilippus* derives a sex pheromone from the ingested alkaloids and also transmits alkaloids to the female, that on turn deposes alkaloid in the eggs.

4. Costs and benefits of chemical protection

4.1. Introduction

As we have seen there are many different ways, sometimes quite efficient, by which plants protect themselves to herbivores. Nevertheless, no plant escapes from herbivory completely, and the majority of species show heavily damaged populations every now and then. This has led to the conclusion that there are factors that limit the protective po-

tential of plants. The limiting factor is often referred to as cost, although there is no clear definition of the "cost of chemical defense". We will use the following definition: "The cost of a chemical defense system comprises every negative effect that the presence of a chemical defense system may have on the growth, survival and reproduction of the protected plants". The cost of chemical defense may have the following components:

1. Direct costs, i.e. the use of energy and nutrients to produce and transport the protective compounds and to maintain their level.
2. Indirect costs, i.e. damage to the plant or the cost of mechanisms to avoid damage, as well as damage to pollinators, natural enemies of the herbivores and symbionts.
3. Non-functional costs: The effects of alleles linked and/or in linkage disequilibrium with one or more of the alleles promoting the production of protective substances.

Direct en indirect costs can be looked upon as pleiotropic effects of the genes responsible for the protective mechanism. In that sense they are unavoidable, although natural selection will tend to minimize them. Non-functional costs can be avoided by rearrangement of the genome in the population, but this is a slow process. One would expect this type of cost in connection with defence systems that are recently acquired by a species.

4.2. Quantitative versus qualitative differences

Whatever the nature of the cost may be, one should be able to measure it as a difference in growth, survival or reproduction related to a difference in strength of the defense system. Polymorphic species offer the best opportunities for this kind on study but the expectation is that differences are small in these species. On the other hand, as polymorphic systems are often evolutionary young, the non-functional cost could be important. In nonpolymorphic species one would expect a negative correlation between the level of the defense system and growth, survival or reproduction. As a correlation does not establish a functional link, the nature of the dependence should be studied separately, for instance by manipulating experimentally the level of protective substances.

4.3. Experimental studies on the cost of chemical defense

Hanover (1966) has found that in *Pinus monticola* growth rate is negatively correlated with α-pinene and total terpene content. The variation in these traits is quantitative, with very high estimates of heritability. The heritability of growth rate is low, as expected. It would appear that terpene content is limited in *Pinus monticola* by the negative effect on growth rate. Van den Bergh & Matzinger (1970) have found a negative correlation between nicotin content and leaf production in 10 tobacco varieties. Coley (1986) has found a negative correlation between tannin content and leaf production in *Cecropia peltata*. In her experiments the genetics of tannin content were not studied and phenotypic variation was used for the regression analysis. 27% of the variation in leaf production could be accounted for by the phenotypic variation in tannin content. Berenbaum *et al.* (1986) have shown that in *Pastinaca sativa* the variation in resistance to the parasite *Depressaria pastinacella* was for 75% attributable to the variation in four furanocoumarins. The furanocoumarins show a negative genetic correlation with potential seed production. The authors argue that in this case cost and benefit of the protective system are equal and consequently an "evolutionary stalemate" has been reached between plant and parasite. The conclusion seems premature as not all the stages in the life cycle, but only growth and reproduction of adult plants were studied.

Cates (1975) studied growth and palatability to slugs and snails in *Asarum caudatum*. He divided the plants in two groups "palatable" and "unpalatable" on the basis of damage score in the field. Clones of the selected plants showed significant differences in the area of leaf discs eaten by the slug *Ariolimax columbianus*. There was also a difference when powder of dried leaves suspended in agar was offered to the slugs. The unpalatable morph had lower seed production, lower dry weight and later flowering time than the palatable morph. Cates concludes that in localities with many slugs, selection favors genotypes investing in protection, whereas in areas with low slug density selection acts in the opposite way. Unfortunately the nature of the protective substances is unknown and the heritability of the difference in palatability was not studied.

Simms & Rausher (1987) studied the cost of resistance of *Ipomoea purpurea* to the herbivore *Chaetocnema confinis*. They found no correlation between seed production and resistance. However the mechanism of resistance was not studied and the published data leave open the possibility that there was no heritable difference in resistance in the partial diallel studied.

Surprisingly, only a few studies on cost and benefit have been published on the species most suitable: *Lotus corniculatus* and *Trifolium repens*, both polymorphic for a system with proven value for the protection against herbivores. Jones (1970, 1978) studied root growth, salt tolerance and seed production in cyanogenic and acyanogenic plants. He found differences in some, but not in all populations (see Kakes (1990) for a review of the results).

In *Trifolium repens*, Foulds & Grime (1972) and Dirzo & Harper (1982) found that acyanogenic plants produce more flowers than cyanogenic plants. Kakes (1989) has studied the cost of defense in *Trifolium repens* in detail in a B2 segregating for Ac and Li. He found a pronounced difference in flower production associated with linamarin/lotaustralin. Plants containing these substances produce only half of the flowers and seeds compared to plants lacking cyanoglucosids. By determining the caloric value of the flowers he obtained a minimum estimate of the energy required to produce the "excess" of flowers of the acyanogenic plants. As the biochemical pathway of the cyanoglucosides is known as well as the cost to produce the sugar and aminoacids used as precursors, a reasonable precise estimate can be made of the cost of producing one mole of linamarin/lotaustralin. It turns out that the minimum estimate of the flower production cost is about 26 times the energy required to produce the cyanoglucosids present in the leaves of the plants. The discrepancy can be explained in several ways (See also the introduction to section 4 of this paper). Firstly the energy for the transport and maintenance of the cyanoglucosides may be much higher than the actual production costs. Very little is known about the cost of transport to the vacuole, but it is unlikely that transport alone is responsible for the discrepancy found. The maintenance cost mainly depends on the turnover of the cyanoglucosides. Again very little is known about that, but the 26 fold difference would mean that the cost of maintenance per day would be accounted for by the daily renewal of approximately 20% of the amount present at any moment. It is not very likely that a substance in the vacuole would have such a high turnover rate. Secondly, the indirect cost could be high. It is difficult to understand however, why only flower production, and not vegetative growth would be affected. The third possibility is that not the level of cyanoglucosides, but a gene or genes linked to Ac would cause the difference in flower production. It is very difficult to test which factor or combination of factors is responsible for the difference, but we should keep in mind the following:

1. Whatever the cause may be, the difference constitutes the cost of chemical defense.
2. The argument of the opponents of the cost-benefit theory that the amount of energy and nutrient locked in chemical defense substances is small compared to the total amounts used in the metabolism of the plant is not a generally applicable argument.

5. Conclusion

The field of the study of secondary plant substances is vast, spanning the distance from molecular biology to ecosystems. One cannot expect with the present state of knowledge a unified theory on the functions of secondary compounds. There can be no doubt however that they play an important role in the complex interactions between organisms on the level of individuals, populations and species. Consequently the study of the functions of secondary plant substances will in the future continue to contribute to our understanding of the web of life. The gains will not come easily however. The requirements for a relevant and efficient study in this field are not easily met in the prevailing scientific (and political) climate. The more important requirements and the barriers that prevent us to fulfil them are: The study of the functions of secondary substances requires a broad education, encompassing not only plant and animal biology on all levels of complexity, but also biochemistry and mathematics. However, the relatively short and strongly regulated curriculum of our universities hardly allows this kind of education, even for the most gifted and dedicated of our students. Research in plant-animal interactions requires a close cooperation between researchers specialized in these kingdoms. However, the growing specialization of individual scientists and the accompanying compartimentalization of the research organisations hinders the free and informal exchange of ideas. Moreover, emphasis on the planning of research, valuable in itself, may easily promote bureaucratic straitjackets, stifling projects that demand cooperation between disciplines. The study of functions of secondary compounds is essentially a study of populations. Although laboratory experiments are a necessary adjunct, the observation of natural populations with all their complex interaction is indispensable. Such studies are only valuable however if they are sustained for a long time. The properties of populations can only be understood as the result of selective and stochastic processes acting over many generations. At the same time the influence of short term but erratically occurring events can be enormous. Long term projects are not very popular with scientific planners and even researchers who are aware of the need for long term projects are apprehensive of the long delay of results in the form of publications.

Yet, if the conditions are far from optimal, there is no reason to be gloomy. The historical excursions in this paper and many other examples to be found in the scientific literature show that progress in biology is in a large part not dependent from optimal, or even moderate conditions. An open mind, perseverance and a keen naturalist's eye have enabled researchers in the past to progress with very limited material means and there is no reason that the present generation will fare less, in any field but certainly not in the exciting and promising study of the secondary plant compounds.

References

Allard, R. W., 1988. Genetic changes associated with the evolution of adaptedness in cultivated plants and their wild progenitors. J. Hered. 79: 225—238.

Belotti, A. C. & Schoonhoven, L. M., 1978. Mite and insect pests of cassava. Ann. Rev. Entomol. 23: 39—67.

Berenbaum, M. R., Zangerl, A. R. & Nitao, J. K., 1986. Constraints on chemical coevolution: Wild parsnips and the parsnip webworm. Evolution 40: 1215—1228.

Berg, P. v.d. & Matzinger, D. F., 1970. Genetic diversity and heterosis in Nicotiana III Crosses among tobacco introductions and flue cured varieties. Crop Sci. 10: 437—440.

Bleiler, J. A., Rosenthal, G. A. & Janzen, D. H., 1988. Biochemical ecology of canavanine-eating seed predators. Ecology 69: 427—434.

Bove, C., Conn, E. E., 1961. Metabolism of aromatic compounds in higher plants II Purification and properties of the oxynitrililase of Sorghum vulgare. J. Biol. Chem. 236: 207—236.

Bruyn, G. H. de, 1971. Étude du caractère cyogénique du manioc (Manihot esculenta Crantz). Thesis Landbouwhogeschool Wageningen.

Cates, R. G., 1986. The interface between slugs and wild ginger: Some evolutionary aspects. Ecology 56: 391—400.

Coley, P. D., 1986. Costs and benefits of defense by tannins in a neotropical tree. Oecologia 70: 238—241.

Collinge, D. B. & Hughes, M. A., 1982. In vitro characterisation of the Ac locus in white clover. Arch. Biochem. Biophys. 218: 38—45.

Conn, E. E., 1980. Cyanogenic compounds. Ann. Rev. Plant. Physiol. 31: 433—451.

Dawson, C. D. R., 1946. Tetrasomic inheritance in Lotus corniculatus. Journal of Heredity 42: 49—72.

Dawson, R. F., Christmann, D. R., Solt, M. L. & Wolf, A. P., 1960. The biosynthesis of nicotin from nicotinic acid. Chemical and radiochemical yields. Arch. Biochem. Biophys. 91: 144—150.

Dickenmann, R., 1982. Cyanogenesis in Ranunculus montanus s.l. from the Swiss Alps Ber. Geobot. Inst. E.T.H. 49: 56—75.

Dirzo, R., Harper, J. L., 1982. Experimental studies on slug-plant interactions IV The performance of cyanogenic and acyanogenic morphs of Trifolium repens in the field. Journ. Ecol. 70: 119—138.

Dussourd, D. E., Ubik, K., Harvis, C. Resch, J., Meinwald, J. & Eisner, T., 1988. Biparental endowment of eggs with acquired plant alkaloid in the moth Utetheisa ornatrix. Proc. Natl. Acad. Sci. USA 85: 5992—5996.

Ellis, W. M., Keymer, R. J. & Jones, D. A., 1977. The effect of temperature on the polymorphism of cyanogenesis in Lotus corniculatus. Heredity 38: 339—347.

Ernst, W. H. O., 1987. Scarcity of flower color polymorphism in field populations of Digitalis purpurea L. Flora 179: 231—239.

Ford, E. B., 1964. Ecological genetics London, New York.

Foulds, W. & Grime, J. P., 1972. The response of cyanogenic and acyanogenic phenotypes of Trifolium repens to soil moisture supply. Heredity 28: 181—187.

Hanover, J. W., 1966. Genetics of terpenes. I Gene control of monoterpene levels in Pinus monticola. Heredity 21: 73—84.

Harborne, J. R., Mabry, T. J. & Mabry, H., 1975. The flavonoids. Chapman and Hall, London.

Harborne, J. B., 1976. Functions of flavonoids in plants. In: T.W. Goodwin (ed.) Chemistry and biochemistry of plant pigments. 2nd ed. Vol. 1. Academic press London pp 736—776.

Harborne, J. B., 1980. Plant Phenolics. In: E.A. Bell & B.V. Charlwood (eds.) Encyclopedia of plant physiology new series vol. 8: Secondary plant products. Springer Berlin pp. 329—395.

Hughes, M. A. & Stirling, J. D., 1984. A study of dominance at the locus controlling cyanoglucoside production in Trifolium repens. Euphytica 3: 477—483.

Hughes, M. A., Stirling, J. D. & Collinge, D. B., 1982. The inheritance of cyanoglucoside content in Trifolium repens. Biochem. Genet. 22: 139—151.

Johnsson, L. M. V. & Schram, A. W., 1987. Regulation of flavonoid biosynthesis in higher plants, in particular Petunia hybrida. Plant Physiol. (Life Sci. Adv.) 6: 9—21.

Jones, D. A., 1970. On the polymorphism of cyanogenesis in Lotus corniculatus III Some aspects of selection. Heredity 25: 633—641.

Jones, D. A., 1978. Characteristics of cyanogenic and acyanogenic white clover plants. Soil and Crop Sci. Soc. of Fla. Proc. 38.

Jones, D. A., 1972. Cyanogenic glucosides and their function. In: Phytochemical ecology (ed.) J.B. Harborne Acad. Press.

Kakes, P. Properties and functions of the cyanogenic system in Angiosperms. Euphytica 48: 25—43.

Kakes, P., 1989. An analysis of the costs and benefits of the cyanogenic system in Trifolium repens. Theor. Appl. Genet. 77: 111—118.

Kossel, A., 1891. Ueber die chemische Zusammensetzung der Zelle. Archiv für Physiologie 14: 181—192.

Nahrstedt, A., 1985. Cyanogenic compounds as protecting agents for organisms. Pl. Syst. Ecol. 150: 35—47.

Nass, H. G., 1972. Cyanogenesis: Its inheritance in Sorghum bicolor, Sorghum sudanense, Lotus and Trifolium repens — A Review. Crop Sci. 12: 503—506.

Nayar, A. & Fraenkel, G. L., 1963. The chemical basis of the host selection of the mexican bean beetle Epilachna varivestis. Ann. Ent. Soc. Am. 56: 174—178.

Parsons, J. & Rothschild, M., 1964. Rhodanese in the larva and pupa of the common blue butterfly (*Polyommatus icarus* (Rott.)), Lepidoptera. Ent. Gaz. 15: 58—59.

Robiquet & Boutran-Charlard. 1830. Ann. de Chim. 44: 352.

Rosenthal, G. A., 1983. The adaptation of a beetle to a poisonous plant. Scientific American 249: 164—171.

Rosenthal, G. A., 1986. Biochemical insight into insecticidal properties of L-canavanine, a higher plant allelochemical. J. Chem. Ecol. 12: 1145—1156.

Rothschild, M., Alborn, H., Stenhagen, G. & Schoonhoven, L. M., 1988. A strophantidin glycoside in siberian wallflower: A contact deterrent for the large white butterfly. Phytochemistry 27: 101—108.

Simms, E.L., Rausher, M. D., 1987. Costs and benefits of plant resistance to herbivory. Am. Nat. 130: 570—581.

Smit, J. D. G., Urbanska, K. M., 1986. Rhodanese activity in *Lotus corniculatus* s.l. J. Nat. Hist. 20: 1467—1476.

Stahl, E., 1888. Pflanzen und Schnecken. Jena Z. Naturwiss. 22: 557—684.

Swain, T., 1977. Secondary compounds as protective agents. Ann. Rev. Plant Physiol. 28: 499—501.

Till, J., 1987. Variability of expression in white clover (*Trifolium repens*). Heredity 59: 265—271.

Tjon Sie Fat, L. A., 1979. Contribution to the knowledge of cyanogenesis in Ranunculaceae. Proc. Koninkl. Nederl. Acad. Wetensch. 82C: 197—209.

Tollstein, L. & Bergstrom, G., 1988. Headspace volatiles of whole plants and macerated plant parts of *Brassica* and *Sinapis*. Phytochemistry 27: 2073—2077.

Underhill, E. W., 1980. Glucosinolates. In: E.A. Bell & B.V. Charlwood (eds.) Encyclopedia of plant physiology new series vol. 8: Secondary plant products. Springer Berlin Heidelberg, New York. pp. 329—395.

Urbanska, K., 1982. Polymorphism of cyanogenesis in *Lotus alpinus* from Switzerland. Berichte des Geobotanischen Institutes der Eidg. Techn. Hochschule. Stiftung Rubel 49: 35—56.

Waal, R. de, 1942. Het cyanophore karakter van witte klaver, *Trifolium repens* L. Thesis Landbouwhogeschool, Wageningen, The Netherlands.

Wiering, H. & Vlaming, P. de, 1984. In: K.C. Sink (ed.) Monographs on Theoretical and applied genetics 9. Springer-verlag Berlin Heidelberg, New York. pp. 49—67.

Witthohn, K. & Naumann, C. M., 1987. Cyanogenesis, a general phenomenon in the lepidoptera? Journal of Chemical Ecology 13: 1789—1809.

Author's Address
P. Kakes
Department of Ecology and Ecotoxicology
Vrije Universiteit
De Boelelaan 1087
1081 HV Amsterdam
The Netherlands

Forest floor vegetations are affected by phenolic substances released from decomposing leaf litter from canopy species (Photograph: A. T. Kuiters.)

CHAPTER 22

Phenolic substances in forest leaf litter and their impact on plant growth in forest vegetations

A. T. KUITERS

Abstract. The role of phenolic secondary substances released from decaying plant litter and entering the forest soil system is discussed. Low molecular-weight phenolic acids have been found to leach in high amounts from various types of forest leaf litter. Different leaching patterns were established between litter from coniferous trees and litter from deciduous trees. An overview is given about the physiological impact of phenolic acids on plant development. By affecting ectomycorrhizal fungi, phenolics may influence the establishment and growth of tree seedlings. Polymeric phenols in forest leaf litter, i.e., hydrolyzable and condensed tannins, have a large influence on nutrient cycling in forests and thereby on the growth conditions of plants. They play a decisive role in the development of the humus profile. In this respect, two distinctive forms of humus formation, i.e., mor and mull humus type, are discussed. The allelopathic potential of species by the action of low molecular-weight phenolic acids, may be influenced by the humus form prevailing under the vegetation. Finally, the implications of monomeric and polymeric phenols for vegetation processes in forests are discussed.

1. Introduction

Secondary plant substances occur in high amounts in plant tissue and chemical very diverse groups are involved, ranging from simple low molecular-weight substances to complicated polymeric macromolecules. They have no direct role in primary metabolism but are essential constituents of plant defense systems against herbivory or pathogens. This is discussed in another chapter (P. Kakes).

During senescence and the subsequent decomposition of dead plant material, secondary metabolites are released from plant tissue and enter the soil system. Outside the plant they may have large effects on the functioning of other plants and micro-organisms. There is a substantial amount of literature providing evidence that secondary metabolites play an essential role in altering plant growth conditions (see review Rice 1984). In this respect phenolic substances are one of the most important groups. Phenolic compounds occur in plant material as free compounds in vacuoles, mostly as glycosides, or esterified to cell wall components. The phenolic substances in plant tissue represent a chemically very heterogenous group of compounds. Two main groups are distinguished: a) monomeric phenols: derivatives of benzoic and cinnamic acids; b) polymeric phenols: e.g., hydrolyzable and condensed tannins. In this chapter, the role of monomeric and polymeric phenolic substances in forest vegetations from the temperate zone will be discussed in more detail.

2. Low molecular-weight phenolic acids and the growth of forest-floor species

2.1. Release of phenolic acids from forest leaf litter

The top litter-layer of forest soils is an important source of phenolic substances. Freshly fallen leaves are rich in water-soluble phenolics, that are released by leaching during the early phases of decomposition. Moreover, large amounts are washed

J. Rozema and J. A. C. Verkleij (eds.), Ecological Responses to Environmental Stresses, 252—260.
© *1991 Kluwer Academic Publishers.*

from the canopy by rainwater. Figure 1 shows the amounts of water-soluble phenolic substances in leaf and needle litter from several common deciduous and coniferous tree species at three dates during the first period after litter fall. The amounts of water-soluble phenols differs substantially between tree species. Freshly fallen litter from deciduous tree species contains considerable amounts of water-soluble phenols, that are rapidly leached during the first period after leaf fall (from October to February), whereas evergreen tree species as Scotch pine, Norway spruce and Douglas fir have only low amounts of water-soluble phenols in their needles. The water-impermeable layer around freshly fallen needles prevents phenols to be leached.

The spectrum of phenolic acids released by leaching from leaf litter is more or less similar for different tree species (Kuiters & Sarink 1986). There is a group of commonly occurring phenolics (see Fig. 2), including both benzoic acid derivatives (e.g., gallic, p-hydroxybenzoic, vanillic, gentisic, syringic acid) and cinnamic acid derivatives (e.g., ferulic, caffeic, p-coumaric, o-coumaric acid). However, the concentrations of the released compounds differ largely between litter types.

Phenolics may also be released from litter material by the breakdown of lignin polymers. This degradation process occurs rather slowly and the organisms involved in lignin biodegradation are mainly white-rot fungi producing ligninases (Harvey et al. 1987).

2.2. Adsorption of phenolic substances by the solid soil phase

The phenolic substances, released from the top-litter layer, are transported to the underlaying hu-

mus layer of forest soils with the percolation water. In the soil environment they are subjected to various processes that reduce their concentration in the soil solution, including metabolization by micro-organisms, sorption by the solid soil phase and polymerization reactions.

Phenols may be adsorbed to mineral soil particles, e.g., iron- or aluminium oxides. Number and position of the hydroxyl-substitutions on the benzene ring have a large impact on this process. Ortho-substituted compounds (e.g., protocatechuic acid, catechol) are more strongly adsorbed than other compounds. Shindo & Kuwatsuka (1976) observed that the adsorption of phenolic acids is mainly determined by the organic matter content of the soil. The pH has within the range found in most forest soils (3.0–6.5) only a small effect on these sorption phenomena. Adsorption to the organic soil layers results into a decrease of phenol concentrations with increasing depth.

Another important phenomenon is the property of phenolic acids to form complexes with metal-ions, although this process becomes only significant beyond pH 5.0. The soluble phenolic substances may increase metal-concentrations (e.g., Fe, Al) in the equilibrium soil solution by complexation. Consequently, metals become more available for uptake by plant roots and behave more mobile through the soil profile. They are involved in podzolization processes of soils (Bruckert 1970).

2.3. Biological metabolization

Microbial breakdown may also decrease concentrations of phenols in the soil solution. Especially the lower molecular-weight phenolic acids are easily metabolized by soil micro-organisms that may use them as energy source (Haider & Martin 1975). This occurs especially in soils with high microbial activity, i.e., mull humus (Bruckert 1970). As soil microbiological activity is seasonally variable, the concentration of phenolic compounds in the soil solution also shows a distinct pattern (Jalal & Read 1983; Kuiters 1987).

Phenolic compounds are also involved in polymerization reactions. Extracellular polyphenol-oxidases catalyze these condensation and polymer-

Table I. Mechanisms of action of phenolic acids on the development of higher plants (after Rice 1984).

cell extension	chlorophyll synthesis
cell division	photosynthesis
membrane permeability	respiration
nutrient uptake	enzyme activity
protein synthesis	

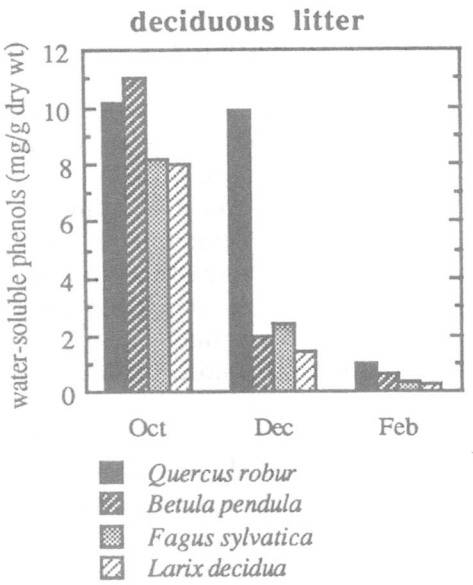
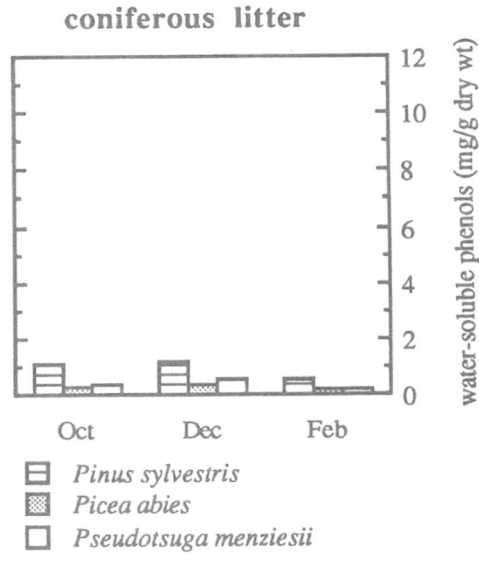

deciduous litter   coniferous litter

■ Quercus robur
▨ Betula pendula
▦ Fagus sylvatica
▧ Larix decidua

⊟ Pinus sylvestris
▦ Picea abies
☐ Pseudotsuga menziesii

Fig. 1. Water-soluble phenols in leaf and needle litter from several deciduous and coniferous tree species, determined at three times during the first phase of decomposition (Kuiters & Sarink 1986).

ization reactions (Filip & Preusse 1985). With mono- and polysaccharides phenolic compounds are polymerized to humus polymers, i.e., fulvic acids and finally to the more condensed humic acids and humins (Schnitzer & Khan 1978). The actual phenol concentrations in the soil solution is at any moment determined by processes of release (leaching from litter and decomposition of plant constituents, especially lignins) and processes that decrease phenol concentrations, i.e., adsorption, microbial breakdown or polymerization reactions.

From the previous paragraphs it will be clear that the actual concentration of phenolic acids in the soil solution is strongly dependent of soil conditions (humus content, humus type, clay content, pH), soil depth, season and vegetation composition. Rice (1984) reviewed the literature data at this point and stated that the concentration of low molecular-weight phenolic acids in forest soil solutions seldom exceeds a value of 0.1 mM.

2.4. Physiological responses of higher plants to phenolic acids

Effects of phenolic substances on higher plants, especially of low-molecular weight phenolic acids are well documented in literature (see for review, Rice 1984). Most investigations refer to weed and crop species but data for forest-floor species are also available. Table 1 summarizes current knowledge of the mechanisms by which phenolic acids may affect the development of higher plants. The proposed mechanisms include most major plant processes, including mineral nutrition, growth regulation, photosynthesis and water balance. Effects of phenolic acids occur throughout the life cycle of a plant, although radicle growth appears to be more sensitive than germination (Von Lioba & Schütt 1987; Kuiters 1989). Root development in the early seedling stage is highly influenced by phenolics. The physiological response of plants to phenolic acids is mostly not linear over a concentration range. Many phenolics are inhibitory at millimolar concentrations, but stimulate at micromolar concentrations, as illustrated in Fig. 3.

The mineral nutrition of plants exposed to phenolic acids may be affected by changes in root membrane permeability, due to their lipophylic properties, resulting in an efflux of ions such as phosphate or potassium (Glass 1973). Experiments with herbaceous woodland plants (Kuiters 1987) revealed that nutrient uptake can be inhibited or enhanced, depending on phenolic acid concentration and test species (Fig. 4). The translocation of nutrients within the plant may be changed as well. In general, grass species are less inhibited by phenolics than dicotyledonous species (Kuiters 1987).

There is an extensive amount of literature, reporting upon bioassays with extracts of living or dead plant material (Rice 1984). Bioassays are a useful tool for the determination of the potentially phytotoxic effects of plant extracts and for testing the sensitivity of certain plant species for specific compounds. However, the relevance of most bioassays as reported in literature is questionable from an ecological point of view. Methods of extraction are often not adequate (grinding of plant material, strong extractants), the concentrations used are much too high or the test species are not relevant. Apart from that, bioassay experiments with plant leachates have only a restrictive relevance as the effects of leachates or compounds may be totally different when tested under natural soil conditions.

Plant growth experiments on soils after addition of intact leaf litter material are more realistic. The soluble organics are gradually released and detoxification mechanisms in the soil, such as microbial breakdown or adsorption, are allowed to occur. Experiments of this type, with leaf litter of several deciduous and coniferous tree species revealed that freshly fallen litter (November) may severely inhibit plant growth, as illustrated in Fig. 5. From the chemical analyses of the test plants it was concluded that addition of fresh litter material resulted into a temporary net immobilization of nitrogen. This effect of nitrogen depletion may initially mask effects of phenolic substances, and occurs especially where litter material of high C/N ratios is applied. When the experiments are carried out with leaf litter material, picked from the forest soil at the start of the vegetation period (April) after leaching has proceeded, both inhibitive and stimulative effects may be observed, dependent on the type of forest litter and the plant species tested (Fig. 5).

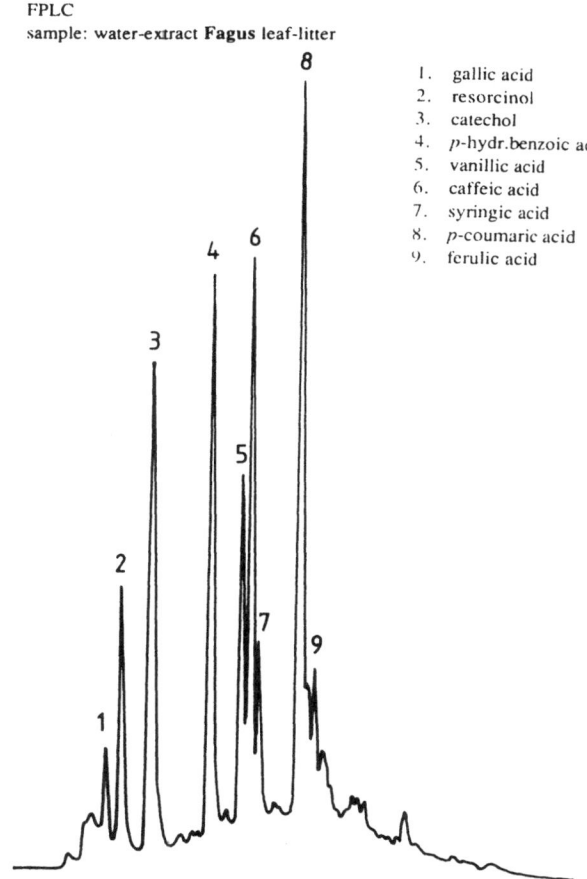

FPLC
sample: water-extract **Fagus** leaf-litter

1. gallic acid
2. resorcinol
3. catechol
4. p-hydr.benzoic acid
5. vanillic acid
6. caffeic acid
7. syringic acid
8. p-coumaric acid
9. ferulic acid

Fig. 2. Phenolic acids in an aqueous extract of freshly fallen leaves of *Fagus sylvatica.* (reversed phase chromatography; Kuiters 1987)

2.5. Effects on ectomycorrhizal fungi and litter-decomposers

There is an increasing amount of evidence that organic substances originating from forest floor litter also affect the activity of ectomycorrhizal fungi and this may have consequences for the establishment of tree seedlings. Rose *et al.* (1983) observed both positive and negative effects of litter leachates of several shrub and tree species on mycorrhizal fungi, depending on litter type, leachate concentration and fungal species. Olsen *et al.* (1971) found that aqueous extracts of aspen leaves (*Populus tremula*) strongly inhibited the growth of several mycorrhizal fungi of the genera *Boletus,* whereas

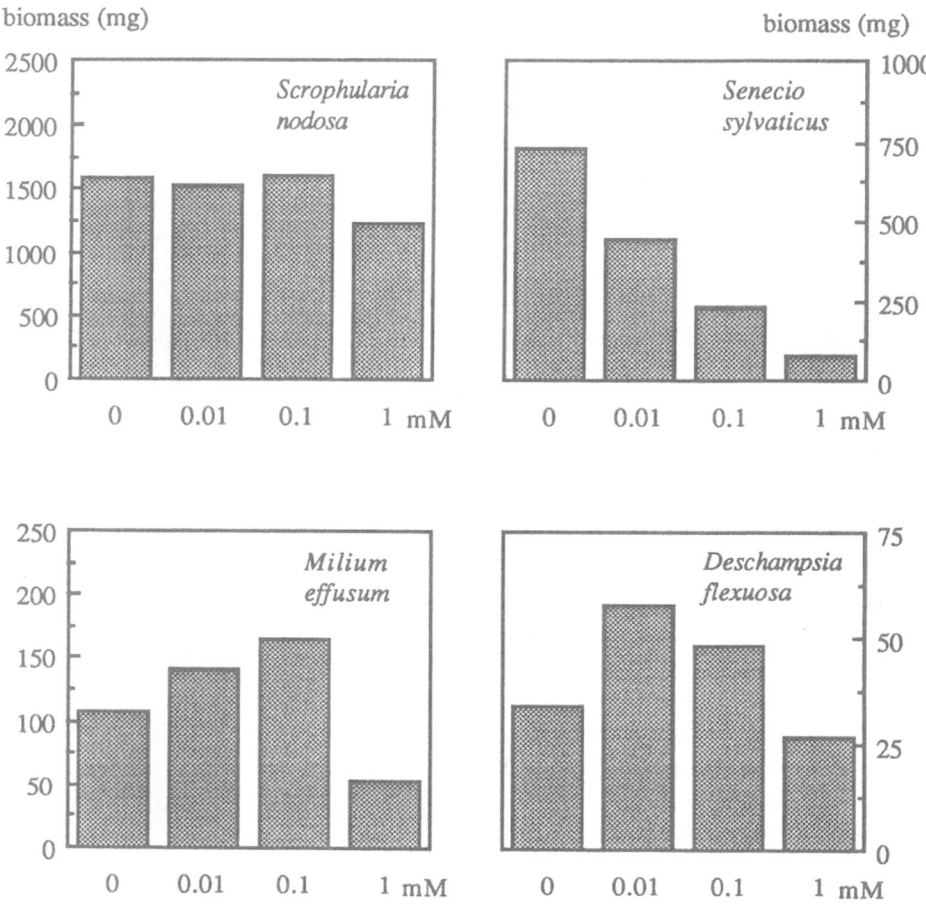

Fig. 3. Biomass of four herbaceous forest plant species grown during four weeks in a nutrient solution with a mixture of phenolic acids added in increasing concentrations (Kuiters 1987).

litter-decomposers of the genera *Marasmius* were stimulated. This difference between two types of fungi, i.e., ectomycorrhizas and litter-decomposers, might partly be explained by the fact that the latter species produce extracellular phenol-oxidases (Olsen *et al.* 1971). Amino acids and carbohydrates released from decaying plant litter may stimulate growth of mycorrhizal fungi, but there are strong indications that phenolic substances have negative effects above a certain threshold concentration.

It is well known that most litter-decomposers are found relatively specific on certain kinds of litter. *Collybia dryophila* is found on broad-leaved litter, whereas *Marasmius androsaceus* prefers coniferous needles (Hering 1982). Little information is available about the biochemical background of these relations, but future research may reveal that food preference of litter-decomposers is closely related with secondary plant chemistry.

3. Polymeric phenols and the formation of mull and mor humus

3.1. Polyphenols and humus formation

Forest tree species may have high polyphenol contents in their leaves (Bate-Smith 1962). These polymers, with hydrolyzable and condensed tannins

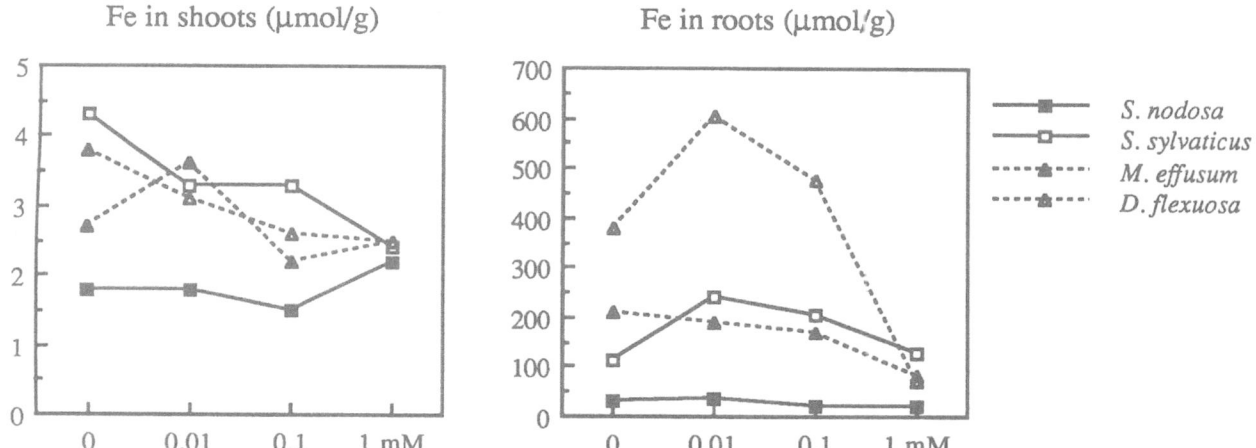

Fig. 4. Fe concentration in shoots and roots of plants grown during four weeks in a nutrient solution with a mixture of phenolic acids in increasing concentrations (Kuiters 1987).

(proanthocyanidins) as most important representatives, are generally produced in higher amounts than the monomeric phenolic compounds. They are part of the quantitative defense system of forest trees (Haslam 1988). In contrast to phenolic acids, they are not easily leached from litter. During senescence of plant tissue hydrolyzable tannins, which are polymers of gallic acid and sugar moieties, form insoluble complexes with cell proteins. The condensed tannins occur presumably mainly as insoluble compounds complexed with polysaccharides in the cell wall structures and have especially a structural role (Haslam 1988).

The breakdown of litter and the subsequent process of humification may follow two distinctive patterns of decay, leading to different kinds of humus profile development, indicated by mull and mor humus profiles (as two extremes of a continuum). The quality of the litter material is often a decisive factor in humus profile development (Duchaufour 1977). If polyphenol-protein complexes occur in high amounts, it renders litter material highly unpalatable for soil faunal species. These complexes are suggested to have a deterrent effect on detritus-feeding soil animals. Many fungal litter decomposers can tolerate these conditions and are able to decompose these complexes by the production of polyphenol oxidases. In typical mor soils, a primary phase of microbial decay precedes litter de-

composition by the soil fauna (Hering 1982). Mor humus is formed when litter material is not mixed through the mineral soil and distinct organic layers are build up. Boreal forest soils are often typical examples of mor humus profiles.

Mull humus is formed when the litter material is continuously mixed through the mineral soil by the activity of soil animals. At mull sites, fresh litter has much lower concentrations of phenolic substances and is more palatable for the soil fauna, especially earthworms. The microbial decay comes here in a later stage of decomposition, when litter material is removed form the soil surface and has become largely a mass of worm casts, mixed with the inorganic soil particles. Brown forest soils, as typical examples of mull humus profiles, are maintained in a constant state of horizon mixing.

Dwarf shrub species of the Ericaceae are well known for their mor humus formation. Other examples are species of the genera *Vaccinium, Rhododendron, Sarothamnus* and *Pteridium*. Coniferous species as *Picea* and *Pinus* form only at nutrient-poor soils or under cool climatic conditions (boreal forests) mor humus profiles. Beech (*Fagus*) forms mor humus at nutrient-poor and dry soil conditions. Most deciduous tree species form mull humus profiles (Miles 1985). Brown forest soils are the result. Litter from plants grown on acidic, nutrient-poor sites is generally more rich in polyphe-

257

litter collected in November

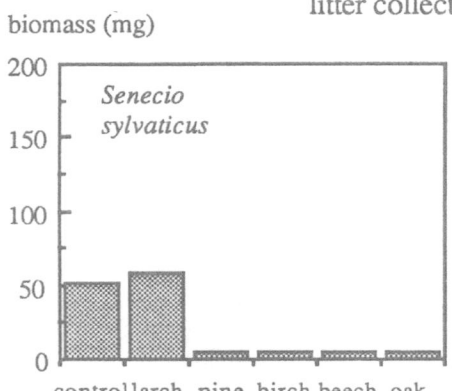
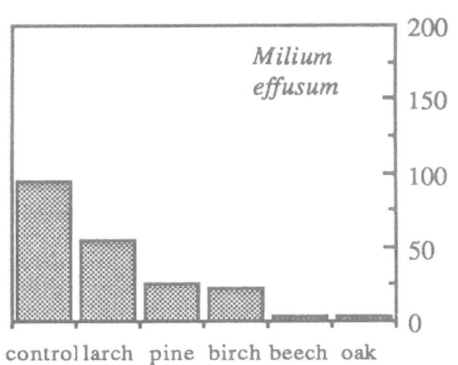

litter collected in April

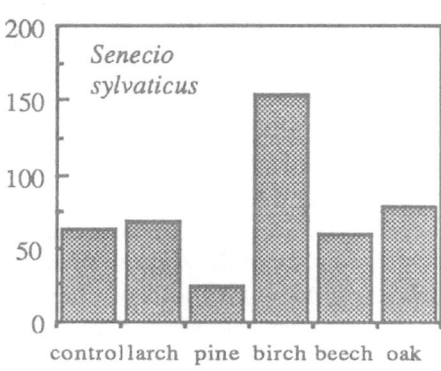
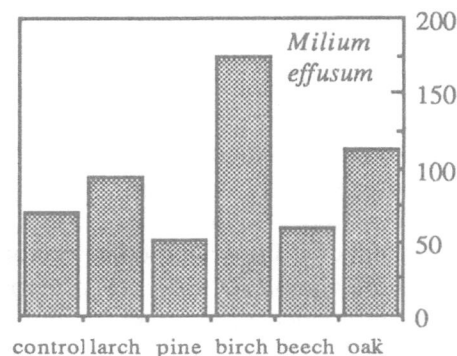

Fig. 5. Biomass of two herbaceous forest plant species grown during 12 weeks on a soil mixed with fragmented leaf litter of different tree species. Freshly fallen litter collected in November (upper part of the figure); litter collected in April (lower part). The control treatments had not received any litter material (Kuiters *et al.* 1986).

nols. As mineralization rates at mull sites proceed at a much higher level, the availability of nutrients is here much higher than at mor sites. The polyphenol content of litter is thus an important factor regulating nitrogen cycling. Humification is one of the processes where the effects of phenolic compounds on the growth conditions of plants is most pronounced.

3.2. *The nitrification hypothesis*

Rice & Pancholy (1972) hypothesized that phenols in plant litter have also a profound effect on nitrifi-

cation rates. Bioassay experiments revealed that the activity of micro-organisms involved in nitrogen mineralization, i.e., *Nitrosomonas* and *Nitrobacter,* is negatively influenced with increasing concentrations of phenolic substances. They suggested that during the maturing of ecosystems plants of later succession stages have generally higher contents of polymeric phenols (e.g., tannins) in their tissue. Inhibition by these substances of nitrification results in relatively higher amounts of ammonium nitrogen and lower amounts of nitrate nitrogen in these soils. As ammonium is less easily leached from the soil this finally results into a conservation of nitrogen in these mature forest ecosystems. This topic is still controversial in literature.

4. Implications for vegetation processes

4.1. Nutrient cycling in forests

Phenolic secondary substances in plant litter have an important impact on the processes of humification in forest systems. As described in paragraph 3.1, the content of phenolics in plant litter, especially of polyphenols (hydrolyzable and condensed tannins) largely influences the humus type that is formed (mull, moder or mor). The high content of polyphenol-protein complexes in mor-forming plant litter seriously delays decomposition processes (Horner *et al.* 1988). This has large implications on rates of element cycling in forests. At mor-humus sites mineralization rates are much lower than at mull-humus sites and this may seriously inhibit plant productivity.

Little is known about the content of secondary metabolites in plant roots. As root litter production in forest systems may be as high as above ground litter production (Swift *et al.* 1979), secondary metabolites in root litter may also have a large effect on decomposition processes, an important area of future research as pointed out by Horner *et al.* (1988).

4.2. Allelopathic potential of plant species

The type of humus profile prevailing under a certain forest vegetation may also have a large influence on the allelopathic potential of plant species. Under mor humus conditions, monomeric phenolic acids, with potentially phytotoxic properties are only slowly metabolized, as microbial activity in these soils is relatively low (Bruckert 1970; Swift *et al.* 1979). The influence of phenolic acids on plant growth will occur especially at mor humus profiles. This may explain why species typical for mor sites, such as *Pinus sylvestris*, *Picea abies*, *Calluna vulgaris*, *Deschampsia flexuosa*, *Vaccinium myrtillus*, *Pteridium aquilinum* and *Rubus* spp. are often found to have allelopathic properties.

The allelopathic potential of (tree) species may also be influenced by vitality. In this respect the findings of Von Lioba & Schütt (1987) are striking. They found that the foliage of diseased beech trees released more substances with phytotoxic activity towards forest floor species than healthy beeches. In view of the proceeding forest decline in western and central Europe this may have a large impact on forest floor vegetations and needs further attention in future research.

4.3. Regeneration failure in forest stands

In forestry, problems with natural regeneration or re-afforestation of stands after clear-cutting have often been reported. Becker & Drapier (1984) have extensively reported about the problems of natural regeneration in stands of *Abies alba* Mill. in forest stands of central Europe. Among the factors preventing successful establishment of tree seedlings, phenolic substances have often been mentioned (Fisher 1980; Rice 1984). As described in paragraph 2.5, these problems may be attributed to the effects of phenolic substances on ectomycorrhizal colonization. Tree seedlings are often dependent on the activity of ectomycorrhizal fungi for the uptake of nutrients. When successful colonization of roots by mycorrhizal fungi is prevented, the establishment of tree seedlings will be hindered. Direct allelopathic effects of leachates from tree litter (Becker & Drapier 1984) or from herbaceous plant litter (Schütt *et al.* 1975) on tree seedlings have been observed as well. In this respect, the interference among plants (and fungal species) by phenolic substances released from decaying plant litter may have effects on vegetation succession in forest systems.

In summary, it can be concluded that there is a growing amount of literature, providing evidence that the ecological significance of secondary plant substances, e.g., phenolic substances, is even manifest after their release from living or dead plant parts. They have a profound effect on humification processes and subsequently on nutrient cycling in forest systems. The lower molecular weight compounds are often rapidly released from decaying plant litter and enter the soil environment, where they are involved in the complex biochemical inter-actions between herbaceous and woody plant species and the fungal community in forest soils.

References

Bate-Smith, E. C., 1962. The phenolic constituents of plants and their taxonomic significance. J. Linn. Soc. (Bot.) 58: 95—173.

Becker, M. & Drapier, J., 1984. Rôle de l'allélopathie dans les difficultés de régénération du sapin (*Abies alba* Mill.). I. Propriétés phytotoxiques des hydrosolubles d'aiguilles de sapin. Acta Œcologica Œcol. Plant. 5: 347—356.

Bruckert, S., 1970. Influence des composés organiques solubles sur la pédogenèse en milieu acide. Ann. Agron. 21: 725—757.

Duchaufour, P., 1977. Pédologie. 1. Pédogenèse et classification. Masson, Paris.

Filip, Z. & Preusse, T., 1985. Phenoloxidierende Enzyme — ihre Eigenschaften und Wirkungen im Boden. Pedobiologia 28: 133—142.

Fisher, R. F., 1980. Allelopathy: A potential cause of regeneration failure. J. Forestry 6: 346—350.

Glass, A. D. M., 1973. Influence of phenolic acids on ion uptake. I. Inhibition of phosphate uptake. Plant Physiol. 51: 1037—1041.

Haider, K. & Martin, J. P., 1975. Decomposition of specifically ^{14}C labelled benzoic and cinnamic acid derivatives in soil. Soil Sci. Soc. Am. Proc. 39: 657—662.

Harvey, P. J., Schoemaker, H. E. & Palmer, J. M., 1987. Lignin degradation by white rot fungi. Plant Cell Environ. 10: 709—714.

Haslam, E., 1988. Plant polyphenols (syn. vegetable tannins) and chemical defense — a reappraisal. J. Chem. Ecol. 14: 1789—1805.

Hering, T. F., 1982. Decomposing activity of basidiomycetes in forest litter. In: J.C. Frankland, J.N. Hedger & M.J. Swift (eds.) Decomposer basidiomycetes. Their Biology and Ecology. Cambridge University Press, Cambridge. pp. 213—225.

Horner, J. D., Gosz, J. R. & Cates, R. G., 1988. The role of carbon-based plant secondary metabolites in decomposition in terrestrial ecosystems. Am. Nat. 132: 869—883.

Jalal, M. A. & Read, D. J., 1983. The organic acid composition of *Calluna* heathland soil with special reference to phyto- and fungitoxicity. II. Monthly quantitative determination of the organic acid content of *Calluna* and spruce dominated soils. Plant Soil 70: 273—286.

Kuiters, A. T., 1987. Phenolic acids and plant growth in forest ecosystems. Ph.D. Thesis. Free University, Amsterdam.

Kuiters, A. T., 1989. Effects of phenolic acids on germination and early growth of herbaceous woodland plants. J. Chem. Ecol. 15: 467—479.

Kuiters, A. T. & Sarink, H. M., 1986. Leaching of phenolic compounds from leaves and needles of several deciduous and coniferous trees. Soil Biol. Biochem. 18: 475—480.

Kuiters, A. T., Van Beckhoven, K. & Ernst, W. H. O., 1986. Chemical influences of tree litters on herbaceous vegetation. In: J. Fanta (ed.) Forest dynamics research in Western and Central Europe. Pudoc, Wageningen, pp. 103—111.

Miles, J., 1985. The pedogenetic effects of different species and vegetation types and the implications of succession. J. Soil Sci. 36: 571—584.

Olsen, R. A., Odham, G. & Lindeberg, G., 1971. Aromatic substances in leaves of *Populus tremula* as inhibitors of mycorrhizal fungi. Physiol. Plant. 25: 122—129.

Rice, E. L. & Pancholy, S. K., 1972. Inhibition of nitrification by climax ecosystems. Am. J. Bot. 59: 1033—1040.

Rice, E. L., 1984. Allelopathy. Second Edition. Academic Press, New York.

Rose, S. L., Perry, D. A., Pilz, D. & Schoeneberger, M. M., 1983. Allelopathic effects of litter on the growth and colonization of mycorrhizal fungi. J. Chem. Ecol. 9: 1153—1162.

Schindo, H. & Kuwatsuka, S., 1976. Behavior of phenolic substances in the decaying process of plants. IV. Adsorption and movement of phenolic acids in soils. Soil Sci. Plant Nutrition 22: 23—33.

Schnitzer, M. & Khan, S. U., 1978. Soil organic matter. Elsevier, Amsterdam.

Schütt, P., Schuck, H. J., von Sydow, A. & Hatzelmann, H., 1975. Zur allelopathischen Wirkung von Forstunkräutern. Forstw. Club 94: 43—53.

Swift, M. J., Heal, O. W. & Anderson, J. M., 1979. Decomposition in Terrestrial Ecosystems. Blackwell, Oxford.

Von Lioba, P. & Schütt, P., 1987. Auswaschung phytotoxischer Substanzen aus Blättern kranker und gesunder Buchen — Schaden an der Bodenflora. Eur. J. For. Path. 17: 356—362.

Author's Address
A. T. Kuiters
Department of Ecology & Ecotoxicology
Biological Laboratories
Free University
P.O. Box 7161
1007 MC Amsterdam
The Netherlands

Ecological field course in the tropical savanna ecosystem of Botswana. (Photograph: J. Schroten.)

CHAPTER 23

Plant responses to human activities in the tropical savanna ecosystem of Botswana

T. TIETEMA, D. J. TOLSMA, E. M. VEENENDAAL & J. SCHROTEN

Abstract. In this chapter, views are expressed regarding the impact of human activities, such as grazing of livestock and wood harvesting on the savanna ecosystem of Botswana.

Grazing by livestock affects the species composition, productivity and microclimate of the grass layer in the savanna ecosystem. It alters the competitive balance between trees and grasses in favour of the trees. Competition for water and nutrients, as well as food preference of the various grazers, are used to explain the process of bush encroachment. In a degraded savanna the remaining grasses may profit from the favourable environmental conditions underneath the trees.

Wood in Botswana is harvested for fuel, fencing and building material. Harvesting affects the population structure and species composition of the natural woodlands. Scarcity of wood is felt, especially around the larger settlements. As plantations of trees have been shown ineffective, research is focused on management of the natural woodland. Particularly regeneration through coppicing and through seedlings is being studied.

It is concluded, that ecological research in Botswana should continue to focus on sustained productivity of natural resources and on rehabilitation of the environment where it has been damaged.

1. Introduction

Although the physical environment and the biological potential primarily determine the type of vegetation growing in a given habitat, human activities have been, in many cases, the decisive factor in establishing what kind of vegetation is now present in many locations.

This is also the case in Botswana. Two examples of the impact of human activities on the environment of Botswana will be discussed in this chapter. The first example deals with some aspects of overgrazing. Extensive cattle grazing initially changed the species composition of the grass layer and later allowed for an explosive development of thorn bushes in the rangeland.

In the second example some biological aspects of the use of wood for firewood and construction purposes will be discussed in conjunction with the possibilities of a sustained provision of these wood resources.

1.1. Physical environment

The climate of Botswana is semi-arid with an average rainfall varying from less than 250 mm year^{-1} in the south-western part of the country, to over 650 mm year^{-1} in the extreme north. Rain falls mostly in summer (October–April), and for the largest part of the country the average rainfall is between 300 and 550 mm year^{-1}. This rainfall is highly variable in space and time; the co-efficient of variation for the seasonal rainfall is roughly between 30 and 45%, the more erratic rainfall coinciding with lower average annual rainfall (Fig. 1). The variation within one season is even higher and the coefficient of variation can be over 100%, which means that in each season the chance of an

J. Rozema and J. A. C. Verkleij (eds.), Ecological Responses to Environmental Stresses, 262—276.
© *1991 Kluwer Academic Publishers.*

unpredictable dry spell occurring is high (Arntzen & Veenendaal 1986; Bhalotra 1985). Average maximum temperatures in summer are between 30 and 35° C, but they can exceed 40° C, and the soil surface temperature can be up to 70° C. In winter the day temperatures are mild, but the nights are cold, with an average monthly minimum in the coldest month (July) of 2 to 7° C; night frosts down to − 8° C are common in most of the country. The potential evapotranspiration is high, up to 2000 mm year^{-1}. Together with the low rainfall this means that vegetation faces a shortage of water during most of the year.

About two-third of Botswana is covered by sand, but in the eastern part the soil is more loamy and stony, and shallow around hills. The natural fertility is low (Sims 1981; Tolsma *et al.* 1987; Huntley & Walker 1982), with nitrogen and especially phosphorus only available in very low concentrations, often limiting plant growth.

2. The reaction of vegetation to grazing pressure in Botswana's grazing lands

The vegetative cover of most of Botswana can be described as savanna. This is a vegetation with mainly C_4 grasses and a woody layer. The actual density and productivity of grasses and woody plants is determined by the amount of rainfall, the availability of nutrients, the occurrence of veld fires and the number and activity of herbivores.

In the past over the whole of Botswana and still at present in many parts of the country (about 35%), large numbers of indigenous herbivorous species occur. Each of these species has its own food preference. Most herbivores change their grazing area continuously and many show annual migratory patterns (especially in the drier parts of the Kalahari).

Cattle were introduced into Botswana presumably around 2000 years ago, but their number remained low until the beginning of this century. Since 1900 the number of cattle has risen dramatically (Fig. 2), except in times of drought. Developments such as borehole technology and im-

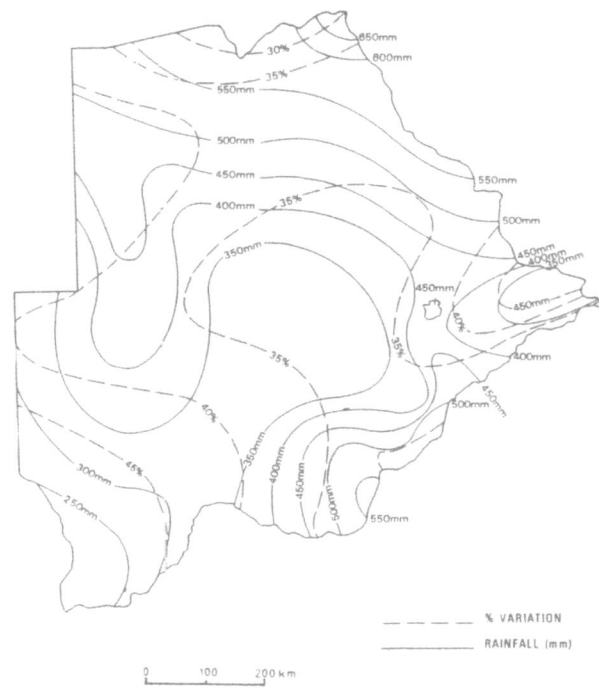

Fig. 1. Rainfall patterns in Botswana (Bhalotra 1985).

proved veterinary facilities helped the rapid growing human population to acquire more cattle. Grazing areas, which so far had been inaccessible (Western Kalahari, Western Ngamiland), could now be used for grazing.

The introduction of large numbers of cattle as a predominantly grazing species into the savanna ecosystem has caused enormous changes in the vegetation. Cattle are mainly grass consumers and are concentrated around artificial boreholes (especially during the dry winter). Rotational grazing is not common in Botswana.

The average carrying capacity is estimated at 10–15 ha per head of cattle for Botswana (Tietema 1984). During times of drought, due to the lower production of grasses, this figure should be higher. In 1984 almost all grazing areas suffered from overgrazing. In the small districts overgrazing has already been a common phenomenon since approximately 1930.

The effects of overgrazing depend on factors such as climate, soil, existing vegetation and num-

263

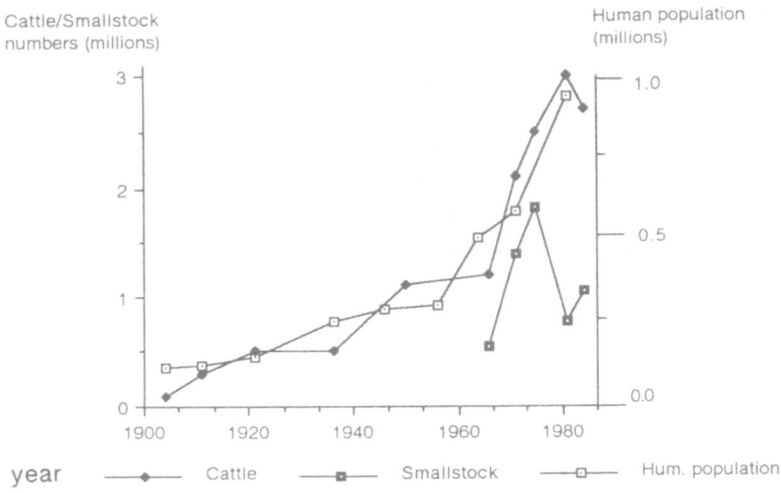

Fig. 2. Development of human, cattle and small stock populations in Botswana.

ber of animals per unit area. In the two following sections some of the main changes, caused by overgrazing, in the grass and the tree layer of Botswana's savanna will be discussed.

2.1. *Effects of overgrazing on the grass layer*

2.1.1. *Changes in the grass layer under grazing stress*

Grazing and trampling have a tremendous impact on the grass layer in the savanna ecosystem of Botswana. In an undisturbed savanna ecosystem the biomass production in the grass layer can reach a level between 10 and 15 ton ha^{-1} year^{-1}. Due to heavy overgrazing this production level can go down to less than 1 ton ha^{-1} year^{-1}. In the process of overgrazing also the species composition is changed. The following generalized picture can be given:

The process of degradation of the range starts with the reduction in density of the most palatable perennial species (e.g., Eastern Botswana hardveld: *Themeda triandra, Panicum* spp., *Anthephora pubescens, Brachiaria nigropedata;* Western Botswana sandveld: *Digitaria milaniana, Stipagrostis uniplumis*). At the same time generally less palatable perennials (e.g., *Aristida* spp., *Eragrostis* spp.) and annuals (e.g., *Tragus berteronianus,*

Chloris virgata, Enneapogon spp., *Urochloa* spp.) increase (Van Vegten 1982).

If the grazing pressure continues at high levels, a reduction of these more grazing resistant species can be expected and the grass layer becomes characterized by unpalatable often poisonous herbs and shrubs (e.g., *Tribulus* spp., *Dipcadi* spp., *Gnidia polycephala, Dichapetalum cymosum*). The total vegetation cover in the herb layer decreases. The change in species composition and the large reduction in biomass production is accompanied by drastic changes in the physical and microclimatological conditions in the grass layer. Soil surface temperatures of the exposed soils can rise to high levels (over 70° C on sunny days) causing rapid drying out of the surface after rains and thus impeding successful establishment through seeds. Wind and water erosion result in a loss of soil nutrients and a change in soil structure. The tendency of the more loamy soils to form crusts impedes the burial of seeds, root penetration (DLFRS 1985) and enhances their washing away and losses through predation. Such a degraded environment is considered an inhospitable environment for plants, except for the few species that survive and thrive at the expense of the majority.

264

2.1.2. *Annuals and perennials*

In many habitats, including arid ones, annual grasses colonize disturbed sites rapidly and are slowly replaced by perennials in succession (Jackson & Roy 1986). In Botswana the number of annuals also increases in the early stages of range degradation and the relative success of some species (e.g., *Tragus berteronianus*) may lead to the conclusion that under grazing stress annuals have adaptive advantages over perennials. A short life cycle helps to avoid both grazing and drought stress. In addition the small size and decumbent growth pattern, the apparent heat resistance of its leaves and a high reproductive effort with a short interval between flowering and seed production can help to explain the success of species like *Tragus berteronianus* in colonizing open spaces (Ernst & Tolsma 1988).

However, some perennials may also show adaptations that make them successful under grazing stress. They may be unpalatable (e.g., *Aristida* spp.) or even spiny (*Odyssea paucinervis*); or they may contain chemical constituents (*Cymbopogon plurinodis* "turpentine grass"). Some of the most successful perennial species seem to be characterized by a good horizontal growth performance (notably *Cynodon dactylon, Odyssea paucinervis*), a low shoot/root ratio, drought resistance and nutrient efficiency. Although the "vegetative strategy" may be more vulnerable to grazing stress as such, at least the vulnerable seedling stage can be avoided. As a number of perennials continue to be successful in severely degraded ranges their adaptations should be studied carefully, especially in relation to the possibilities for range restoration.

2.2. *Effects of overgrazing on the tree layer*

The savanna ecosystem can be described as a mixture of grass and woody species. According to the "two layer" hypothesis of Walter (1971) these two groups of species have access to two different sources of water: grasses are more efficient than woody species in extracting water from the topsoil, while woody species have an exclusive access to water from the sub-soil. The actual density of the grass and woody species in the "natural" savanna (this is before the introduction of cattle) is determined by the average rainfall: with an average rainfall of less than 200 mm year^{-1} all the water is used by grass (grassland); above 200 mm year^{-1} water becomes available for woody species (savanna and dry woodland). After the introduction of cattle the grasses are the first to disappear (see 2.1). More water, light and nutrients become available to other species, not only to unpalatable or even poisonous herbs as *Tribulus terrestris* and *Datura stramonium,* but also to seedlings of shrub species, which rarely survive under closed canopies of grass. Because of the disappearance of palatable grasses, cattle will change their diet, first to other grasses and then to other plant species, herbs as well as shrubs.

Given enough time, the last group has the possibility of growing above the grazing height of cattle. *Acacia* species have two advantages over other species: thorns and the ability to fix nitrogen. The nutrient content of the soil is very low, especially in nitrogen and phosphorus. Tolsma *et al.* (1987b) found that the nitrogen content in the leaves of encroaching *Acacia* species is about twice as high as in non-encroaching *Acacia* species. The encroaching species are all shallow rooted. Due to the higher root biomass in the topsoil they are capable of fixing more nitrogen through root nodules. The high nitrogen content in the leaves of the encroaching *Acacia* species makes them of value as a stockfeed. Also the pods and seeds of *Acacia* species have not only a very high nitrogen, but also a very high phosphorus content (Tolsma *et al.* 1987b).

Especially during the dry season *Acacia* seeds form a large part of the food intake. The consumption of *Acacia* seeds can play a role in their encroachment. In the past emphasis has been placed on the difference between dehiscent and indehiscent pods of *Acacia* spp. in the diet of game. For example according to Gwynne (1969) and Lamprey (1967) seeds from indehiscent pods can be found in dung, while seeds from dehiscent pods are not found in dung. However, Table I shows that not only seeds from indehiscent pods are found in cattle dung but also seeds from the dehiscent pods of *Acacia hebeclada*. Although the pods of this species split when they are ripe, nearly all the seeds remain in the pod. Seeds from species with de-hiscent pods and a soft coat are not found in cattle dung, although they occur in the investigated areas (only *A. karroo,* a species with dehiscent pods and seeds with a hard coat, does not occur). Thus it is not the (in)dehiscence of the pod that is important, but whether the seeds remain in the pod and the hardness of the seedcoat.

The consumption of *Acacia* seeds with a hard seed coat results in increased dispersal, they germinate better after passage through the alimentary tract of the animal and the seed ends up in a nutrient rich environment. The chance that a seed is found by cattle is larger for seeds from indehiscent pods because the whole pod falls to the ground. Also, the number of seeds per pod is higher for species with indehiscent pods (e.g., 15.5 for *A. erioloba,* compared to 2.5 in *A. erubescens*). Seeds from dehiscent pods fall to the ground during a longer time span and are therefore not easily found.

The germination success of the seeds not only depends on their passage through the alimentary tract, but also on their quality. Seeds of Leguminosae are frequently infested by beetles of the family Bruchidae. A comparison between encroaching and non-encroaching *Acacia* species (Table I) shows that the infestation percentage lies between 12% (*A burkei*) and 80% (*Dicrostachys cinerea*). Species which can encroach, such as *A. erubescens* and *A. tortilis* can be as infested as non-encroaching species.

In explaining the change of vegetation, the importance of increasing cattle densities is emphasized and just as important, if not more so, is the disappearance of the indigenous game. A group of herbivore species with different food preferences is replaced by two species only: cattle and goats. The first group includes species capable of destroying *Acacia* seeds because of their smaller bite size (e.g., Impala, up to 95% of the ingested (hard coated) seeds are destroyed during passage (Jarman 1976). In other words: the change towards bush vegetation is not necessarily caused by an increase in the number of consumed *Acacia* seeds by cattle; what is important is the larger percentage of consumed seeds which passes through the digestive system undamaged.

2.3. The relation between trees and grasses

The interaction between trees and grasses in the savanna is generally being discussed in the context of competition for water and nutrients (Walter 1971, Tolsma *et al.* 1987b). Bush clearing has been studied as a method of improving the production of the grass layer in the savanna in Botswana (APRU, 1979).

On the other hand, certain grass species turn out to be associated with the tree canopy (e.g., *Panicum maximum; Digitaria pentzii*) (Barnes 1982) and the removal of the tree layer will result in their disappearance. Some research was done on the close association between *Panicum maximum* in

Table I. Characteristics, infestation by Bruchid beetles (N = 500 seeds) and occurrence in cattle dung of the seeds of *Acacia* species and *Dicrostachys cinerea.* (deh = dehiscent; ind = indehiscent; n.f. = not found; ± = s.d.).

Species	Pods	Seedcoat	% Infested	Nr. seeds/dropping	Encroaching
A. burkei	deh	soft	12	n.f.	no
A. erubescens	deh	soft	18	n.f.	yes
A. fleckii	deh	soft	0	n.f.	yes
A. mellifera	deh	soft	22	n.f.	yes
A. karroo	deh	hard	15	n.f.	no
A. hebeclada	deh	hard	38	95 ± 132	yes
A. tortilis	ind	hard	32	208 ± 417	yes
A. nilotica	ind	hard	35	10 ± 20	yes
A. erioloba	ind	hard	52	52 ± 98	yes
D. cinerea	ind	hard	80	n.f.	yes

closed and open canopy sites and grassland (Bosch & Van Wijk 1970, Kennard & Walker 1973) but no conclusive data on the nature of this association was produced, with the exception of a marked higher survival of *P. maximum* seedlings under shade (Kennard & Walker 1973). Under trees, favourable conditions in terms of lower soil surface temperatures, increased nutrient availability and soil moisture, especially during the seedling stage, can be expected.

In severely degraded rangeland in south eastern Botswana there is a strong association between tree cover and the remaining grass cover. Although the protection from grazing through the thorny branches of the bushes and trees may play a role, the more favourable microclimate might be more important. In Botswana this aspect has until now not been studied in great detail. For the management of these degraded ecosystems it is important to understand the relationship between shaded areas and recovery of the grass layer. It may be that in the degraded environment of Botswana trees should be seen more as facilitators of grasses than as their competitors.

3. The response of woodland to human exploitation

In Botswana three main types of savanna woodland can be distinguished (Fig. 3):
1. The Open Savanna Woodland, covering most of the Kalahari sands.
2. The Acacia Woodland, that is found in most of the eastern hardveld.
3. The Mopane Woodland, which covers most of the northern part of the country.

Next to these three main woodland types, there are three more woodland types which, although they cover smaller areas, are of importance. These are the Zambesi Teak forests, in the north; the hill forests, in the south east, and the Riverine forests, covering the fringes of the non-perennial rivers throughout the country and the edges of the river systems in the Okavango Delta.

The people of Botswana have free access to most of these woodlands and are permitted to collect wood or anything else for subsistence purposes

Fig. 3. The distribution of woodlands in Botswana ([   ] Open Woodland Savanna; [\\] Acacia Woodland; [////] Mopane woodland).

from the natural woodlands. The wood collected from the woodlands is used for firewood, fencing and building material and for the carving of utensils and ornaments. Next to these uses roots and bark for dyeing are also collected, together with a whole variety of forestry products for food and preparing traditional medicines.

3.1. The quantities of wood harvested from the woodlands

Information on the quantities of wood harvested from the woodlands in Botswana is very scarce. Only the average amount of firewood harvested per head of the population has been quantified reasonably and can be set at approximately 500 kg person^{-1} year^{-1} (Jelenic & Van Vegten 1981; ERL 1985).

The wood needed for fencing and building has only been quantified spasmodically and no countrywide figure is available.

However, a probable conservative estimate, for the Modubwana lands in south east Botswana, indicates that a family farming in the traditional way

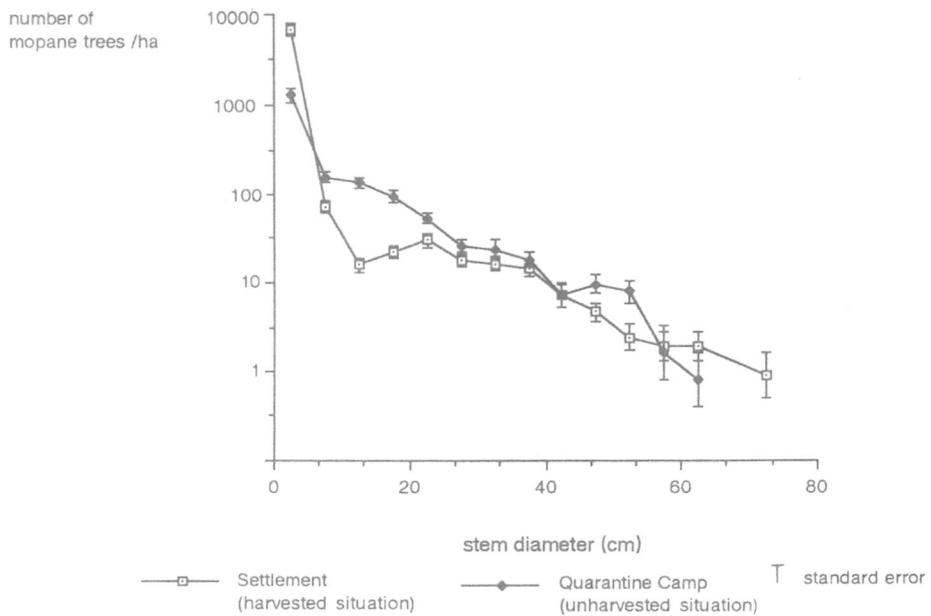

number of
mopane trees /ha

stem diameter (cm)

———□——— Settlement
(harvested situation)

———◆——— Quarantine Camp
(unharvested situation)

⊤ standard error

Fig. 4. Structure of the *Colophospermum mopane* population in Dukwe forest.

might use annually 1.5 times more wood per person, just for maintenance of the bushfence around their arable field, than they use for firewood. In this figure the wood needed for erection of the bushfence is not included (Tietema & Geche 1987). In Dukwe (Central North Botswana) the wood needed annually, to maintain the stacked woodfence around the vegetable garden only (15 × 15 m), was already equivalent to half the national average per capita needs for firewood. (Tietema *et al.* 1988). Next to this also bushfences were used around the arable fields.

The amount of wood needed for the construction of houses varies with the type of design. In northern Botswana houses are basically build of wood, with a plaster made of a mud and cowdung mixture. In Dukwe, the wood needed in these houses was estimated at 5 ton per house of 3.5 by 4.0 meters (Tietema *et al.* 1988). In the south eastern part of Botswana, however, the amount of wood used in the construction of houses is considerably less, as people use mud walls, without a wooden reinforcement.

It will be clear that it is only possible to make rough estimates of the amount of wood needed

annually to allow the population to continue its present way of life. Nevertheless, taking these constraints in mind, the national annual need of wood per capita can roughly be estimated at 1.5 ± 0.5 ton person^{-1}. With a population of approximately 1 million people, this would result in an annual national wood consumption of 1.5 ± 0.5 million ton of wood. All this wood is harvested from the natural woodlands.

3.2. The effect of woodharvesting on the woodlands

The effect of wood-harvesting on a woodland is determined by the way in which the harvesting is done. In Botswana people are, at least initially, selective in their way of harvesting. The species and sizes harvested have to match certain criteria in conjunction with the anticipated use of the wood. As a consequence of this selective way of harvesting firstly the population structure of the woodland will change and secondly the species compositon.

3.2.1. The effect on the population structure
The majority of wood uses require wood of be-

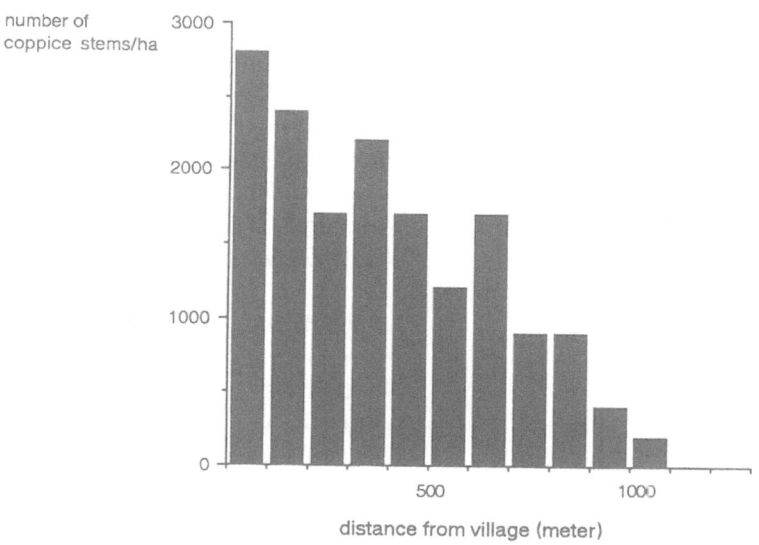

number of coppice stems/ha

distance from village (meter)

Fig. 5. Change in coppice shoot frequency in relation to the distance from Morwa village (south east Botswana, near Mochudi).

tween 5 and 20 cm diameter. Consequently this is the size class that will disappear first when people start to harvest a previously not harvested woodland. Fig. 4 shows this for the Mopane woodland in Dukwe where the harvesting of wood had been going on for 9 years at the moment that the measurements were taken. Another effect wood harvesting has on the population structure of the woodland, is the increase in the number of coppice shoots amongst the smaller size classes. Usually the number of coppice shoots decreases with decreasing wood cutting intensity; which in most cases means with increasing distance from the villages (Fig. 5).

The result of this way of wood harvesting is, that the recruitment of trees into size classes larger than 20 cm diameter decreases, while at the same time the natural dying off of older trees continues. This creates a situation in the major villages, in which finally none, or at least hardly any, mature indigenous trees are left.

Also the regrowth of coppice after cutting is not complete. Around 10% of the stumps die after cutting, hence a certain amount of natural regeneration is needed to ensure that there will be a woody vegetation in any harvested area. The environmental factors affecting this generative regeneration, grazing/browsing, soil type and rainfall, in the end

determine the appearance of a heavily harvested area.

On the heavily overgrazed sandy soils of the Kalahari sandveld, regeneration is usually negligible, initially resulting in villages with big trees only; all the rest having been cut down and the coppice and seedlings browsed and trampled to the point of extermination (e.g., Kang & Maun). If no action is taken, then these villages will develop a situation where no trees occur in the entire settlement (e.g., Tshabong).

On the more loamy soils of the eastern hardveld the situation is less desperate, as there is at least a vigorous regeneration of thorny tree species. The final appearance here is of villages with hardly any big trees, but surrounded by a well developed thorn-bush scrub (e.g., Molepolole, Mochudi).

3.2.2. *The effect on the species composition*
Alongside the changes in the population structure also changes in species composition occur in a harvested woodland. The changes in species composition depend on the type of wood that is predominantly harvested in a given area. In arable land areas, where wood is being used for bush fencing, there is a shift towards non thorny species, because of the permanent harvest of young thornbush trees

269

for the fencing of arable fields (Tietema & Geche 1987).

The time needed to effectuate a change in the species composition in a woodland is much longer than the time needed to create a marked change in the population structure of a woodland. For example in Dukwe, after 9 years of intensive wood cutting, the population structure had changed dramatically (Fig. 4), whereas in the species composition no significant change had taken place.

3.3. Ways of supplying the population with the wood needed

In order to supply a population with the wood needed, two ways are open to the policy makers. The first is to plant extensive stretches of land with trees; the second way is to make use of the production potential of the still existing but allegedly low yielding indigenous woodlands.

3.3.1. Wood production in plantations

Production of wood in plantations was the way to solve the African fuelwood crisis, according to the perceptions of the late seventies (Heermans & Minnick 1987). Big plantation schemes were started in the Sahel countries, in which mainly *Eucalyptus* species were planted.

The same also happened in Botswana, where between 1972 and 1981 a number of medium sized *Eucalyptus* plantations were started. At the moment of planting a production rate of 9–18 ton ha^{-1} year^{-1} was anticipated, which compared favourably with the 1–2 ton ha^{-1} year^{-1} that the indigenous woodland was estimated to yield (Nickerson 1984). However, the anticipated production of the *Eucalyptus* plantations did not materialize, as after growth periods of 8–12 years the actual production rate turned out to be only 1.25–1.46 ton ha^{-1} year^{-1} (Tietema *et al.* 1987). On top of this low growth rate also a high incidence of die off of trees (up to 95%) occurred in plantation blocks older than 8 years (Tietema *et al.* 1987). This die off was presumably caused by drought, not an unusual situation in Botswana.

The poor performance of *Eucalyptus* in combination with the high establishment costs of planta-tions (around US $ 1000 ha^{-1}) means that *Eucalyptus* plantations cannot be considered a viable option in solving the firewood crisis. The only future for plantations will be in growing tree species artificially, which have high demand, and economic value, and which cannot be obtained from the natural woodlands in sufficient quantities (e.g., *Spirostachys africana*).

3.3.2. Woodlandmanagement and natural regeneration

Another way of addressing the problem of how to supply the wood needed, is through an organized policy of harvesting natural woodlands. If this possibility is to be pursued, then it will be necessary to know how much wood can be harvested, what the rate of regrowth of the woodland will be after cutting and to what extent regeneration takes care of replacement.

3.3.2.1. The amount of wood available from Botswana's natural woodlands.
Similar to the wood consumption, also the information on the amount of wood available from the natural woodlands in Botswana is very scanty. Actual measurements of the quantity of wood available have only been made occasionally. Information only exists for a number of small sections of woodland in eastern Botswana and for the Mopane woodland in Dukwe, central north of Botswana (Table II).

However, if this scanty amount of information is

Table II. Woody biomass available from natural woodlands in Botswana.

Woodland type	standing biomass (ton fresh weight ha^{-1})
Acacia tortilis woodland	14[1]—30[2]
A. karroo — *A. erubescens* woodland	60[3]
Mopane woodland unharvested	80[4]
Mopane woodland harvested	63[5]

[1] Recolonization on a 20 year old abandoned arable field.
[2] A full grown stand.
[3] A dense bushencroachment stand near a borehole.
[4] The *Colophospermum mopane* component only, which is 70% of the total basal area at ankle height.
[5] As above, with *C. mopane* consisting of 82% of the total basal area at ankle height.

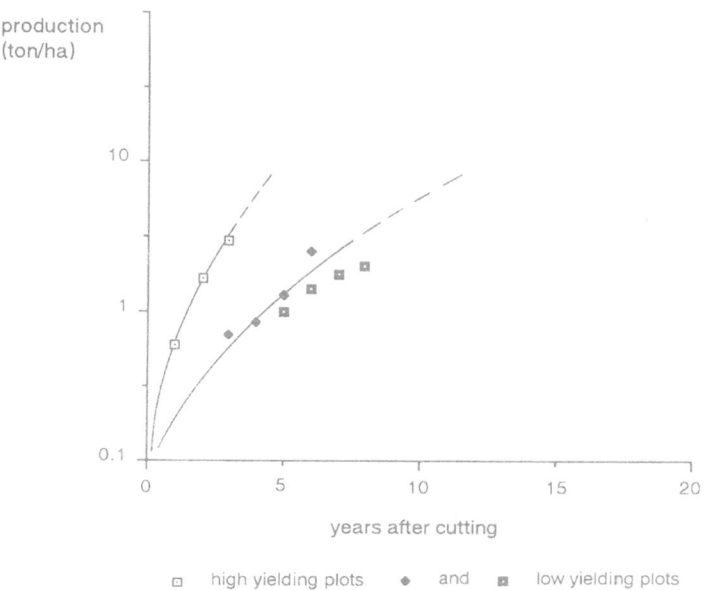

production
(ton/ha)

Fig. 6. The regrowth of natural *Acacia* woodland near Morwa (south-east Botswana, near Mochudi) after clear cutting.

combined with Landsat data, then it seems reasonable to assume an average standing fresh woody biomass of approximately 20 ton ha^{-1} throughout the country. With firewood being on average 50% of the total fresh biomass, and with an average moisture content of the wood of only 30%, this means an availability of dry firewood of on average 7 ton ha^{-1}.

Countrywide this results in a standing crop of approximately 450 million ton of firewood, which if compared with the annual consumption figure of 1.5 million ton means, that at a countrywide basis enough fire wood is available. However, the population is not evenly distributed over the country and this results in big discrepancies between production and demand, especially near the larger settlements in the eastern part of the country.

3.3.2.2. The rate of regrowth of the natural woodlands after cutting. Actual measurements of the regrowth of indigenous woodland only exist for two locations in the Acacia woodlands of south-eastern Botswana. These measurements only cover the period of 1982 to 1988, during which period the rainfall was significantly below average and extrapolations to obtain long term production rates have

to be considered tentative. Nevertheless, within this constraint it is already clear that the indigenous woodlands in a coppicing system are producing much more than the anticipated 1 ton ha^{-1} year^{-1}. At present extrapolations allow a production forecast of between 5 and 10 ton ha^{-1} year^{-1} in coppicing cycles of between 10 and 20 years (Fig. 6).

This conclusion for south-eastern Botswana was supported by the study on the Mopane woodland in Dukwe (Tietema *et al.* 1988) and is in line with similar results obtained from the Sahel countries (Gibson 1988 pers. comm.).

3.3.2.3. Generative regeneration. Recruitment of young trees is the result of a chain of processes in which each link is essential. First adult trees must produce seeds, then the seeds must survive hazards such as seed predation and extreme temperatures. In many tree species dormancy develops first, and has to be broken before germination. After germination the seedling, and later on the young plant has to survive until maturity. All this takes place in a hot and dry climate, with unpredictable dry periods, and in a nutrient-poor soil. A tree has to produce a large number of seeds to ensure that at least one tree survives the long and hazardous peri-

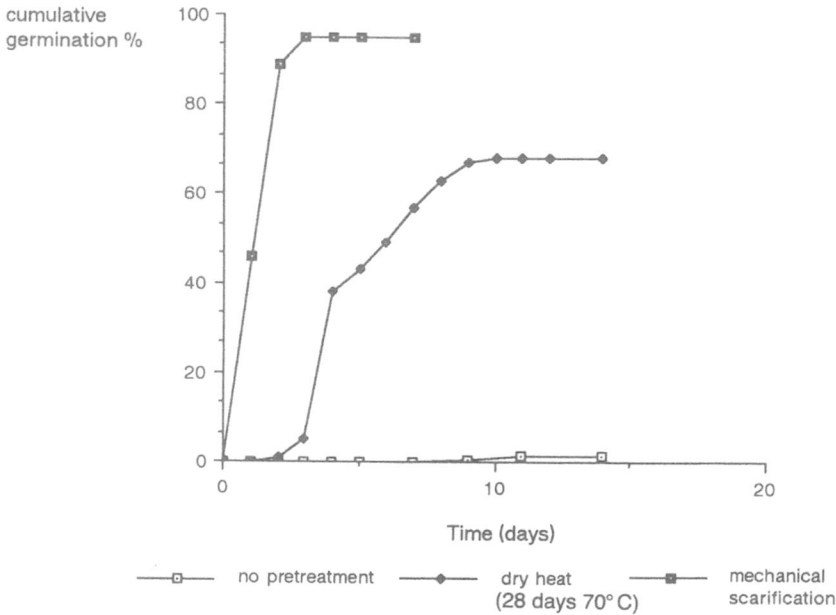

cumulative germination %

Time (days)

—□— no pretreatment   —◆— dry heat (28 days 70° C)   —■— mechanical scarification

Fig. 7. Germination (at 30° C) of *Acacia tortilis* seeds after different pretreatments.

od during which most of the potential new trees die. In dry years many species produce hardly any seeds. This also depends on the timing of flowering. Some species always flower just before or at the beginning of the rainy season, e.g., *Acacia mellifera*. If the subsequent rains are too late or insufficient, many developing fruits are aborted and the seed production for that year fails. Other species flower after each rainfall event, e.g., *Acacia tortilis*. Given the unpredictability of the rains, this spreads the risk of a total failure of the annual seed production. For both types of trees the low nutrient content in the soil poses a limit to the number of seeds produced (Tolsma *et al.* 1987b). Escaping from seed predators seems to be very difficult, as seeds from many species are often heavily infested with seedpredators (see Table I, for infestation of *Acacia* seeds with one group of seedpredators). Infestation occurs in spite of plant metabolites found in seeds that are assumed to avert seed predators. The dependence of a good seed crop on the rainfall may incidentally act as a defence against seed predators: in a wet year with a high seed production following some dry years with low seed production the suddenly large number of

seeds may cause seed predator satiation and hence result in a larger number of seeds escaping predation. To ensure germination at a suitable time, many species have developed dormancy mechanisms. As dormancy is not broken in all seeds at the same time, it also ensures germination of only part of the seeds at each rainfall period, thereby spreading the risk. For instance, germination of the total seed crop after a rainfall event could lead to the loss of all the seeds, if a prolonged dry period were to follow. Environmental factors important in breaking of dormancy are seed predation by large mammals, heating by solar irradiation and by veld fires (Bewley & Black 1982). For almost all savanna tree species little is known about the germination process, but germination often takes longer than might be expected in view of the high temperatures. Seeds are capable of much quicker germination. This is shown by the fact that artificial pretreatments can speed up the germination process by days (Fig. 7). This delay in germination seems a waste of time: the sooner a seedling can start growing after the rain, the better. However, the delay in germination might be a precaution against small rainfall events: in that case quick germination

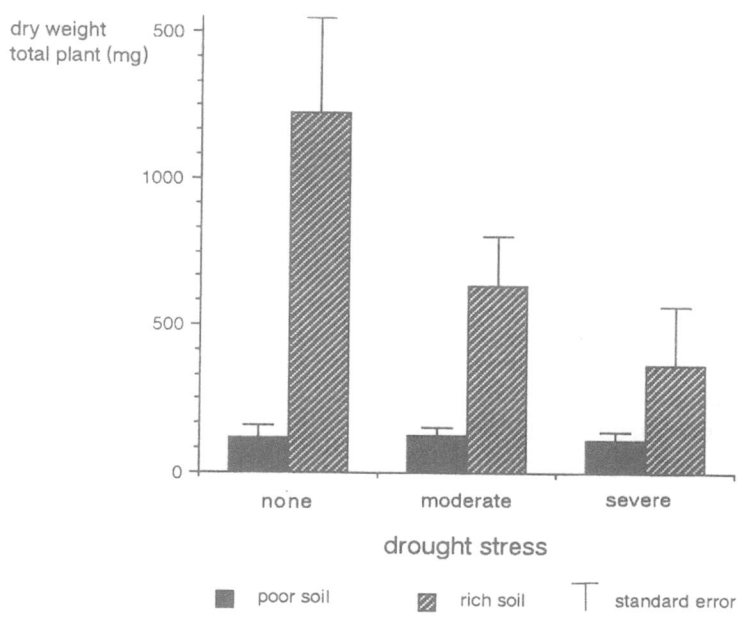

Fig. 8. Water and minerals as limiting factors for the growth of nine weeks old *Acacia tortilis* seedlings (The nutrient concentration in the rich soil was approximately 4 × higher than in the poor soil).

would result in death of the seedling, because of water shortage. A small seedling has to cope with several stresses at a time. First, the risk of a long dry period is always present, and even if the summer has enough well-spaced rain, the seedling must be able to survive the dry winter period, often 4 or 5 months without any significant rain at all. Many species put initially a lot of effort in the production of a well-developed root system, resulting in a low shoot/root ratio. Only after a few years growth rate of a shoot increases. High soil temperatures also pose a problem for the seedling, especially in open areas. To a certain extent this can be countered by a high transpiration rate, but this will deplete the often scarce water resources all the sooner. Therefore seedlings often only survive in shadowed areas or in small depressions where the soil remains moist for a longer time. A third constraint for growth is the poor soil. Except in very dry years, not the amount of water, but the limited supply of nutrients from the soil may be growth-limiting (Fig. 8; Penning de Vries & Van Keulen 1982). Some species benefit from symbionts, e.g., Leguminosae with *Rhizobium*-bacteria for more nitrogen, many

species with mycorrhyzal fungi for more phosphorus (Bowen 1980). But hardly anything is known about how quickly a seedling starts symbiotic relationships, and at which stage this enhances the growth rate.

3.3.2.4. Effects of overgrazing and woodcutting on generative regeneration. Overgrazing and woodcutting can occur separately e.g., around cattleposts overgrazing takes place, but not necessarily excessive woodcutting. Especially around villages both processes do occur at the same time. As already mentioned in section 2, overgrazing, initially, leads to a change in species composition in the grass layer and later to the emergence of spots of bare soil (see 2.1). In later stages of overgrazing the soil becomes increasingly bare, resulting in rising maximum soil temperatures, which affect seedling survival and growth both directly and indirectly. In severely degraded areas also soil erosion takes place, especially early in the rainy season, when the annual grasses have not yet established themselves. This results in effects such as washing away of the soil around seedlings or covering them, or in the wash-

273

ing away of seeds to small depressions. In the latter case, this might in itself be advantageous for germination and growth, as these places tend to stay moist for longer periods, and also more clay and silt is accumulating. In the advanced stages of overgrazing, owing to the lack of grasses and herbs, cattle and goats change their diet from grasses to trees and seedlings (APRU 1985). This results in a higher mortality amongst seedlings and young trees, and in more damage to adult trees.

Woodcutting leads to reduced seed production, as mainly (parts of) adult trees are cut. Also, like overgrazing, it leads to less soil cover, causing more extreme micro-climatic conditions of the soil, and thereby a higher mortality of seedlings and young plants. All this will be detrimental to regeneration. However, the susceptibility to these stresses differs among species. To a certain extent some species even benefit from the changes caused by overgrazing and woodcutting; see for example 2.2 on bush encroachment.

All in all the result is initially a shift in the species composition, resulting ultimately in the appearance of the tree vegetation around villages as described in 3.2.

4. Conclusion

When early this century the numbers of people and cattle in Botswana started to increase dramatically, a series of changes in the savanna vegetation was initiated. The increase in cattle numbers primarily caused changes in the grass layer. These changes involved the replacement of the more palatable perennial grass species by less palatable annual ones, followed by the encroachment of unpalatable herbs and thorn trees. Initially this development was predominantly seen around the cattle watering points, but later they occurred at a much wider scale.

The increase in the human population of Botswana, which occurred alongside with the increase in cattle numbers, resulted in an increased demand for wood products. The effect of this increased demand of wood was initially most noticeable in the woody vegetation around the settlements, but

has at present also started to spread out over bigger areas, especially in the eastern part of Botswana. The effects of overexploitation of the woodlands are in principle quite similar to those of overgrazing in the grass layer. First a change in population structure of the woodland, followed by a change in species composition much later. In cases where overgrazing and overharvesting of wood occur next to each other, a dense thornbush scrub can develop.

Obviously these changes in both grassland and woodland are the ultimate results of grazing/harvesting practices that do not take into account how much can be harvested from a certain environment without damaging the existing situation in that environment.

If one accepts the fact that the human population of Botswana uses environment for its subsistence, then the question arises, whether that is possible without damaging the environment. The answer to this question depends in the first place on what one considers to be damage done to the environment. If no alterations should occur in the environment, then probably no or only a very limited use of environment can be made. If, however the use of environment should be sustainable, and resulting changes in environment are acceptable, within this sustainability, then a more realistic situation emerges.

From the effects human activities have on both the grasslands and the woodlands, it can be concluded that presently no sustained harvest situation exists, but a situation of plain overharvesting. This situation applies for all areas surrounding villages and boreholes. The underlying cause of this overharvesting problem is the rapid expansion of the human population, and also the number of cattle. This calls for a change in the organization of subsistence harvesting of environment. People, due to their increasing numbers and concentration in huge villages can not continue to collect what they need in the traditional way, as this way of use of environment creates locally overexploited areas, with related problems like soil denudation and soil erosion, lack of firewood and no grass.

The conclusion is hence, that the use of environmental resources by the population has to be regulated in order to avoid an expansion beyond control of the presently emerging environmental problems. However, this conclusion poses the great practical problem of lack of information, on which one could base a kind of realistic environmental resources harvesting regulation system. The legal framework to do such an exercise is available, but the knowledge how to fill the framework does at present only exist in a few scattered areas. Consequently, this poses two main research questions to the ecologists in Botswana. The first one is how much can one harvest from any component of environment and how do we have to organize that harvest in such a way that the harvest can be sustained. The second one is, how can we repair the damage already done to the environment, and avoid a further expansion of the degraded areas.

The answer to these two questions has to be found in studying the production ecology and population dynamics of the more important products obtained from the environment and in studying the possibilities of natural or artificial rehabilitation of degraded areas by means of reintroducing the vegetation that had disappeared.

Acknowledgements

The authors wish to thank Mrs M. Vink for typing the main body of the text and Prof Dr C. Wilkinson for his comments and his assistance in improving the English text.

References

A.P.R.U., 1979. Livestock and range research in Botswana. Ministry of Agriculture, Gaborone, Botswana.

A.P.R.U., 1985. Livestock and range research in Botswana 1983—1984. Ministry of Agriculture, Gaborone, Botswana.

Arntzen, J. W. & Veenendaal, E. M., 1986. A profile of environment and development in Botswana. N.I.R./University of Botswana. Gaborone, Botswana, I.E.S., Free University Amsterdam.

Barnes, D. L., 1982. Management strategies for the utilisation of Southern Africa savanna. In: B. Huntley & B.H. Walker (eds.) Ecology of Tropical Savannas. Ecological Studies 42. Springer Berlin, Heidelberg, New York pp. 626—656.

Bewley, J. D. & Black, M., 1982. Physiology and biochemistry of seeds in relation to germination, Vol. 2. Springer-Verlag, Berlin, Heidelberg, New York.

Bhalotra, Y. P. R., 1985. Rainfall maps of Botswana. Department of Meteorological Services, Ministry of Works and Communications, Gaborone, Botswana.

Bosch, O. H. J. & Van Wyk, J. J. P., 1970. Die invloed van die bosveldbome op die productiwiteit van *Panicum maximum*. 'n Voorlopige verslag. Proc. Grassld. Soc. Sth. Afr. 5: 69—74.

Bowen, G. D., 1980. Mycorrhizal roles in tropical plants and ecosystems. In: P. Mikola (ed.) Tropical Mycorrhiza Research. Clarendon Press, Oxford, pp. 165—190.

DLFRS, 1985. Soil properties, crusting and seedling emergence. Final report Phase III, Vol. 5. Ministry of Agriculture/Division of Agricultural Research, Sebele, Botswana.

ERL — Energy resources limited, 1985. Study of energy utilization and requirements in the rural sector of Botswana. Ministry of Mineral Resources and Water Affairs, Gaborone, Botswana.

Ernst, W. H. O. & Tolsma, D. J., 1988. Dormancy and germination of semi-arid annual plant species, *Tragus berteronianus* and *Tribulus terrestris*. Flora 181: 243—251.

Gwynne, M. D., 1969. Notes on the nutritive values of *Acacia* pods in relation to *Acacia* seed distribution by ungulates. East Afr. Wildl. J. 7: 176—178.

Heermans, J. & Minnick, G., 1987. Guide to Forest restoration and management in the Sahel, based on case studies at the National Forests of Gueselbodi and Gorou — Bassounga, Niger. Ministry of Hydrology and the Environment, Land use and Planning Project, Niamey, Niger.

Huntley, B. & Walker, B. H., (eds.) 1982. Ecology of Tropical Savannas. Ecol. Studies 42. Springer-Verlag, Berlin, Heidelberg, New York.

Jackson, L & Roy, J., 1986. Growth patterns of Mediterranean Annual and Perennial Grasses under simulated Rainfall Regimes of Southern France and California. Oecol. Plant. 7: 191—212.

Jarman, P. J., 1976. Damage to *Acacia tortilis* seeds eaten by impala. East Afr. Wildl. J. 14: 223—225.

Jelenic, N. E. & Van Vegten, J. A., 1981. A pain in the neck. The firewood situation in South-West Kgatleng, Botswana. NIR Research Notes No. 5. Gaborone, Botswana.

Kennard, D. G. & Walker, B. H., 1973. Relationships between tree canopy cover and *Panicum maximum* in the vicinity of Fort Victoria. Rhod. J. Agric. Res. 11: 145—153.

Lamprey, H. F., 1967. Notes on the dispersal and germination of some tree seeds through the agency of mammals and birds. East Afr. Wildl. J. 5: 179—180.

Nickerson, R. A., 1984. The need for fuel-wood plantations in Eastern Botswana, including a draft proposal for implementation. Annual Journal of the Forestry Association of Botswana 1984, pp. 57—64.

Penning de Vries, F. T. W. & Van Keulen, H., 1982. La production actuelle et l'action de l'azote et du phosphore. In: F.T.W. Penning de Vries & M.A. Djitéye (eds.) La productivité des pâturages sahéliens. PUDOC, Wageningen, pp. 196—226.

Sims, D., 1981. Agroclimatological information, crop requirements and agricultural zones for Botswana. Land Utilisation Division, Ministry of Agriculture, Gaborone, Botswana.

Tietema, T., 1984. Production ecological backgrounds of overgrazing. In: Anonymous. Proceedings of the Workshop on Overgrazing. Ministry of Agriculture/Ministry of Local Government and Lands, Gaborone, Botswana (Annex III, pp. 1—3).

Tietema, T. & Geche, J., 1987. A quantitative determination of the amount of wood needed for the erection of bush fences around arable fields in Botswana. Annual Journal of the Forestry Association of Botswana 1986/87, pp. 19—25.

Tietema, T., Kgathi, D. L. & Merkesdal, E., 1988. Wood production and consumption in Dukwe. A feasibility study for a woodland management and plantation scheme. NORAD/ NIR, Gaborone, Botswana.

Tietema, T., Merkesdal, E., Kgafela, S. & Maembolwa, J., 1987. The productivity of *Eucalyptus* plantations in Botswa-na. The case of the Molopolole airstrip plantation. Annual Journal of the Forestry Association of Botswana 1986/87, pp. 91—96.

Tolsma, D. J., Ernst, W. H. O. & Verweij, R. A., 1987a. Nutrients in soil and vegetation around two artificial waterpoints in Eastern Botswana. J. Appl. Ecol. 24: 991—1000.

Tolsma, D. J., Ernst, W. H. O., Verwey, R. A. & Vooys, R., 1987b. Seasonal variation of nutrient concentrations in a semi-arid savanna ecosystem in Botswana. J. Ecol. 75: 755— 770.

Van Vegten, J., 1982. Increasing stock numbers on deteriorating rangeland. In: R.K. Hitchcock (ed.) Botswana's First Livestock Development Project and its future Implications. Symposium Proceedings. N.I.R./University of Botswana. Gaborone, Botswana, pp. 98—107.

Walter, H., 1971. Ecology of tropical and subtropical vegetation. Oliver & Boyd, Edinburgh.

Authors' Address
T. Tietema, E. M. Veenendaal & J. Schroten*
National Institute of Development Research and
Documentation
University of Botswana
Private Bag 0022
Gaborone
Botswana

* Present address
Department of Ecology and Ecotoxicology
Biological Laboratory
Free University
De Boelelaan 1087
1081 HV Amsterdam
The Netherlands

D. J. Tolsma*
Thusano Lefatsheng
Private Bag 00251
Gaborone
Botswana

Agricultural settlement in the Punjab, Pakistan. Drinking place for buffaloes. The reservoir contains rainwater and is also used for washing. (Photograph: J. Rozema.)

CHAPTER 24

Agricultural problems of saline arable land in Pakistan

J. ROZEMA

Abstract. Salinity represents a serious threat to irrigated arable land. It has been estimated that about one-third of the irrigated land of the world is now salt-affected. In addition, vast areas of potentially fertile land cannot be used for the cultivation of conventional crops because of an excessive salt content of the soil.

Crop growth is not only depressed by excess salt but also by the poor quality of the structure of the saline soil. Under certain conditions chemical amendment (application of gypsum) may improve the physico-chemical quality of arable land but only locally and at high costs. For the arid and semi arid conditions in Pakistan it can be calculated that using the amount of water the river Indus transports through the flood plains, only a limited area can be properly irrigated and used for agriculture without salination. Therefore an extensive area of salt affected land will always remain and be abandoned by the farmers. During the last two decades biosaline research aims at cultivation of salt adapted plants on saline arable land.

The perspective of conventional screening and breeding techniques to obtain salt tolerant crops is discussed. Also advanced biotechnological approaches (callus and cell culture, meristem culture, plant regeneration, and protoplast fusion) have improved the possibilities to increase salt tolerance in plants. Fodderbeet is a domesticated cultivar of the coastal halophyte *Beta vulgaris*. Cultivars of the fodder beet were found fairly salt tolerant and even improved growth of the leaves and beet was found on saline land in field trials as compared with non-saline land.

Further field studies are being carried out to test the usefulness of this salt tolerant crop on salt affected land in Pakistan combined with other salt tolerant crops (Chenopods, grasses, legumes), to be used in rotation schemes.

In addition, since the mechanism of salt tolerance is still incompletely understood, fundamental research is required to see how higher plants may grow well under saline conditions.

1. Salinity and agriculture

In many parts of the world, arid and semi-arid zones in particular, possibilities for food production are reduced because of conditions of drought or soils may have become salinated. Estimations by FAO (Massoud 1974) indicated that 350 million hectares of the total surface of terrestrial ecosystems, is more or less affected by salt. This is a low estimate. High estimates amount to 950 millions of hectares of saline land. Saline arable land often relates to inadequate application of irrigation techniques. Reeve & Fireman (1967) estimated that, as a result of this, about one third of the irrigated arable land is now salt affected. There are naturally saline soils such as salt marshes at the coastal fringes of continents and along the borders of estuaries, and inland salines (e.g., Neusiedlersee, Austria; Dead Sea, Israel; Great Salt Lake, Utah, USA) (Fig. 1) may occur where evapotranspiration exceeds precipitation. As a result, soluble salts accumulate at the soil surface. Man-made salination of (arable) land may occur as a result of irrigation practices in semi-arid and arid areas. Salinized arable land is well known from Northern Africa, the Middle East, (the Eufrat and Tigris Valleys in Iraq), the delta of the Indus (Pakistan) and the Ganges (India), and in the San Joaquin Valley in California (Fig. 1). In Pakistan irrigation of the Indus Valley has led not only to salination of soils,

J. Rozema and J. A. C. Verkleij (eds.), Ecological Responses to Environmental Stresses, 278—288.

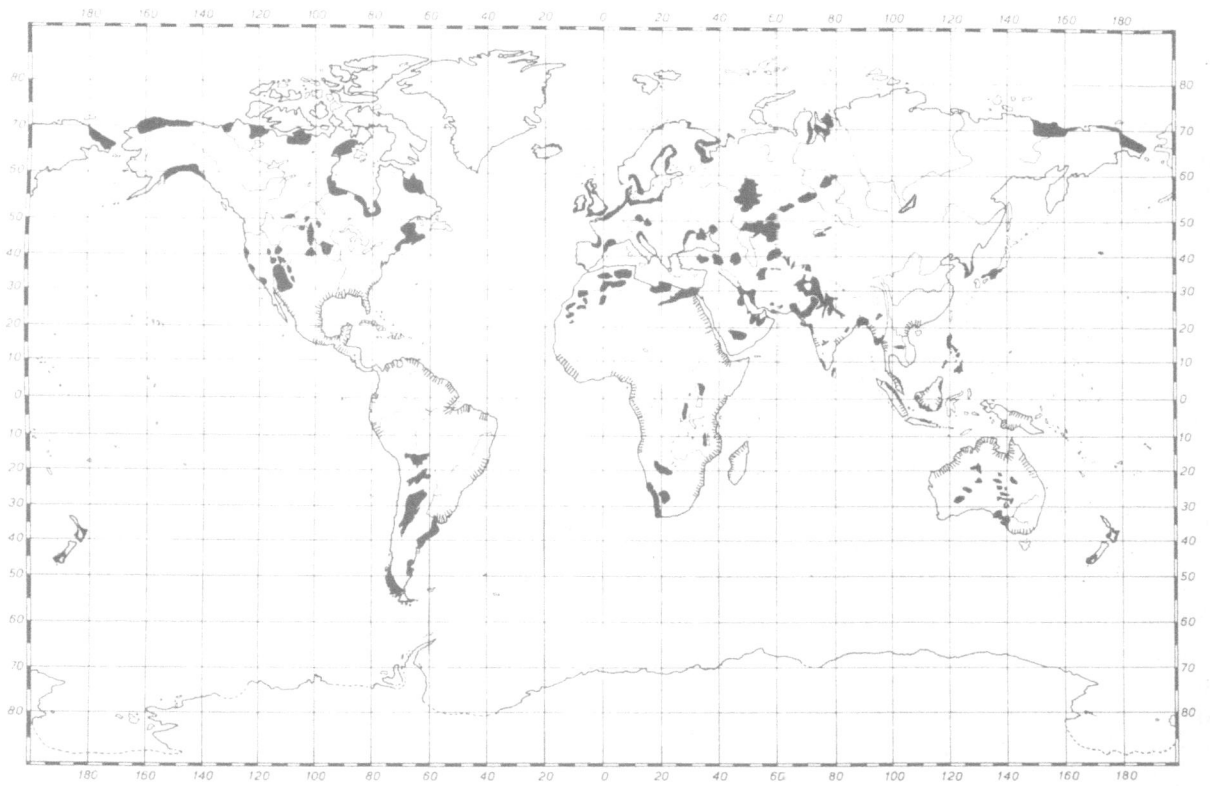

Fig. 1. Saline areas of the world. Black areas represent coastal salt marshes or salinized inland soils. Dashed coastline indicates mangrove areas. Modified after Chapman (1975) and Ponnamperuma (1984).

but also to waterlogging conditions, as a result of the gradually rise of the groundwater table.

2. Plant growth in saline soils

Most terrestrial plant species are not capable of growing well at a salt concentration in a nutrient solution of 50 mM NaCl or higher (Waisel 1972). Only a small group of halophytic plant species, that possesses a highly specialized type of physiology and morphology shows good growth under saline conditions, up to the salinity of sea water, i.e., 500 mM NaCl (Epstein 1985).

Even at this salt concentration, that occurs in all seas and oceans, occupying about 80% of the surface of the earth, most naturally halophytic plant species demonstrate some depression of growth, compared to growth at non-saline conditions (Rozema *et al.* 1985). From a physiological point of view this implies that most terrestrial halophytic higher plant species do not require seawater levels of salinity to reach an optimal growth rate. Under field conditions halophytes may be characterized as plant species, whose presence is dominant in saline areas. Apart from physiological adaptations, it is the success in the competitive interference with glycophytes that explains the abundance of halophytes on saline or brackish soils (Rozema *et al.* 1988). When salinity occurring in arable land exceeds 4 mmhos/cm (40 m eq/l), growth of many crop plants is reduced. The land therefore renders no longer useful for conventional agricultural practice and may become abandoned.

279

3. Saline soils and salinated arable land

South Asia is a region with a vast area of salinated soils (Fig. 1, Table I). Within this region Pakistan and India in particular face the problem of extensive areas with salt affected arable land. In Pakistan the area with salt affected arable land is still continuously growing with a rate of about 250 acres per day. It has been estimated that in total an area of more than 600,000 acres is useless for agricultural purposes (Sandhu & Qureshi 1985). About 80% of the irrigated land in Pakistan is more or less affected by salinity (Toeniessen 1984). Salt affected arable land in Pakistan implies serious problems to the agricultural system and economy of the country. Saline soils are abundant in the Sind and Punjab areas. About 81.0% of salt affected soils of the Punjab is saline sodic and the rest (19.0%) is saline. For Sind about 50% of the salinated arable land is saline sodic and the other half is saline. The agricultural problems of salt affected arable land are twofold. Firstly, high concentrations of soluble salts produce harmful effects on most agricultural plant species. Secondly, increased soil salinity badly affects the structure of the arable soil.

In saline soils, the content of soluble salts exceeds 0.1% of the dry weight of the soil, depressing the yield of crop plants. However, the soil characteristics may not be affected. Salts in saline soils consist of a mixture of sodium, magnesium, calcium as cations and chloride and sulfate as anions.

Sodium and chloride are the dominating ions. Alternatively, the U.S. Salinity Laboratory Staff (1954) defines saline soils as soils with an electrical conductivity (EC) of the saturation extract exceeding 4 millimhos/cm (about 40 m eq/l) and an exchangeable sodium percentage (ESP) less than 15; sodic soils (non-saline alkali soils) may technically be defined as soils with an ESP greater than 15 and with an EC less than 4 millimhos/cm. The high ESP modifies the physical and chemical properties of the soil. Clay particles in the soil become dispersed. The dispersed clay colloids block the pores in the soil, reducing the hydraulic conductivity of the soil and leading to poor aeration.

Saline alkali soil refers to soils with an ESP > 15 and EC > 4 mmhos/cm. The properties of these soils depend on the ratio of soluble salts and exchangeable sodium. Properties of sodic soils may be amended by the application of $CaSO_4$ or $CaCl_2$ improving permeability, drainage and aeration (Ahmad et al. 1988).

In Table II physico-chemical characteristics of non-saline and saline arable land in the North-Western Frontier Province in Pakistan is presented. The Sodium Adsorption Ratio (SAR) of the soil refers to the ratio:

$$SAR = \frac{Na^+}{\sqrt{\frac{(Ca^{2+} + Mg^{2+})}{2}}}$$

where the concentration of the cations can be given

Table I. Distribution of the area (million hectares) of saline soils on the land surface of the continents of the world. The percentage of the total is given in brackets.

Region	Strongly saline		Moderately saline		Total	
Europe	0.5	(0.4)	1.5	(0.6)	2.0	(0.6)
Africa	16.5	(14.5)	37.0	(15.6)	53.5	(15.5)
Australia	16.6	(14.6)	0.8	(0.3)	17.4	(5.0)
Mexico and Central America	0.24	(0.2)	1.72	(0.7)	1.96	(0.6)
North America	0	(0)	120	(5.2)	6.2	(1.8)
South America	10.5	(9.2)	58.9	(24.8)	69.4	(20.1)
North and Central Asia	22.5	(19.7)	69.2	(29.2)	91.7	(26.5)
South Asia	47.2	(41.4)	36.1	(15.3)	83.3	(24.1)
South East Asia	0	(0)	20.0	(8.4)	20.0	(5.7)
Total	114.04	(100.0)	237.22	(100.0)	345.06	(100.0)

From Massoud (1974); Chapman (1975) and Ponnamperuma (1984).

in milli-equivalents per liter. In the salt-affected soil the SAR, assessing relative abundance of Na^+ is high and, correspondingly the Exchangeable Sodium Percentage.

4. Saline arable land in Pakistan

The Indus basin of Pakistan is one of the large fertile river plains with the potential of providing food for human consumption. The river Indus contains water from snowmelt and precipitation of the Himalaya Mountains and this is being used for irrigation practices. In Pakistan the total area of salinized land consists of 14 million acres of arable land. The increased salinity, waterlogging and the changed chemical and physical characteristics make conventional crop production impossible. The soils of salt-affected arable land have increased salinity and in addition to this, the soils are impermeable, poor in organic matter and low in biological activity (Malik *et al.* 1986).

The complete solution of the problem of salinated waterlogged arable land requires an expensive energy and water consuming drainage system to remove excess ground water and salt accumulated in the surface layer of the soil. If it is possible to apply vertical drainage to lower the ground water table, the problem arises where to dispose the saline waste water that has been pumped out. Also, complete reclamation of salinated land requires replacement of the accumulated sodium by calcium. This can be done by application of powdered gypsum from formations of rocks in Pakistan (Ahmad *et al.* 1988). The costs for engineering, management and energy required makes that these improved drainage techniques will have only local importance, and on relatively small areas.

In Pakistan, like in many other semi-arid and arid countries, the area of salinated arable land is infact still continuously growing at a rate of 250 acres per day (Sandhu & Qureshi 1985). In some areas of Pakistan farmers have moved to other parts in the country because agricultural production is no longer possible. In other parts of Pakistan the problem of salt affected soils seriously reduces the income of the rural population.

Irrigation of the river plains of the Punjab has extended the area for agricultural use markedly. In a recent study Hendrickx (person. comm. 1989) concluded that the salinity problem of the arable land in Pakistan can only partly be solved with improved drainage and irrigation practices. From model calculations it is shown that the water supply by the river Indus allows proper irrigation of a only limited area of Pakistan, avoiding the severe problems of salinization. This implies however, that a vast area of saline arable land will remain and the search for biosaline exploitation becomes even more important.

5. Crop growth in saline arable land

Plant species in general, including crop plants, differ with regard to their tolerance to salinity. It may also be that crop plants are tolerant to salinity at one growth stage but sensitive during another stage of the life cycle. In general, plants are most sensitive to salinity during seedling growth. When salinity is increased, growth will be suppressed, for sensitive species already at a low salt concentration, for tolerant species at a high salinity. Other environmental factors may seriously affect salt tolerance. Most crops tolerate greater stress under cool and humid conditions compared to hot and dry conditions (Waisel 1972). Plants grown on infertile soils may seem more salt tolerant than crops cultivated in a fertile medium, because fertility, rather than salinity appears to be the primary factor limiting growth. Fertilization will increase yields of saline soils proportionally more than of non-saline soils. Relative salt tolerance of crops may be based on several criteria. Firstly, salt tolerance during germination and emergence, this is in fact based on survival of early seedlings. Secondly, salt tolerance of older plants may be based on a decrease of growth rate or yield. Barley and cotton have a high relative salt tolerance with a 50% yield reduction at 18 dS/m and 50% seedling emergence at about 20 dS/m. Sugarbeet is equally salt tolerant based on the yield reduction criterion, but 50% emergence reduction is already obtained at 9 dS/m, which implies that germination and emergence is more salt sensitive than of cotton and barley. Rice and bean are among sensitive crops (50% yield reduction at 3.6 dS/m) (Maas 1986). Using a threshold and slope parameter Maas (1986) has ranked numerous herbaceous and woody crops according to their salt tolerance. The threshold represents the maximum allowable salinity without yield reduction below that for non saline conditions, and slope is expressed by the percent yield decrease per unit increase in salinity beyond the threshold. The electrical conductivity is measured of an extract of a saturated soil paste, at 25°C with units of deci-siemens/meter (1 dS/m = 1 mmhos/cm).

6. Possibilities to improve salt tolerance of crop species

When high soil salinity in salt-affected fields obstructs the growth of many conventional crop plants, one may look for new salt tolerant crop species. The yield of salt tolerant crops on the salinated land must then be at least high enough for extensive agricultural units.

In theory, several possible ways are open to achieve salt tolerance or increased salt tolerance of crop species.
1. Screening and selection for salt tolerance in existing races and cultivars of crop species.
2. Breeding for salt tolerance using crosses between non-tolerant and salt-tolerant crop plant species.
3. Cell culture and recombinant DNA methods to improve salt tolerance of plants.
4. Domestication of natural halophytic plant species to salt tolerant crop species.
5. In a biosaline type of agricultural system wild halophytic plants are being cultivated extensively in salt affected land and used for example as forage crop (Malcolm & Swaan 1985; Ahmad & San Pietro 1986).

Genetic selection for salt tolerance assumes the existence of genetic variability for salt tolerance. Therefore the question arises: what is the genetical background of salt tolerance. At the same time it is desirable to know whether salt tolerance of plants may be recognized not only from growth under saline conditions, but also from physiological parameters (Epstein 1985; Meiri & Plaut 1985).

The limited genetic variability of salt tolerance in various crop plants led Mudie (1975) and Tal (1985) to conclude that longterm cultivation of these crops in a non-saline environment has resulted in a loss of genes for salt tolerance from the gene pool. However, in a screening program for salt tolerance in rice, many salt tolerant races have been obtained (Ponnamperuma 1984). Yeo & Flowers (1984) hypothesized that the existence of environmental stress will lead to reduction of genetical variation. Gene diversity for salt tolerance may therefore be much lower in natural halophytes than in non-stressed glycophytic plant species. Irrespective of the degree of genetic variability, there seems to be general agreement on the favourable perspective of improvement of salt tolerance of crop plants through a selection and breeding program. However, knowledge of the genetic control of salt tolerance is poor. It seems not likely that a one gene relationship with total salt tolerance will be demonstrated (Shannon 1984). The polygenic basis of salt tolerance will imply that transfer of salt tolerant genes will require at least three crosses rather than one or two.

From a plant physiological point of view salt tolerance is regarded as a characteristic of the whole plant and adaptations at the cellular level may only represent part of the complete set of mechanisms that we call salt tolerance.

7. Conventional screening and breeding for improved salt tolerance

Through conventional screening, breeding and genetic engineering it is attempted to improve salt tolerance of wheat, cotton, rice, barley, sugar cane, pulses and *Brassica* species (Shannon 1985, Abdullah 1987a,b). Wheat (*Triticum aestivum*) (cf. Gorham *et al.* 1985) is particularly important because it is a staple food in many irrigated valleys such as the Indus basin of Pakistan. In addition to the exploitation of existing variation in tolerance to salt in cultivars and species of wheat, it is being attempted to incorporate genes from wild salt tolerant grasses, *Aegilops squarrosa* and the sand couch *Thinopyrum bessarabicum* (Gorham *et al.* 1985). This sand couch is native to sand dunes of the Black Sea in the Soviet Union, and the species is resistant to exposure to high salinity levels. The hybrids of sand couch and wheat have inherited the dominant salt tolerant genes of sand couch and these are expressed in the hybrid wheat. In hybrid wheat enhanced tolerance to salt seems to be contributed by a single chromosome of *T. bessarabicum*. This finding suggests that for wheat only a limited number of genes have a major effect on tolerance to salt. For wheat breeding programs it is important to aim at improval of both salt tolerance and flooding tolerance because salt affected land is often also waterlogged (Forster 1988).

Cultivation of Fodderbeet under saline conditions

A screening and breeding program to test and improve salt tolerance and flooding tolerance of fodderbeet is now being carried out. Using seed of a variety of cultivars it was found at the National Agricultural Research Center in Islamabad that germination and seedling growth of the fodder beet, *Beta vulgaris* spp *vulgaris* represent stages of the life cycle of beet that are more susceptible to salt injury than growth of the adult plant (Abdullah

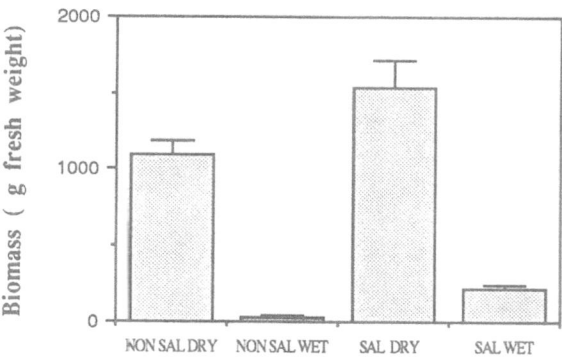

Fodderbeet *Beta vulgaris* cv majoral

Fig. 2. Growth (g fresh weight root and shoot) of Fodderbeet (*Beta vulgaris*) cultivar majoral under non-saline and saline as well as dry and waterlogged conditions. Average value of 4 plants with standard error of the mean. Growth period of the plants was 5 months, in an experimental field.

Table II. Physico-chemical characteristics of non-saline and saline soil used for the cultivation of Fodderbeet cultivars at Mardan, North-Western Frontier Province, Pakistan. SAR = Sodium Adsorption Ratio of the soil. ESP = Exchangeable Sodium Percentage (Fodderbeet experiments at Mardan, N.W.F.P.).

Soil properties	Non-saline	Saline
pH	7.82	8.85
EC_e dS m^{-1}	1.23	5.17
CO_3^{2-} meq/l	–	–
HCO_3^- meq/l	1.0	2.0
Cl^- meq/l	10.0	35.0
$Ca^{2+} + Mg^{2+}$ meq/l	16.5	8.25
Na^+ meq/l	1.74	81.0
K^+ meq/l	0.4	5.0
SAR	0.61	40.0
ESP	2.16	38.2

1987a). In growth experiments in the greenhouse and on an experimental field of the Department of Ecology and Ecotoxicology, Amsterdam growth of fodder beet cultivars appeared not to be reduced but on the contrary, to be stimulated by a salinity level of 50–100 mM NaCl in the soil moisture (Fig. 2). However, under waterlogged conditions, both in non-saline and saline treatment, growth of the fodder beet is severely reduced. Successful cultivation of fodderbeet in salt affected waterlogged soils requires not only salt tolerance of cultivars but also flooding tolerance. In addition to this varieties of fodderbeet (*Beta vulgaris* spp. *vulgaris*) are being crossed with the maritime progenitor *Beta vulgaris* ssp. *maritima,* the sea beet.

Growth of Fodderbeet on saline arable land in Pakistan

Growth of four varieties of the fodderbeet (Monoval, Monored, Majoral and Polygroeningia) has been studied at the area of Mardan and D. I. Khan (Ratta Kulachi) in the North West Frontier Province in Pakistan in the winter period (1988/1989). Most of the area is canal irrigated. Low lying arable lands are salt-affected by salinity and sodicity due to seepage and the crop productivity (sugarcane)

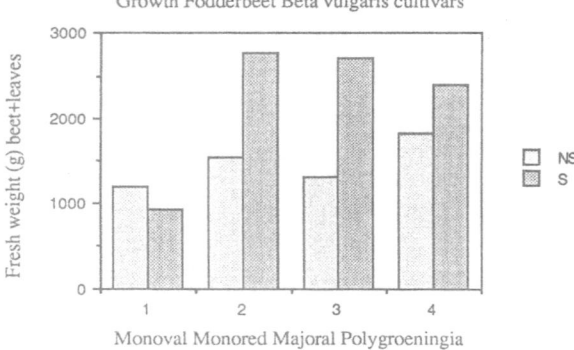

Growth Fodderbeet Beta vulgaris cultivars

(y-axis: Fresh weight (g) beet+leaves; legend: NS, S; x-axis: 1 2 3 4 / Monoval Monored Majoral Polygroeningia)

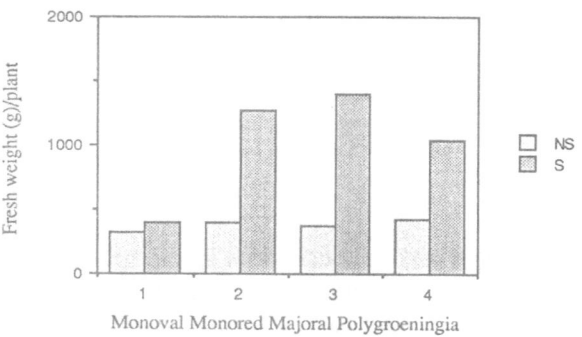

Growth Fodderbeet Beta vulgaris/leaves

(y-axis: Fresh weight (g)/plant; legend: NS, S; x-axis: 1 2 3 4 / Monoval Monored Majoral Polygroeningia)

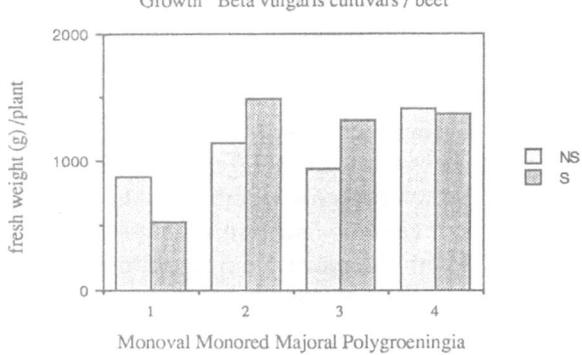

Growth Beta vulgaris cultivars / beet

(y-axis: fresh weight (g) /plant; legend: NS, S; x-axis: 1 2 3 4 / Monoval Monored Majoral Polygroeningia)

Fig. 3. Growth four cultivars of the fodderbeet grown in non-saline and saline soil (See Table I) at the field station Mardan, North Western Frontier. Province, Pakistan 1987.

has been reduced. The salt-affected arable land covers about 1.264 millions hectares of the North West Frontier Province. (Niazi & Abdullah 1989). A comparison has been made between growth of these cultivars under non-saline soil and saline soil (Table II). Growth of the varieties Monored,

Majoral and Polygroeningia was markedly increased at saline soil conditions. Only in the case of cultivar monoval there was a slight reduction of fresh weight per plant (Fig. 3). The content of chlorophyll a and b increased with fodderbeet plants of all four varieties grown in saline soil.

As is the case for most halophytic plant species (Rozema 1975), the germination of fodderbeet is sensitive to salinity (Abdullah 1987a). With an increase of the electric conductivity from 4 to 20 d S m^{-1}, the germination percentage of fodderbeet seed decreased from 80% to 30%. Both the studies

Table III. Relative salt tolerance of crops based on maximum allowable salinity (threshold) and percent yield reduction per unit increase in salinity above the threshold (slope) salinity is measured as electrical conductivity of a saturnated soil extract and expressed as $dS.m^{-1}$. Modified after Maas (1986) and Malik *et al.* (1980).

	threshold	slope
Salt tolerant crops		
Barley, *Hordeum vulgare*	8.0	5.0
Cotton, *Gossypium hirsutum*	7.7	5.2
Sugarbeet, *Beta vulgaris*	7.0	5.9
Wheat, *Triticum aestivum*	8.6	3.0
Bermuda grass, *Cynodon dactylon*	6.9	6.4
Kallar grass, *Leptochloa fusca*	20	22.3
Date palm, *Phoenix dactylifera*	4.0	3.6
Medium tolerant crops		
Soy bean, *Glycine max.*	5.0	20.0
Cowpea, *Vigna unguiculata*	4.9	12.0
Sorghum, *Sorghum bicolor*	6.8	16.0
Medium sensitive crops		
Sugar cane, *Saccharum officinarum*	1.7	5.9
Corn, *Zea mays*	1.7	12.0
Broadbean, *Vicia faba*	1.6	9.6
Berseem, *Trifolium alexandrium*	1.5	5.7
Cucumis, *Cucumis sativis*	2.5	13.0
Lettuce, *Lactuca sativa*	1.3	13.0
Tomato, *Lycopersic esculentum*	2.5	9.9
Sensitive crops		
Rice, *Oryza sativa*	3.0	12.0
Bean, *Phaseolus vulgaris*	1.0	19.0
Carrot, *Daucus carota*	1.0	14.0
Grapefruit, *Citrus paradisi*	1.8	16.0
Orange, *Citrus sinensis*	1.7	16.0
Plum, *Prunus domestica*	1.5	18.0

on the experimental field in Amsterdam and the studies on saline arable land in Pakistan indicate good growth of fodder beet. It appears that fodder beet cultivars are more salt tolerant than many other crops, even under the hot conditions in Pakistan (Maas 1986). Beet and leaves of Fodderbeet have considerable value for cattle feeding. Raising of cattle implies the production of milk and meat.

Kallar grass (*Diplachne (Leptochloa)fusca*) has also successfully been grown on saline sodic soils of Pakistan (Malik *et al.* 1986). Kallar grass is used as fodder crop and grown well during the summer. However, there is no growth of Kallar grass during the winter period. The dead grass may be adequate as fodder for buffaloes but not for goats, in which marked losses in weight and fertility occur.

Therefore the alternate cultivation of Kallar grass and Fodderbeet on saline land in the summer and winter time seems to be very promising.

Also other crops can be applied in a rotation scheme with Fodder beet and Kallar grass: legumes like Chickpea (Abdullah 1987), Egyptian clover (*Trifolium alexandrinum*), salt and flooding tolerant *Juncus* species, salt tolerant chenopods for forage and oil seeds, and a variety of *Brassica* species.

A problem in this respect seems to be the negative coupling of increase of salt tolerance and the growth rate: the coastal beet is tolerant to seawater salinity (500 mM NaCl) but grows slowly. Less salt tolerant cultivars of the domesticated fodder beet exhibit a much higher growth rate.

8. Perspective of biotechnology for the salt tolerance of crops: cell culture and recombinant methods

Cell culture techniques provide several advantages for use in selection for salt tolerance (Hanson 1984; Gamborg *et al.* 1986). The environment and growth medium are well defined. Complications related to differences in the response to salinity due to the developmental stage of intact plant do not exist. Cell culture makes study of cellular physiology of salt tolerance easily possible. And of course, millions of cells can be screened for salt tolerance. There are also difficulties associated with cell cul-

ture: regeneration from selected cells, and the decrease of the capacity for regeneration with time. Cell culture techniques for selection of salt tolerance have been applied to *Nicotiana tabacum, Capsicum annuum, Citrus sinensis, Oryza sativa* and *Medicago sativa* (Kochba *et al.* 1982; Hasegawa *et al.* 1986).

In our laboratory successful callus and cell suspension cultures of fodder beet have been obtained. Also regeneration from callus culture is possible. Studies for selection of salt tolerant cell lines are in progress. There has been limited success for this biotechnological and genetic engineering approach. Possibly this relates to the discrepancy that may exist between salt tolerance at the cellular level and that of the whole plant (Hasegawa *et al.* 1986, Mcloy 1987, Rozema *et al.* 1986, 1987). Nevertheless further extension of cell culture and recombinant methods is needed, combined with improved physiological and genetical knowledge of salt tolerance to help to understand adaptations to a saline environment and the possibilities for genetic engineering (Cheeseman 1988).

9. Domestication of natural halophytes and the biosaline approach

In Pakistan a group of scientists from the Nuclear Institute for Agriculture and Biology (NIAB), Faisalabad, have successfully developed the cultivation of Kallargrass (*Leptochloa fusca*). Kallar grass, having C-4 photosynthesis is a highly salt tolerant perennial grass, which grows well even under waterlogged conditions (Malik *et al.* 1986). Kallar grass may provide 50 tons of biomass per hectare when irrigated with brackish water. These and other attributes make it a suitable crop to be introduced into salt affected land.

In Australia, establishment and cultivation of halophytic shrubs has been made possible by development of the niche seeding techniques. In this technique difficulties caused by salinity, high temperature and waterlogging are reduced and is now in commercial use. Grazing experiments and farmer experience have shown that halophyte shrub pastures (*Atriplex amnicola, A. undulata, Halosar-*

cia pergranulata, Maireana brevifolia, Paspalum vaginatum and Puccinellia ciliata) provide useful fodder during the autumn and winter feed supply gap (Malcolm 1986).

In sandy inland deserts in Pakistan comprising of Thal, Cholistan and Tharparkar and in a coastal sandy belt ranging from Karachi to Gwadar, scarcity of water has made the land barren. Irrigation of these sandy soils with saline water has made possible the cultivation of various salt tolerant plant species such as the grasses *Leptochloa fusca, Panicum turgidum* and *Sporobolus arabicus,* the trees *Eucalyptus camaldulensis, Azadirachta indica* and *Casuarina equisetifolia* (Ahmad *et al.* 1986). The shrub *Prosopis juliflora* proved to grow well under irrigation with saline underground water (Khan *et al.* 1986). Following this biosaline approach, that is the production of (plant) biomass using saline water for water and nutrient supply, plant material is gained that can be used for fodder, fuel and wood. On salt flats of the United Arab Emirates, the halophyte *Salicornia* from saline inland and coastal sites in the U.S. is now grown as a crop, for the time being mainly to provide fodder for livestock in arid lands (Charnock 1988). A selected strain bred from germplasm in the U.S. known as *Salicornia* oilseed selection (SOS-7), may be grown to obtain vegetable oil. The perennial halophytic herb *Kosteletzkya virginica* produces seed resembling millet, and contains 25% protein and 15% oil. Irrigation with seawater (25% salinity) may give rise to a yield of 1500 kg per hectare (Gallagher 1985). Domestication of wild halophytes to forage crops or vegetable crops seems to be a promising outlook.

Domestication of natural halophytes and the development of crop varieties with a high growth rate and tolerance to seawater salinity will be an important challenge for both plant breeders, geneticists and physiologists.

The irrigation systems, particularly in the developing countries have no adequate drainage and it is unlikely that the problem of salt affected arable land will reduce on a short term. Bio-saline production of plant biomass represents a good perspective and support for the agricultural economy of arid and semi-arid areas. The scientific part of biosaline production is not easy and deserves the help of western knowledge of plant science and plant biotechnology.

Acknowledgements

The author is indebted to Ms. D. Hoonhout for typing the manuscript and to Mr. M. B. Meyer for correction of the English text. The seed material of the cultivars of fodder-beet was kindly provided by Zwaan en de Wiljes, Scheemda, The Netherlands. The Center for Development Cooperation Services of the Vrije Universiteit is acknowledged for financial support for the visit of institutes in Pakistan. The author is indebted to Dr Zaib-un-Nisa Abdullah and Mr Banaras Niazi of the Stress Physiology group of the National Agricultural Research Center, Islamabad for the data on growth of fodder beet at Mardan, N.W.F.P.

References

Abdullah, Zaib un Nisa, 1986. Physiology and biochemistry of plants under saline environment. Ph.D. Thesis University of Karachi: pp. 208.

Abdullah, Zaib un Nisa, 1987a. Physiology of Fodderbeet Cultivation. In: Pakistan Agricultural Research Council. Annual. Report. pp. 178—181.

Abdullah, Zaib un Nisa 1987b. Potential of Chickpea and constraints to its production in Pakistan. In: Adaptation of chickpea and pigeonpea to abiotic stresses. Proc. Consultants workshop/Icrisat, India pp. 143—148.

Ahmad, R., Ismail, S. & Khan, D., 1986. Use of highly saline water for irrigation at sandy soil. In: R. Ahmad & A. San Pietro (eds.) Prospects for biosaline research. Shamir Printing Press, Karachi, pp. 389—414.

Ahmad, M., Niazi, B. H. & Sandhu, G. R., 1988. Effectiveness of gypsum, HCl and organic matter for the improvement of saline sodic soils. Pakistan J. Agric. Res. 9: 373—378.

Ahmad, R. & San Pietro, A., 1986. (eds.) Prospects for biosaline research. Shamir Printing Press, Karachi, pp. 587.

Chapman, V. J., 1975. The salinity problem in general, its importance and distribution with special reference to natural halophytes. In: A. Poljakoff-Mayber & J. Gale (eds.) Plants in saline environments. pp. 7-24.

Charnock, A., 1988. Plants with a taste for salt. New Scientist 3: 41—45.

Cheeseman, J. M., 1988. Mechanisms of salinity tolerance in plants. Plant Physiol. 87: 547—550.

Epstein, E., 1985. Salt tolerant crops: origins, development, and prospects of the concept. Plant & Soil. 89: 187—198.

FAO-UNESCO, 1973. Irrigation, drainage and salinity, An international source book. Hutchinson, London.

Forster, B., 1988. Wheat can take on more than a pinch of salt. New Scientist 3: 43—45.

Gallagher, J. L., 1985. Halophytic crops for cultivation at seawater salinity. Plant and Soil 89: 323—376.

Gamborg, O. L., Ketchum, R. E. B., Nabors, M. W., 1986. Tissue culture and cell biotechnology for increased salt tolerance in crop plants. In: R. Ahmad & A. San Pietro (eds.) Prospects for biosaline research. Shamir Printing Press Karachi, pp. 93—100.

Gorham, J., McDonnell, E., Budrewicz, E. & Wyn Jones, R. G., 1985. Salt tolerance in the *Triticeae:* growth and solute accumulation in leaves of *Thinopyrum bessarabicum*. J. Exp. Bot. 36: 1021—1031.

Hanson, M. R., 1984. Cell culture and recombinant DNA methods for understanding and improvening salt tolerance of plants. In: R. Staples (ed.) Salinity tolerance in plants: strategies for crop improvement. pp. 335—359.

Hasegawa, P. M., Bressan, R. A. & Hauda, A. K., 1986. Cellular mechanisms of salinity tolerance Hort. Science. 21: 1317—1324.

Khan, D., Ahmad, R. & Ismail, S., 1986. Case history of *Prosopis juliflora* plantation at Makhan coast raised through saline water irrigation. In: R. Ahmad & A. San Pietro (eds.) Prospects for biosaline research. Shamir Printing Press Karachi, pp. 559—585.

Kochba, J., Ben-Hayim, G., Spiegel, Roy, P., Saad, S. & Neuman, H., 1982. Selection of stable NaCl-tolerant callus cell lines and embryos in *Citrus*. J. Plant Physiol. 106: 111—118.

Maas, E. V., 1986. Salt tolerance of plants. Applied Agricultural Research 1: 12-26.

Malcolm, C. V., 1986. Rainfed halophyte forage production on salt affected soils. In: R. Ahmad & A. San Pietro (eds.) Prospects for biosaline research. Shamir Printing Press Karachi, pp. 542—551.

Malcolm, C. V. & Swaan, T. C., 1985. Soil mulches and sprayed coatings and seed washing to aid chenopod establishment on saline soil. Aust. Ranged. J. 7: 22—28.

Malik, K. A., Aslam, Z. & Naqvi, M., 1986. Kallar grass, a plant for saline land. Ghulamali Printers, Lahore, pp. 93.

Massoud, F. I., 1974. Salinity and alkalinity as soil degradation hazards. FAO/UNDP. Expert consultation on soil degradation. Rome.

McLoy, T. J., 1987. Tissue culture evaluation of NaCl tolerance in *Medicago* species: Cellular versus whole plant response. Plant Cell Reports. 6: 31—34.

Meiri, A. & Plaut, Z., 1985. Crop production and management under saline conditions. Plant and Soil 89: 253—271.

Mudie, P. J., 1974. The potential economic uses of halophytes. In: P.J. Reimold & W.H. Queen (eds.) Ecology of Halophytes. Academic Press, New York pp. 565—597.

Niazi, B. H. & Abdullah, Z., 1989. Physiology of fodderbeet cultivation in Pakistan under saline conditions. National Agricultural Research Centre 30 pp.

Ponnamperuma, F. N., 1984. Role of cultivar tolerance in increasing rice production on saline lands. In: R.C. Staples & G.H. Toeniessen (eds.) Salinity tolerance in plants: strategies for crop improvement. John Wiley, New York pp. 255—271.

Reeve, R. C. & Fireman, M., 1967. Salt problems in relation to irrigation. Agronomy 11: 988—1008.

Richards, L. A., 1954. Diagnosis and improvement of saline and alkali soils. U.S. Dept. Agric. Hand Book No. 60.

Rozema, J., 1975. The influence of salinity, inundation and temperature on the germination of some halophytes and non-halophytes. Oecol. Plant. 10: 341—353.

Rozema, J., 1990. Growth, water and ion relations of monocot and dicot halophytes; a unified concept. Aquat. Botany (in press).

Rozema, J., Arp, W., Diggelen, J., van, Kok, E., Fanger, A. M. & Letschert, J., 1986. Comparative ecophysiology of the water relations of salt resistant monocotyledonae and dicotyledonae. In: R. Ahmad & A. San Pietro (eds.) Prospects for biosaline research. Shamir Printing Press Karachi, pp. 101—114.

Rozema, J., Arp, W., Diggelen, J. van, Kok, E. & Letschert, J., 1987. An ecophysiological comparison of measurements of the diurnal rhythm of the leafelongation and changes of the leaf thickness of salt-resistant dicotyledonae and monocotiledonae. J. Exp. Bot. 38: 442—453.

Rozema, J., Bijwaard, B., Prast, G. & Broekman, R., 1985. Ecophysiological strategies of coastal halophytes from fire dunes and salt marshes. Vegetatio 62: 487—497.

Rozema, J., Scholten, M. C. T., Blaauw, P. A. & Diggelen, J. van., 1988. Distribution limits and physiological tolerances with particular reference to the salt marsh habitat. In: A.J. Davy, M.J. Hutchings & M.J. Watkinson (eds.) Plant Population Ecology, Blackwell, Oxford pp. 137—164.

Sandhu, G. R. & Qureshi, R. H., 1985. Salt affected soils of Pakistan and their utilization. Reclam. Reveg. Res.

Shannon, M. C., 1984. Breeding, selection and the genetics of salt tolerance. In: R.C. Staples & G.H. Toeniessen (eds.) Salinity tolerance in plants: strategies for crop improvement. John Wiley, New York, pp. 231—254.

Shannon, M. C., 1985. Principles and strategies in breeding for higher salt tolerance. Plant and Soil. 89: 227—241.

Stavarek, S. J., & Rains, D. W., 1984. Cell culture techniques: selection and physiological studies of salt tolerance. In: R.C. Staples & G.H. Toeniessen (eds.) Salinity tolerance in plants: strategies for crop improvement, John Wiley, New York, pp. 321-334.

Tal, M., 1985. Genetics of salt control in higher plants: theoretical and practical considerations. Plant and Soil. 89: 187—198.

Toeniessen G. H., 1984. Review of the world food situation and the role of salt tolerant plants. In: R.C. Staples & G.H. Toeniessen (eds.) Salinity tolerance in plants: Strategies for crop improvement. pp. 399—413.

USDA, 1954. Salinity laboratory staff: saline and alkali soils. U.S. Dept. Agriculture Handbook no. 60. U.S. Government Printin Office, Washington, DC. pp. 160.

287

Waisel, Y., 1972. Biology of halophytes. Academic Press pp. 395.

Yeo, A. R. & Flowers, T. J., 1984. Mechanisms of salinity resistance in rice and their role as physiological criteria in plant breeding. In: R.C. Staples & G.H. Toeniessen (eds.) Salinity tolerance in plants: strategies for crop improvement. pp. 151—169.

Author's Address
J. Rozema
Department of Ecology and Ecotoxicology
Vrije Universiteit
De Boelelaan 1087
1081 HV Amsterdam
The Netherlands

288

Infestation of faba beans in Syria by the parasitic weed *Orobanche crenata*. (Photograph: M. J. van Hezewijk.)

CHAPTER 25

Genetic variability in *Orobanche* (broomrape) and *Striga* (witchweed) and its implications for host crop resistance breeding

A. H. PIETERSE & J. A. C. VERKLEIJ

Abstract. An overview is given of what is known of genetic variability in the economically most important species of the parasitic weeds *Orobanche* (broomrape) and *Striga* (witchweed). Data are included on morphotypes, chromosome numbers, iso-enzyme patterns and biotypes differing in virulence to their hosts. It has been concluded that there is ample variability and that, especially in the autogamous species, an important part of the variability is genetically determined. As a consequence it was concluded that in breeding programmes for tolerant/resistant crops, apart from available defense mechanisms in the hosts, the genetic variability of the parasites should be taken into consideration. The population ecology of weedy *Orobanche* and *Striga* species has been discussed in connection with their genetic variability and it has been hypothesized that adaptation to agro-ecosystems may have brought about a significant effect on their survival strategy. In order to obtain a better understanding of the evolution from wild parasitic plants into agressive parasitic weeds as well as the genetic resource in wild hosts regarding tolerance/resistance mechanisms, it has been proposed to carry out comparative studies with weedy biotypes of *Orobanche* and *Striga* species in natural vegetation.

1. Introduction

Species of *Orobanche* (broomrape) and *Striga* (witchweed) cause serious losses in the yields of important crop plants in sub-tropical and tropical regions (Pieterse 1979, Pieterse & Pesch 1983, Parker & Wilson 1986, Musselman 1982, 1987a). Broomrapes (Family: Orobanchaceae) and witchweeds (Family: Scrophulariaceae) are root parasites which take up nutrients and water from their hosts. *Striga* species may also bring about a toxic effect. The control of these parasitic weeds is very difficult, especially because the small seeds, which are shed in large numbers, in general only germinate if exposed to germination stimulants. These germination stimulants occur in root exudates of host plants (and certain non-host plants).

Orobanche is mainly a problem in the Middle East, in areas around the Mediterranean and in South-eastern Europe. *O. crenata* Forssk. attacks faba bean (*Vicia faba* L.), lentil (*Lens culinaris*

Medik.), chickpea (*Cicer arietinum* L.), vetch (*Vicia sativa* L.) and other leguminosae. *O. ramosa* L. and *O. aegyptiaca* Pers., which are closely related, parasitize mainly upon tomato (*Lycopersicon esculentum* Mill.), hemp (*Cannabis sativa* L.), tobacco (*Nicotiana tabacum* L.), lentil and some other crops. *O. cumana* Wallroth (= *O. cernua* Loefl.) causes extensive losses in sunflower (*Helianthus annuus* L.), especially in the Soviet Union. *O. minor* Sm., which represents a taxonomically complex group of variants found throughout the world, is less important. However, it can be troublesome on clover species and sporadically on some other crops. The distribution of the *Orobanche* species of economic importance is shown in Fig. 1 (a–e).

Striga is an important pest in the semi-arid regions in Africa, especially in the Sahel. *S. hermonthica* (Del.) Benth. and *S. asiatica* (L.) Kuntze (which is also a problem in India and a small area in the eastern part of the United States where it has been introduced in or shortly before the 1950s)

J. Rozema and J. A. C. Verkleij (eds.), Ecological Responses to Environmental Stresses, 290—302.
© *1991 Kluwer Academic Publishers.*

attack sorghum (*Sorghum vulgare* Pers.), pearl millet (*Pennisetum typhoides* (Burm. f.) Stapf & Hubbard), maize (*Zea mays* L.) and other grasses. *S. gesnerioides* (Willd.) Vatke attacks cowpea (*Vigna unguiculata* (L.) Walp.). The distribution of the *Striga* species of economic importance is shown in Fig. 2 (a–c). A main approach which is followed to tackle these parasitic weed problems is breeding for resistant or tolerant crops. Up to now the most promising results have been obtained with sunflower, sorghum, cowpea and vetch (Aggarwal *et al.* 1984, 1986; Gil *et al.* 1984; Aggarwal 1987; Cubero 1986; Ramaiah 1987, 1988; Cubero *et al.* 1988). In maize, hybrids have been produced which are tolerant to *S. hermonthica*, i.e., the plants are infected but the parasite does not cause much damage (Kim *et al.* 1985; Kim 1988).

In spite of the fact that large scale breeding programmes are carried out at various institutes, little attention has been paid to the genetic variability and population ecology of the parasites. However, it may be expected that the parasites are able to overcome resistance or tolerance (or at least certain types), especially when their genetical variability is high and resistance or tolerance of the host is determined by one or a few genes. This is illustrated by the infestation of resistant sunflower cultivars in the Soviet Union, which were bred in the beginning of this century, by new biotypes of *O. cumana* with a different physiology (Cubero 1986; Vrânceanu *et al.* 1986). In this context it should also be noted that different biotypes of *Orobanche* and *Striga* species have been reported which attack different hosts.

In the present paper an overview is given of what is known of genetic variability of economically important *Orobanche* and *Striga* species and how this is related to the population ecology of these parasites. It is considered to be a basis for further research, especially in connection with breeding programmes for resistant or tolerant crops.

2. Orobanche and Striga species of economic importance

At the (first) International Symposium on Parasitic Weeds, which was held in Malta in 1973, a paper was presented on the identification of economically important *Orobanche* and *Striga* species (Hepper 1973). Subsequently, Ramaiah *et al.* (1983) published an identification handbook on *Striga* with illustrations in colour. More recently Musselman (1986, 1987b) published extensive taxonomic reviews on *Orobanche* and *Striga*.

The genus *Orobanche* is divided into four Sections, i.e., Gymnocaulis, Myzorrhiza, Trionychon and Orobanche. The number of species is ca. 100, however, the taxonomy of the genus is complex and needs clarification. The *Orobanche* species of economic importance are within the Section Orobanche (in which the bracteoles are absent and the calyx is 1–2 teethed): *O. cumana*, *O. crenata* and *O. minor*, and within the Section Trionychon (characterized by two bracteoles and a four teethed calyx): *O. ramosa* and *O. aegyptiaca*.

Distinguishing factors of *Orobanche* species of economic importance:

– *O. cumana*. Height up to ± 50 cm; inflorescence unbranched; flowers up to 2 cm long, pale with dark blue-purple corolla lobes.
– *O. crenata*. Height up to ± 100 cm; inflorescence unbranched; flowers up to 3 cm long, colour white with purple markings, very fragrant.
– *O. minor*. Very variable. Height up to ± 70 cm; inflorescence unbranched; flowers up to 2.5 cm long, colour pale yellow tinged with pink, red or brown.
– *O. ramosa*. Height up to ± 30 cm; inflorescence usually branched; flowers up to 1 cm long, colour pale to bright blue, occasionally white; filaments not densely haired.
– *O. aegyptiaca*. Height up to ± 40 cm; inflorescence usually branched; flowers up to 2 cm long, colour pale to bright blue; filaments densely haired.

The taxonomy of *Striga* is much more simple than that of *Orobanche*. There are ca. 30 species of which three are of economic importance. Their distinguishing factors are:

– *S. hermonthica*. Height up to ± 50 cm; leaves

291

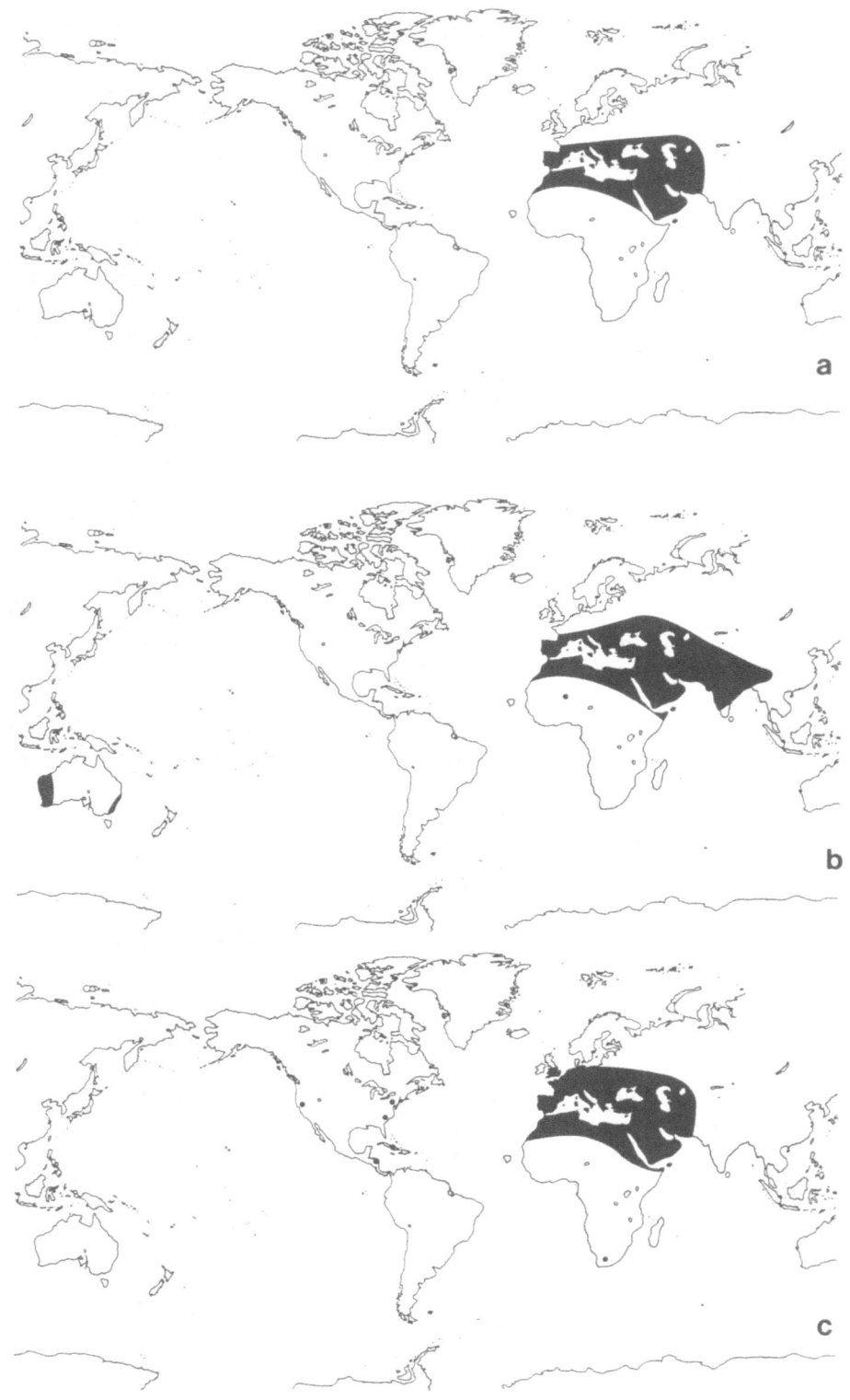

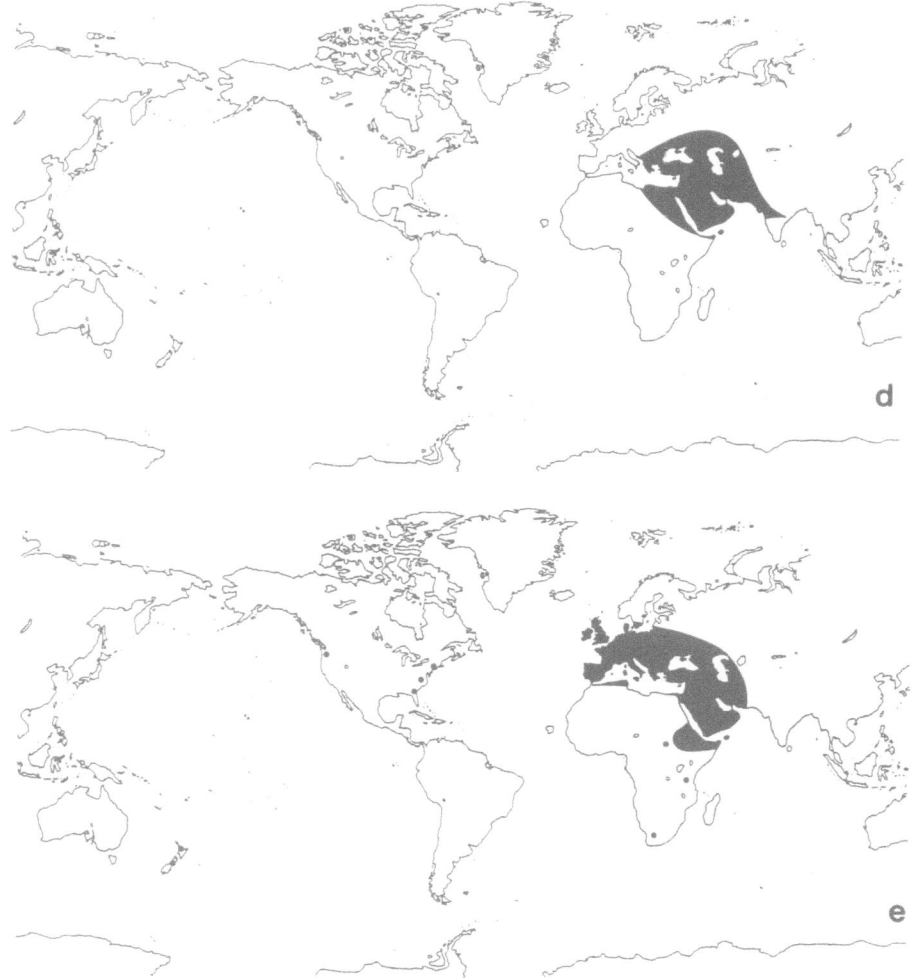

Fig. 1. a–e. General distribution of: a. *O. crenata.* b. *O. cernua.* c. *O. ramosa.* d. *O. aegyptiaca.* e. *O. minor.*

green, 2–6 cm long; flowers up to 1.5 cm long, colour bright pink; corolla tube about equal to calyx in length (± 1 cm); calyx ribs 5.

– *S. asiatica.* Plants extremely variable; height up to ± 30 cm; leaves green, 2–6 cm long; flowers up to 1 cm long, colour red, white or yellow; calyx ribs more than 5.

– *S. gesnerioides.* Plants extremely variable; height up to ± 15 cm; usually densely branched; leaves scale-like, rarely expending 0.5 cm in length; flowers up to 0.5 cm in length, variable in colour: cream-white/mauve/pink/purple; calyx ribs 5.

3. Breeding systems

A major factor influencing the genetic variability of flowering plants is the breeding system, i.e., whether the plants are autogamous (self-pollinators), allogamous (cross-pollinators) or a combination of both (mixed mating species) (Loveless & Hamrick 1984). Inbreeding, whether from autogamy or a restricted gene flow in mixed mating species, will increase the potential for genetic differentiation. As a consequence, the pollination system of *Orobanche* and *Striga* species may have an important effect on their variability.

293

Fig. 2. a–c. General distribution of: a. *Striga hermonthica*. b. *Striga asiatica*. c. *Striga gesnerioides*.

In 1982 Musselman *et al.* presented an overview of the breeding systems of agronomically important *Orobanche* and *Striga* species. Both autogamous and allogamous species occur. In *Orobanche* the species *O. cumana*, *O. minor* and *O. ramosa* are autogamous. *O. crenata,* on the other hand, was reported to possess three patterns of pollination: obligate allogamy, facultative autogamy and obligate autogamy. In general, however, it seems that cross pollination is the most common system in this species. *O. aegyptiaca* is also an outcrosser although there appears to be a form of facultative autogamy if pollen vectors are absent. Cross-pollination in *Orobanche* species is achieved through large hymenoptera, especially bumble bees.

In *Striga* the species *S. asiatica* (see also Nickrent & Musselman 1979) and *S. gesnerioides* are autogamous while *S. hermonthica* is allogamous (probably an obligate outcrosser). Pollination is effected by various insects, including butterflies and moths.

4. Morphotypes

In all species which are discussed in the present paper there are distinct morphotypes and in many cases these appear to have a genetic basis. The differences which are mentioned in the literature are especially connected with corolla morphology and plant size (Musselman *et al.* 1979; Musselman & Parker 1981; Musselman 1984). In *O. ramosa* and *O. aegyptiaca* there is the complication that certain morphotypes are more or less intermediate between the two species.

In *O. crenata* there are marked colour differences in the corolla varying from white-yellowish to white with purple markings. When the corolla has purple markings the penduncle is generally dark purple (a light coloured corolla is commonly connected with a yellowish penduncle). In Spain the morphological variations in this species are so large that it was suggested to distinguish at least two sub-species (Moreno *et al.* 1979). However, in general it may be concluded that this species is morphologically less variable than the other *Orobanche* species.

A very complex species is *O. minor,* but it is not clear to what extent this is due to environmental conditions (incl. different host plants) or genetic variability.

As far as *Striga* is concerned the morphotypes are especially conspicuous in *S. gesnerioides*. Musselman (1984) distinguished four main types:

a) a type parasitizing *Euphorbia* which is characterized by cauline branching, dark purple flowers and dark red stems.
b) a type parasitizing *Convolvula, Ipomoea, Meremmia* and *Jaquemontia,* characterized by branched stems and pinkish flowers.
c) a type parasitizing cowpea with branched stems and bluish flowers.
d) a type parasitizing tobacco with unbranched stems and bluish flowers.

In addition, Ralson *et al.* (1987) reported that in Botswana the following morphotypes can be recognized which appear to be host-specific:

a) a type parasitizing *Ipomoea* with short internodes and yellow flowers.
b) a type parasitizing *Indigofera* and *Pteridiseus* with short internodes and light-pink to deep-purple flowers.
c) a type parasitizing *Tephrosia* with medium internodes and light pink flowers.
d) a type parasitizing *Rhynchosia* and *Tephrosia* with long internodes, stems occasionally red pigmented and large light-pink flowers.

The branching pattern varied considerably and each type could be found unbranched and branched.

In *S. asiatica* there are biotypes which differ in the colour of the flowers (white, red, pale-pink, red with yellow or yellow) and in plant size. In this species it has also been possible to discern differences in seed surface features of different geographic origin (Musselman & Parker 1981).

Musselman (1979) and Musselman *et al.* (1979, 1985) have presented detailed descriptions of corolla variations in *S. hermonthica*. These were especially connected with colour as well as the morphology of the lobes. White flowered forms are occasionally seen at very low frequency (Ramaiah *et al.* 1983).

5. Chromosome numbers

The chromosome number of the *Orobanche* and *Striga* species of economic importance are presented in Table I. Differences in chromosome numbers within species have only been reported for *O. cumana* and *S. asiatica*. In *O. cumana* n = 12 and n = 17 has been observed (Musselman 1986). In *S. asiatica* Musselman *et al.* (1988) found n = 19 in plants from Africa whereas Kondo (1973) and Musselman *et al.* (1988) reported n = 12 chromosomes in plants from the United States, and Kumar & Abraham (1941) n = 20 for plants from India.

Moreno *et al.* (1979) observed meiotic anomalies and some variation in the basic number of chromosomes in about 20% of *O. crenata* plants from populations in Spain. However, it is unknown whether these aberrations persist in the populations.

6. Estimation of the genetic variability by means of iso-enzyme markers

Iso-enzyme analyses have been carried out in *O. crenata, O. aegyptiaca, S. hermonthica, S. asiatica* and *S. gesnerioides*. However, these studies were only concerned with a few populations and therefore the results can only give initial information.

In *O. crenata* and *O. aegyptiaca* a preliminary study has been conducted with plants form a single population collected in an area near Aleppo in Syria from faba bean and lentil (Verkleij *et al.* 1986). The results suggest a relatively high intraspecific genetic variation and a marked genetic difference between the two species. A more detailed study on *O. crenata* has been carried out recently with two populations, one which was also collected from the Aleppo region, the other from the coastal area in Syria (both from faba bean) (Verkleij *et al.* 1990). It appeared that there was a very large variability within the populations, which could be ex-

plained on the basis of the presumably allogamous pollination system. On the other hand it was remarkable that there were no significant differences between the iso-enzyme patterns of the two populations, in spite of the fact that the distance between the sites is more than 100 km and the areas are separated by mountains.

In *S. hermonthica* three populations have been examined for iso-enzyme patterns, i.e., one population form pearl millet in Burkina Faso and two populations from sorghum from respectively Burkina Faso and the Sudan (Musselman *et al.* 1988). The intra-populational diversity of the iso-enzyme patterns was relatively high. In general it could be concluded, after carrying out statistical tests, that the results were in conformance with the obligate outcrossing habit of the species. Furthermore the results indicated a greater genetic distance between populations from different countries than from different hosts.

In *S. asiatica* two populations from the United States have been analyzed (Werth *et al.* 1984). All individual plants tested appeared to be monomorphic. This lack of genetic diversity can be explained by the fact that *S. asiatica* has recently been introduced into the United States with a few seeds

Table I. Chromosome numbers in *Orobanche* and *Striga* species of economic importance.

Species	N	
Striga asiatica	19[a]	(Africa)
	12[a,b]	(USA)
	20[c]	(India)
Striga gesnerioides	20[a]	(Africa, India)
Striga hermonthica	32[a]	
Orobanche crenata	19[d]	
Orobanche cumana	12[d]	
	19[d]	
Orobanche aegyptiaca	19[e]	
	12[d]	
Orobanche minor	19[d]	
Orobanche ramosa	12[e]	

[a] Musselman *et al.* (1988).
[b] Kondo (1973).
[c] Kumar & Abraham (1941).
[d] Musselman (1986).
[e] Hambler (1954, 1956).

(founder population) and the predominantly auto-gamous breeding system of the species. In the population of *S. asiatica* in the United States there is at least one deviating morphotype, a yellow flowered plant (about 0.01% of the plants exhibit this phenotype, the others are red flowered) but apparently this deviation is not connected with the enzymes which were tested (at least not in such a way that there were differences in the analyzed iso-enzyme patterns).

The iso-enzyme study with *S. gesnerioides* was concerned with populations from three sites in Burkina Faso and three sites in Niger, all from cowpea (Shawe *et al.* 1988). Within the populations there were distinct differences in iso-enzyme patterns, with the exception of one population from Burkina Faso. Differences in iso-enzyme patterns between the populations were variable, one population (from Niger) showed a very different set of iso-enzyme patterns. On a geographical basis there was not much variation, however, the populations from Burkina Faso were slightly more similar to each other than to the populations from Niger.

7. Biotypes differing in virulence to their host(s)

7.1. Orobanche cumana
Various biotypes of *O. cumana* have been recorded which differ in their virulence to sunflower cultivars. The literature has recently been reviewed by Cubero (1986) and Vrânceanu *et al.* (1986).

During the end of the 19th century sunflower in the Soviet Union was heavily infested by a so called "A" biotype of *O. cumana*. A large scale breeding programme was started and during the period 1910–1920 new sunflower cultivars were selected which proved to be highly resistant. However, these cultivars became eventually infested by another *O. cumana* biotype which was called "B". Breeding for resistance to biotype "B" proved to be extremely difficult, especially as it appeared that "B" consisted of different populations which differed in the degree of virulence. As a consequence multiple resistance had to be built in to the various "B" types.

Other biotypes of *O. cumana* on sunflower (which were distinctly different from A and B) were subsequently reported from Bulgaria, Romania, Yugoslavia and Spain (Cubero 1986).

7.2. Orobanche crenata
Variability within this species in connection with virulence to specific hosts has recently been reported by Radwan *et al.* (1988). The number of *Orobanche* shoots per host plant (different varieties of faba bean were tested) varied when the host plants were exposed to *Orobanche* seed from different collection sites. These differences in aggressiveness were more pronounced when, rather than counting the number of *Orobanche* inflorescences, parasitic effects on host growth and development were examined. However, the relationship between host genotype and specific *Orobanche* population appeared to be dependent upon environmental conditions and it was concluded that the utility of host genotypes to differentiate between different biotypes of *O. crenata* is rather limited. A previous study of Cubero and Moreno (1979) on this matter with different populations of *O. crenata* and various inbred lines of faba bean had not been conclusive. Although there was a great variation in agressiveness it appeared that other factors, such as differences in the time of emergence of the parasite, could also play a role.

7.3. Orobanche minor
This species represents a morphologically very complex group of plants. The taxonomy, however, has not been studied in detail. It is not clear whether the different forms are linked to specific hosts. Musselman & Parker (1982) have shown that there is some host specificity within the species. A biotype from lettuce parasitized on tobacco, red clover (*Trifolium pratense* L.) and white clover (*Trifolium repens* L.), white biotypes from red and white clover never emerged on lettuce (*Lactuca sativa* L.). Moreover Parker (1986) reported that in Ethiopia a population which attacked *Xanthium* did not parasitize tobacco, in spite of the fact that the *Xanthium* plants occurred in tobacco fields.

7.4. Orobanche aegyptiaca
In this species there is some evidence of host spe-

cialization. For example Singh & Pavgi (1975) observed that *O. aegyptiaca* from cruciferous hosts did not attach sunflower and Beilen (1948) reported that a special form of *O. aegyptiaca* ("*O. muteli*"), in contrast to the "normal" form, did not parasitize cucurbits.

7.5. Orobanche ramosa

Vinogradov *et al.* (1981) reported differences among *O. ramosa* populations as far as virulence to tomato was concerned. Similar observations have been made regarding hemp (Kolyadko 1973). In addition there is a report by Musselman & Parker (1982) that two populations of *O. ramosa,* from respectively tobacco and tomato, did not or in very low numbers develop on lettuce, which normally is a good host.

7.6. Striga hermonthica

In this species it has been demonstrated that there are biotypes which are specific to sorghum or to pearl millet. Evidence was first presented by Parker & Reid (1979) who also showed that this specificity is based on a response to different germination stimulants. Subsequently, Ramaiah (1984) reported the occurrence of biotypes which attack both sorghum and pearl millet and biotypes which are mainly specific to sorghum. However, this author did not observe a hundred percent specificity. It should also be taken into consideration, that the samples studied had been collected from a narrow range of latitudes.

The existence of intra crop-specific biotypes (i.e., biotypes which are specific to a particular cultivar of a crop) has been suggested but the experimental evidence is not conclusive (Vasudeva Rao & Musselman 1987).

7.7. Striga asiatica

In India *S. asiatica* (the white flowered form) occurs on sorghum in predominantly sorghum growing areas. However, in general it does not parasitize pearl millet in the sorghum zone and sorghum in the pearl millet zone. An exception is formed by a small area in Andhra Predesh where both crops are attacked on a large scale (Vasudeva Rao & Musselman 1987). It was demonstrated by Bharathalakshmi & Jayachanda (1979), who worked with *S. asiatica* collected separately from sorghum and pearl millet from Karnataka (an area in India where mainly sorghum is grown), that in this area there are two markedly different biotypes and that the host specificity is connected with the germination stimulant(s). Vasudeva Rao (1984), who collected seed form *S. asiatica* parasitizing sorghum and pearl millet in a predominantly sorghum growing area in Andhra Pradesh, found that cross inoculation on sorghum and pearl millet (pearl millet infested with sorghum *Striga* and sorghum infested with pearl millet *Striga*) only led to an infestation of sorghum. This suggests that the pearl millet *Striga* in this area somehow has adapted itself to sorghum.

7.8. Striga gesnerioides

It has clearly been demonstrated that different biotypes of this species occur which vary in their host specificity to different varieties of cowpea (Aggarwal *et al.* 1986, 1987). Cowpea varieties which had been developed by the International Institute of Tropical Agriculture (IITA) appeared to be resistant to *S. gesnerioides* in Burkina Faso (Aggarwal *et al.* 1984, 1986). However, when these varieties were grown in Niger and Nigeria they were found to be susceptible (Aggarwal 1987). In addition there is ample evidence that different biotypes occur on different host plant species, at least in many instances. *S. gesnerioides* has a very wide host range of predominantly broad-leaved species, including members of Solanaceae, Convolvulaceae, Leguminosae and Euphorbiaceae. Musselman & Parker (1981) found in greenhouse experiments a strong host specificity which was to a large extent correlated with a variation in morphological characteristics.

8. Discussion

It may be concluded that there is ample variability in the *Orobanche* and *Striga* species which are discussed in the present paper and that an important part of the variability seems to be genetically determined. The amount of the genetic variability varies to a large extent, mainly due to the various pollination systems. In the autogamous species distinct biotypes have developed which may parasitize specific hosts. In allogamous species there is also host specialization, as is shown by the occurrence of sorghum and pearl millet biotypes in *S. hermonthica*. However, based on iso-enzyme studies in allogamous species the differences between populations from different areas as well as from different hosts are relatively small, whereas the intrapopulational variability is relatively large. As a consequence it may be assumed that in *S. hermonthica*, at least in the populations which have been analyzed for iso-enzymes, there is no correlation between iso-enzyme marker genes and genes involved in host plant-parasite interaction.

The question why different pollination systems have evolved could be connected with the spread and distribution of the host plant(s) populations. According to Solbrig (1976) cross-pollination is theoretically the favoured breeding system for flowering plants but self-pollination became profitable for species with small populations and clumped distributions. As far as *Orobanche* and *Striga* are concerned, it is conceivable that self-pollination became profitable for species which parasitize hosts with an irregular distribution.

In natural vegetations there has presumably been a co-evolution of both the populations of parasitic plants and their hosts. This interaction between the genetic diversity of host and parasite populations has probably led to a certain level of resistance in the host plant populations, thus precluding a total eradication by the parasite. In this context it could be noted that populations of the semi-parasitic plant *Rhinanthus angustifolius* show an optimal growth after establishment, however, the populations subsequently decline and persist at a low density (ter Borg 1985). As host-parasite mechanisms are in general characterized by a wide variety of polymorphic genes which brings about a regulatory effect on the infection (O'Brien & Evermann 1988) this mechanism could also play a role in the interactions between parasitic plants and their hosts.

The question arises to what extent these genetic interactions have been influenced by the domestication of certain host plants. Most likely, selection of the host plants as well as monocropping has led to an adaptation of the parasites. As a consequence it may be assumed that biotypes have evolved which specifically occur in agro-ecosystems. Indications for this hypothesis are manyfold but experimental evidence is lacking. For example, in western Africa *S. asiatica* is abundant on wild grasses in natural savannas but rarely if ever occurs on crops, in contrast to southern Africa, India and the United States (Pieterse 1988 (pers. obs.)). On the other hand, *S. hermonthica* is never found in native grasslands in western Africa although it is a major pest of cereal crops in this area (Musselman *et al.* 1988). According to Musselman & Hepper (1986) *S. hermonthica* is native to eastern Africa and these authors suggested that it was spread with sorghum. It should also be taken into account that, although *O. crenata* and *O. aegyptiaca* frequently occur on wild plants, this is mainly in the vicinity of arable land. Whether these species, or at least the biotypes which occur on crops, are still able to survive in natural vegetation, is uncertain. Other possible examples of genetic adaptations to agro-ecosystems are the biotypes of *S. gesnerioides* which occur on cowpea or tobacco and the biotypes of *O. cumana* which are specific to sunflower.

The adaptation of parasitic plants to agro-ecosystems may have had a significant effect on their survival strategy. Although it cannot be ruled out that in the past man has selected for tolerance in local crop varieties, the aggressiveness of weedy parasites to their host crops is in general very high. This could be explained by the fact that for its survival in agro-ecosystem a parasite does not necessarily depend on the survival of its host, as man will provide new hosts during following seasons. In other words selection of more aggressive biotypes is not counterbalanced by a decrease in host plant population. Further it may be presumed that the parasites have adapted themselves to the special ecological conditions in agro-ecosystems which are amongst others characterized by changes in water relations and soil structure.

Apart from available defense mechanisms of host plants the genetic variability of the parasites should be taken into consideration in connection with breeding programmes. The recent work of

Radwan *et al.* (1988) has demonstrated that *O. crenata* from different geographical regions varied in aggressiveness to different faba bean lines. The suggestion of these authors that different "low level" tolerance mechanisms occur in faba bean and that *O. crenata* biotypes differ in their degree of aggressiveness to these mechanisms, could be of practical importance for the breeding programmes.

It is obvious that the genetic diversity of most crop plants is relatively small. The chance to find differences in tolerance/resistance will be much greater when crop plants can be crossed with wild relatives or when genes involved in the mechanism of resistance of wild relatives could be transferred by means of genetic engineering techniques. Up to now little attention has been paid to the genetic resource in wild species. However, in the opinion of the present authors much could be learnt by studying the response of wild relatives of host crop plants to *Orobanche* and *Striga* species. For a better understanding of the evolution from wild parasitic plants into aggressive parasitic weeds which attack crop plants it would be useful to carry out comparative studies with weedy biotypes of *Orobanche* and *Striga* species in natural vegetations.

Acknowledgements

The authors like to thank Prof. Dr L. J. Musselman and Dr S. J. ter Borg for critically reading of the manuscript.

References

Aggarwal, V. D., 1987. Latest in cowpea *Striga* research at IITA. In: H.Chr. Weber & W. Forstreuter (eds.) Parasitic Flowering Plants. Marburg. pp. 27—36.

Aggarwal, V. D., Haley, S. D. & Brockman, F. E., 1986. Present status of breeding cowpea for resistance to *Striga* at IITA. In: S.J. ter Borg (ed.) Proceedings of a workshop on biology and control of *Orobanche*. LH/VPO, Wageningen. pp. 176—180.

Aggarwal, V. D., Muleba, N., Drabo, I., Souma, J. & Mbebe, 1984. Inheritance of *Striga gesnerioides* resistance in cowpea. In: C. Parkers, L.J. Musselman, R.N. Polhill & A.K. Wilson

(eds.) Proceedings of the Third International Symposium on Parasitic Weeds. ICARDA/International Parasitic Seed Plant Research Group, Aleppo. pp. 143—147.

Beilen, I. G., 1948. *Orobanche mutelii* F. Schulz and specialization. C.R. Acad. Science, URSS 61: 943—944.

Bharathalakshmi & Jayachandra, 1979. Physiological variations in *Striga asiatica*. In: L.J. Musselman, A.D. Worsham & R.E. Eplee (eds.) Proceedings of the Second Symposium on Parasitic Weeds. North Carolina State University, Raleigh. pp. 132—143.

Borg, S. J. ter, 1985. Population biology and habitat relations of some hemiparasitic Scrophulariaceae. In: J. White (ed.) The population of vegetation. Dr. W. Junk Publishers, Dordrecht. pp. 463—484.

Cubero, J. I., 1986. Breeding for resistance to *Orobanche* and *Striga*: a review. In: S.J. ter Borg (ed.) Proceedings of Workshop on biology and control of *Orobanche*. LH/VPO, Wageningen, The Netherlands. pp. 127—139.

Cubero, J. I. & Moreno, M. T., 1979. Agronomic control and sources of resistance in *Vicia faba* to *Orobanche crenata*. In: D.A. Bond, G.T. Scarascia-Mugnozza & M.H. Pulsen (eds.) Some current research on *Vicia faba* in Western Europe. Commission of the European Communities, Luxembourg. pp. 41—80.

Cubero, J. I., Pieterse, A. H., Saghir, A. R. & Borg, S. J. ter, 1988. Parasitic weeds on cool season food legumes. In: R.J. Summerfield (ed.) World Crops Cool Season Food Legumes, Proceedings of the International Food Legume Research Conference on Pea, Lentil, Faba Bean and Chickpea, Spokane, Washington, USA, 6—11 July 1986. Kluwer Academic Publishers, Dordrecht, The Netherlands. pp. 549—563.

Gil, J., Martin, L. M. & Cubero, J. I., 1984. Resistance to *Orobanche crenata* Forssk. in *Vicia sativa* L. II. Characterization and genetics. In: C. Parker, L.J. Musselman, R.M. Polhill & A.K. Wilson (eds.) Proceedings of the Third International Symposium on Parasitic Weeds. ICARDA/International Parasitic Seed Plant research group, 7—9 May 1984, Aleppo, Syria. pp. 221—229.

Hambler, D. J., 1954. Cytology of the Scrophulariaceae and Orobanchaceae. Nature 174: 838.

Hambler, D. J., 1956. Further chromosomes counts in Orobanchaceae. Nature 177: 438—439.

Hepper, F. N., 1973. Problems in naming *Orobanche* and *Striga*. In: Proceedings European Weed Research Council (First) Symposium on Parasitic Weeds, Malta. pp. 9—17.

Kim, S. K., Khadr, F., Parkinson, V., Fashemisin, J. & Efron, Y., 1985. Maize breeding for *Striga* resistance. In: Proceedings OAU/FAO Workshop on *Striga* in Africa, held in Yaounde, Cameroon 23—27 September 1985. pp. 58—74.

Kim, S. K., 1988. Breeding for *Striga* tolerance in maize and development of a field infestation technique. Paper presented at the IITA/ICRISAT Workshop, held at the International Institute of Tropical Agriculture, Ibadan, Nigeria, 22—24 August 1988.

Kolyadko, I. V., 1973. The resistance of hemp hybrids to

branched broomrape (in Russian). Len i Konoplya 1, 27—28.

Kondo, K., 1973. The chromosome numbers of *Striga asiatica* and *Triophyllum peltatum*. Phyton 3, 1—2.

Kumar, L. S. S. & Abraham, A., 1941. Cytological studies in Indian parasitic plants. I. The cytology of *Striga*. Proceedings of the Indian Academy of Sciences, Section B, 14, 509—516.

Loveless, M. D. & Hamrick, J. L., 1984. Ecological determinants of genetic structure in plant populations. Ann. Rev. Ecol. Syst. 15, 65—95.

Moreno, M. T., Cubero, J. I. & Martin, A. 1979. Meiotic behavior in *Orobanche crenata*. In: L.J. Musselman, A.D. Worsham & R.E. Eplee (eds.) Proceedings of the second International Symposium on Parasitic Weeds, held 16—19 July 1979 at Raleigh, North Carolina, USA. pp. 73—78.

Musselman, L. J., 1979. Fertility and floral patterns in some species of *Striga* (Scrophulariaceae). National Geographic Society Research Reports 1979, 487—491.

Musselman, L. J., 1982. Parasitic weeds of arable land. In: W. Holzner & N. Numata (eds.) Biology and ecology of weeds, Chapter 16, Dr. W. Junk, The Hague, The Netherlands. pp. 175—185.

Musselman, L. J., 1984. Taxonomic problems in *Striga* with particular reference to Africa. In: C. Parker, L.J. Musselman, R.M. Polhill & A.K. Wilson (eds.) Proceedings of the third International Symposium on Parasitic Weeds. ICARDA/International Parasitic Seed Plan Research Group, 7—9 May 1984, Aleppo, Syria. pp. 53—57.

Musselman, L. J., 1986. Taxonomy of *Orobanche*. In: S.J. ter Borg (ed.) Proceedings of a Workshop on biology and control of *Orobanche*, LH/VPO, Wageningen, The Netherlands. pp. 2—10.

Musselman, L. J. (ed.), 1987a. Parasitic weeds in agriculture. Volume I *Striga*. CRC Press, Inc., Boca Raton, Florida, USA.

Musselman, L. J., 1987b. Taxonomy of witchweeds. In: L.J. Musselman (ed.) Parasitic weeds in agriculture. Volume I *Striga*. CRC Press. pp. 3—12.

Musselman, L. J. & Hepper, F. N., 1986. The witchweeds (*Striga, Scrophulariaceae*) of the Sudan republic. Kew Bull. 41: 205—.

Musselman, L. J. & Parker, C., 1981. Surface features of *Striga* seeds (Scrophulariaceae). Adansonia Ser. 2, 20: 431—437.

Musselman, L. J. & Parker, C., 1982. Preliminary host ranges of some strains of economically important broomrapes (*Orobanche*). Economic Botany 36, 270—273.

Musselman, L. J., Nickrent, D. L., Mansfield, R. A. & Ogborn, J. E. A., 1979. Field notes on Nigerian *Striga* (Scrophulariaceae). SIDA 8: 196—201.

Musselman, L. J., Parker, C. & Dixon, N., 1982. Notes on autogamy and flower structure in agronimically important species of *Striga* (Scrophulariaceae) and *Orobanche* (Orobanchaceae). Beitr. Biol. Pflanzen 56: 329—343.

Musselman, L. J., Bhrathalakshmi, Safa, S. B., Knepper, D. A., Mohamed, K. I. & White, C. L., 1988. Recent research on the biology of *Striga asiatica*, *S. gesnerioides* and *S. hermonthica*. Paper presented at the IITA/ICRISAT *Striga* Workshop held at the International Institute of Tropical Agriculture, Ibadan, Nigeria, 22—24 August 1988.

Nickrent, D. L. & Musselman, L. J., 1979. Autogamy in the American strain of witchweed *Striga asiatica* (Scrophulariaceae). Brittonia 31: 253—256.

O'Brien, S. J. & Evermann, J. F., 1988. Interactive influence of infectious disease and genetic diversity in natural populations. TREE 3: 254—259.

Parker, C., 1986. Scope of the agronomic problems caused by *Orobanche* species. In: S.J. ter Borg (ed.) Proceedings of a Workshop on biology and control of *Orobanche* LH/VPO, Wageningen. The Netherlands. pp. 11—17.

Parker, C. & Reid, D. C., 1979. Host specificity in *Striga* species — some preliminary observations. In: L.J. Musselman, A.D. Worsham & R.E. Eplee (eds.) Proceedings of the Second Symposium on Parasitic Weeds. North Carolina State University, Raleigh, N.C., USA. pp. 79—90.

Parker, C. & Wilson, A. K., 1986. Parasitic weeds and their control in the Near East. FAO Plant Prot. Bull. 34: 83—98.

Pieterse, A. H., 1979. The broomrapes (Orobanchaceae) — a review. Abstracts on Tropical Agriculture 5(3): 9—35.

Pieterse, A. H. & Pesch, C. J., 1983. The witchweeds (*Striga* spp.) — a review. Abstracts on Tropical Agriculture 9(8): 9—37.

Radwan, M. S., Abdalla, M. M. F., Fischbeck, G., Metwally, A. A. & Darwish, D. A., 1988. Variation in reaction of faba bean lines to different accessions of *Orobanche crenata*, Forsk. Plant Breeding 101: 208—216.

Ralston, D. M., Riches, C. R. & Musselman, L. J., 1987. Morphology and hosts of three *Striga* species (Scrophulariaceae) in Botswana. Bull. Mus. natn. Hist. nat., Paris, 4e ser., 9, 1987, section B, Adansonia, no. 2: 195—215.

Ramaiah, K. V., 1984. Physiological specialization of *Striga hermonthica* and crop specificity. In: C. Parker, L.J. Musselman, R.J. Polhill & A.K. Wilson (eds.) Proceedings of the Third International Symposium on Parasitic Weeds. ICARDA/International Parasitic Plant Research Group, 7—9 May 1984, Aleppo, Syria. pp. 58—65

Ramaiah, K. V., 1987. Breeding cereal grains for resistance to witchweed. In: L.J. Musselman (ed.) Parasitic Weeds in Agriculture, Volume I *Striga*. CRC Press. pp. 227—242.

Ramaiah, K. V., 1988. Breeding for *Striga* resistance on sorghum and millet. Paper presented at Striga Workshop, 22—24 August 1988, IITA, Ibadan, Nigeria.

Ramaiah, K. V., Parker, C., Vasudeva Rao, M. J. & Musselman, L. J., 1983. *Striga* identification and control handbook. ICRISAT Information Bulletin, No. 15. 52 pp.

Safa, S. B., Jones, B. M. G. & Musselman, L. J., 1984. Mechanisms for outbreeding in *Striga hermonthica* (Scrophulariaceae). New Phytologist, 96: 299—305.

Shawe, K. G., Ingrouille, M. & Stewart, G., 1988. Isoenzyme survey of *Striga gesnerioides* (Willd). Vatke on cowpea in West Africa. Cowpea-Striga project. 3rd year report. ODA-Birkbeck College. 60 pp.

Singh, S. L. & Pavgi, M. S., 1975. Observations on seed germination of *Orobanche*. Science and Culture 41: 296—297.

Solbrig, O. T., 1976. On the relative advantages of cross- and self-fertilization. Ann. Missouri Bot. Gard. 63: 262—276.

Vasodeva Rao, M. J. & Musselman, L. J., 1987. Host specificity in *Striga* spp. and physiological "strains". In: L.J. Musselman (ed.) Parasitic weeds in agriculture, Volume I *Striga,* CRC Press. pp. 13—25.

Vasodeva Rao, M. J., 1984. Patterns of *Striga* resistance to *Striga asiatica* in sorghum and millets, with special reference to Asia. In: E.S. Ayensu, H. Doggett, R.D. Keynes, J. Marton-Lefevre, L.J. Musselman, C. Parker & A. Pickering (eds.) Striga: Biology and Control. ICSU Press, Paris, France. pp. 93—112.

Verkleij, J. A. C., Jansen, J. & Pieterse, A. H., 1986. A preliminary study on isoenzyme variation in *Orobanche crenata* and *O. aegyptiaca,* In: S.J. ter Borg (ed.) Proceedings of a Workshop on biology and control of *Orobanche,* LH/VPO, Wageningen, The Netherlands. pp. 154—159.

Verkleij, J. A. C., Egbers, W. S. & Pieterse, A. H., 1990. Allozyme variations in populations of *Orobanche crenata* Forssk. from Syria. In: K. Wegman & L.J. Musselman (eds.) Recent Progress in Orobanche Research. In press.

Vinogradov, V.A., Mironov, E. K. & Sarychev, Y. F., 1984. Degree of field resistance to *Orobanche* in tobacco mutants, hybrids and varieties. Plant Breeding Abstracts 54: 3464.

Vrânceanu, A. V., Pirvu, N., Stoenescu, F. M. & Pacureanu, M., 1986. Some aspects of the interaction *Helianthus annuus* L. *Orobanche cumana* Wallr. and its implications in sunflower breeding. In: S.J. ter Borg (ed.) Proceedings of a Workshop on biology and control of *Orobanche,* LH/VPO, Wageningen, The Netherlands. pp. 181—189.

Werth, C. R., Riopel, J. L. & Gillespie, N. W., 1984. Genetic uniformity in an introduced population of witchweed (*Striga asiatica*) in the United States. Weed Science 32: 645—648.

Authors' Addresses
A. H. Pieterse
Royal Tropical Institute
Rural Development Programme
Mauritskade 63
1092 AD Amsterdam
The Netherlands

J. A. C. Verkleij
Department of Ecology and Ecotoxicology
Free University
De Boelelaan 1087
1081 HV Amsterdam
The Netherlands

Species Index

Abies alba 259
Acacia 265, 271
  burkei 266
  erioloba 266
  erubescens 266, 270
  fleckii 266
  hebeclada 265, 266
  karroo 266, 270
  mellifera 266, 272
  nilotica 266
  tortilis 266, 270, 272, 273
Acer pseudoplatanus 13
Achillea 2
Aegilops squarrosa 284
Agrostis 45, 185
  canina 42, 44, 46
  capillaris 3, 8, 203, 204
  castellana 42, 44, 45, 46
  delicatula 42, 44
  giganthea 120
  stolonifera 3, 46, 68, 70, 86, 107, 143
  tenuis 8, 9, 13, 46
Aira caryophyllea 142
  praecox 142, 143
Allium ursinum 138
Althaea officinalis 105
Anagallis arvensis 241
Anemone nemorosa 138
Anthephora pubescens 264
Anthoxanthum odoratum 3, 13
Archanara 174
  geminipunctata 173
Ariolimax columbianus 246
Aristida 264, 265
Armeria 110
  maritima 107
Artemisia 186, 189
  maritima 91, 105, 106, 107, 110, 120, 121, 188
Asarum caudatum 246
Aster 185, 187

tripolium 63, 65, 67, 68, 69, 70, 104, 105, 106, 113, 120, 121, 128, 129, 186, 222, 224, 225, 227
Atriplex 106
  amnicola 286
  littoralis 63, 68, 69, 105, 126
  prostrata 104, 120, 121
  undulata 286
Avena 223
Azadirachta indica 287

Beta vulgaris 279, 284, 285
Betula pendula 140, 254
Boletus 255
Bouteloua gracilis 121
Brachiaria nigropedata 264
Brassica 238, 243, 286
Bratrachospermum moniliforme 166

Calamagrostis epigejos 88
Calluna vulgaris 201, 202, 259
Canavalia ensiformis 236
Cannabis sativa 290
Capsicum annuum 286
Carex 55, 80, 81
  elata 55
  pseudocyperus 53, 55, 57
  remota 53
Casuarina equisetifolia 287
Cecropia peltata 246
Centaurium littorale 78, 79, 80, 82, 88
  pulchellum 78, 82, 83, 86
Cerastium semidecanarum 142, 143
Ceratium hirundinella 166
Chaetocnema confinis 246
Chaetophora incrassata 166
Chaetosphaeridium globosum 157
Chamaenerion angustifolium 139
Chamobates cuspidatus 215
Chara aspera 161, 162, 166, 167, 168
  canescens 161
  connivens 161

contraria 161, 162, 167
globularis 149, 150, 161, 162, 166, 167, 168
hispida 161, 166, 167, 168
vulgaris 150, 161, 166
vulgaris var. longibracteata 161
Cheiranthus x allionii 243
Chloris virgata 264
Chrysopyxis 157
Chrysophaera botryoides 158
Cicer arietinum 290
Cirsium palustre 137
Citrus paradisi 285
  sinensis 285, 286
Cladophora glomerata 150, 166
Cochlearia officinalis 105
Collybia dryophila 256
Colophospermum mopane 268, 270
Convolvula 295
Crotalaria 245
Cucumis sativus 285
Cymbopogon plurinodis 265
Cynodon dactylon 265, 285

Dactylis glomerata 201
Danaus gilippus 245
Datura 9
  innoxia 14, 15
  stramonium 265
Daucus carota 285
Depressaria pastinacella 246
Deschampsia caespitosa 14
  flexuosa 139, 256, 257, 259
Dichapetalum cymosum 264
Dicranochaete reniformis 157
Dicrostachys cinerea 266
Digitalis purpurea 137, 139, 140, 241
Digitaria milanjeana 264
  pentzii 266
Dinobryon 157
Dioclea megacarpa 236
Dipcadi 264

Subject Index